PRAISE FOR

GRAPES

— OF THE —

HUDSON VALLEY

AND | OTHER COOL CLIMATE REGIONS OF THE UNITED STATES AND CANADA

PRAISE FOR THE SECOND EDITION

This second edition of *Grapes of the Hudson Valley and Other Cool Climate Regions of the United States and Canada* unlocks the fascinating stories and lesser-known agricultural traditions of our New England region. Here in Salem, Massachusetts, this book has us looking at our museum's grape arbors with fresh eyes and new insights.

— **ROBIN PYDYNKOWSKI,** Peabody Essex Museum, MA

Steven Casscles shines a well-deserved light on these breeding pioneers as well as the hybrid varieties they meticulously crafted. A fluid combination of history, viticulture, and hands-on knowledge, this library staple is perfect for those looking to learn more about who set the stage for the modern Northeastern wine industry.

— **ALLISON DUSH,** Director of Guest Experience and Education, Massachusetts Horticultural Society, MA

This is a book all New England and New York wine grape growers, wine enthusiasts, grape breeders, and horticultural history buffs will want on their bookshelves. Stephen Casscles is a thorough historical researcher and a viticulturalist par excellent. His writing is well researched and is written in a style that is both easy to read and informative. I was facinated with the detailed historical perspective of the New England grape breeders dating back to the early 1800 ... a valuable contribution.

— **WIN COWGILL,** Professor Emeritus, Rutgers, the State University, NJ

J. Stephen Casscles's *Grapes of the Hudson Valley* is enriched in this second edition by two new chapters, one on Edward Staniford Rogers and the other on a number of nineteenth-century New England breeders, including the legendary John Moore and Ephraim Bull. The new chapters make the book all the more a treasure trove, not only for the breeder and grower but for anyone interested in the history of the region's grapes, the methods of cultivation and improvement, and the growers themselves, as well as the new varieties they produced.

— **DANIEL J. KEVLES,** Professor of History Emeritus, Yale University, CT

Grapes of the Hudson Valley is reference material gold. It honors, and adds to, the work of generations among the vines while making the case for viticulture in the northeast today, and into the future. Stephen Casscles's research effort is admirable and it matches the rigor with which he tends his own vines, both of which lead to the great benefit of we, the readers.

— **TODD TRZASKOS,** Vermont Wine Media; contributing editor, *thecorkreport.us*

PRAISE FOR THE FIRST EDITION

Grapes of the Hudson Valley is essential not just for winegrowers in the Hudson Valley, but equally for the Champlain Valley, Vermont, and Quebec. We've waited a long, long time for such a book; it is marvelous to be able to say that the wait was well worth it!

— **GEORGE GALE,** author, *Dying on the Vine: How Phylloxera Changed the World of Wine*

A valuable guide for viticulturists not only in the Hudson Valley, but in cold climate grape production regions globally. Stephen Casscles unashamedly gives the reader a full picture of what interspecific hybrid grapes are, and why they are such an important resource for wine grape viticulture. Let me just say that *Grapes of the Hudson Valley* will be one book I will refer to often.

— **WILLIAM H. SHOEMAKER,** retired fruit and vegetable horticulturist, University of Illinois, IL

Grape breeders, grape growers, and grape and wine enthusiasts will want this book on their shelf.

— **CARL CAMPER,** Chateau Stripmine, WA

Stephen Casscles's *Grapes of the Hudson Valley* takes a fascinating approach to presenting information about grape varieties grown in the Hudson Valley. Not only does it present excellent information about northern grape varieties, it also answers the question, "Where did these varieties come from?"

— **TOM PLOCHER,** viticulturalist, grape breeder
author, *Northern Wineworks: Growing Grapes and Making Wine in Cold Climates*

It's rare to see a book related to wine cover brand new territory, but this is exactly what Stephen Casscles does in *Grapes of the Hudson Valley*. Nowhere are these grapes more important to the local wine industry than in the Northeast, and Casscles gives them the respect and breadth of analysis that they deserve. A must-read for anyone looking to produce wine in a cool climate or just interested in what makes our region of the world so unique. Bravo!

— **RICHARD OLSEN-HARBICH,** winemaker, Bedell Cellars, Long Island, NY

GRAPES
— OF THE —
HUDSON VALLEY

AND | **OTHER COOL CLIMATE REGIONS OF THE UNITED STATES AND CANADA**

SECOND EDITION

J. STEPHEN CASSCLES

FLINT MINE PRESS

ISBN 978-0-9982545-1-7

Library of Congress Control Number: 2022950116

Editor: Robert Bedford
Book design: Linda Pierro
Illustrations: Scott Krontilik

Published by Flint Mine Press
PO Box 353, Coxsackie, NY 12051
www.flintminepress.com

Printed and bound in the United States of America.
First edition June 2015.

This book is dedicated to my family and friends,
who for the past thirty-five years have helped me
to cultivate grapes, make them into wine, and listened,
or pretended to listen, to my musings on
grape growing and wine making.

I would particularly like to dedicate this book to my
paternal grandparents, Joseph Levi Casscles and Rose Casscles,
who introduced me to grape cultivation and
who brought me with them to visit their many old-time,
fruit-grower friends who made their living
in the Hudson Valley growing grapes and other fruits
during the early twentieth century.

CONTENTS

Mystery Red. *Grapes from a chance seedling growing on the author's Cedar Cliff farm in Athens, NY.*

Foreword

BY KEVIN ZRALY

I have known of Steve Casscles and his grape growing and winemaking expertise for over 30 years. I have been to his home vineyard, tasted many of his experimental older wines and most recent vintages. He has also made wine from my vineyard in the Hudson Valley, and as the chief winemaker at Hudson-Chatham Winery every year I have the opportunity to taste all of his new blends.

His new book, GRAPES OF THE HUDSON VALLEY is a work of passion just like his grape growing and winemaking. His enthusiasm is contagious and his knowledge and experimentation of cool climate grapes is unmatched. You will find out why some grapes work and others don't, where you should be planting grapes, and why. There has never been a book written like this and I am sure you will agree with me that this is a complete history of Hudson Valley winemaking. I never knew that there were so many grapes, whether it be *vinifera*, *labrusca*, or hybrids, that can grow in cool climate regions. This is without question the seminal guide.

Steve and I have a lot in common. I was born in the Hudson Valley and have been a lifelong resident and this is where I began my own personal passion with wine. We both planted grapes in the early '70s. Steve has been a success and I am still "trying." The first vineyard I ever visited was Benmarl Winery located in Marlboro, NY, which for me was life-changing. Steve had a great relationship with the owners of Benmarl, the Miller family, and worked in the cellar there for many years in the '70s and '80s.

His fascinating chapter on the history of the Hudson Valley, which includes old photographs and historical pencil sketchings, brought back many fond memories of my early days studying wine. As a history major in college this may be the reason why I made wine my career. I have always envisioned that the Hudson Valley would be the new "Napa Valley."

Steve includes what I would call a "simple" guide to winemaking without being overly detailed. In another chapter he gives credit to the pioneers who made it all possible.

This is not just another book on wine. It is both a technical book and a fun read, a rarity in wine writing.

Kevin Zraly
Author, *Windows on the World Complete Wine Course*
Educator of the Windows on the World Wine School, NYC
2011 James Beard Lifetime Achievement Award Winner

Preface
BY ERIC MILLER

I was delighted and excited when Steve Casscles said he was writing a book about viticulture in the Hudson Valley, home of my family's Benmarl Vineyards in Marlboro, New York. Although my first real contact with wine growing occurred several years before, while living in Burgundy, it was at Benmarl that I was first faced with the terrifying and truly important questions about *terroir*. A word I'm sure you will know a great deal more about after reading GRAPES OF THE HUDSON VALLEY.

Steve and I go back about 40 years and thousands of gallons of wine. Miles of trellised vines. Lifetimes of idle conversations pruning on the steep and snow-blanketed hills where he came to work for us as a young man. Even then, in his teens, Steve's curiosity and enthusiastic spirit made me feel like our vineyard was the most important thing in his life. We read the manuals and trained under the direction of the likes of Nelson Shaulis – whose grape-growing experiments in the Finger Lakes revolutionized vineyard practices around the world – but I'm sure we destroyed tons of cropping potential in the early days of our "artful" pruning. Slowly, however, we began the process of interpreting and adapting other regions' experts to find out what really worked in our corner of the Hudson Valley.

Eventually Steve Casscles grew up and became a lawyer who worked for various New York state senators, but his heart never really left the New York wine industry. His intelligent insights are easy to find in what may be a dozen key bills affecting wine and spirits in New York State. Along the way he planted his own vineyard, wrote and proposed remedies to improve the wine industry in the Hudson Valley, and became an accomplished winemaker. It should come as no surprise then, that he has written this important work that is a record of the *terroir* of the Hudson Valley, a log of the people who worked there, and a view of what they have done in this region. And Steve gives us a special gift of personal perspective by weaving in his own family's long history in the agricultural theater of the Valley.

Like all fine grape growing regions, the Hudson Valley is margined by the pitfalls of climate, soil, grape variety, cultural practices and winemaking decisions. In GRAPES OF THE HUDSON VALLEY the author not only shows it as a wildly beautiful place, but wisely points out its deadly extremes and even where to find them – and he is eminently qualified to do so after so many years of visiting vineyards and wineries, asking questions and observing, since long before he was of drinking age!

What I enjoyed most about GRAPES OF THE HUDSON VALLEY are the pages spent on the grape varieties Steve feels are well-suited to the Valley. Not content to give just their growing characteristics, he indulges in generations of history of these varieties' parentage and, in many cases, the people who created them. In this way each variety really comes alive. Woven into pages of straightforward facts are Steve's impassioned personal views about those age-old grape varieties we accept as mainstream, verses the newer hybrids and interspecific crosses that may hold a delicious and viable future for wine growers and American wine drinkers alike.

You will see that Steve is not only an historian but he is progressive. He has a broad appreciation and unbiased view of growing wine. He obviously enjoys and admires the success of traditional *vinifera* wines from the Hudson Valley and other cool climates. He also has suggestions for using a lab and current technology. And by citing the non-traditional Hungarian, American, French, and German varieties as potential for Hudson Valley soils and climate, he intimates that untraditional winemaking techniques should be considered. And Steve goes on to cite some of his own practice.

I have asked myself many times why the Hudson Valley has not become the private wine cellar of Manhattan. Perhaps this book will open peoples' minds to the means and potential for viticulture here. GRAPES OF THE HUDSON VALLEY is both charming and informative and anyone thinking about planting vines in their back yard or sinking millions into a vineyard and winery should have it under their belt, and in their library.

Eric Miller
Winemaker and author, *The Vintner's Apprentice*

Preface
to the Second Edition
BY CARLO DEVITO

Unhappy plants, whether a house plant or in a farmer's field, are easy to spot. There is nothing sadder than to witness an unhappy plant. Lean, bereft of foliage, and struggling for life. Many a grower has seen such sad sights. In 2007, this is what we faced as we started our new vineyard and winery, Hudson-Chatham Winery in Ghent, New York. We had four acres, and more than 20 rows of unhappy vines. We had Chancellor and a mixed variety of muscat grapes. The Chancellor, normally a hardy grape, was struck with downy mildew three years running and produced little to no fruit. And our muscats were anemic. It was just sad. Our supposedly hardy hybrids were not happy. Endless rows of sad, sickly plants.

It was then that Steve Casscles came to visit. Steve immediately told us what was wrong. The Chancellor was in a depression where water and cold air pooled. A tough spot for any grape, but especially Chancellor. The muscats we had chosen, he said, would not make it. Steve replaced the Chancellor and some of the muscats with Chelois, a grape certainly not in vogue at the time. And then replaced some of the other rows with the Middle Hope clone of Baco Noir, Burdin, Chambourcin, and LeColonel. All of these were obscure hybrids, except for the Baco Noir. That season and the next, the plants flourished, big lines of lush green leaves, swelling, bulging with life. For the first time, we had to hedge or rows, a drudgery in more established vineyards, but a godsend to us.

I use this story to illustrate the genius of Steve Casscles. He knows the many stories behind the breeding of the grapes, and their parentage and what they are like to grow firsthand. When you read Steve's books and articles, for the first time you begin to understand that certain hybrids were bred (not just to fight phylloxera) but to grow in specific kinds of soil, while others were meant to grow in wetter circumstances, and yet others were bred to live in drier soils. And of course, temperatures being what they are in the Northeast, many were bred for winter hardiness and resistance to fungus disease. Steve's book goes far, far beyond the scant nursery notes offered by most providers. Steve's book is a grower's absolute must have for these types of cool-climate hybrid grape varieties.

If you are a grower, in the cooler climates of North America, Europe or Northeast Asia, the reasonable thing is to grow hybrids. Heck, even Bordeaux has broken down, and now allowed a select few hybrid grapes to be grown. But why is it advantageous to grow hybrid grapes? Several reasons. First, they grow on their roots, and some argue that this produces a more authentic wine. Many hybrid grapes are more fungus disease resistant. They require fewer sprays in the vineyard. That means less chemicals in the air and soil to break down, fewer hours in tending these vines, less fossil fuel emissions, and a smaller carbon footprint. Not the least of which, the wines produced are something other than Cabernet Sauvignon or Chardonnay. These grapes offer up a great new fruit profile, and make new, exciting, and deliciously unique wines.

Steve was also one of the people who showed that one could make true, quality wines with hybrid grapes. Steve has been involved in winemaking since he was a teenager, where he worked as a cellar rat at Benmarl Vineyards under tutelage of Mark and Dene Miller and their sons Eric and Kim for many years. Steve combined his knowledge of grape growing, as well as involving the same care and skill it took to make great vinifera wines. We pulled leaves, we dropped fruit, aged the wines in older French and American oak, to make, in Steve's words, "soft, supple, and approachable, dry reds."

Hudson-Chatham benefited from two other individuals – Dominique DeVito and Bryan Van Deusen. Dominique was a unique sales and marketing force, whose palate we relied upon for honest feedback. Bryan Van Deusen, who worked for us for ten years, as Tasting Room Manager, then General Manager and Winemaker, soaked up much of Steve's recipes and advice. He became a talented winemaker in his own right. In the second half of the winery's first 15 years, Bryan helped elevate the wine program to new heights! The wines we made as a team with Steve at Hudson-Chatham Winery broke the glass ceiling of what red hybrid wines could do. Our Baco Noirs and Chelois earned rave reviews. Our Middlehope Baco Noir was the first dry red hybrid wine to break the 90-point ceiling. Our wines were featured in several *Edible Communities* magazines, *Forbes*, *Wine Enthusiast*, *The New York Times*, *The Washington Post*, *Times Union*, *New York Post*, *New York Daily News*, *Sommelier* magazine, *Beverage Media*, *Hudson Valley* magazine, *Valley Table*, and featured by countless bloggers and influencers. Our Baco Noir and Chelois were served in some of the best restaurants in New York, including Rouge Tomate, where they were selected by Pascaline Lepeltier. The wines were regulars on the wine list at the Culinary Institute of America in the Hudson Valley. The culmination of our winery was when it was named Best Winery in the Hudson Valley in 2019.

Most importantly, Steve is a historian. For more than two centuries, the Hudson Valley was one of the largest grape growing regions in the country. Hundreds of tons of grapes were delivered from the banks of the Hudson River to the city's piers, on Hudson sloops and later steamships, with announcements of their arrival in many of the city's newspapers. Many great hybridizers called the region home. And Steve has brought back many of their grapes in his Hudson Valley Heritage series of wines at Milea Estate Vineyards.

Steve has also brought forth, in this new edition, exciting new information on all the great grape breeders of New England. Did you know the Concord grape was developed in Massachusetts? And Steve also has extensive new histories and information on many of the Rogers hybrids, as well as others.

And Steve isn't just a dithering historian, with his nose pressed between the pages of old, musty, dusty books. Steve has one of the best, largest, and privately owned experimental vineyards in the Northeast. I remember touring his vineyard one season, along with his small and steadfast black dog, Nixon, and he proudly stated, "I now have all the Rogers varieties planted here, as well!" That's the thing about Steve, he's actually grown most of the grapes he's written about here. And developed a few of his own, including Lynnwood, Annie Noir, and Pallmer. Steve is also involved with the CANR Experimental Vineyard at SUNY Cobleskill that is monitoring, evaluating, and ultimately recommending cool climate grape varieties for those in the Upper Hudson Valley, and Mohawk and Schoharie Valleys.

And just as well, Steve has spoken throughout the Northeast United States, at the Eastern Winery Exposition, American Wine Society, the Peabody Essex Museum, the Massachusetts Horticultural Society, Tower Hill Botanic Garden, Queens Botanical Gardens, and Berkshire Botanical Garden, and other prestigious venues. He is also regularly published by the Arnold Arboretum (*Arnoldia*), University of Massachusetts Amherst – Stockbridge School of Agriculture, and Rutgers University, and supplies heritage grape varieties for vine exhibits for botanical gardens such as Bartram's Garden in Philadelphia, the Horticultural Society at Elm Bank, in Wellesley, and the Peabody Essex Museum homes and gardens in Salem, Massachusetts.

The first addition of this book was a landmark book to detail the genetic history, cultivation and winemaking attributes of these cool climate grape varieties, and the wines that they produce. This second edition is so much more as it covers the rarely covered New England grape breeders, their grape varieties, and the significance of these New England heritage grape varieties. Congratulations to Steve for another significant advancement for American horticulture and cool climate viticulture.

Carlo DeVito
Author and winemaker

Fruit Growing Areas
of the Hudson Valley

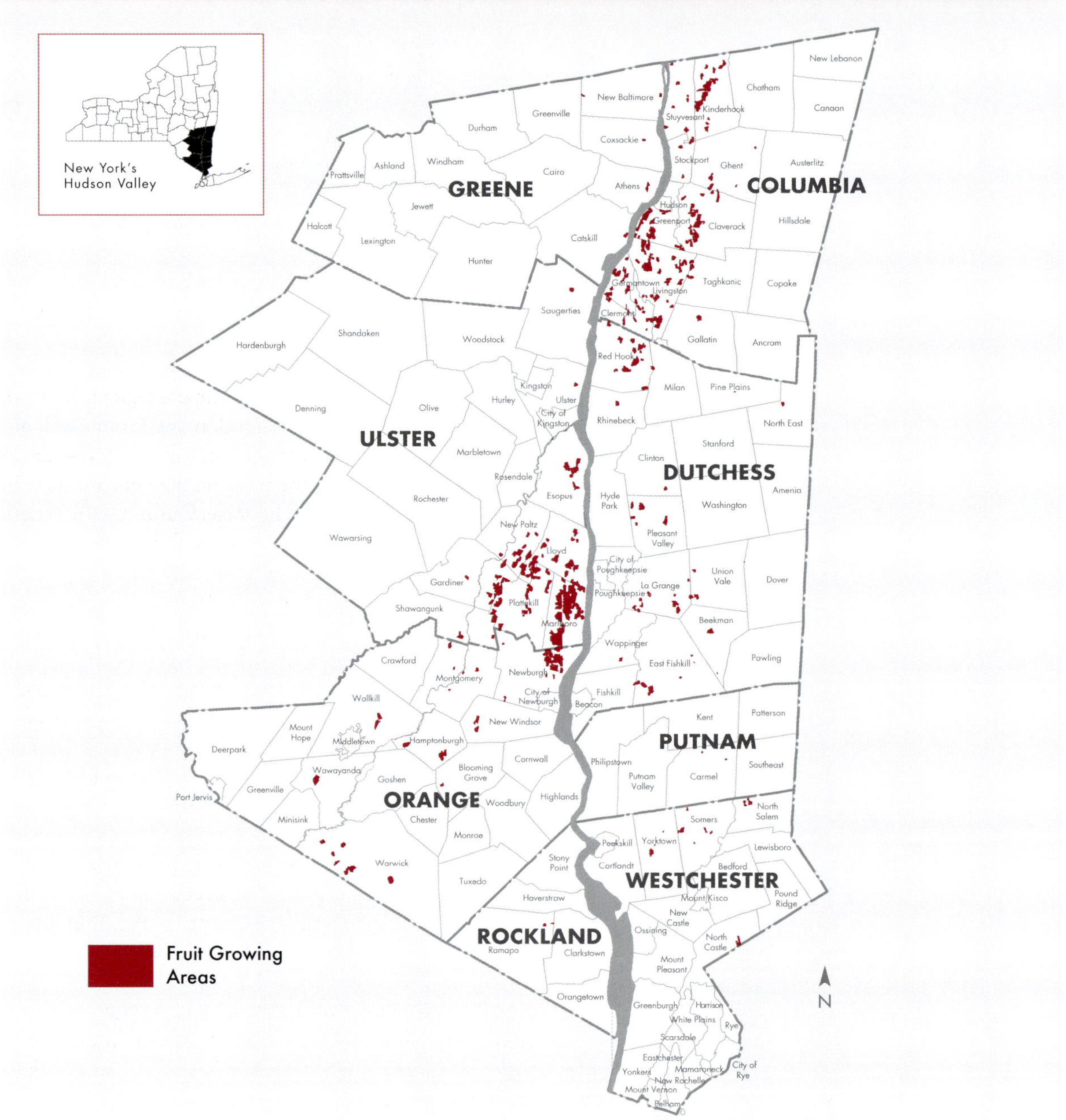

Source: Adapted from The New York State Senate Task Force for Hudson Valley Fruit Growers. Action Plan Final Report. Albany, NY: December 2004.

Introduction

This book has three separate but reinforcing goals. First, it identifies those grape varieties suitable for cultivation in the Hudson Valley for winemaking and, in some circumstances, table use. These varieties may also be suitable for cultivation in other cool climate areas of the northeastern United States, the Midwest, the Pacific Northwest, and Canada. Second, the book will specify the kinds or styles of wine that can be produced from these grapes. This may, at some points, lead to a discussion of various winemaking techniques. However, I tried not to dwell too much on how to make wine as this topic has been covered extensively in other texts (see the end notes).

Third, the book discusses the genetic makeup of the grapes commonly used for winemaking in the Hudson Valley and other cool climate regions. Many do not realize that all grapes—from the "noble" Chardonnay, Pinot Noir, and Cabernet Sauvignon to the "lowly" *labrusca* varieties such as Concord—are all hybrids of one sort or another. Like people, there are no pure grapes, they are all hybrids. This book will outline their genetic heritage. Outlining the genetic makeup of these varieties should help the reader to more fully understand the growing characteristics and winemaking capabilities of these varieties.

This book is unique as the grapes discussed are not organized in alphabetical order or by the color of the grape, the region where it came from, its degree of winter hardiness, or its disease-resistant qualities, but by the hybridizer who either purposefully bred or found these grape varieties in his or her travels along the road or in the gardens of acquaintances. By organizing the presentation of interspecific hybrid grapes by hybridizer, we can begin to understand the goals and work of each hybridizer, as well as the qualities of each individual grape hybrid. Also included is a brief biography of each hybridizer and where he or she did their work.

This book was organized so that we may better understand, in their totality, the individual grape varieties, their genetic history, the context in which they were created, and their strengths and weaknesses, both in the field and in the cellar.

This book is a practical field guide for those who want to successfully grow grapes in the Hudson Valley and other cool climate regions. It is also meant to assist those who want to successfully vinify the fruit of the vine into wine for commercial sale or their own personal consumption. I have not seen a book for the eastern part of the United States that is both a practical field guide for growing specific grape varieties as well as a guide for making such grapes into wine. In my discussions with other grape growers in cool climate areas, they too found that there were few satisfactory books that comprehensively covered the many grapes that can be profitably grown in cool climate areas and that also detailed their attributes in the cellar. Those who are interested in growing grapes for the first time will also find this book useful.

As I conducted research on local grape varieties and those who hybridized them, I realized that there were no books that accurately presented the rich history of grape growing in the Hudson Valley or that placed the Valley in the correct historical context and discussed its significance within the history of the American wine industry. Hence, the scope of the book grew to address these shortcomings. I hope that you will enjoy reading about the many titans of the American grape and wine industry who came from the Hudson Valley as much as I enjoyed researching them.

After embarking on this project, I explored my own family history in the Hudson Valley and found, to my surprise, that my ancestors were closely associated with the leaders of the Valley's grape and wine industry. My great-great-great-great-great-grandfather, James Albert Casscles, operated a sawmill with his three sons on Croton Point from the mid- to late 1700s to the early 1810s, at about the same time when the Underhill family operated a gristmill there and were the first to establish a vineyard on Croton Point and hybridize their own new grape varieties.

From Croton Point, my ancestors moved across the Hudson River around 1820 to Caldwell's Landing (also known as Jones Point) and to nearby Tomkins Cove, Rockland County, New York, where they manufactured bricks and fished for a living. Meanwhile, across the river at Croton Point, the Underhills were also manufacturing bricks. Five miles to the north, at Iona Island on the Hudson River, Dr. Charles W. Grant was establishing his own vineyard and undertaking hybridization work on grape varieties such as Iona.

In 1871, along with William Springstead, William Tomlins, and George King, the Casscles family established the Episcopal Church at Caldwell's Landing and The House of Prayer at Tomkins Cove. Both of these churches remain standing at Jones Point and at Tomkins Cove, and Episcopal services continue to this day at the Chapel of St. John the Divine at Tomkins Cove.

Sometime in the 1870s, my great-great-grandfather, Alonzo Palmer Casscles, and his family moved to Marlborough (Marlboro), New York, some 25 miles north of Tomkins Cove, to establish a small nine-acre fruit farm on Highland Avenue to grow grapes, red currants, raspberries, and some tree fruits. This farm was located directly across the street from the farm and nursery owned by Andrew Jackson Caywood, the famous grape breeder. That farm is now called Benmarl Vineyards, and is reputed to be the oldest continuously farmed vineyard in the United States. The Casscles raised three more generations of fruit farmers there who knew and worked with Caywood's descendants, the Wardells.

In 1951, my grandfather, Joseph Levi Casscles, sold the family farm in Marlboro and moved to the farm of his wife's parents—John and Mary Sabo—located one-and-a-half miles to the south on U.S. Route 9W. That is where I grew up and where I still maintain a small three-quarters of an acre vineyard. I got my interest in growing grapes from my grandfather, and worked at Benmarl Vineyards on and off from 1973 to 1986. I happily count the members of the Miller family, the then owners of Benmarl Vineyards, as my friends.

The family of my paternal grandmother, Rose Louise Sabo, came from the small village of Cseke or Csike, near Sátoraljaújhely, in northern Hungary, near where the famous Tokaji dessert wines are produced. The Sabos (or Taylor in Hungarian) married into the Hungarian families of Sedlak and Horaz in the Hudson Valley. In 1905, in light of the tensions brewing between Russia and the Austro-Hungarian Empire, and since they lived on the border of these two antagonistic countries, they wisely decided it was time to leave their homes for Roseton, New York, far away from the fighting that became World War I, to a place that specialized in the manufacture of bricks.

I have been growing grapes at my vineyards in Middle Hope, New York, since 1976, and in Athens, New York, since 1990. Also, I am the winemaker at the Hudson-Chatham Winery, in Ghent, New York. As for my day job, I have worked happily in the New York State Senate for the past twenty-eight years as an attorney, where I have written over twenty-one laws that relate to the production, sale, and distribution of wine and distillants in this state.

As a teenager, when I first began to think more seriously about growing grapes as an avocation, I looked to others for advice on the kinds of grapes that could be profitably grown in the Hudson Valley and the kinds of wine they made. I talked extensively with my grandfather, Joseph Levi Casscles; my then employers, Mark Miller and his sons Eric and Kim Miller of Benmarl Vineyards; Robert (Bob) Pool at the New York State Agricultural Experiment Station in Geneva, New York; and Philip Wagner at Boordy Vineyards and Nursery. I particularly appreciated the advice, guidance, and time that Philip Wagner and Eric Miller gave me. Philip Wagner's advice came at a critical time in both my life and his.

In 1989, I had just purchased, with my wife Lilly Casscles, 12 acres of land in Athens, New York, in the Hudson Valley, where I planned to establish a vineyard and fruit farm. During this period, I was actively looking for new and old grape hybrids that could be grown at my farm, now named Cedar Cliff, to make quality wines. I relied on my old favorites that were grown at Benmarl, such as Baco Noir, Chelois, Maréchal Foch, Léon Millot, and Chambourcin. To these I wanted to add newly developed grape hybrid varieties. However, Philip Wagner suggested that instead of looking to the newer varieties being developed by Cornell University and others, I should look back to the past seventy years of breeding activity for those hybrids developed in France by Joanny Burdin, Pierre Landot, Albert Seibel, and members of the Seyve family—Bertille Seyve Sr., Bertille Seyve Jr., and Joannes Seyve.

Philip knew the Albany, New York, area, near where my farm is located, because he had worked in public relations at General Electric, in Schenectady, New York, before going to the *Baltimore Sun* in the early 1930s. His firsthand knowledge of our Hudson Valley weather was of great assistance in advising me on the grape varieties that could grow here.

As I mentioned above, the early to mid-1990s was a critical time for both Philip Wagner and me. While I was now beginning to embark on my new career to identify and grow quality grapes to be made into wine, Philip was winding down his own long and distinguished grape-growing and winemaking career. His wife Jocelyn was ill and he wanted to wrap up his nursery business affairs. In the years after 1991, I purchased a few of Philip's favorite hybrids. I was particularly surprised and honored in the mid-1990s, when I had ordered no more than five different varieties, and he sent me, at no charge, over twelve varieties with a note that it would be best to pass them on to someone who was younger and able to pursue the cultivation of these grapes and evaluate their merits. Of his favorites, the Burdin hybrids stood out with grapes such as B.6055, B.11402, B.8753, B.4672, and B.5201, and several Landot and Seibel hybrids. I have been evaluating Wagner's favorite varieties ever since, and find that most of them are good both in the field and in the cellar.

Philip Wagner helped to change my views on grape hybridization and the selection of appropriate grapes. He believed that many of the varieties developed in France in the early to mid-twentieth century had not received a fair shake. The critical evaluation of these grapes was hindered starting from the tense period preceding World War I to after World War II in the late 1940s. These varieties needed more careful consideration before being dismissed in favor of the newer grape varieties that were being developed from the 1970s onward.

This book reintroduces the general public and growers to grape varieties that have been forgotten or nearly forgotten. I think that the reader will see the merits of some of the grapes profiled in this book, and hopefully the book will provide solid guidance to both the experienced and novice grape grower alike in our cool climate areas of North America.

As a member of a family that has been cultivating grapes in the Hudson Valley for some 150 years and as someone who has been growing grapes and making wine from them since 1976, it is with special pride and obligation that I write this book to document a portion of the history of grape growing and winemaking in the Hudson Valley. I hope to share some of the knowledge I have gleaned from my own experiences during the past thirty-five years, and to pass on to others the knowledge that has been conveyed to me by older, much more experienced growers and winemakers who have successfully made a living in this tough but rewarding profession. Some of these recollections are from individuals who worked in the wine and grape industry or who knew people who worked in the industry well before World War I. This book is dedicated to them.

J. Stephen Casscles
Athens, New York
January 30, 2015

Introduction to the Second Edition

It has been seven years since the first edition of *Grapes of the Hudson Valley and Other Cool Climate Regions of the United States and Canada* was published in 2015. The first edition sold out more than two years ago, but with the setbacks of the coronavirus pandemic (COVID-19), it has taken some time to bring this second edition to print.

In this edition, there is an added section containing two chapters dedicated to New England grape breeders. I was initially attracted to the New England heritage grape varieties because of my personal interest in eastern Massachusetts—I worked and lived in Boston for three years while at Northeastern Law School (1981-1984), and have vacationed with my family in the Rockport/Gloucester area of Massachusetts for the past thirty years. During our summer vacations on the North Shore and my many trips to the Concord and Groton areas, and on sojourns to the botanical gardens at the Arnold Arboretum, Massachusetts Horticultural Society, Tower Hill Botanical Garden, and the gardens at the Peabody Essex Museum, I became acquainted with the places that these New England breeders lived and worked.

Coupled with my own experiences growing heritage grape varieties, these places instilled in me the desire to include the fascinating, but untold, history of New England grape breeders in a future edition of this book. As I conducted my research on the New England breeders, I found that there were many connections between their work and the work of Hudson River Valley heritage grape breeders, nursery owners, horticultural societies, and horticultural writers. The two additional chapters included in this second edition presents the resultant creations of these breeders and their relationships with the Hudson Valley.

As with the first edition, the goal is to detail the numerous cool climate grape varieties that are suitable for cultivation in the Hudson River Valley, New England, and other cool climate regions of the United States and Canada. I have also noted those cool climate grapes which are gaining popularity in Northeast Asia or which could be successfully grown there. Related to the first point, this book not only details the cultivation attributes of such varieties, but includes information on the kinds of wine that they produce.

The second goal is to incorporate details about the lives of these cool climate grape breeders, including those newly added New England breeders, and their hybridization goals, and to include information to help the reader understand how their lives influenced their work and the grape varieties they created.

What is different between the publication of the first and second editions of this book is the radically different political and perceived environmental and climatic conditions faced by our world today. While it has only been eight years from the publication of the first edition, it is clear now that our planet is more rapidly warming than we had first thought and future weather patterns will not only have increased incidences of warmer temperatures, but that we will be experiencing more violent weather patterns and higher precipitation levels. This makes the content and context of this book more relevant today, as large commercial growers are now actively searching for more resilient grape varieties that can be grown profitably in this new warmer and wetter climatic reality, and who may wish to establish vineyards further north than had been possible before the advent of global warming.

The first fourteen chapters of the first and second editions of this book are virtually the same. However, it has come to our attention that through DNA testing, some of the genetic heritages that were reported in the first edition needed to be updated. Hence, revisions in the second edition have been made to the genetic make up of the grape varieties Catawba, Concord, Chambourcin, and Vignoles. It is important to note that while many had thought that Concord was a pure *labrusca* variety, it turns out that it is an off-spring of Catawba, whose pollen parent is the *vinifera* variety Sémillon. This means that Concord possesses significant *vinifera* genetic heritage, and consequently, all of the many hybrid offspring of Concord reported in this book also have *vinifera* genetic heritage. It is important to document the true genetic heritage of Catawba, Concord, Chambourcin, and Vignoles since they are major cool climate grape varieties.

This second edition adds two new chapters to detail my work over the past fifteen years with New England cool climate heritage grape varieties developed by Rogers, Bull, Clough, Crehore, Allen, Moore, and others. I was attracted to these New England heritage varieties, especially those bred by E. S. Rogers, because he was one of the first in North America to successfully breed a number of quality grape varieties that were winter hardy, fungus disease resistant, productive, and produce quality wine and table grapes. The work of Rogers that began in 1851 was followed and replicated by Hudson Valley breeders such as J. H. Ricketts, the Underhill family, and others. Working with and evaluating these heritage grape varieties during these past years, I can now share my thoughts on how these grape varieties perform in cool climate areas.

Since the publication of the first edition, I have also become more aware of the burgeoning grape and wine industries of South Korea and Japan. The northeast sections of the United States and Asia share similar climates, native local flora, and grape varieties that can be cultivated successfully. Many of the cool climate grape varieties covered in this book have applications to the expanding grape growing regions of South Korea and Japan. Already, we share several grape varieties that we commonly grow, such as Delaware, Campbell Early, Verdelet, Riesling, Cabernet Franc, and others. It is my hope that this book will foster more communication and cooperation between our two grape growing regions.

The work does not stop here. For the past thirty years, I have been simultaneously cultivating and analyzing several of the lesser known French-American hybrids as well as some chance seedlings that I have come across at my farm Cedar Cliff in Athens, New York. Among these promising cool climate grapes are Baco Blanc, Bell, Bertille-Seyve 3408, Boordy White Rogue, Burdin 4672, Captivator, J.S. 12.428, Seibel 10.868, Seibel 13.047, Seibel 13.680, and Valarian, as well as chance seedling varieties that I have identified and named Annie Noir, Lynwood Noir, and Pallmer.

It is my hope to report upon these promising "new" varieties for the benefit of cool climate grape growers and winemakers in North America and Northeast Asia in a third edition of this book, should such an opportunity present itself.

With all of this said, I wish to you all, cheers!

J. Stephen Casscles
Athens, New York
February 6, 2023

Acknowledgments

This book, like most books, was not written by any one person. It is the outgrowth of shared interests and decades of working in the field and discussing the topics of grape growing, making wine and consuming wine over meals with friends, neighbors and associates in the Hudson Valley and in other cool climate areas of North America. It is also the result of our shared interests in putting down on paper the scarcely documented rich history of grape growing in the Hudson Valley and the unique place that this Valley has had on the development of America's wine industry.

With that said, I would like to first thank and acknowledge, my grandparents Joseph L. Casscles and Rose Casscles, and my parents Joseph P. Casscles and Janet M. Casscles for their guidance, inspiration and historical perspective which eased me into this fascinating field, and to my sisters and brother – Lorraine Casscles, Karen Williams and Brian Casscles – and my Uncle Paul and Aunt Elaine Zamenick.

A special thanks goes to my fellow winemakers and partners in crime for many years, first and foremost to my own family – my wife Lilly Casscles, sons J. Benjamin and Noah Lovick Casscles, and my daughter Grace Anne Elizabeth Casscles – who have helped me grow grapes and make wine for over twenty years and suffered through many a trying growing and harvest season, thanks again; and to John and Barbara Bellucci, Eric and Lee Miller, and the entire Miller Family, Mark and Dene Miller, and Kim and Mary Miller at Benmarl Vineyards, Carlo and Dominique DeVito at Hudson-Chatham Winery, Tim Biacalana of Riverview Winery, and my wine making buddies Marty Wallace, Nick Cahill, Dylan and Dawson DeVito and the boys from UNICO in Marlboro, New York; Cesar Baeza of Brotherhood Winery, Kevin Zraly, the late Philip M. Wagner of Boordy Vineyards; growers and home winemakers that I have learned so much from, some of whom have passed on, Lawrence Bealli, Ida Martini; and my neighbors in Middle Hope, Florence Carlino and her son Matty Carlino, and John Hudelson.

Also, a special thanks to those I have worked with and from whom I have obtained great insights into fruit growing and the history of fruit growing in the Hudson Valley: the Greiner family (David, and sons Mark and Eric) and Jack Baldwin; Richard Fino, Howard Quimby, Vito and Joel Truncali, Ernie Borchert, Johnny Corrado, and Lois and Amy Hepworth; and my fellow Hudson Valley winery proprietors Dick Eldridge of Brimstone Hill Winery, Michael Migliore of Whitecliff Vineyards, Mark Stopke of Adair Vineyards, and Doug Glorie of Glorie Farm Winery; and to my bosses, New York State Senator

William J. Larkin (R-Cornwall), Senator Jeffrey D. Klein (D-Bronx) and the late Senator Jess J. Present (R-Bemus Point).

I would also like to give an overall thanks to the many people who shared their experiences, photographs and background material with me: Robert Fakundiny, New York State Geologist, Doug Moorhead of Presque Isle Wine Cellars, Jim Trezise of the New York Wine & Grape Foundation, Hudson Cattell of *Wine East*, Bob Pool, Richard Vine, Carl Camper of Chateau Stripmine, George Gale, University of Missouri-Kansas City, Patrick Pierquet of Ohio State University at Wooster, William H. Shoemaker, Clifford Ambers, Lon J. Rombough, Mark Hart, Kenton Erwin, Tom Plocher, and Bob Bedford; and to Dr. David A. Rosenberger, Stephen A. Hoying, Michael Fargione and Peter J. Jentsch of the Hudson Valley Laboratory and Agricultural Experiment Station; Julie Suarez, Assistant Dean, College of Agriculture and Life Sciences, Cornell University; Dr. Bruce I. Reisch and Timothy Martinson of Cornell University; Peter Hemstad of the University of Minnesota; and Aaron Bouska of the New York Botanical Gardens.

There are many people who provided me with valuable research and background information and personal photographs which helped to bring the book to life as a celebration of those who worked in this field. I would like to thank Bernard Matabos, Maite Laberiotte, Chairman of the Cultural Center of Pays D'Orthe, the Pommies Family, Daniel Dufau, Mayor of Belus; Barbara Mathe, Museum Archivist and Head of Special Collections and Gregory Raml of the American Museum of Natural History and the keepers of papers of the John Burroughs Society, Pat Favata and Johanna Porr, Historical Society of Newburgh Bay & the Highlands, Mary McTamaney, City of Newburgh Historian, Dorothy R. Pezanowski, Village Historian of the Village of Croton-on-Hudson, and the Croton-on-Hudson Historical Society, Marc Cheshire, Croton Friends of History, U.S. Congressman Chris Gibson's Office and Constituent Services Representative Kathy Fallon, Diane Wunsch, Special Collections, National Agricultural Library, Donald "Doc" Bayne, "the Mayor of Iona Island," Irving Geary, Vice President of the Minnesota Grape Growers Association, Patrick Raftery, Librarian, Westchester County Historical Society, Cathy Roy of the Niagara Falls Public Library, Ontario Canada, Sue Smith of the Palisades Interstate Park Commission, Gan-Yuan Zhong, PhD, United States Dept. of Agriculture, Agricultural Research Service, and Michael P. Fordon, Library Coordinator, Frank A. Lee Library, NYSAES, Cornell University.

A note of personal thanks to those who offered advice and continued to cheer me on during this more than ten-year-long process to bring this book to fruition, with a special thanks to Dawa Jung for her interest, time and proofing the manuscript, and to Teresa Rossi, Karen Bennett-Geel, Chris Klaeysen, and Gabriel Paniza for their encouragement and help.

Without the work of editors, designers, and proofreaders, this book would not be in its current form and far less readable; a thank you to Scott Krontilik, who created many of the illustrations in the book, and to Valerie Ahwee, our able copyeditor. I would like to thank my friends and publishers Bob Bedford and Linda Pierro, whose expertise, dedication, and attention to detail was fundamentally important in all stages of this process. Without their involvement this book would not have been produced.

Acknowledgments
for the Second Edition

This second edition, like the first edition, was not written by one person. It is the outgrowth of shared interests and decades working in the field and discussing the topics of grape growing, making wine, and consuming wine with friends and colleagues. I would like to thank and acknowledge the following people for their assistance and knowledge that helped make this edition possible.

A special thanks goes to my winemaking pals who make our wines year in and year out, and who have now specialized in the production of New England heritage wines, including, Marty Wallace, Michael Swanson, Vinnie Adamo, Doug and Carol Malloch, Louis Bush, my sons Ben and Noah Casscles, Laura Stigter Carino, Bill Sproat, Peter McLaren, and my wife Lilly Casscles who often provides us with the sustenance needed during our long winemaking days.

I wish to thank the staff at the University of Massachusetts Amherst, Stockbridge School of Agriculture and UMass Extension Fruit Program, Professor Emeritus Dr. Wesley Autio, Professor Sonia Schloemann, and Dr. Elsa Petit; Professor Emeritus Win Cowgill, Rutgers University and Editor of the Horticultural News, New Jersey State Horticultural Society; the staff at the Arnold Arboretum, Jamaica Plain, Massachusetts, Dr. Michael Dosmann, Jonathan Damery, Editor of *Arnoldia*, and Lisa Pearson; Dan Lipcan and Robin Pydynkowski of the Peabody Essex Museum, Salem, Massachusetts; Dr. Leslie Judd, Professor David Campbell, and Jed and Jaime Radcliff of SUNY at Cobleskill, Fermentation Sciences Program; Allison Dush, Massachusetts Horticultural Society; Todd Trzaskos, Vermont Grape and Wine Council; Susan Lacerte and Rebecca Wolf, Queens Botanical Garden, Queens, New York; Andrew Farmer, Northeastern Vine Supply; David Falchek and Jim Rink, American Wine Society; Carlo DeVito, Dominique DeVito, and Bryan Van Deusen, the owners of the Hudson-Chatham Winery (2007-2020), where we made some spectacular wines from French-American hybrids and heritage Hudson Valley and New England Heritage grape varieties; the entire staff at the New York Wine and Grape Foundation; the staff at the Tower Hill Botanical Gardens, Boylston, Massachusetts; the staff at the Berkshire Botanical Garden, Stockbridge, Massachusetts; and a special shout out to my colleagues and partners-in-crime at Milea Estate Vineyards, Staatsburg, New York—Barry and Sang Milea, Bruce Tripp, and Edward Evans—who have graciously allowed me to make wines for Milea's Hudson Valley Heritage Grape Project, including these Hudson Valley and New England heritage wines that are covered in this book.

And finally, many thanks to my editors and publishers Robert Bedford and Linda Pierro, for all of their help and guidance to make this second edition all that it can be.

Hardiness Zones

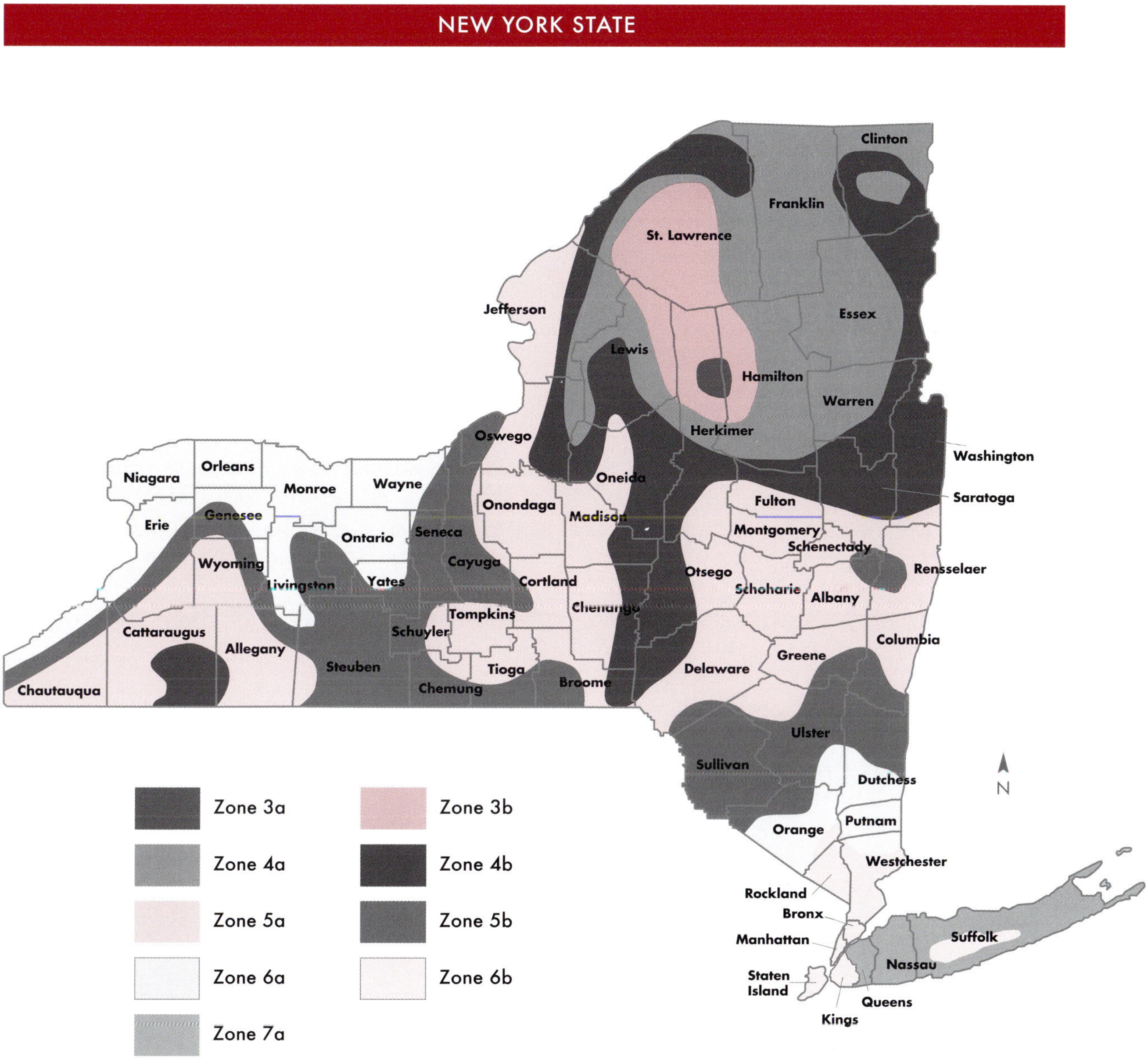

Source: USDA Agriculture Research Service, New York Plant Hardiness Zone Map.

How to Use This Book

Each major grape variety detailed in this book contains a key to its growing characteristics—color, harvest dates, winter hardiness, fungal disease resistance, vigorousness, productivity—and its wine quality, based mainly on my own observations in the field and in the cellar. The variety's parentage is also included.

While other guides have tended to rank varieties with a numeric rating system, this book adopts a letter grading system which should be clearer and more self-explanatory for the reader, hence an **A** is better than a **B**, and a **B** is better than a **B-**, etc.

An explanation of harvest dates in the Hudson Valley, and a key to winter hardiness terms are explained below.

While all varieties contained in this book can survive in the Hudson Valley, some will need additional work to be grown successfully, while others will be virtually indestructible in the field.

For those readers in other cool climate growing regions, they merely need to compare the general growing conditions in their regions to the Hudson Valley. Having a single reference point should benefit all readers no matter their location.

HARVEST DATES IN THE HUDSON VALLEY

Very early	Late August to around Labor Day
Early	Labor Day (September) to one week after Labor Day
Late early	September 10th to 20th
Early mid-season	September 15th to 25th
Mid-season	September 20th to 30th
Late mid-season	September 30th to October 5th
Early late	October 5th to October 15th
Late	October 15th to 25th
Very late	After October 25th

GRAPE COLOR

Red   Pink   White   Red/Black

WINTER HARDINESS

Very hardy	Nearly indestructible in all parts of the Hudson Valley, surrounding foothills and nearby Catskill and Berkshire Mountains, even if soils are not optimum well-drained soils.
Hardy	Generally winter hardy and can produce a solid commercial crop even after a very harsh winter. Can tolerate most soil types including wetter sites.
Medium hardy	Will sustain some cold damage in harsh winters. Should be grown only on good well-drained soils, but will crop on wetter sites.
Medium tender	Needs to be grown in the warmer parts of the Hudson Valley on the best, well-drained soils. Even so, it will generally sustain some cold damage in the harshest of winters.
Tender	Needs optimum growing sites, optimum well-drained fruit soils, in the warmest parts of the Hudson Valley that are sheltered from damaging winds, but will sustain cold damage in the coldest of winters.

NAMED VARIETY

HYBRIDIZER DESIGNATION

GRAPE COLOR

PARENTAGE

HARVEST DATE

SEYVAL BLANC (S.V. 5-276)

Seyval Blanc is one of Bertille Jr.'s first early successes in hybridizing. This white wine grape was developed in 1921.[48] Galet and Morton maintain that Seyval Blanc is a cross of S.5656 × S.4986 (Rayon d'Or),[49] while the Geneva *Vineyard and Cellar Notes* maintains that the cross may have been S.4995 × S.4986.[50] It is agreed, though, that Rayon d'Or (S.4986) is one of its parents and shares this parent with Vidal Blanc. This grape, after its popularity began to grow, became known in France as Seyval Blanc,[51] the name being the contraction of Seyve and Vallier.[52] Seyval Blanc has been used to breed Cayuga White, Chardonel, La Crosse, Melody, and St. Pepin.

PARENTAGE
lincecumii, rupestris, vinifera

HARVEST DATE
Mid-season to late mid-season

A A– C A A

The grape is adaptable to different regions and climates and is grown throughout the eastern United States, northern France, and England.[53] It is a versatile grape that can be made into many different wine styles, such as very fruity, semi-dry Germanic whites; Sancerre-like and even semi-loose, depending on the clone and the soil that it is grown on, with berries of medium size. The compactness of the clusters can lead to berry splitting at harvest time, particularly if it rains before harvest. This berry splitting leads to bunch rot. Seyval grows on a standard-sized growth habit and should be spur pruned to reduce the risk of overcropping and to balance its crop load to the quantity of grapes that the vine can fully ripen. Seyval can be subject to poor fruit set if the spurs are pruned too short and overcropping if pruned too long.[54]

KEY TO ICONS AND RATING SYSTEM

WINTER HARDINESS	FUNGAL DISEASE RESISTANCE	VIGOROUSNESS	PRODUCTIVITY	WINE QUALITY
A+ Very Hardy	A Slightly susceptible	A Very vigorous	A+ Very productive	A+ Very high
A Hardy	B Moderately susceptible	B Vigorous	A Productive	A High
B Medium hardy	C Very susceptible	C Moderately vigorous	B Moderately productive	B Medium
C Medium tender	D Extremely susceptible		C Low productivity	C Low
C– Tender				

CHAPTER ONE

A Short History of Viticulture in the Hudson Valley

OVER A CENTURY AGO, the prominent horticulturalist, writer, and educator Dr. Ulysses Prentiss Hedrick stated that "the Valley of the Hudson has more reason to be called the birthplace of American viticulture than any other of the grape-growing districts of this country." The confluence of propagating new and old grape varieties, cultivating grapes for both the table and wine production, and extensive breeding operations by more than several nationally recognized breeders to create new superior varieties for the domestic table grape and wine industry can lead one to maintain that the Hudson Valley is truly the birthplace of American viticulture.[1]

In the twentieth century, U. P. Hedrick (1870–1951) was singularly one of the most important chroniclers of the American grape and wine industry. He was a horticulturalist, director of the New York State Agricultural Experiment Station at Geneva, New York, and wrote landmark books such as: *The Grapes of New York* (1908), *Manual of American Grape Growing* (1919), *A History of Agriculture in the State of New York* (1933), *Fruits for the Home Garden* (1944), and *Grapes and Wines from Home Vineyards* (1945).[2]

It is here in the Hudson Valley where America's oldest commercial winery—the Blooming Grove Winery, opened in 1839—continues to operate today as the Brotherhood Winery in Washingtonville, Orange County. Further, Benmarl Vineyards, one of the oldest continuously farmed vineyards in America, is located in Marlboro, Ulster County.

Vineyards in the Hudson Valley are generally found in the geological division known as the Taconic Province, a broad valley that extends from Pennsylvania across New Jersey, taking in Orange County and parts of Ulster, Dutchess, Columbia, and Greene counties before extending into Massachusetts. The rocks of this geological division are shales, slates, schists, and limestones. The soils are derived from these rocks. The grape lands, for the most part, are those areas in which there is a preponderance of coarse shale or slate fragments, with finer particles of clay or gravelly loams. This district, unlike others in New York State, is more or less hilly, with some vineyards located in broad, gently undulating plains, some on knolls or hollows of only a few acres, and others on steep slopes.[3] Until the late 1940s, the common practice was to plant vineyards up and down a hill to negate the need for terracing and to accommodate the steep terrain and shallow soils in some places.

An October Scene on the Hudson. *Gathering grapes in a commercial vineyard in the mid-Hudson Valley. Etching from Harper's Weekly, October 26, 1867.*

The Hudson Valley's climate changes rapidly as one goes north up the Hudson River and away from the river's shores. This variation in temperature is partly due to the diversity in physical features that separate the land from the river and the proximity of the Catskill and Berkshire mountains. In the fruit-producing areas of the Hudson Valley, summer temperatures are high owing to the position of vineyards between mountain ranges and to the warm southerly winds that prevail in the summer. In winter, the winds can be more northerly out of Canada and temperatures can be relatively low, which makes the culture of cold-sensitive grape varieties more difficult. Overall, however, the influence of the river, which is really a broad tidal estuary or fjord, is most favorable to growing fruit crops.[4]

The lowlands of the Hudson Valley receive somewhat less rainfall when compared to the rest of New York because when moisture is carried inland from the Atlantic Ocean, it falls as rain over the Hudson Highlands and the mountains and highlands of New England. Another desirable characteristic of the Hudson Valley's rainfall patterns is that, on average, July rains are more plentiful than in the rest of the state, but reduced in September and October during the critical harvest season. This relatively light rainfall in the fruit's maturing months is more marked in the Hudson Valley than in other New York grape-growing regions.[5]

Early Settlers in the Hudson Valley

Some of the earliest observations and reports by European explorers, such as Henry Hudson and Johannes Megapolensis in the early- and mid-1600s, was that the Hudson Valley and other areas of the eastern seaboard from Nova Scotia to the Carolinas were very favorable environments for the cultivation of grapes. This was evidenced by the number of wild grapevines growing in the Hudson Valley and most of the eastern seaboard. These early explorers found many grapes "as good and sweet as in Holland" and in such quantities that they inspired the launch of viticulture in the newly settled Hudson Valley. In 1646, Johannes Megapolensis stated that "if people would cultivate the vines they might have as good wine here as they have in Germany or France. I had myself last harvest a boat-load of grapes and pressed them. As long as the wine was new it tasted better than French or Rhenish Must, and the color of the grape juice here is so high and red that with one wine glass full you can color a whole pot of white wine." These very overoptimistic reports and statements, as well as those issued by the Dutch government, were used to entice Europeans to settle in the Hudson Valley.[6]

Henry Hudson's *Half Moon* on the Hudson River. *Early exploration of the region prompted the cultivation of grapes in the Hudson Valley by seventeenth- and eighteenth-century settlers.*

It was clear that the Hudson Valley was climatically a very suitable environment for the cultivation of fruit. Before the arrival of Europeans, the Native Americans supplemented their diet of game and fish with wild fruits that grew where orchards and vineyards now stand. These native fruits included various species of plums, grapes, raspberries, blackberries, dewberries, cranberries, gooseberries, huckleberries, and blueberries. In 1647, Peter Stuyvesant, the Dutch colonial governor, set out a farm he called the *Bouwerie* on the site of the present Bowery on Manhattan Island. On this farm he planted many fruits, and grafts from his own orchards were sent up the Hudson River to establish new orchards and vineyards.[7]

European colonists, upon arriving in the Hudson Valley, strove to make wine from imported and native fruits to help ease the burden of frontier life. The French Huguenots were the first to systematically cultivate grapes in the Hudson Valley for wine. Facing religious persecution in France, the Huguenots, also known as French Calvinists, a Protestant sect, began leaving France in large numbers in the seventeenth and eighteenth centuries. Many of those who came to America ultimately settled in Ulster County around De Paltz, now called New Paltz, and the Wallkill River Valley after 1677.

They brought cuttings of their European vines with them to establish new vineyards in their adopted land. Unfortunately, as was the case across the American colonies, these vines died from attacks of the soil-borne root louse called phylloxera, which killed the soft roots of European grapevines, also known as *vinifera* grapes.[8]

Packing grapes for market. *A sorting table used to trim and pack table grapes for market in the late nineteenth century, near Marlboro-on-the-Hudson. Illustration circa 1884.*

The few vines that survived phylloxera were killed by the Hudson Valley's harsh winters (much harsher than today's winters) and native fungal diseases such as black rot, downy mildew, and powdery mildew, which attacked both the fruit and leaves of those few struggling plants that remained. The Huguenots could not succeed with European or *vinifera* grape varieties, so they turned to cultivating wild grapes for fresh fruit and for the production of limited amounts of wine. The wine made from these native grapes, mostly *labrusca* and *riparia* varieties, was not similar to European-type wines at all and were generally considered to be unpalatable, so the Huguenots began to cultivate apples instead of grapes to produce hard cider as their main alcoholic beverage.

From the colonial era until the 1820s, the most common cultivation practice for growing orchard crops and grapes in the Hudson Valley was a slightly managed control of wild fruit varieties that were already growing on the farm or in lots behind houses in the villages. Often these fruits were barely pruned or not pruned at all. Certainly pesticides and insecticides were not applied. Neighbors shared cuttings of grapes and scions of fruit trees to establish new orchards and vineyards, but such varieties were not identified or named. Further, many new orchard fruit trees and grapes were propagated from the seedlings of locally grown varieties. There was little mention of specific names for these cultivars of fruits. An apple was an apple, and a grape was a grape.[9]

Practically all of the fruit grown in the Hudson Valley was used for the production of jams, jellies, preservatives, sauces, or alcoholic beverages for home consumption. There were very few large-scale commercial fruit farms dedicated to producing fruit that could be harvested, transported, and sold in urban areas. The economics for fruit cultivation began to change slowly after 1800 as evidenced by the increasing number of commercial fruit farms being established. Much of this change came about due to advances made by local horticulturalists; writers who disseminated new information on horticultural issues; the work of newly formed agricultural and horticultural societies; the expansion in the size and number of commercial fruit nurseries that sold clearly identified new grape cultivars; and more efficient modes of transportation to deliver grapes to urban markets.

Early Nurseries and Viticulturists

One of the first nurseries in New York was the Linnaean Botanic Garden, established by William Prince in Flushing, Long Island, a few years before the American Revolutionary War. This nursery was operated by four successive generations of the Prince family. For approximately 100 years, the Princes sold fruit plants to many farmers and townspeople who cultivated fruit on back lots throughout the Hudson Valley.

The Princes's collection of fruit trees, grapes, and small fruits included every hardy variety of fruit plant that could be obtained in the United States and

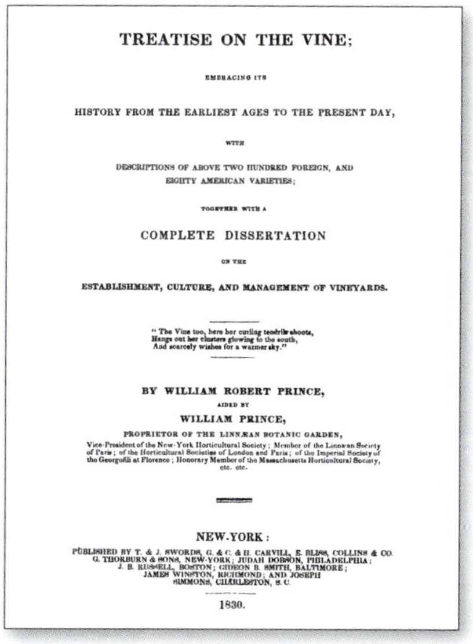

TREATISE ON THE VINE;

EMBRACING ITS

HISTORY FROM THE EARLIEST AGES TO THE PRESENT DAY,

WITH

DESCRIPTIONS OF ABOVE TWO HUNDRED FOREIGN, AND EIGHTY AMERICAN VARIETIES;

TOGETHER WITH A

COMPLETE DISSERTATION

ON THE

ESTABLISHMENT, CULTURE, AND MANAGEMENT OF VINEYARDS.

"The Vine too, here her curling tendrils shoots,
Hangs out her clusters glowing to the south,
And scarcely wishes for a warmer sky."

BY WILLIAM ROBERT PRINCE,

AIDED BY

WILLIAM PRINCE,

PROPRIETOR OF THE LINNÆAN BOTANIC GARDEN,

Vice-President of the New-York Horticultural Society; Member of the Linnæan Society of Paris; of the Horticultural Societies of London and Paris; of the Imperial Society of the Georgofili at Florence; Honorary Member of the Massachusetts Horticultural Society, etc. etc.

NEW-YORK:

PUBLISHED BY T. & J. SWORDS, G. & C. & H. CARVILL, E. BLISS, COLLINS & CO. G. THORBURN & SONS, NEW-YORK; JUDAH DOBSON, PHILADELPHIA; J. B. RUSSELL, BOSTON; GIDEON B. SMITH, BALTIMORE; JAMES WINSTON, RICHMOND; AND JOSEPH SIMMONS, CHARLESTON, S. C.

1830.

Treatise on the Vine, 1830. *One of the first horticultural manuals on grape growing and vineyard management was written by William Robert Prince whose Long Island, New York, nursery supplied fruit and grape plants to many Hudson Valley growers.*

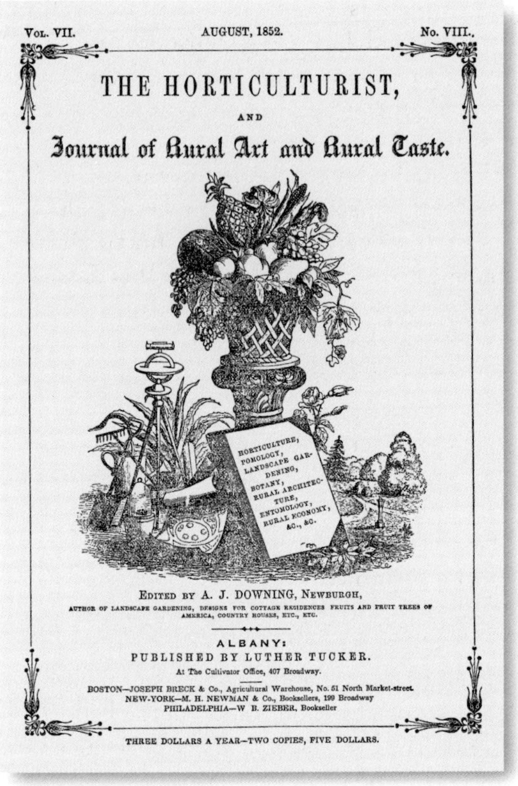
Vol. VII. AUGUST, 1852. No. VIII.

THE HORTICULTURIST,

AND

Journal of Rural Art and Rural Taste.

HORTICULTURE, POMOLOGY, LANDSCAPE GARDENING, BOTANY, RURAL ARCHITECTURE, ENTOMOLOGY, RURAL ECONOMY, &C., &C.

Edited by A. J. DOWNING, Newburgh,

AUTHOR OF LANDSCAPE GARDENING, DESIGNS FOR COTTAGE RESIDENCES FRUITS AND FRUIT TREES OF AMERICA, COUNTRY HOUSES, ETC., ETC.

ALBANY:

PUBLISHED BY LUTHER TUCKER.

At The Cultivator Office, 407 Broadway.

BOSTON—JOSEPH BRECK & Co., Agricultural Warehouse, No. 51 North Market-street.
NEW-YORK—M. H. NEWMAN & Co., Booksellers, 199 Broadway
PHILADELPHIA—W B. ZIEBER, Bookseller

THREE DOLLARS A YEAR—TWO COPIES, FIVE DOLLARS.

The Horticulturalist, 1852. *This important national journal contained articles on landscape design and rural architecture, integrating fruit cultivation into country estates and gardens.*

Europe. Further, the Princes's nursery was one of the first American nurseries to successfully breed grape varieties. One of their most significant contributions to fruit cultivation in the Hudson Valley, and to the nation as a whole, was standardizing the naming and use of established named grape varieties instead of using seedling grapes or cuttings of local grape varieties of unknown origin, or which had an unproven ability to be grown successfully.

William Prince, of the second generation of Princes, wrote one of the first treatises on fruits in New York, which was published in 1828. His son, William Robert Prince, wrote the *Treatise on the Vine*, published in 1830, and, with his father, published the *Pomological Manual* in 1831.

Another patron of horticulture was Dr. David Hosack (1769–1835). Besides his vocation in practicing and teaching medicine, Hosack was interested in botany, and in 1801 established the 20-acre Elgin Botanic Garden in the northern part of Manhattan Island. This public botanical garden was sold to the State of New York in 1810. After 1825, he built another large botanical garden north of Hyde Park, at his 700-acre Hudson River estate, which he purchased from his associate Dr. Samuel Bard. At his home and gardens, he entertained Hudson Valley writers, painters, and naturalists such as Washington Irving, Samuel F. B. Morse, and William Cullen Bryant. Hosack was well known in Europe and, through these acquaintances, obtained many new fruits from European orchards, which were eventually distributed to fruit growers in the Hudson Valley.[10]

In addition to Dr. Hosack, gentleman farmers such as Chancellor Robert Livingston of Clermont and Edward Livingston of Montgomery Place near Barrytown, New York, had Hudson River estates, where they established formal gardens, orchards, vineyards, and arboretums at which they or their staff worked to advance horticulture.

In the early part of the nineteenth century, local horticultural societies were being established so that fruit growers had a forum in which they could exchange information and ideas on new fruit varieties and on new fruit-cultivation techniques. The New York Horticultural Society, the first such organization in the U.S., was organized in 1818 by Dr. Hosack, André Parmentier, Michael Floy, and William Wilson. In 1829, the Albany Horticultural Society was founded by Jesse Buel, Esq., the noted pomologist, nurseryman, and horticultural writer from Albany.

Andrew Jackson Downing and his brother, Charles Downing of Newburgh, and Jesse Buel did much to disseminate new plant material and spread the knowledge of cultivating grapes and other fruits through their nurseries and published writings on horticultural issues. These individuals, unlike the Livingstons or Dr. Hosack, were not landed gentry who had vast financial resources to pursue their horticultural interests as a hobby, but made their living writing about horticultural issues or by selling plants at their nurseries.

A. J. Downing was, in fact, a nationally renowned American landscape designer and writer who was one of the founders of the *The Horticulturalist* and the author of *A Treatise on the Theory and Practice of Landscape Gardening: Adopted to North America* (1841), *Cottage Residences* (1842), *The Fruits and Fruit Trees of America* (1845), and *The Architecture of Country Houses* (1850). He was the editor of *The Horticulturalist* and one of the chief contributors to this important magazine on fruit cultivation and landscaping before the Civil War. In his writings, he often incorporated the cultivation of grapes into his numerous rural and urban landscape designs.[11]

After A. J. Downing's untimely death in 1852 at the age of thirty-six, his older brother Charles continued his extensive work in horticulture and communicated with national horticultural experts and growers by operating the family's nursery business, accepting and shipping new grape varieties throughout the United States, and contributing numerous articles to *The Horticulturalist*. In his communications with fellow nurserymen and commercial growers, he offered advice, and identified and secured new grape varieties. He often attended horticultural shows across the eastern United States.

Charles Downing, through his largely unsung leadership and practical knowledge of new grape varieties and other fruits, strongly influenced the practice of horticulture on a national level by publishing the greatly expanded second edition of the book *The Fruit and Fruit Trees of America*. This book incorporated, documented, and recognized many new grape varieties after his brother's death in 1852. It advised nurserymen across the country on which new grape varieties to collect, propagate, and sell for home use, landscaping, and for commercial vineyards.[12] Both Charles and A. J. Downing are buried at the Cedar Hill Cemetery in Middle Hope.

Jesse Buel was another significant force who advanced horticultural issues in New York. After serving as a journeyman printer on newspapers in New York State, in 1797, he began to publish his own newspapers in Lansingburg, Troy, Poughkeepsie, and Kingston, culminating in the founding of the *Albany Argus* in 1813. He was heavily involved in Whig politics and became the official printer for the New York State government.

Jesse Buel, Esq. *Like many other Hudson Valley horticultural leaders in the early nineteenth century, Buel had other interests that supported his agricultural pursuits and writings.*

At the age of forty-three, Buel turned his attention to improving pomology and horticultural practices and agriculture in general. He believed that farming needed to be more sustainable, and that soils should be systematically improved by better farming practices. He encouraged the study, identification, and implementation of new scientific farming practices such as using deep plowing, manures, green crops, and crop rotation to improve soil fertility and productivity. To that end, he identified and disseminated information on these practices and created his own 85-acre experimental farm just west of Albany along what is now Western Avenue.

In addition to founding the Albany Horticultural Society in 1829, Buel became the secretary of the State Board of Agriculture, pushed for the establishment of a state agricultural school when he was in the State Assembly, and helped to establish the New York State Agricultural Society in 1832. In 1834, he began to publish what would become the nationally popular farm journal *The Cultivator*, which was the official organ of the New York State Agricultural Society. In 1839, he published the *Farmer's Companion*, which embodied his philosophy on scientific farming concepts.[13]

The First Commercial Vineyards

Some of the first commercial vineyards in the Hudson Valley date back to before 1830, when Robert Underhill and his two sons, Richard T. Underhill and William A. Underhill, planted a vineyard of Catawba and Isabella grapes at Croton Point, Westchester County This vineyard eventually covered approximately 75 acres. For some years, this large vineyard supplied the New York City market with grapes.[14]

The Underhills cultivated grapes on Croton Point for the next sixty years until 1873. After that date, the Underhill descendants lost interest in grape cultivation and breeding.[15] More can be found on the Underhill family's grape and breeding activities in Chapter Eight of this book.

In 1829, Rufus Barrett of New Paltz started to ship small quantities of Isabella grapes to New York City. Barrett lived in a community that was settled 150 years earlier by French Huguenots, who, as previously noted, planted native grape varieties after unsuccessfully experimenting with European *vinifera* varieties. Barrett may have obtained his inspiration and knowledge of

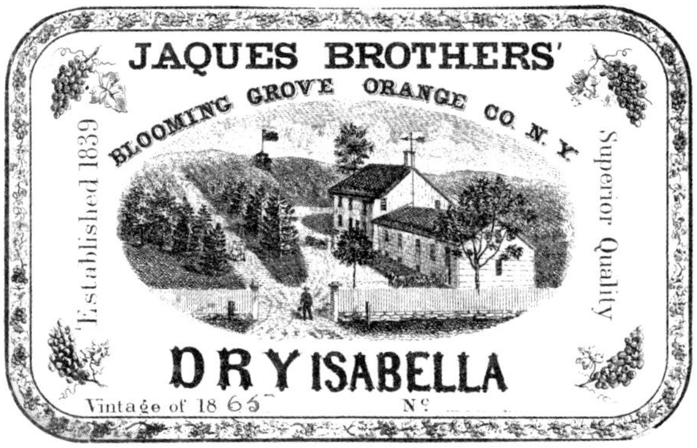

Jaques Brothers' Winery. *Early Isabella wine label from Orange County, New York. Circa mid-1860s.*

cultivating grapes from the descendants of these early French settlers.[16]

In 1809, nineteen-year-old John Jaques, by trade a shoe and boot maker of Scottish heritage, settled in Little York, later called Washingtonville. It is here, in 1839, he produced his first vintage, founding what would become Brotherhood Winery, the oldest continuously operating winery in America.

Jaques first established a store at the corner of North and Main streets in 1824 to sell his shoes, groceries, and general merchandise. At the back of his store, he grew a half acre of grapes, which were shipped by Hudson River sloops to New York City for sale. Due to the high cost of transporting grapes and the declining grape prices, Jaques began to press his grapes for juice and wine production. By 1835, he purchased a large parcel of land on a high ridge further up North Street, where he began to cultivate grapes. He planted approximately 10 acres of Catawba and Isabella grapes—the common varieties of the day—which were purchased from the Linnaean nursery on Long Island, operated by William Prince.[17]

Jaques established a large juice- and winemaking facility that included long hand-dug cellars, over which were built several buildings for wine production. These ancient caves, which look like those found beneath old wineries in Europe, may be the largest wine-storage tunnels in North America.[18] The Jaques Winery, also known as the Blooming Grove Winery, produced its first vintage in 1839, and wine has been made there ever since. Jaques sold some of his first wines to the First Presbyterian Church, where he was an elder. Thereafter he continued to prosper by selling sacramental wine, which later helped to keep Brotherhood Winery in operation during Prohibition.[19]

Simultaneously, across the Hudson River in Amenia, in Dutchess County, a utopian community called the "Brotherhood of New Life" established its own grape growing and winemaking operation at one of its communes. Led by Thomas Lake Harris, the Brotherhood made wine in the Hudson Valley between 1860 and 1864. They believed that wine had divine and miraculous powers and Harris's wine became commercially popular. In 1863 the commune, along with their agricultural and winemaking operations were relocated to Brocton, on Lake Erie, in Chautauqua County.[20]

In the late 1850s, John Jaques turned his business over to his three sons, John Jr., Oren, and Charles, who, in 1858, renamed the winery to the Jaques Brothers Winery. It remained in the Jaques family until 1886, when the last surviving brother, Charles, sold the vineyards and winery to New York City wine merchants J. M. Emerson and Sons. Jesse Emerson and his son Edward were New York City sales representatives for the Brocton-based Brotherhood wines, noted above, and when the Brotherhood commune splintered and relocated to the west coast, the Emersons consolidated the two operations at the Washingtonville property under the name of the Brotherhood Wine Company, which is still in operation today.[21]

Today, the Brotherhood Winery is capable of producing 500,000 gallons of wine a year, and has established a new one-acre vineyard at nearly the same site as its old vineyards to grow heirloom grape varieties that were developed in the Hudson Valley from the 1860s to the 1890s. The original century-old vines located there were removed in the late 1950s, due to age and to the increasing popularity of new grape varieties.[22] Once again, Brotherhood can not only claim to be the oldest continually operated winery in the country and has grapes grown on one of the oldest vineyard sites in the country.

Another early vineyard of Isabella was planted by William T. Cornell near Clintonville (probably Clintondale, Ulster County) in 1845. Mrs. Cornell and Mrs. William A. Underhill, of Croton Point, Westchester County, were sisters, so Cornell's vines presumably came from the Underhill vineyard at Croton Point. Andrew Jackson Caywood, an important grape breeder after the Civil War was a brother-in-law of W. T. Cornell. His interest in grape cultivation and grape breeding was probably due to the influence of both the Cornell and Underhill families. More information on A. J. Caywood's grape-breeding activities in the Hudson Valley is in Chapter Eight.[23]

Advances in Grape Growing and Breeding

From the two decades before the Civil War to the 1890s, the Hudson Valley was one of the major centers for the breeding of new grape varieties in the United States. Local breeders tried to enhance grape quality for both table consumption and wine production. In their quest to develop better-quality grapes, they cross-pollinated Old World *vinifera* grapes with native *labrusca* and *riparia* grape varieties. In doing this, however, their disease resistance tended to be lower than that of the hardy *labrusca* and *riparia* parents. This was unlike the breeding programs in Texas, Massachusetts, and France, where the "hardy" gene pool used for parents tended to be *labrusca*. However, these "hardy" *labrusca* parents also tended to be the better-quality *labruscas*, such as Delaware or Iona, which already had some *vinifera* in their heritage.

Famous Hudson Valley grape breeders such as A. J. Caywood (Poughkeepsie and later Marlboro), Dr. Charles William Grant (Iona Island), James H. Ricketts (Newburgh), Richard T. Underhill and his nephew Stephen W. Underhill (Croton Point), among many others, developed some of the nation's most important new grape varieties. Quality grape varieties, with local names such as Croton, Downing, Dutchess, Iona, Irving, Empire State, Quaissaic, Poughkeepsie, Senasqua, Ulster, and countless others were all developed in the Hudson Valley. These varieties became the genetic building blocks for a new generation of American varieties, such as Moore's Diamond and Cayuga White, both here in the United States and Europe.

By the mid- to late-1800s, William Kniffin, of Clintondale, experimented with new techniques to cultivate and prune grapes. Kniffin developed new grape-pruning systems for native American *labrusca* grape varieties. Prior to Kniffin, during the early years of American grape cultivation, the training systems used were most likely European. The vines were kept well headed back and were trained to stakes of varying heights. However, the growth habit of our *labrusca* varieties needed to be trained to give the fruit and foliage more light to reduce fungal diseases and hasten fruit ripening. Also, a new system needed to be developed to regulate bearing wood, to maximize fruit production that would not lead to the overproduction of unripe fruit, and to control the height of the main trunk.[24] Kniffin's grape-pruning systems included the Hudson River Umbrella, and the Four-Arm and Two-Arm Kniffin Systems, which are still extensively used today. (As an aside, I learned that William Kniffin is a member of my extended family in the Hudson Valley.[25])

In addition to grape hybridizers, nurserymen, and horticulturalists, the Hudson Valley had more than a few large commercial growers of grapes and other fruits. These growers helped to advance the knowledge of grape and other fruit cultivation. They pursued the profitable growing of such fruits in the region.

One such person was Robert Livingston Pell, Esq. (1813–80) who, from 1854 to 1880, had a 1,200-acre fruit and mixed-use farm along the Hudson River in Ulster Park, in the town of Esopus.

This magnificent country estate called Pelham Farm, was sometimes referred to as Cliffwood. On the farm was a 70-square-foot mansion built of brick in the Roman style with columns and extensive piazzas. The property also had 10 miles of graveled roads and paths, fifteen picturesque bridges, ten manmade lakes, drain tile, and stone under-drains. This farm was reported to

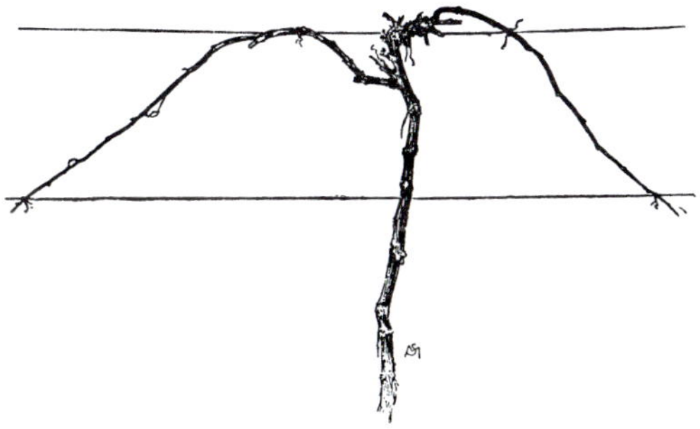

Early example of a Kniffin trellis. *Grape pruning and training systems that were developed in the Hudson Valley in the mid- to late-1800s are still used throughout the United States today.*

have, at its height, approximately 200 acres of Newtown Pippin apples, 50 acres of Isabella grapes, 80 acres of potatoes, significant acreage in raspberries, currants, peaches, and strawberries, and a 400-foot cold frame or greenhouse that contained forty-five European grape varieties. The farm also had sheep, some thoroughbred horses and cattle, and fish ponds.

Gathering Grapes in the Vineyard at Marlboro-on-the-Hudson. From Frank Leslie's Illustrated Newspaper, October 1879.

It was a profitable operation with a dock on the Hudson River, called Pell's Dock, which was used to ship produce by steamboat to New York City. On this wharf was a large stone building, covered with slate and a ventilated roof, to sweat apples for shipment to Europe. Mr. Pell grew Isabella grapes because they ripened much later than the more commonly grown, earlier-ripening varieties such as Concord, Hartford, or Iona, so there was less competition and higher prices for the Isabella grapes when shipped to market.

In addition, Pell was, for many years, the president of the Farmers Club and the prestigious American Institute. Pell attended meetings of state and local pomological associations, and contributed many papers on fruit culture. Further, he experimented with new types of grape varieties and new horticultural techniques.[26]

Farming in the Hudson Valley: John Burroughs, a Case Study

The famous naturalist writer John Burroughs (1837–1921) was another viticulturalist who resided in the Hudson Valley. Burroughs, a leader of the early U.S. conservation movement, worked with President Theodore Roosevelt and John Muir on national conservation issues and to establish the United States National Park System. Further, he associated with prominent writers such as Walt Whitman and Ralph Waldo Emerson, and industrialists such as Henry Ford, Harvey Firestone, and Thomas Edison.

Detailing Burroughs's farming activities in the Hudson Valley will help to give an accurate picture of the local grape industry from the 1870s to the 1920s. Raised in Roxbury, New York, John Burroughs was a schoolteacher who taught in the Hudson Valley towns of Olive, Marlboro, and Highland Falls from 1854 to 1864.[27] He began to write nature essays and poems for national magazines such as the *Atlantic Monthly*, *Century*, and *The Nation,* and later published numerous books. In 1864, Burroughs relocated to Washington, D.C., to secure a federal government position. He was hired that year as a clerk with the U.S. Treasury Department. Ultimately he became a federal bank examiner who audited banks in the Hudson Valley from 1873 to 1886.[28] This job required quite a bit of travel each week throughout the region to examine banking operations. Enamored with the Hudson Valley, in 1873, Burroughs purchased nine acres of land and an old Dutch farmhouse in West Park, Ulster County, which became his home and base of operations. The land had been an old fruit farm, which he later called Riverby, and was located near the banks of the Hudson River, just above the huge icehouse owned by the Knickerbocker Ice Company. After purchasing the land, Burroughs, with hired help, cleared the old farm lands and reestablished a mixed-fruit farm, which was common at that time in the Hudson Valley. While John Burroughs favored apples for consumption, the farm he established grew grape varieties such as Delaware, Niagara, Concord, Worden, and Moore's Early, as well as raspberries, red currants, strawberries, and some apples.[29]

Burroughs, who had a background in dairy farming, had a lot to learn about fruit farming. He turned to pamphlets from the U.S. Department of the Interior and books on practical fruit farming, written by local authors such as the Rev. Edward Payson Roe, a Presbyterian minister who lived in Cornwall-on-Hudson, about 20 miles south of West Park.[30]

At that time the Hudson Valley was (and is still) characterized by relatively small, intensively cultivated farms because of the region's high population density and the competition that local farms face from other urban and suburban land uses. However, Hudson Valley soils are very suited for fruit cultivation and are highly productive so that such small farms can still be economically profitable.

Burroughs left government service in 1886, primarily to devote more time to his writing.[31] Expanding his fruit farm from nine to 17 acres in 1886 enabled him to earn sufficient income to replace the federal job, and have free time from October to April to devote to travel and writing. In the spring of 1888, Burroughs planted an additional 2,400 grapevines, 2,000 red currants, and 2,000 hills of raspberries. In the fall of that year, after observing the success of his neighbors and the high price of table grapes, not wine grapes, he planted mostly white Niagara grapes, with a few rows of Concord and Delaware. To package his table grapes, he built a packing house to crate his crop for shipment to Boston and New York City.[32] Burroughs son Julian later recollected that only after the vineyard was established and his father began to ship grapes did he name the place Riverby. The Riverby Vineyards' logo, an oval with a bunch of grapes in the middle and the address below, was stamped on the lid of every crate or basket of grapes.[33]

During the 1890s, Burroughs's vineyards, like most fruit farms in the Hudson Valley, were very profitable. He continued to plant and evaluate a few new experimental grapes each year in search of better grape varieties, and liked the Winchell the best.[34] As Burroughs got older, he, along with Julian, constructed in 1895 an Adirondack-style cabin of wood and bark called Slabsides, on land just west of Riverby near the Black Creek, which he had recently purchased.

As the elder Burroughs aged, and as his travels remained extensive, Julian came back to live on the farm in 1902 with his new wife Emily. Julian first

John Burroughs and his vineyard at Riverby. *Looking east to the Hudson River in West Park, New York. Circa 1890s.*

helped tend the vineyard and celery patch, eventually becoming the primary caretaker of the vineyard. He was able to earn a living and build a nice home of his own overlooking the vineyards.[35] In 1903 President Theodore Roosevelt visited Burroughs at Riverby, as did Walt Whitman.[36]

However, by 1916, Julian was forced to move out of his home on the farm and live one mile north on the estate of Standard Oil tycoon Oliver Hazard Payne, who employed him as the superintendent of his estate. Julian and his wife continued to manage the farm at Riverby and made additional income renting out their home there, but they could not financially sustain themselves solely on farm income derived from the 17-acre Burroughs farm.[37]

The Riverby home is still standing on the east side of U.S. Route 9-W, where a bookstore devoted to Burroughs's writings is located next door.[38]

Mixed-Fruit Farming and Early Grape Cultivation

From the 1880s until the 1940s, the layout of mixed-fruit farms in the Hudson Valley often included vineyards with rows 10–12 feet apart, with red currants planted beneath the grapes. Raspberries, strawberries, or some vegetable crop would be cultivated between grape rows. At the end of every third grape row might be an apple, cherry, or pear tree.[39]

One of the economic reasons for the establishment of mixed-fruit farms was to spread the cultivation and harvesting work throughout the year instead of concentrating farmwork into a few months. Such work could then be done cheaply by family members, relatives, and a few hired hands instead of retaining many workers, who were generally not available for the harvest of a single fruit crop. The farm year would start in February or March, when trees and vines were pruned, then fruit crops would be hoed and cultivated from April until June. The sequence of harvesting crops was strawberries in June; currants in early July; cherries and raspberries in late July and August; grapes, pears, and apples in September until October; followed by home-improvement projects until Christmas. Then in February and March, the cycle would begin again.

On those farms that concentrated on grapes, often the rows would go up and down the hill rather than be terraced against the hill because of the thin soils and outcroppings. To minimize erosion, the space between the rows was very narrow—about 4 feet—just wide enough for a horse to go through to cultivate the ground. Due to the configuration of the vineyards and the narrow aisles, the horse remained an important beast of burden on the farm until the late 1940s and was not easily replaced by the gas-powered tractor.[40]

From the 1870s until the 1950s, it was common for merchants, craftsmen, and other industrial workers over the age of fifty-five, such as John Burroughs, who were nearing the end of the first stage of their work lives to "retire" and start a second career in the Hudson Valley as a small-fruit grower. Relatively inexpensive farms under 15 acres were available, and the capital costs for operation were low. Also, it got the grandparents out into the country and established a place for their children and grandchildren to go to for recreation or to construct a home to live.[41]

New York State Governor and then President Franklin D. Roosevelt also had a fruit farm on his Hudson River estate in Hyde Park, Dutchess County, where he grew apples, small fruits, and grapes. To relax from the rigors of public life, he would often visit neighboring farmers on both sides of the river to discuss the latest growing techniques that could be used on his own orchards and vineyards.

The Hudson Valley Grape Industry

The economics of grape cultivation changed for the worse in the Hudson Valley from 1900 to 1905 because of increased competition from very productive California growers, who had increasingly better access to large eastern urban markets due to cheaper rail transport; increased fungal disease from the high concentration of grapes cultivated in the Hudson Valley; lower crop yields due to the Valley's shallow soils and more difficult terrain; and no corresponding increase in the quality of grapes grown since the varieties sold were overwhelmingly older table grape varieties.[42] However, in today's grape market, which has a greater emphasis on wine grapes, the lower yield per acre in Hudson Valley vineyards is offset by the increased quality and higher price of these wine grapes. Hence, the economics for grape cultivation are now much more favorable for local growers than it was in the early 1900s.

Over a century ago, when the Hudson Valley grape industry was at its height, the region was New York State's third largest grape-growing district. In 1890,

"Marlboro Thirds" fruit baskets. *In the early days of the Hudson Valley's fruit industry, produce was shipped in round wooden baskets which were manufactured locally and intended to be returned to the grower.*

there were approximately 13,000 acres of grapes in the Hudson Valley, but by 1900, that acreage had shrunk by at least half when a considerable number of old vineyards, which had been planted with too many or worthless varieties or had been set out in places that were not suitable for vineyards, were eliminated.[43]

As noted earlier, the high concentration of vineyards led to the widespread dissemination of fungal diseases such as black rot and, to a lesser extent, other fungal diseases that growers could not suppress. Further, many grape growers were forced to shift to other fruit or agricultural crops in reaction to the Banking Panic of 1907, which caused a large downturn in the economy. At the same time, significant technological advances and economic forces were radically changing how fruit crops were transported, refrigerated, stored, and marketed. However, this substantial reduction in vineyard acreage was reported to have made the industry stronger and more prosperous going forward.[44]

The Hudson Valley was also facing strong competition from other more productive grape-growing areas such as the Finger Lakes in upstate New York, and California. The combination of the significant national economic downturn due to the panic of 1907, the stiff competition for the sale of grapes, the spread of grape fungal diseases, and the relatively low productivity of Hudson Valley soils and steep terrain, which increased the cost of producing table grapes, encouraged local growers to switch to other fruit crops such as currants, gooseberries, and raspberries, and tree fruits such as apples, pears, peaches, and cherries.

In 1908, an estimate of grape acreage by county was: Columbia, 865 acres; Dutchess, 448 acres; Orange, 865 acres; and Ulster, 4,021 acres for a total of 6,199 acres.[45] In the early 1900s there were only two or three commercial wineries in the Hudson Valley, so approximately 90 percent of all locally grown grapes were sold as table grapes or to those who made wine in small quantities for home consumption.[46] While a substantial portion of the grapes grown were for table use, in the first half of the twentieth century, a sizeable amount of grapes were sold to be processed into wine by local residents or by those who came from the New York City area to purchase grapes for wine and grape juice.[47]

After the first real consolidation of the Hudson Valley's grape-growing industry before 1910, Concord (as in other grape-growing regions in the state) led the list of most popularly grown grapes, followed by (in descending order by acreage) Delaware, Niagara, Worden, Moore Early, Bacchus, Pocklington, Campbell's Early, Hartford, and Vergennes, after which were many lesser known Ricketts and Rogers hybrid varieties grown in blocks of an acre or less. Even after this first consolidation, the Hudson Valley had a very diverse selection of grape varieties.[48]

Revitalization of the Grape Industry after World War I

Some European immigrants from Italy and Central Europe, after first establishing residence in New York City, later moved to the Hudson Valley to establish new fruit farms or to help revitalize older ones. An example of this revitalization of fruit farming in the Hudson Valley and the creation of new businesses that used locally grown fruit was the Bolognesi family.

Alessandro Bolognesi and his family came to America from Bologna, Italy, around 1900. Alessandro was a relatively experienced businessman and well-off English-speaking Italian immigrant who worked within the growing American-Italian community as a *banchista*, one who facilitated Italian immigrants' transfer of money and savings to those back home, exchanged money, notarized and prepared legal documents, and provided other financial services. He was also a U.S. agent for the Navigazione Generale Italianna, whose steamships traveled between New York and Italy. Ultimately, Alessandro established a financial services company called A. Bolognesi & Co., which included his eldest son Aldo.

Possibly motivated by his boyhood recollections of working at his uncle's vineyard in Italy, or his search for new business or investment opportunities to support himself and his children, in 1904 Alessandro Bolognesi purchased a rundown farm, in Highland to establish a new productive vineyard. The site is located on a beautiful high bluff that borders the west bank of the Hudson River and overlooks the city of Poughkeepsie. He then devoted the resources needed to revitalize the farm and to establish the Hudson Valley Wine Company.

Bolognesi's winery was bonded in 1904 and his first wine was produced in 1907. The vineyards and wine production were greatly expanded until Alessandro's death at the age of seventy-six in 1924. In a wise business move, the winery was licensed to make sacramental wine in 1919. During Prohibition, which lasted from 1919 until 1933, the winery was permitted to continue producing and selling sacramental wine during the Roaring Twenties and through the Great Depression until Prohibition was repealed in 1933. Wines under such names as Loyola, Aquinas, and St. Benedict were exempt from the provisions of Prohibition.

A. Bolognesi & Co. went bankrupt shortly after World War I, in 1920. By then, son Aldo Bolognesi was listed as an investment banker who used the farm to raise racehorses. It seems that Aldo lost most of his money during the stock market crash of 1929, so he devoted himself full time to the wine business. After the death of patriarch Alessandro in 1924 and that of his wife in 1926, Alessandro's four children, none of whom had children of their own, operated the winery until 1968. Daughter Ada kept the house for the entire family, Aldo handled the paperwork, Alfred was the winemaker, and Alphonso managed vineyards. In 1968, when Aldo died, his widow sold the business to Monsieur Henri Wines, Ltd., which was owned by the Feinberg brothers.[49]

Until the late 1970s, the vineyards at Hudson Valley Wine Company consisted entirely of *labrusca* wine grape varieties such as Delaware, Catawba, Iona, and Bacchus, and French-American hybrids such as Baco Noir and Aurora. In the 1970s, it still had the same Italian winemaker, Sam Williams, who was born on the estate and whose father was one of the vineyard managers for the farm.[50] Unfortunately, the winery continued its slow decline to the late 1970s and is now closed. However, due to its prime location, it is hoped that this premium wine-production facility may one day be revived.

Fruit Growing after 1930

Even as late as the 1930s, the number of grape varieties grown in the Hudson Valley was far greater than in other parts of the country. This was to be expected since so many new varieties were bred here and disseminated to local growers for their opinions before being released to the public. Further, the Valley's proximity to many metropolitan markets meant that it could support the cultivation and sale of a wide variety of exotic or unique grape varieties. Since grape cultivation was gradually declining, no new vineyards were established and no newer varieties were planted, leaving only the older vineyards with many of the older exotic varieties.

There was little uniformity in the packaging in which local fruit was shipped and the manner in which it was packed. One reason was because the Hudson Valley

Hudson Valley Wine Company. *The Bolognesi family's Italianate-style wine processing facility on their 325-acre farm in Highland, Ulster County, produced a full line of wines, including Chablis and Sparkling Burgundy.*

was so close to large urban markets, and the shipping facilities were unsurpassed in the state. Most grapes were packed in the popular, disposable "Climax" baskets in either the four- or eight-pound standard size. Some growers, like the Wardells of Marlboro, packed two or three varieties into one basket to give it a range of color and quality.[51]

At the time, the shipping facilities along the river were exceptional. Most grapes were shipped by boat down the Hudson River to New York City or used locally for jellies, preserves, and wine. When shipped by boat, the fruit was loaded late in the evening when it was cool and it reached its destination early the next morning. This method of transportation was also used for short shelf life fruits such as, strawberries, raspberries, blackberries, and currants grown in the Valley.[52]

Direct rail connection to New England was also good, so large shipments went eastward by rail while smaller quantities went inland and south. During Prohibition, trucks were frequently used to transport fruit because they also had the added benefit of providing a way to ship local growers' illegally produced liquor to the city.[53] By the 1930s and 1940s, truck transport became a major method of shipping fruit to Manhattan and Boston, and by the 1940s, had supplanted riverboats as the most common method for shipping fruit.[54]

My grandmother, Rose L. Casscles, was a part-time independent fruit broker who assembled shipments from various growers from Middle Hope, Marlboro, and Highland to be transported by truck to Boston and New York City markets. It was a tough, stressful job because the growers always thought the broker could get them higher prices and the purchasers were always looking for lower prices from competing brokers. Further, fruit crops are more perishable than other agricultural crops so returning from the city with an unsold load of fruit was not a viable option.[55]

From 1919 to 1945, when America was in the throes of Prohibition, the Great Depression, and then World War II, there was a revolution in our country's transportation system that radically reduced the cost of shipping agricultural products, including grapes. This led to an avalanche of shipments of grapes from California to the East to the detriment of the Hudson Valley's grape industry. As a result, local grape cultivation declined even further after the 1920s and farm acreage shifted to the cultivation of other fruits, especially tree fruits such as apples.

The Casscles family in the Hudson Valley. *Top: The author's paternal great-great-grandparents, Alonzo Palmer Casscles and Rachel Berean Casscles in Marlboro, circa 1920s. Bottom: The author's great-grandmother Mary Sabo (left), at the Sabo farm in Middle Hope, circa 1945.*

What did invigorate the Hudson Valley's fruit-growing industry in the 1900s to the 1940s was the influx of many Italian immigrants who purchased small fruit farms, grew grapes for the production of wine for home consumption, and who raised their families here.[56] The Lofaro, Porpiglia, Truncali, Bealli, Fererra, Troncillito, Corrado, and many other families settled in the Marlboro and Highland areas, and joined the fruit growers of mostly English descent—the Quimby, Weed, Nicklin, Lyons, Hepworth, Barnes, Mackey, Wardell, Quick,

Bailey, Berrian, Purdy, DuBois, Greiner, Borchert, Conway, Welsh, Cosman, and Casscles families—whose forebears settled in this area in the eighteenth and early nineteenth century.

From the 1940s to the 1960s, growers gradually ceased producing fruit on small diversified 10- to 30-acre farms. Those who stayed in farming consolidated smaller farms into larger holdings of up to 100- to 300-acre farms. Ultimately these small diversified farms, which had produced a mix of grapes, berry crops, currants, gooseberries, and various tree fruits, were supplanted by much larger farms that concentrated solely on apples.[57]

High Tor Vineyards. *Located in Rockland County, High Tor was one of the first vineyards in the United States to cultivate French-American hybrids to produce commercially available wines.*

The Growth of Local Wineries

In this environment, from the middle of the 1950s to the 1980s, small wineries began to be established in the Valley. In 1949, Everett Summer Crosby purchased an old farm and the historic Van Orden house near the top of High Tor, in Rockland County, a craggy mountain about which Maxwell Anderson wrote his prize-winning play of the same name. Crosby, who lived in New York City, was the head of the Voice of America and wrote scripts for radio.

Crosby planted High Tor Vineyards in 1951 and by 1954 was selling wine made completely from French-American hybrids, one of the first to do so in New York State. He was also one of the first winemakers in the Hudson Valley to grow and use French-American hybrids, such as Aurora, Baco Noir, Chancellor, and Seyval Blanc at his winery.

His four-year-old Rockland Red was rich in tannin and perfectly balanced with good bouquet, and his white was dry and fragrant. He produced a rosé as fine as any made from *vinifera*. Crosby sold High Tor in 1971 to the restaurateur Richard Voigt, the owner of Peppermill Restaurants in Connecticut, whose winemaker was the Episcopal priest Father Thomas Lee Hayes.[58] The farm and winery continued to produce wine until 1976, after which less labor-intensive non-fruit crops were grown until the 1980s, when the property was sold to the Scenic Hudson Land Trust to become part of a nature preserve that is now known as High Tor State Park.

In the end, Crosby proved that he could produce quality wines just 20 miles north of the George Washington Bridge. High Tor Vineyards was able to produce and sell out of wines such as Rockland White, Rockland Red, and Rockland Rosé, which were made from French-American grapes. Crosby acted as an inspiration and mentor to other aspiring local vintners, including Mark Miller and his son, Eric Miller.

A new winery, Benmarl Vineyards, was next established by Mark and Dene Miller and their family in 1967. The name Benmarl comes from the early Gaelic words *ben*, meaning "hill," and *marl*, which describes the vineyard's slatey soil.[59] Mark Miller and his family had moved onto a portion of the historic Caywood property in Marlboro to establish a new vineyard. In 1957, when Mark Miller was shown the property by William Wardell, the grandson of A. J. Caywood (the aforementioned grape hybridizer, who bred grapes such as Dutchess, Ulster, and Walter) some of Caywood's original hybrids were still growing in the old vineyards on the hill. Most of the vineyards were planted with Delaware, the Caywood hybrids (which were mostly Delaware hybrids), and Dutchess.[60] Some of these grapes continue to grow today at my vineyards in Middle Hope and in Athens.

In 1962, Mark Miller, a prominent magazine illustrator, moved his family to Europe to chase the few remaining illustration jobs that English and French magazines offered. The demand for illustrators in the United States had declined due to the increased use of photography in magazine production. By 1967, the Millers moved back to the Hudson Valley.[61]

They constructed a winery in the late 1960s, designed by Dene Miller, and then built three large additions to expand the winery during the 1970s. Along with his wife Dene, and sons Eric and Kim, by 1973 Mark Miller gradually planted over 80 acres of French-

American hybrid grapes at their Benmarl farm, the Hampton Estate Vineyard on Hampton Road, and Mt. Zion Vineyard, all in Marlboro. The grapes planted in order of importance were Baco Noir, Seyval Blanc, Chelois, and Chancellor at Benmarl; Seyval Blanc, Vidal, Vignoles, and Verdelet at Hampton Estate Vineyard; and Chelois and Villard Noir at Mt. Zion Vineyard.

They produced their first vintage from these grapes in 1967 and commercial wines under the Benmarl label starting in 1972. I had the opportunity to work in these vineyards and the winery during and after high school and college from 1973 to 1986. In 2006, the Millers sold the winery to Victor Spaccarelli Jr. and his family, who continue to operate it today.

The establishment of both High Tor and Benmarl Vineyards, in addition to the continuous operation of wineries such as Brotherhood Winery, Hudson Valley Wine Company, Kedem Winery, Mandia at Clintondale, and Marlboro Imperial Cellars, led others to establish new wineries in the 1970s: Brimstone Hill, Cagnasso Winery, Cascade Mountain, Chateau Georges, Clinton Vineyards, Cottage Vineyards, Eaton Vineyards, Millbrook Vineyards, North Salem Vineyards, Tom Clark's Vinifera Vineyards, Walker Valley Vineyards, and West Park Vineyards.

In subsequent decades, approximately twenty-five more wineries were established, including Applewood Winery, Clearview Vineyard, Demarest Hill Winery, Palaia Vineyards, Pazdar Winery, Riverview Winery, Silver Stream Winery, Warwick Valley Winery & Distillery in Orange County; Adair Vineyards, Baldwin Vineyards, Cereghino Smith, El Paso Winery, Glorie Farm Winery, Magnanini Farm Winery, Robibero Winery, Stoutridge Vineyard, and Whitecliff Vineyard in Ulster County; Alison Vineyards and Oak Summit Vineyards in Dutchess County; Hudson-Chatham Winery and Tousey Winery in Columbia County; Brookview Station Winery in Rensselaer County; and others in the foothills of the Hudson Valley such as BashaKill Vineyards in Sullivan County, Windham Vineyards in Greene County, Elk Hill Winery in Albany County, and Prospero Winery in Westchester County.

Farm Winery Act of 1976

The establishment of so many small farm wineries in the Hudson Valley and across New York State was encouraged by the enactment of the Farm Winery Act on June 4, 1976. (Chapter 275 of the Laws of 1976), which established a new reduced fee for a farm winery license. Under this Act, a New York farm that grew grapes could be licensed as a farm winery that could produce and sell its wines wholesale and retail only if New York–grown grapes were used. This new license allowed farm wineries to operate outside of the three-tier system of distribution of alcohol that had been established in 1934 after the repeal of Prohibition. In essence, the small farm winery could produce wine and then sell it directly to the consumer without going through a wholesaler.

This revolutionary new law was cosponsored in the State Senate by Sen. Richard E. Schermerhorn of Cornwall, who represented the west bank of the mid-Hudson Valley. However, much of the advance work on this bill was done in the late 1960s by Sen. D. Clinton Dominick of Newburgh. Further, Mark and Dene Miller of Benmarl Vineyards and John S. Dyson of

Benmarl Vineyards. *Owner and winemaker Mark Miller and his newly planted vines on the site of the former nineteenth-century Caywood vineyards in Marlboro, overlooking the Hudson River. Circa early-1970s.*

Millbrook—first as commissioner of the New York State Department of Agriculture and Markets in 1975, and then as commissioner of the New York State Commerce Department—lobbied the State Senate and Assembly vigorously to ensure that this bill was enacted into law.

The bill was enthusiastically signed into law by Governor Hugh L. Carey, who became an important supporter and promoter of New York–produced wines. In recognition of Mark and Dene Miller's efforts in advocating for the enactment of the Farm Winery Act of 1976, Benmarl Vineyards received the first issued bonded farm winery license, license number 1.[62]

In 2002, Sen. William J. Larkin Jr., of Cornwall-on-Hudson, was appointed chairman of the New York State Senate Task Force for Hudson Valley Fruit Growers. I was the counsel assigned to this Task Force. After conducting roundtable meetings in Highland and at the Glynwood Center in Cold Spring, Putnam County, to which Hudson Valley growers and wineries were invited, the Task Force developed an action plan to encourage local viticulture and the production of local wines and other alcoholic beverages produced by local fruits. Local Hudson Valley senators John Bonacic, Vincent Leibell, Thomas Morahan, Stephen Saland, and James Seward, in addition to Chairman William J. Larkin Jr., all were appointed to the Task Force. Task Force members introduced over forty-five bills in the New York State Legislature to implement their recommendations, which were outlined in the *Action Plan Final Report*, dated December 2004.[63]

Twenty-two of the forty-five recommendations that were introduced as bills were ultimately signed into law. These statewide bills recognized and encouraged the development of micro-wineries; reduced license and annual fees for wineries; facilitated wine tastings and the sale of local wines at charitable events and at farm stands; greatly expanded the farm distillery provisions and the ability to sell distilled products on site; enhanced the marketing of locally produced wines and other agricultural products; expanded the Shawangunk Wine Trail; repealed a law that prohibited the cultivation of black currants in New York State; and provided technical and financial assistance to expand or replace vineyards by the Urban Development Corporation.

This is an exciting time in the Hudson Valley's grape-growing and winemaking history because, unlike California, which has relied on a few select *vinifera* grape varieties, we are working with many new grape varieties and turning them into new and exciting wines. These grapes include *labrusca* varieties, French-American hybrids, newer hybrids from Minnesota and Geneva, and *vinifera* varieties.

As it did after the Civil War until 1900, the Valley is beginning to accumulate a critical mass of people, money, and talent needed to establish a truly nationally recognized wine region that can produce wines of exceptional quality.

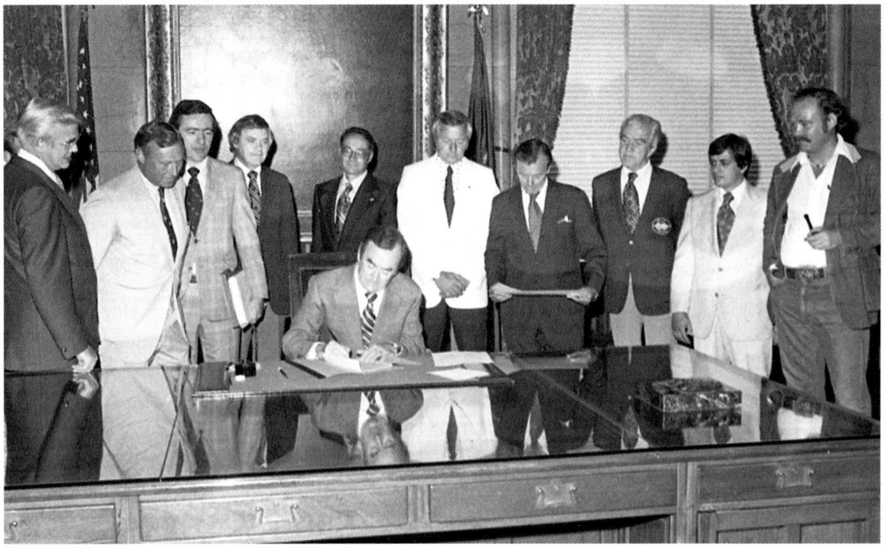

Governor Hugh L. Carey signing the Farm Winery Act, June 1976.
Left to right: Senator Jess J. Present (Chautauqua); Assemblyman Melvin N. Zimmer (Onondaga); Assemblyman Rolland E. Kidder (Chautauqua); Assemblyman Daniel B. Walsh (Cattaraugus); Assemblyman George J. Hochbrueckner (Suffolk); Governor Carey; Senator Richard E. Schermerhorn (Orange); Mark Miller (Benmarl Vineyards); and far right, Walter S. Taylor (Bully Hill Vineyards).

NOTES

1. U. P. Hedrick, *The Grapes of New York. Report of the New York Agricultural Experiment Station for the Year 1907*. Department of Agriculture, State of New York. Fifteenth Annual Report, Vol. 3 – Part II. (Albany, N.Y.: J. B. Lyon Company, 1908), 90.

2. The importance of the work of Dr. Ulysses Prentice Hedrick (1870–1951) over his long and productive life cannot be overemphasized. His work as a horticulturalist at the New York State Agricultural Station at Geneva and, most importantly, his writings, continue to be the basis for how the cultivation of grapes was and is viewed and practiced by growers and academicians. In a manner similar to the Hudson Valley horticulturalist and nurserymen Andrew Jackson Downing and his brother, Charles Downing, Hedrick's writings helped to consolidate and disseminate the idea that there should be a new scientific method to cultivate grapes. Further, these writers chronicled the many existing and new grape varieties being grown commercially or by hobbyists. Even today, *The Grapes of New York* is used by practicing grape breeders across the country as a primary reference guide when selecting older grape varieties to breed new, more disease-resistant and cold hardy grape varieties with superior qualities for the table or the cellar.

3. Hedrick, *Grapes of New York*, 88.

4. Ibid., 89; and many years of personal observation by myself and area growers.

5. Hedrick, *Grapes of New York*, 89.

6. Noted as "a man of scholarship, piety, energy and good sense," Rev. Johannes Megapolensis Jr. was one of the earliest Dutch ministers in New Netherlands and the Hudson Valley, and preached to the Dutch colonists, as well as the Indians, as early as 1643. He was minister of Rensselaerswijck from 1642–49, and then New Amsterdam/Manhattan from 1649 until his death in 1669. His account of the Mohawk Indians comes from various letters he sent to his friends, and was reprinted in Dutch and English several times. Here are his observations on the local grapes:

"Grapevines also grow here naturally in great abundance along the roads, paths, and creeks, and wherever you may turn you find them. I have seen whole pieces of land where vine stood by vine and grew very luxuriantly, climbing to the top of the largest and loftiest trees, and although they are not cultivated, some of the grapes are found to be as good and sweet as in Holland. Here is also a sort of grapes which grow very large, each grape is as big as the end of one's finger, or an ordinary plum, and because they are somewhat fleshy and have a thick skin we call them Speck Druyven [what we now call hog-grapes]. If people would cultivate the vines they might have as good wine here as they have in Germany or France. I had myself last harvest a boat-load of grapes and pressed them. As long as the wine was new it tasted better than any French or Rhenish Must, and the color of the grape juice here is so high and red that with one wine-glass full you can color a whole pot of white wine." Source: J. Franklin Jameson, ed., *Narratives of New Netherland 1609–1664* (New York: Charles Scribners Sons, 1909), 168–80.

7. U. P. Hedrick, *A History of Agriculture in the State of New York. The New York State Agricultural Society*. (Albany, N.Y.: J. B. Lyon Company, 1933), 380.

8. Patricia Edwards Clyne, *Hudson Valley Faces and Places* (New York: Overlook Press, 2005), 219–20.

9. Hedrick, *History of Agriculture*, 380–82.

10. Ibid., 382; Robert Wilson Hoge, "A Doctor for All Seasons: David Hosack of New York," ANS Magazine, Spring 2007, volume 6, number 1 (New York: American Numismatic Society). It is interesting to note that despite his record of achievements in both medicine and horticulture, Dr. David Hosack is best remembered as the Hamilton family doctor, and thus his service as attending physician in the historic duel between Alexander Hamilton and Aaron Burr, in Weehawken, New Jersey, in 1804.

11. Clyne, *Hudson Valley Faces*,174–79.

12. Hedrick, *History of Agriculture*, 393–95.

13. Ibid., 120, 321, 336, 350, 385, 418.

14. Hedrick, *Grapes of New York*, 89.

15. Ibid., 226.

16. Ibid., 89.

17. Ibid.

18. Leon D. Adams, *The Wines of America* (Boston: Houghton Mifflin, 1973), 122.

19. Ibid., 123.

20. Thomas Pinney, *A History of Wine in America, From the Beginnings to Prohibition, vol. 1* (Berkeley: University of California Press, 2007), 333.

21. Robert Bedford, *The Story of Brotherhood, America's Oldest Winery* (Coxsackie, N.Y.: Flint Mine Press, 2014), 19–29; Edward J. McLaughlin III, *Around the Watering Trough: A History of Washingtonville, N.Y.* (Washingtonville, N.Y.: Washingtonville Centennial Celebration, Inc., 1994), 116-122. Note that McLaughlin was the official Washingtonville historian and great-great-grandson of John Jaques. See also Clyne, *Hudson Valley Faces*, 220–21.

22. Adams, *Wines of America*, 122.

23. Hedrick, *Grapes of New York*, 90.

24. Ibid., 91.

25. Conversation with Joseph Levi Casscles, my grandfather, in the late 1970s.

26. R. G. Pardee, "Iona, Pelham, and Wodenethe," *The Horticulturalist and Journal of Rural Art and Rural Taste*, vol. 14 (New York: C.M. Saxton, Barker & Co, 1859), 401–403; also see "The History of Esopus, NY," *Gazetteer and Business Directory of Ulster County, NY*, for 1871–1872 (Syracuse, N.Y.: Hamilton Child, 1871), 80–86.

27. Edward J. Renehan Jr., *John Burroughs: An American Naturalist* (Post Mills, Vt. Chelsea Green Publishing Co., 1992), 35, 52, 70.

28. Ibid., 72, 170.

29. Ibid., 120.

30. Ibid. for more information see Chapter Eight.

31. Renehan, *John Burroughs*, 170, 173.

32. Ibid., 171–72.

33. Julian Burroughs, "Conclusion," in John Burroughs, *My Boyhood, with a Conclusion by His Son Julian Burroughs* (Garden City, N.Y.: Doubleday, Page & Co., 1922), 159–160.

34. Renehan, *John Burroughs*, 173.

35. Ibid., 224.

36. Ibid., 254.

37. Ibid., 288.

38. Renehan, *John Burroughs*, 224.

39. Conversation with Joseph Levi Casscles in the late 1970s.

40. Mark Miller, *Wine—A Gentleman's Game: The Adventures of an Amateur Winemaker Turned Professional* (New York: Harper & Row, 1984), 23, 24; and conversations with Eric Miller in 2011, and with Joseph L. Casscles throughout his life.

41. This is anecdotal. Examples: my great-grandparents John and Mary Sabo left the Roseton, N.Y. brickyards in 1928; Lawrence and Florence Carlino left a fruit and vegetable retail business in Brooklyn, N.Y. in the early 1950s; my grandmother's cousins, the Rudicks, left the foundry business located in northern New Jersey in the early 1950s.

42. Miller, *Wine*, 48.

43. Hedrick, *Grapes of New York*, 88.

44. Ibid., 92. The Panic of 1907, also known as the Bank Panic of 1907 was a nationwide financial crisis which began with a financial recession mid-year, and culminated in the drop in the New York Stock Exchange, and the bankruptcy of two brokerage firms in October. A run on banks and trust companies ensued, and the resulting panic and disruption to the nation's financial markets became the catalyst for the creation of the Federal Reserve banking system. See Ellis W. Tallman and Jon R. Moen, "Lessons from the Panic of 1907," *Federal Reserve Bank of Atlanta Economic Review*, May/June 1990, 2–13.

45. Ibid., 88.

46. Ibid.

47. Conversations with long-established grape growers such as Lawrence Bealli, Jackson Baldwin, Joseph L. Casscles, John Corrado, and the Affuso family.

48. Hedrick, *Grapes of New York*, 90–91.

49. Conversations with Robert Bedford, December 2013.

50. Adams, *Wines of America*, 126.

51. Hedrick, *Grapes of New York*, 91.

52. Ibid.; conversations with those who lived then and shipped fruit—Joseph L. Casscles and the Smalley family.

53. Conversations with Joseph L. Casscles and Laurence Bealli.

54. Conversations with Joseph L. Casscles in the late 1970s.

55. Conversations with Rose Casscles, my grandmother, in the early 1980s.

56. C. M. Woolsey, *History of the Town of Marlborough, Ulster Co., NY* (Albany, N.Y.: J. B. Lyons Company, 1908), 460.

57. William J. Larkin Jr. and J. Stephen Casscles, *The New York State Senate Task Force for Hudson Valley Fruit Growers, Action Plan Final Report* (Albany, N.Y.: New York State Senate Task Force for Hudson Valley Fruit Growers, December 2004), 9–13.

58. Adams, *Wines of America*, 124–25.

59. My conversations with Mark Miller at Benmarl winery in 1974; Adams, *Wines of America*, 125.

60. Miller, *Wine*, 22–24; conversations I've had with Eric Miller, Mark Miller's son, over the years on this subject.

61. Miller, *Wine*, 125.

62. Ibid., 79–82, 177–80.

63. Larkin and Casscles, *Senate Task Force*.

CHAPTER TWO

Benefits of Grape Hybridization

THE CULTIVATION OF *Vitis vinifera* grape varieties has dominated wine and table grape production in Europe and Asia Minor ever since the time of ancient Greece. These *vinifera* grape varieties were wild local grapes that originally came from Asia Minor, around the Caspian Sea. They were selected for cultivation by local growers, and then spread with the expansion of the Greek and Roman empires west across the Mediterranean Sea and then up to northern Europe.[1]

The exclusive use of *viniferas* for the production of wine and table grapes in Europe changed with the discovery of the Americas. Not only did Europeans take *vinifera* grapes to the New World in an unsuccessful effort to cultivate them for the production of fine wines and fruit, but many American grape varieties were sent to Europe for botanical study and evaluation. The importation of America's grape plants to Europe in the late eighteenth and early nineteenth centuries also brought the scourges of phylloxera (*Daktulosphaira vitifoliae*), a soil-borne insect that sucks and destroys the roots and leaves of *vinifera* grapevines, and fungal diseases such as downy mildew, powdery mildew, and black rot. The importation of these diseases had a devastating effect on grape production in Europe.[2] In France alone, from 1860 to 1880, some 2.5 million acres of vineyards were destroyed by phylloxera.[3]

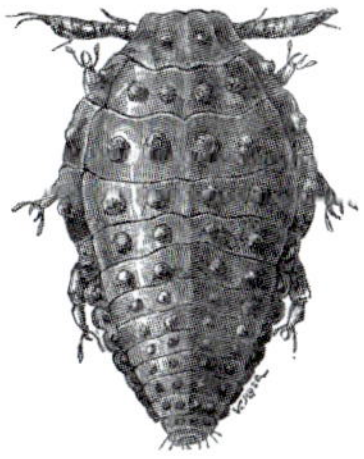

Female adult phylloxera insect. *Enlarged view.*

It cannot be overemphasized how important grape cultivation and wine production were to the nineteenth-century European economy. For many, wine was not just the beverage of choice, but of necessity due to an unreliable source of clean drinking water. Also, at a time before modern food-storage techniques such as canning or freezing, wine consumption was an efficient way to preserve and store fruit and provided necessary calories to enhance the meager diets of most residents on the continent. The body metabolizes the alcohol in wine, which converts to sugar, and supplies important vitamins.[4]

Vinifera grapes were also being transported from Europe to the Americas for cultivation. Initially, colonists attempted to cultivate pure *viniferas*. However, the vines died because the environment was either too warm or too cold, had excessive humidity, or because of the aforementioned fungal diseases such as black rot, downy mildew, powdery mildew, and Pierce's disease, or insects such as phylloxera. From the 1850s to the 1880s, noted Hudson Valley hybridizers such as James H. Ricketts, Dr. Charles W. Grant, the Underhill family, and Andrew Jackson Caywood began to breed *viniferas* with domestic grapes with an eye toward producing high-quality grapes that could thrive in the Hudson Valley, mostly for table consumption, but also for wine production in some limited circumstances.

To produce quality grapes that could thrive in America and in Europe, the best minds from both continents set out to reinvent grape cultivation for the production of wine and table use. There were two schools of thought on the matter. Those who wanted to continue using *vinifera* varieties as they existed before the importation of American grapes, and those who wanted to create new grape varieties that genetically incorporated the disease-resistant genes of selected American varieties into European *vinifera* grape varieties.

The first group, whom I call the Viniferists, had a daunting task. They had to identify suitable American varieties to be used as the basis of new rootstocks to be grafted on to *vinifera* plants to replace the soft fleshy *vinifera* roots and protect them from the soil-borne insect phylloxera. Further, they had to develop new cultivation techniques and, for the first time, apply spray materials and utilize other techniques to protect the vines from imported fungal diseases.

The second group of individuals, whom I call the Hybridists, approached the problem differently by breeding entirely new interspecific grape hybrids to produce wine. These new hybrid grape varieties would incorporate the disease resistance of American varieties with the more horticulturally and oenophologically acceptable characteristics of established native European breeds. They hoped to systemically create new varieties that could withstand phylloxera and fungal diseases. Some of the more famous French hybridizers were Albert Seibel, François Baco, Eugene Kuhlmann, J. F. Ravat, Bertille Seyve, and Joanny Burdin.[5] Up until the early 1950s, it was not clear which philosophy would prevail.

This interesting topic alone is sufficient for its own book, as it combines the issues of politics, economics, sociology, and agriculture, in ways similar to the effect that the Irish potato famine had on the British Empire and the development of the American economy. At the present time, it seems that the Viniferists' position is dominant in most parts of the world. France is the leader of the Viniferists' position because of that country's political and economic need to keep underemployed people in the vineyards to earn some income and to maintain vineyard acreage. Unlike lower yielding *vinifera* grapes, highly productive French-American hybrids were supplying too many grapes for the French economy to absorb. From the late 1940s to the 1970s, the conservative governments of France, which at that time included prominent members from Burgundy and Bordeaux who grew *vinifera*, could not compete economically with these new French-American hybrids, so they banned the cultivation of these new hybrid grapes. In the end, it was politics that did in the French-American hybrids in France. In 1934, 1955, and 1970, the French government enacted progressively more restrictive laws to, at first, prohibit the establishment of new French-American hybrid grape vineyards for selected varieties, and then enacted laws to encourage or force growers to remove existing French-American hybrid vineyards.[6]

While the battle may be over in France for the development and cultivation of new hybrid grapes, grape-growing countries such as Germany, Hungary, and the United States still have extensive grape-breeding programs. These programs are introducing new and exciting varieties such as Regent and Rondo from Germany, Bianca and Reform from Hungary, and the large number of selections from the Geneva/Cornell and Minnesota breeding programs. So ultimately, the battle continues between the Viniferists and Hybridists. In the United States, hybridists continue to work at state university breeding programs in New York, Florida, Illinois, and California. Further, many private citizens are now hybridizing new grape varieties under their own programs, which are more suitable for their climate and growing conditions. As environmental concerns and global-warming issues come to the forefront, hybridization may very well become the wave of the future.

Running counter to the generally held beliefs of the Viniferists—especially those purists who would like to limit production to a few "pure" classic *vinifera* grape varieties, such as Chardonnay, Cabernet Sauvignon, Riesling, or Pinot Noir—all grapes are hybrids. Even the mighty purebred *vinifera* Chardonnay is a naturally occurring hybrid of Pinot Noir and the bulk grape Gouais Blanc.

When all is said and done, however, the wine quality of *vinifera* wines is lauded the world over. Grape breeders in countries such as Germany, Hungary, and the United States who are at the forefront of developing new grape varieties for the twenty-first century should continue to strive to incorporate the superior fruit flavors and acid/sugar/flavor profiles of *vinifera* wines into new breeds that can be grown locally.

In the end, whether a grape is a pure *vinifera* hybrid such as the noble Chardonnay, Cabernet Sauvignon, or Pinot Noir, or an interspecific hybrid such as Baco Noir, Chelois, Dutchess, Seyval Blanc, Rondo, or Bianca, they are all hybrids. The story of wine is really the story of hybridization. As this book will point out again and again, all grapes are hybrids, so people should not look at them negatively. In fact, the development of better hybrids should be encouraged for important economic, social, and environmental reasons (see sidebar).

As the Earth's global population continues to grow to over seven billion people and as global warming may affect the areas and conditions where grapes are grown, the development of new genetically superior grape varieties and other agricultural crops will be critical in enhancing the ecological and economic well-being of the human race worldwide.

FIVE REASONS TO ENCOURAGE HYBRIDS

1. **Environmental**
   Future hybrids could reduce the amount of chemical fertilizers, pesticides, fungicides, or water needed to produce an environmentally sound and economically sustainable grape crop. This will not only help the environment, but will reduce the cost of producing grapes.

2. **Land conservation**
   Future hybrids could increase the tonnage of grapes grown per acre to help reduce the demand for agricultural land or to allow environmentally sensitive lands to revert back to nature.

3. **Better tasting wines**
   New varieties could be developed to produce superior-tasting wines. Further, perhaps varieties could be developed that are capable of producing multiple styles or types of wines to increase the variety's versatility in the cellar to the advantage of both the grower and wine producer.

4. **Cold hardiness**
   New varieties could be developed for greater cold hardiness or increased resistance to late spring frosts to extend grape-growing areas. Further, with the increase in global warming, perhaps breeding resistance to heat or drought will become desirable goals.

5. **Ease of cultivation**
   Hybrids bred for overall ease of cultivation could help to reduce the amount of labor needed to produce a crop, and increase income and leisure time for growers. Varieties also could be developed for resistance to bird or other wildlife damage, susceptibility to berry rupture due to rain, or a change in growth habit to increase the amount of sunlight to grape clusters or to facilitate spray applications.

NOTES

1. Lucie T. Morton, *Winegrowing in Eastern America: An Illustrated Guide to Viniculture East of the Rockies* (Ithaca, N.Y.: Cornell University Press, 1985), 64; see also U. P. Hedrick, *The Grapes of New York* (Albany, N.Y.: Department of Agriculture, State of New York, 1908), 1–4.

2. Thomas Pinney, *A History of Wine in America, From the Beginnings to Prohibition*, vol. 1 (Berkeley: University of California Press, 2007), 9–10.

3. Frank Schoonmaker, *Frank Schoonmaker's Encyclopedia of Wine* (New York: Hastings House, 1969), 251.

4. Hugh Johnson, *Wine* (New York: Simon & Schuster, 1969), 5–11.

5. Morton, *Winegrowing*, 69.

6. Pierre Galet and Lucie T. Morton, *A Practical Ampelography: Grapevine Identification* (1st American ed.) (Ithaca, N.Y. and London: Cornell University Press, 1979), 49.

CHAPTER THREE

Basic Principles of Cool Climate Pruning and Vineyard Management

"Farmerettes" in a Hudson Valley vineyard. *In the wake of labor shortages during World War I, many women from New York City volunteered to work in the Hudson Valley picking fruit and maintaining crops. Circa 1918.*

THIS BOOK DESCRIBES grape varieties that can grow in the Hudson Valley and other cool climate regions, and reports on their basic attributes in the field and in the cellar. While a detailed outline of the basic principles of pruning and vineyard management is beyond the scope of this book, we should at least touch upon some of the general principles of pruning grapes and vineyard management practices in cool climate areas, such as the Hudson Valley. For an excellent layperson's guide on how best to prune grapes and manage a vineyard, I highly recommend reading *From Vines to Wines* by Jeff Cox, *Small-Fruit Culture* by James Sheldon Shoemaker, *A Wine-Grower's Guide* by Philip M. Wagner, and *Manual of American Grape-Growing* by U. P. Hedrick.[1] These books are excellent references for the beginner or advanced vine grower. Other suggested books are included in the bibliography.

The above books all expertly outline how best to establish a vineyard; vineyard soil types and climates to look for and avoid; optimum vine spacing; trellis systems; the first-, second-, and third-year pruning techniques to bring young grape plants into full production; and, finally, once vines are at full maturity, the wide variety of pruning systems that can be used to maximize the production of quality grapes for wine or table use. Pruning systems will vary depending on the grape varieties planted, the climate and vineyard site conditions, the soil type, the desired production level, and the type of wine to be produced.

In describing the grape varieties in this book, I detailed their vegetative growth habit, such as upward, downward, horizontal, bushy, or some variation of these. In addition to the direction of growth, there is the question of the volume of vegetative growth and how that should influence the pruning style. Varieties such as Baco Noir are very vigorous and produce lots of wild and dense horizontal growth. This dense growth, compounded by Baco's large leaf size, inhibits sunlight from penetrating the canopy, so grape clusters get little sun exposure to the detriment of grape quality. The pruning system needed to control the growth habit of Baco, for example, will need to be tailored to that specific grape variety.

Unlike Baco Noir, other varieties such as Chelois and Chambourcin are more outspread, which allows for good sunlight penetration and facilitates the application of sprays to control fungal diseases and insect damage.

Further, their leaves tend to be small to medium in size, which gives them an airy appearance. Other varieties, such as Concord, tend to grow downward and have big thick leaves, which reduce sunlight penetration to the interior of the plant. Each pruning system must be tailored to the growing habits of the variety being pruned.

In sum, the growth habit, canopy density, leaf cover, size and thickness of the leaf, and winter hardiness all influence the choices that a vine grower should weigh when selecting the pruning system to be used for a particular grape variety.

Pruning Systems

There are many different pruning systems that can be utilized to prune grapevines and to manage their growth to produce quality grapes. For grapes in the Hudson Valley, pruning systems include cordon, cane, Kniffin, Kniffin umbrella, Keuka high renewal, Geneva double curtain, Hudson River umbrella, and combinations of these. The books *From Vines to Wines*, *Small-Fruit Culture*, and A *Wine-Grower's Guide* accurately describe each of these pruning systems and recommend specific pruning systems for each variety. Further, they contain illustrations and diagrams to clearly show how to prune these vines and illustrate what the vines look like before and after pruning.

In choosing the proper pruning system, first consider the fertility of the soil to sustain a balanced crop, the soil's water-holding capacity, the climate and prevailing wind (especially winter winds), and the variety to be planted. Further, the previous season's weather effects on the condition of the vines and the prospects for their future growth potential are very important. Look at the extent of the vines' damage due to winter weather, late spring frost, fungus, and the amount of vegetation growth due to precipitation or temperature.

Grape growing in the Hudson Valley, as in most other cool climate regions, can be a challenging endeavor. While most books give the fundamentals of pruning, there are many more considerations to keep in mind to maintain a healthy and successful vineyard. Below are ten points that should be addressed when pruning and maintaining a vineyard in the Hudson Valley and other cool climate regions.

10 POINTS TO CONSIDER WHEN COLD-WEATHER PRUNING

1. **Balance fruit production**

   The fruit produced on a vine should be balanced across the entire trellis to maximize sunlight to the canopy's leaves and fruit. Hence, the pruning system selected should spread out the growth of leaves, canes, and fruit clusters evenly across the trellis and not concentrate growth in one place while leaving other places on the trellis open.

2. **Keep the canopy open**

   Keeping the canopy as open as possible is very important for several reasons. First, it increases the amount of sunlight, and hence energy, that the vine's leaves can absorb. This increases the vine's ability to grow and ripen a crop. Second, the sun dries the vine's leaves and fruit to help inhibit the spread of fungal diseases. Third, it facilitates air circulation through the vine to dry its leaves and fruit. Further, it helps to lower the temperature and humidity of the interior canopy to reduce the spread of disease. Fourth, increased sunshine to the grape clusters enhances the grape crop's quality, maturity, and sugar content. Fifth, increased sunlight to the grape plant's new green canes and buds is important for the productivity of next year's crop.

3. **Think of next year's crop**

   Relative to the "let the sunshine in" theme of the previous point, opening the canopy increases sun exposure to this year's growing green canes, which then become next year's fruiting wood and buds. In each dormant grape bud is the potential for next year's canes, leaves, and grape clusters, so exposing this year's growing green canes to sunlight increases next year's grape productivity, maturity, and quality. A well-developed bud in the first growing season tends to lead to a successful crop in the next season.

4. Bud selection is important

When pruning in the spring, select canes and buds that were fully exposed to the sun to maximize fruit production and quality.

5. Prune according to the vine's strength

This is important. It may seem counterintuitive, but if the vine's growth is weak, prune it more severely. If the vine's growth is vigorous, leave more buds for growth and grape production next year. This is because the vine can handle the extra demands placed on it. When pruning, look to the overall strength of the plant. Leave only a sufficient number of buds to produce a crop that the vine can sustain. The books cited earlier will advise on the number of buds that should be left on a vine based on the amount (i.e., weight) of new wood that is to be pruned in the spring.

6. Prune late

If at all possible, prune in late March instead of earlier in the winter. Late pruning is safer because it is easier to ascertain the amount of winter damage sustained by the vine, and the vine can then be pruned accordingly to compensate for the damage or lack of live buds. Also, late pruning, especially for early budding varieties such as Maréchal Foch and Baco Noir, helps to inhibit early bud break, which can lead to late spring frost damage. This is because if there are more buds on the vine, the plant's sap is divided equally among all the buds, thus discouraging the premature pushing of any one single bud or set of buds, so all of the buds will remain more dormant.

7. All fruit comes from first-year wood

Nearly all fruit grows from buds on the light brown or blonde smooth wood that is produced from buds of the previous year, so in balancing the wood left for the current year's grape production, consider only buds from first-year wood.

8. Size of the cane matters

It seems counterintuitive, but those big, thick, long, strong-looking canes, also known as bull canes, should not be left and relied upon to produce next year's crop. Use average-size canes and buds on such canes—not those bull canes or small, underdeveloped canes—because average canes with short spaces between each node have developed more slowly and consequently are hardier and more resistant to winter injury.

9. There is more than one way to prune for crop size, balance, and quality

A grape variety that produces its grapes on large heavy clusters will produce an ample crop from relatively few buds. If too many buds are left at pruning time, such a vine will overbear, fail to ripen its crop, and can be permanently weakened. However, a variety that produces small clusters should be more lightly pruned so that it can yield a sufficient and balanced crop. It is best, though, to err on the side of leaving too many, not too few, buds. It is easier to prune off excess buds later on. That is why the crop can be further balanced and controlled by late spring pruning to eliminate more canes, spurs, and buds. Further, thinning clusters in the early or late summer to remove and control the vineyard's crop can improve the quality.

10. Pruning is an art, not a science

After pruning for a few years, pruning can be accomplished simply by feel and instinct. In my early days of pruning, I would prune after measuring the weight of each vine's first-year wood (with an actual scale) to determine the number of buds that should be left on the vine. I would weigh the wood, calculate the number of buds that should be left on, and then slowly prune until the correct number of buds were left. After thirty-five years of pruning, I now quickly size up the vine, its past history, what I want the vine to do each year, and then go at it. Pruning a vine should take no longer than five to six minutes. This is why in describing pruning techniques I do not resort to detailed, specific pruning systems, but the considerations that should be weighed when pruning a vine.

Pruning Methods and Training Systems

The purpose of pruning is to maximize the quantity and quality of grapes harvested for winemaking or table consumption. Left to nature's devices, grapevines would produce many short or medium-size spindly canes, with many leaves and few grape clusters. The act of pruning is to alter nature's design and decrease the number of canes, buds, and leaves in order to maximize fruit production, not wood production. Ever see a grapevine in the wild? It generally has a big thick trunk or trunks, is spreading, and has few or no grapes. In the vineyard, the ratio of the mass of vegetation is low when compared to the volume of grapes. And that is the point of having a vineyard—to grow grapes, not branches or leaves.

While it is important to minimize the ratio of second- and third-year wood to grapes produced, it is also important to leave "escape routes" and alternative paths of production so that if a vine suffers from a late spring frost or winter damage, it is still able to produce an economically viable crop.

There are several ways to prune defensively in the Hudson Valley. First, instead of establishing one trunk to support a vine, establish two trunks with separate lateral cordons or spurs so that if one trunk dies, the other trunk can still produce a crop that year. For cold-sensitive varieties such as Riesling or Chardonnay, maintain three trunks to assure a crop. Second, about halfway up each trunk, leave a two-bud replacement spur so that if the cordon or canes above that spur are damaged or die, that replacement spur can supplant the dead cordon or cane.

There are four general systems of pruning: spur pruning, cane pruning, mixed cane/spur pruning, and summer pruning.

Spur Pruning

This consists of pruning back fruitful canes to two to four buds off of a thicker and unproductive second-year wood or older cordon or trunk. Sometimes the thicker older wood of the trunk is formed as a "T" or two back-to-back and upside-down "L's," which together form a "T" with short spurs coming off the older wood. Types of spur pruning include cordon, head training, goblet, and Geneva double curtain.

Cane Pruning

This consists of leaving longer canes of first-year wood with five to ten buds on each cane. The canes grow from a central trunk and spread out along the trellis wire. Cane pruning tends to produce more grapes because there is a higher ratio of fruitful buds to the plant's older wood. The disadvantage is that the growth tends to occur at the end of the cane so that, over time, there are few replacement canes available at the center of the plant for future years, so replacement spurs should be in more central locations of the vine to generate more canes for future fruit production. Types of cane pruning systems include the Hudson River umbrella, two-arm and four-arm Kniffin (which were developed in the Hudson Valley), and Keuka high renewal.

Mixed Cane/Spur Pruning

Mixed pruning is a combination of spur and cane pruning that is used when the vine has severe winter injury or necrosis and the pruner is trying to obtain an ample crop from the vine that year. Further, when used, there tends to be both spurs and a longer cane.

In selecting a pruning system, first look at the growth habit of the vine. If the new first-year canes grow straight upward, then keep the trunks and cordons relatively low so that the new canes can grow up the trellis system that you have selected. If the vine is pruned to a high cordon that goes across the top wire, then all of the new growth will be unsupported above that wire. Also, all of the fruit will be six feet in the air, so the weight will bend and break the canes because there was no support. If the variety is downward-growing, then prune the spurs or canes to the top wire, so that all growth will trail downward and be supported by the chosen trellis system.

Summer Pruning

The initial winter pruning is needed to remove most of the previous year's fruiting and non-fruiting wood. It is done coarsely to balance next year's wood-to-fruit ratio. This is followed by late-spring pruning to lop off the excess first-year wood that the cautious pruner left to ensure that a vine can produce an ample crop that year should a late spring frost occur or cane necrosis afflicts the vine. Summer pruning and cluster thinning are the final steps taken to balance the amount of vegetative growth that a vine produces to mature a quality crop of grapes. I prefer to do cluster thinning before July 4th to balance the crop to what the vine can handle. That way, the vine can channel its energy into maturing the crop that remains. Some wait until mid-August to remove excess clusters because they want to guarantee a large crop. I do cluster thinning earlier in the year and proceed as if the crop that remains will reach maturity.

The remaining part of summer pruning is to hedge the vine's vegetation. This is done by combing the canes by hand so that they are all positioned vertically. This untangles the horizontally growing canes to increase sunlight and air circulation. Sometimes the canes reach down to the ground or grow along the ground. In this case, I then clip the canes with a pair of hedge clippers so that there is at least a two-foot open area all along the bottom of the row. This increases air circulation and reduces the incidence of fungal growth.

The Hudson Valley's summers are humid, so shoot positioning, hedging, and cluster thinning are important in increasing sunlight and air circulation to reduce fungal diseases and damage. Also, from Memorial Day to the Fourth of July, it is important to remove suckers from the trunks of grapes. This shifts growth up to the top wire, where there is more sunlight for optimum maturation of the crop. It also removes suckers at the bottom of the canopy to increase airflow throughout the vineyard.

POPULAR COLD WEATHER TRAINING SYSTEMS

These illustrations feature several of the popular pruning methods used in cold weather climates. The vines are shown after the current fruiting season, indicating where the pruning cuts will be needed the following year.

— Pruning Cuts •••• New Fruiting Cane

High Cordon Spur Renewal: Used for grape varieties that have a lateral or slightly upward cane growth habit such as *riparia* hybrids, like Baco Noir and Bacchus.

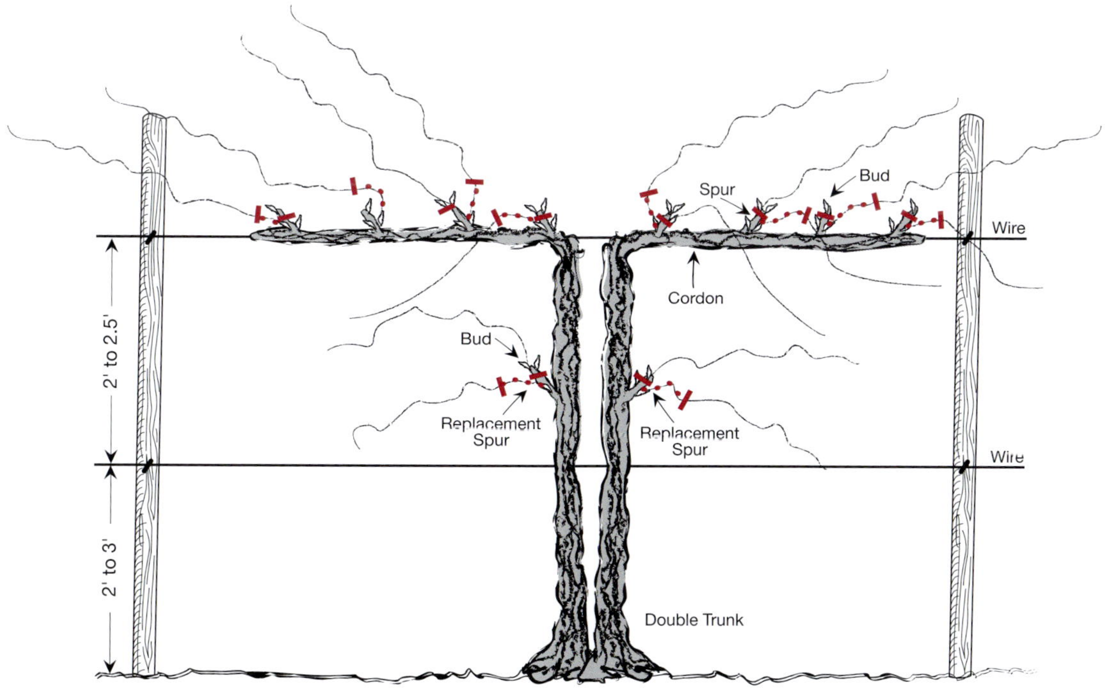

Low Cordon Spur Renewal: Used for grape varieties that have an upward growth habit such as *vinifera* or *vinifera* hybrids, like Chardonnay and B.6055.

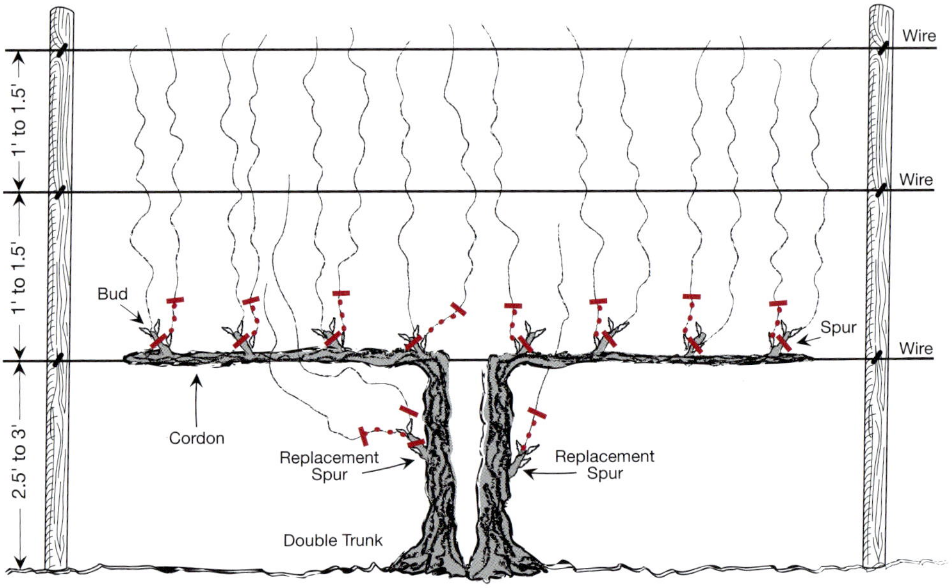

Four-Armed Kniffin: A pruning system developed in the Hudson Valley by William Kniffin that utilizes long canes to bear fruit instead of short spurs. It also spreads out vegetative growth on the top and bottom wire. Used for grape varieties that have a lateral growth habit, such as Le Colonel and Delaware.

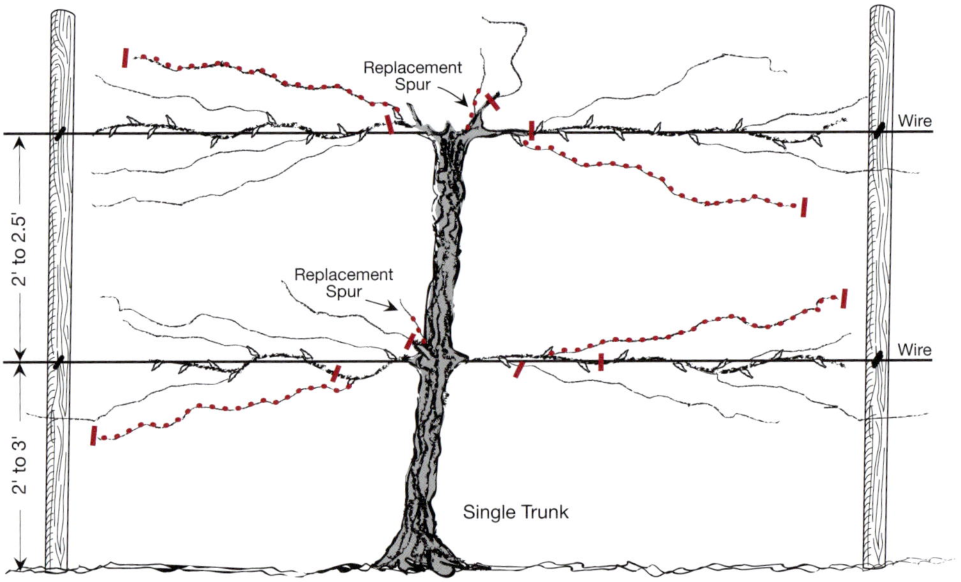

Four-Armed Umbrella Kniffin: A variation of the Four-Armed Kniffin, where the canes are brought over the top wire and tied to secure the canes to the lower wire. Used for grape varieties that have a downward cane growth habit, such as *labrusca* varieties, like Concord. Locally, this is sometimes inaccurately called the Hudson River Umbrella.

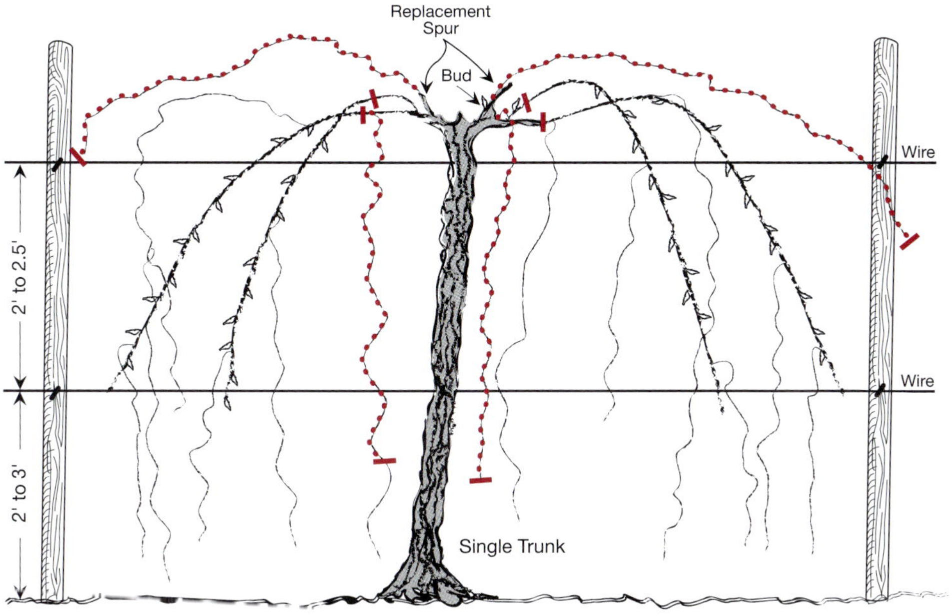

Hudson River Umbrella: A pruning system used for varieties that have an upward and then downward cane growth habit, such as Empire State. The vertical first-year canes are secured to the bottom wire. This system spreads out growth to increase sunlight penetration.

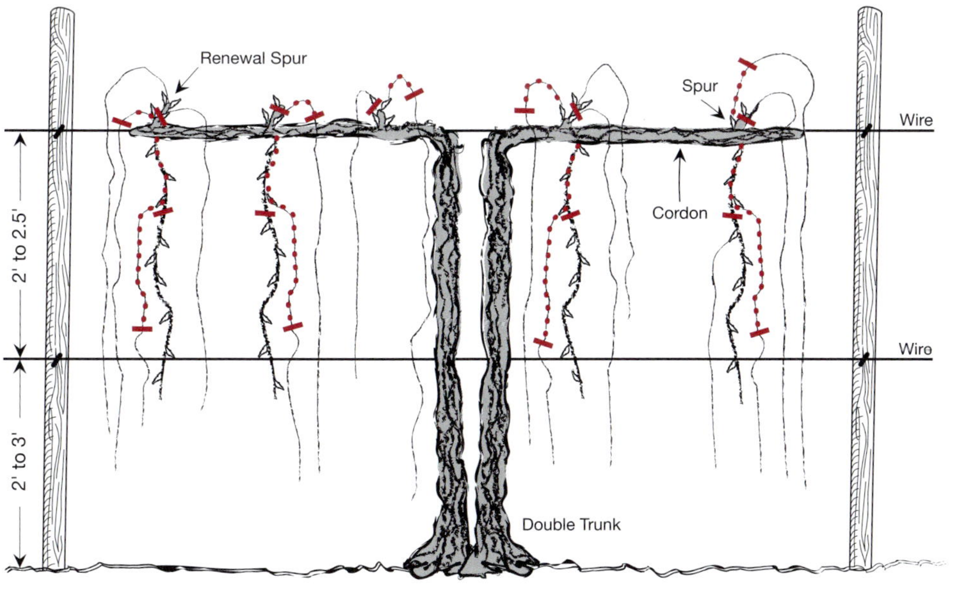

Combination: For vines that have experienced significant winter injury, a combination of spurs and long canes may be used to achieve the sufficient number of buds needed to produce a good crop. The following year, the vine can be converted back to a complete cordon spur renewal system or cane training system.

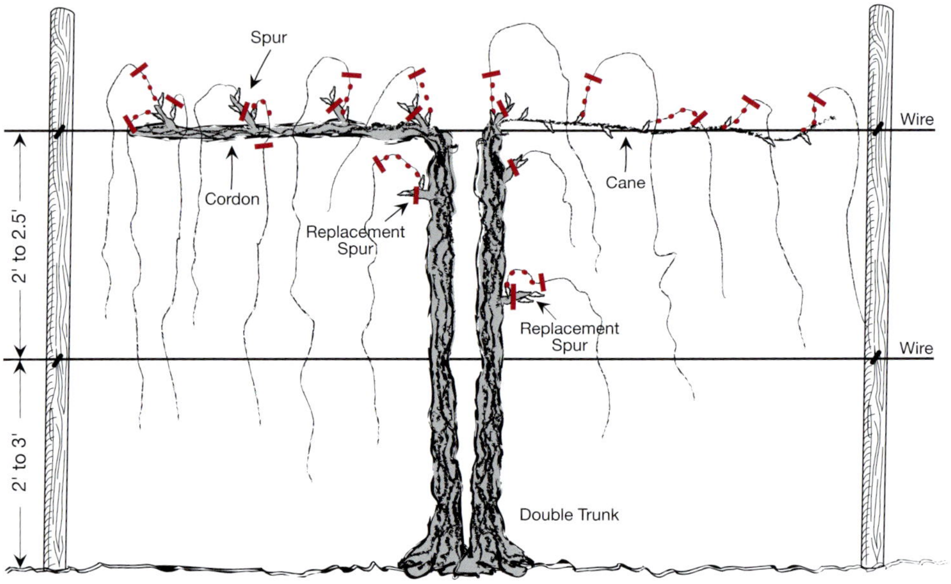

Controlling Diseases in the Vineyard

Fungus diseases that attack grape vines and their fruit thrive in the humid climates of the eastern United States. The control of such fungus diseases and insect populations is of critical importance in establishing a strong and effective vineyard management plan. Fungus disease control is particularly important because fungus damage can wipe out the entire grape crop.

While controlling insect populations is important, insects will not destroy the entire crop. However, an effective insect control program is still a key element in establishing an effective fungus disease program because insects, as they move throughout the vineyard, are vectors that quickly spread fungus diseases.

There are five major fungus diseases: black rot, downy mildew, powdery mildew, anthracnose, and dead arm disease. The books mentioned earlier in the chapter provide very good information on how to identify these diseases in the field, how they overwinter and spread during the next growing season, and how to control their spread.

Note that I use the words "control" or "manage" such fungus disease populations, and not "eliminate" them. While it is conceivable to eliminate such diseases, it takes expensive, toxic fungicides and insecticides to do the job, which can be very damaging to the environment. Three of the most prevalent of these diseases are:

Black rot, the most destructive disease in the Hudson Valley, appears in wet spring months (water activates the release of fungus spores) as small spots on young tender grape leaves and then spreads to still developing green grape clusters. The spores overwinter on dead canes, leaves, and shriveled hard black grape berries and clusters that remain from the previous growing season. Black rot's presence is not generally apparent until the grapes have become infected and begin to prematurely turn color, then become black and shrivel.

Downy mildew also thrives in our humid climate and is particularly destructive to *vinifera* varieties as opposed to *labrusca* and French-American hybrids. Downy mildew attacks all tender young and older green parts of a grape vine, but focuses mainly on its leaves. It forms brownish spots on the upper leaf surface and white patches that produce spores on the leaf's underside.

Powdery mildew, like downy mildew, attaches itself to all new green parts of the vine, especially young shoots and fruit clusters. The disease looks like a white/grayish powder on the young shoots and leaves. When infected, the young shoots lose their color and ultimately drop off. For black rot, downy mildew, and powdery mildew, the fungus only attacks growing green growth and green unripe grape clusters. Once the grape berries begin to change color, (known as veraison), these funguses do not infect the newly pigmented grapes. Hence, as the crop ripens and becomes ready for harvest, the need to apply spray material decreases or is not needed at all.

Controlling Fungus

To control fungus, it is very important to first remove all diseased wood and berries from the vineyard and dispose of it far away from the vineyard, or burn the brush to kill dormant fungus spores. Second, an effective and timely spray material program should be instituted. This means a fungicide/insecticide spray application every ten to fourteen days during the entire growing season, until the grapes begin to turn color. All parts of the vine need to be sprayed, including the brown trunks (since they harbor fungus spores) and the top and undersides of all leaves, shoots and grape clusters.

For varieties that have tight bunched grape clusters, such as Chelois, it is very important to apply the spray to the inside of the cluster before it closes in and becomes tight and unable to accept spray in the interior of the cluster. Otherwise, the interior of the grape cluster becomes a place where fungus diseases can incubate and spread again to the remainder of the vineyard.

The reason for the high number of spray applications is three-fold. First, the spray becomes ineffective after a week or so. Once applied, most fungicides are manufactured to disintegrate after a short period of time so that they do not remain actively toxic, which can damage wildlife. Second, as the vine grows, all new growth, which is especially susceptible to fungus infections, is unprotected by the previous spray application. Third, rain washes off spray material. It is important to apply the spray after a rain and when the leaves are dry so the material can stick to the plant. Another reason to apply after a rain is that fungus spores are released after precipitation, hence the fungicide is very effective in curtailing the spread of the disease after, and not before, a rain shower.

The spray materials used to prevent fungus and insect damage range from "Bordeaux Mixture" (a mixture of copper sulfate, active lime, and water), general fungus/insecticides called fruit or orchard sprays that can be purchased at a local garden shop, powdered sulfur, and organic sprays that can include soaps, and other non-toxic chemicals that hinder the spread of fungus or curtails insect damage. These spray materials are mixed with water and pumped out via a sprayer that disperses the material evenly to all parts of the vine. Such sprayers are available at the same garden centers or larger hardware stores.

Spray Application

Remember, the application of these chemicals are either toxic or, even if organic, designed to kill or curtail fungus and insects, hence they should not be inhaled or come in contact with skin. Further, even organic sprays are fine particulate matter that should not be inhaled. Hence, to minimize contact, the applicator should wear heavy clothes, long-sleeved shirts and pants, gloves, and a facemask for protection.

After spraying has been completed, the spray equipment should be well rinsed with water to avoid clogging, and the water should be disposed in a safe manner and location that is not near streams or other bodies of water. These, and all, chemicals should always be stored in a safe place out of the reach of children. As a final precaution, the applicator should shower after spraying to remove any trace chemicals, and his/her protective clothing should be washed separately.

A final word is needed on the governmental regulation of chemical spray material and its application. Many states have laws on how to conduct spray operations and which spray materials can be used. Also, some states distinguish between commercial spray operators and hobbyists that grow fruit crops. The distinction between the two is generally determined by whether the crop is used for home use or is sold commercially on a large scale. Those who conduct activities that are deemed by state law to be commercial operations may need to be licensed and be required to take a class on how to spray their crop. For hobbyists, it is still not a bad idea to take the class for safety reasons and to reduce the amount of toxic material used, and to minimize possible damage to the environment.

Additional Thoughts on Vineyard Management

There are many suitable books that outline proper vineyard management techniques for cool climate vineyards, including those mentioned earlier in this chapter. Below are a few of the basic practices I utilize that have helped me to maintain a healthier and more productive vineyard. These practices are organic wherever possible and try to minimize the use of chemicals and machinery, yet, while it is easy to understand these basic principles, it is far more difficult to apply them in a timely manner.

Sod and Sod Management

When establishing a vineyard or altering the grass that grows between grape rows, think about what the grass is meant to do. Is it to absorb excess water in the spring and fall so that the grape plants are not standing in water or getting their feet wet? Is it to maintain soil moisture in the summer? Or is it to give the vineyard a neat and pleasing appearance for the benefit of the winery's customers and sales numbers?

The answers to these questions will determine the sod management techniques. If the site is wet, use ryegrasses for sod. Ryegrasses have deep roots, luxuriant growth, and can absorb lots of water. If the site is droughty in the summer due to soil composition or depth, use fescue grasses or other grasses that die in the summer heat, but revive in the spring or fall when conditions are wet again. If the vineyard is near the winery and appearance is important, use a combination of grasses that absorb excess water in the spring and fall, but remain green in the summer.

Mowing

This is a generally neglected topic. Do not mow when the ground is wet because this will lead to soil compaction (the ground's reduced ability to absorb water during the summer months), suppressed root growth, and the loss of soil fertility. Mow so that grass is kept between five and six inches high. There is no need to mow vineyards so that they look like manicured lawns. That takes time, wastes fuel, and compacts the soil. Keeping vineyards mowed leads to better air circulation because undergrowth is kept to a minimum. To retain soil moisture in the summer, mow close to the ground to stunt or kill the grass. This will reduce the grass's consumption of water so that the vines will have more moisture.

Under-canopy Management

It is best to keep the area under an entire vine's row clean of any vegetative growth. This increases the amount of water available for the vine and enhances air circulation to reduce fungal diseases. Vegetative growth can be controlled by applying herbicide or by hoeing (either by hand or by mechanical hoe). Another way, and one that I prefer, is to use stone mulches. Placing lots of small and medium-size stones under the vine in a row decreases weeds, retains soil moisture and heat, and deflects heat toward the vine year-round. While it is labor intensive, I recommend stone mulches wherever and whenever time and material permit. Once a stone mulch is established, it reduces the need for other under-row weed control in the future.

Fertilizers

Unlike vegetable crops, vineyards and orchards are more permanent features, so fertilizers need to be applied around existing trellises and other structures and not plowed in each year since that can damage grape roots. I do not recommend chemical fertilizers such as 5-10-5 or other high-nitrogen synthetic fertilizers because they prompt the growth of weeds and unwanted vegetative growth of vine shoots, and result in little fruit. Further, chemical fertilizers can inhibit the growth of earthworm populations, which are very effective natural tillers of the soil.

I recommend the use of organic fertilizers, such as leaves, municipal compost, grass clippings, mulch, clam or oyster shells. I tend to put such organic matter directly under the vines to feed them and to minimize weed growth. Organic fertilizers, as opposed to chemical fertilizers (i.e., manufactured with oil), have many advantages: they recycle household organic matter (e.g., fruit and vegetable peelings, eggshells, coffee grounds, etc.); improve soil tilth and fertility and reduce soil compaction; improve moisture retention in the summer and water drainage in the wet seasons; and increase the number of earthworms in a vineyard.

Earthworms

Increasing the number and activity of earthworms is also very important for vineyard soil health. Earthworms quickly compost organic material in the soil to increase fertility. Further, they burrow through the soil, which loosens it and increases air and water circulation, thus keeping it light and airy to facilitate root growth. This is particularly important with heavy clay loam soils. When earthworms die, they add more nutrients to the soil as they decompose. Instead of plowing the ground with a tractor, encourage the growth of earthworms to achieve the same effect.

A single acre of cultivated land may contain as many as 500,000 earthworms. Each day, an earthworm deposits approximately its own weight in castings, which is fertilizer that directly conveys nutrients to a vine's roots. Ever notice that an uncultivated forest floor has soft and loose ground, while some cultivated fields can be hard and compacted because of tractors? The difference is the presence of an active community of earthworms.[2]

NOTES

1. Despite the range of dates these books were written, I consider them indispensable and still applicable today. Jeff Cox, *From Vines to Wines: The Complete Guide to Growing Grapes and Making Your Own Wine* (Pownal, Vt.: Storey Books, 1994), 65–104; James Sheldon Shoemaker, *Small-Fruit Culture: A Text for Introduction and Reference Work and a Guide for Field Practice*, 2nd ed. (Philadelphia: The Blakiston Company, 1950), 70–108; Philip M. Wagner, *A Wine-Grower's Guide* (New York: Alfred A. Knopf, 1985), 110–131; U. P. Hedrick, *Manual of American Grape-Growing* (New York: The Macmillan Company, 1919), 61–123.

2. "Understanding Earthworms," Organic Gardening, www.organicgardening.com/learn-and-grow/understanding-earthworms.

CHAPTER FOUR

Considerations When Growing *Vinifera* Grapes in Cool Climate Regions

Bud break on a young vinifera vine. *Experimental Zweigelt vine with Couderc 3309 rootstock growing in a Hudson Valley vineyard.*

THE CULTIVATION of *vinifera* should have a significant part in the future of viticulture in the Hudson Valley. To succeed, cold-tender *vinifera* varieties, which are highly susceptible to fungal diseases, must be grown on optimally selected warm sites with the right location and topography. Well-known names of these classic European wine grapes include Chardonnay, Gewürztraminer, and Riesling for the whites, and Cabernet Franc, Pinot Noir, and Gamay Noir for the reds, as well as some of the Central European *viniferas* covered in this book, such as Grüner Veltliner and Pinot Blanc for the whites, and Dornfelder and Lemberger for the reds. All of these varieties can be grown profitably in the Hudson Valley to produce quality wines.

This book details the genetic makeup and growing characteristics, based in part on their genetic heritage, of individual grape varieties and their potential for wine production. While there are already many excellent books that give detailed information on how to select the appropriate vineyard site, establish a vineyard, grow grapes, and make wine from all types of grapes (including *vinifera*) grown in cool climate regions of the United States,[1] there are very few books or field guides that provide adequate information on the genetic heritage and horticultural characteristics of these grapes. Further, there are few comprehensive field guides for growers and winemakers that address the potential of individual grape varieties to produce wine or that describe the different wine styles that can be made from the same grape.

With that said, this chapter will focus on growing *vinifera* grapes in the Hudson Valley, and by extension, other cool climate regions of the United States. Successfully growing *vinifera* has the same considerations and is similar to growing the more cold-sensitive French-American hybrids such as Villard Blanc or Chambourcin. Further, the same vineyard management practices recommended for French-American hybrids in the previous chapter are suitable for *vinifera*. The choice of a pruning style depends on the growth habit of the variety in question.

The big difference in growing *vinifera*, as opposed to French-American hybrids in the Hudson Valley, is that there is little or no room for error. This means that the right site and the appropriate grape varieties (and clones of those varieties) and rootstocks must be carefully chosen in order to establish a commercially viable *vinifera* vineyard that can consistently produce superior wines.

There are many factors to consider when determining the optimal vineyard site and the right *vinifera* grapes to grow:

General Conditions and Growing Characteristics

The Hudson Valley is close to the northern limit, climatically, to where *vinifera* can be grown on a commercially profitable scale. *Vinifera* grapes are from Europe, so to successfully cultivate them, it is critical to make them as comfortable as possible in this somewhat inhospitable New World climate.

The Hudson Valley, compared to much of Europe, has much colder winter temperatures, warmer summer temperatures with higher humidity, and alien fungal diseases and insects, including phylloxera. Fungal diseases such as downy mildew, powdery mildew, botrytis, and especially black rot thrive here on the East Coast because of the warmer and more humid summer growing conditions. *Vinifera*, much more so than French-American hybrids, are very susceptible to these diseases. A very vigilant spray program must be used with *vinifera* grapes to control these diseases.

Vinifera prefer more moderate climates and do not like wide temperature variations or short-term temperature extremes. The latter can lead to false springs, when vines bud out too early and are damaged by late spring frosts. Trunks can crack in severe winter cold, which can make vines vulnerable to crown gall. Because of the temperature extremes, multiple trunks, canes, and other pruning techniques are needed to ensure that *vinifera* vines survive and can produce a crop each year.

Macro-climate Considerations When Selecting the Location of a Vineyard Site

There is a general range or band of areas that are suitable for growing *vinifera* grapes in the Hudson Valley. A site needs to be fairly close to the shores of the Hudson River or its larger tributaries so that it can benefit from a moderate temperature that such waters provide. As a rule of thumb, a site that is more than 10 to 15 miles away from the river or its major tributaries is probably too far away to grow most *vinifera* varieties on a commercially successful basis.

In the spring and fall, proximity to the Hudson River keeps temperatures above freezing so that a vine or crop will not be damaged by late spring frost or early fall freezes. Also, those vineyards near the Hudson River at lower elevations have a warmer, more maritime climate because moist warm air travels up the river from the south in the spring and fall. For example, during the blizzard of 1996, more than two feet of snow fell in New York City, and 18 inches fell on my vineyard in Athens, 125 miles up the river. However, just six miles further north from my vineyard, in Coxsackie, only three inches of snow fell because Athens on the Hudson River was influenced by the warmer maritime climate of the Atlantic Ocean while Coxsackie was not, as the storm system from the south traveled up the Hudson.

Having even a slightly more moderate maritime climate is a significant benefit in both the spring and fall because temperatures above 28°–32°F and below 28°F can mean the difference between life or death of a vulnerable green growing vine. If a vineyard can survive one cold snap, chances are temperatures will rise again and another more damaging one may occur one or two weeks later. A vineyard's growing season may be extended, on average, by one or two weeks if it was spared frost injury from a single frost episode during the previous fall or in the spring. This is very important for those *vinifera* varieties that need long growing seasons or warmer temperatures to ripen properly.

Another way to determine if a vineyard site is positively influenced by the moderating influences of the Hudson River—and hence more suitable for *vinifera* cultivation—is to look for areas that have morning fog in the spring and particularly in the fall. The colder air from the highlands mixes with warmer air near the river and creates fog. Further away from the river, the air becomes much colder and crystal clear, which is not helpful in warding off late spring or early fall frosts. Due to the uneven topography of vineyard sites along the Hudson River, accurately identifying these macro-climatic conditions that are favorable to vineyard site is very important.

To increase sun exposure on the vines and the soil, southern exposures are best. Also, Hudson Valley sites with eastern exposures are better than western exposures for two reasons. First, the sun rises in the east and dries vineyard leaves faster, thus inhibiting fungal diseases. Second, most frigid winter winds in the Hudson Valley come from Canada from the northwest or west. Establishing a vineyard on eastern slopes of a hill or mountain helps to protect a vineyard from freezing winds that can kill cold-sensitive *viniferas*. An eastern-facing site is better on both the eastern and western sides of the Hudson River. This means that vineyards on the western shore should face the river, and vineyards on the eastern shore should face away from the river, and the site should have an adequate micro-climate to move colder air away from the vineyard to lower areas.

Ask local growers for their opinions about locating a vineyard in their town or portion of that town. They may be able to provide important historical information about fruit-cultivation practices and suitable sites in the area, such as what kinds of fruits have been cultivated in the area previously—if fruits were never cultivated at all, that should raise a red flag—and general information on the length of the growing season, minimum and maximum temperatures for each season, first and last frost dates, prevailing winds by season, and seasonal rain patterns. If fruits were grown in the area, find out if they were cold-hardy tree fruits such as apples and pears, which tolerate damp soils, or cold-sensitive stone fruits, such as cherries and peaches, which prefer warmer soils and spring frost dates that consistently end fairly early.

Micro-climate Considerations When Selecting a Vineyard Site

All the towns that border the Hudson River—from Orange to Albany counties on the west bank of the river and from Putnam to Rensselaer counties on the east bank, as well as the towns that border them further away from the river—will have some suitable lands for *vinifera* vineyards.

Once the town has been selected, the grower needs to evaluate the micro-climate of the vineyard site being considered. The considerations include:

Topography: Is the site located near the crest of a hill or at least halfway up a hill so that cold air and excess water can quickly travel down the slope to minimize frost damage or winter injury? If so, that is the kind of topography that is suitable for a vineyard.

While the average temperature of a vineyard is important to keep in mind, the winter minimum temperature is critical. The minimum temperature is what causes frost damage or vine winter injury. Also, soil water drainage issues need to be considered so that water quickly drains off to lower lands. Wet soils, often located in low lying areas do not drain off quickly, hence they are colder micro-climate sites that are not suitable for grapes, especially *viniferas*. The increased minimum temperatures of a site and good water drainage are often dictated by the topography of the site and the degree of the slope.

The topography can help to determine if slight breezes go through a vineyard and in which direction. While soft breezes in the summer are beneficial to dry off vines and inhibit fungal diseases, they may also bring in frosts that damage vines in the spring, fall, and winter by pushing out warmer air from frost pockets.

So far I have discussed the advantages of sites that are high on a slope, preferably with a southern or eastern exposure. The disadvantages of locating a vineyard in low-lying areas should also be considered: it will have cold air pockets, colder soils, poorer water drainage and airflow, higher humidity, more fungal diseases, and soils that are too rich for the production of quality wines.

Hence, topography is important in determining the temperature of the site and if it is on well drained soils.

Vineyard topography. *Side view of a vineyard site illustrating how low pocket areas collect cold air and vine-damaging frost.*

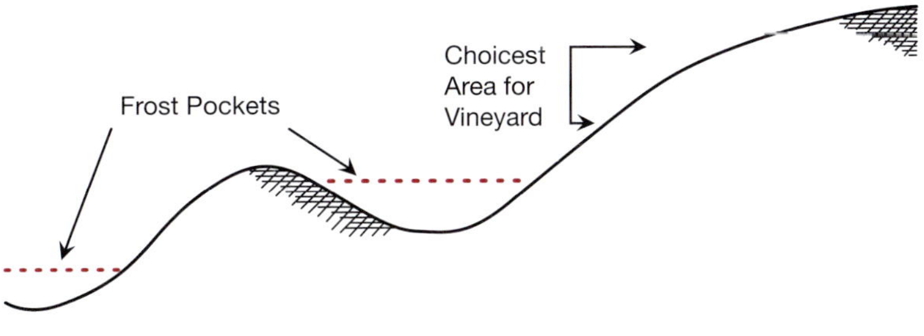

The Hudson Valley and much of the eastern United States have been mapped by the United States Geological Survey. Refer to these maps to look for gentle sloping land that is easy to cultivate and that also allows excess water to drain and cold air to dissipate.

Soil types: It is important to have a site with warmer, friable soil types that retain sufficient water for the vines, facilitate drainage of excess water, and promote root growth. Warmer soils are gravelly silt or clay loams and sandy loams that are high above the floodplain or water table.

Different soil types have varying fertility and pH levels that affect *vinifera* growth. Overly fertile soils lead to excessive vegetative growth, which leads to lower quality herbaceous and watery wines. *Vinifera* tend to be more tolerant of higher pH—that is, more alkaline—soils and those that contain limestone, while French-American hybrids like slightly acidic to neutral soils and *labruscas* do better in more acidic soil.

Markers that can help indicate if a soil is warm and fertile are the color of the soil and the stones in it. Black soils tend to have too much organic matter, which can indicate that they are too fertile, do not drain water well, or are cooler in temperature. Bright red soils tend to have too much clay in them, which impedes water drainage. In addition, red soils tend to have a high iron content, which produces wines with ferric casse, a slight cloudiness in the wine that is difficult to remove without special treatments.

The color of a soil's rocks influences a vineyard's temperature. An abundance of black or dark rocks on the soil's surface efficiently absorb solar heat and reflect it back to the vines in the evening much more readily than sod or lighter colored rocks.

To help identify local soil types, analyze the soil maps published by the United States Department of Agriculture, Natural Resources Conservation Service (NRCS). The soil maps and reports prepared by the NRCS detail the location and class of soils best suited for grape production, provide descriptions of the soil types and their properties, identify optimum agricultural use and management of such soils, and generally indicate the slope of the land.

Soil depth: The site should have well-drained soils of only moderate fertility and should be over 30–40 inches deep. Ideally, underneath this soil, there should be friable rock, such as shale, slate, gravel, or some other water-permeable substrata for good water drainage and not a water-impermeable layer of bedrock, hard pan, or heavy clay. Shallow, compacted heavy clay or silt soils, waterlogged soils, or ones with active springs, both on the surface and in subsoils, should be avoided because they impede good water drainage and severely inhibit soil aeration, which is critical for root growth.

Deep, well-drained soils of only moderate fertility increase the water-holding capacity of the soil and minimize the harm that can occur to vines during a drought. Further, deep soils enable vines to develop the vital root systems required to obtain all the nutrients they need to thrive, which means better production of quality fruit for winemaking, leaf retention during the summer, and vine growth.

Temperature characteristics and length of growing season: It is important to know the history of the vineyard site as it relates to 1) the minimum temperature in the winter; 2) the average temperature during the growing season; and 3) the average time between the last spring frost and the first fall frost. Each of these must be reviewed to determine if a *vinifera* variety will be capable of producing grapes profitably over the long term. If, for example, the number of frost-free days is long enough to grow *vinifera*, but the number of heating degree days is deficient, then *vinifera* will not thrive. Also, the site can have a long enough growing season and sufficient heating degree days, but there may possibly be one or two nights when the minimum killing temperature for a vine is exceeded, and the vine incurs frost damage or severe winter injury. Again, the grower needs to look at all three above criteria to determine if the site can sustain *vinifera* over the long term.

Proximity to wooded areas: When selecting a site for a vineyard, avoid choosing a site that is too close to wooded areas, especially if the tree line blocks sunlight. Wooded areas are breeding grounds for insect pests and fungal diseases; inhibit airflow, which increases the incidence of fungal diseases; are habitat for deer and birds, which devastate grape crops; block sunlight from reaching the vineyard; and compete with the vines for valuable water and soil nutrients.

Human-made climate attributes of a vineyard: These items should be considered after the vineyard site is selected, but they can affect the micro-climate of a vineyard and hence the viability of *vinifera* on marginal sites.

There is a whole science to the subject of whether vine rows should go north-south or east-west and how the width of the space between the rows may affect vineyard soil temperature and the amount of solar energy absorbed by the vines. There are also some differing opinions on which approach should be used relating to the benefits of vine-row orientation and the distance between rows. In my view, for the Hudson Valley at its latitude, vines on north-south-oriented rows tend to absorb more solar energy, which helps to ripen a crop sooner. Vines on east-west-oriented rows tend not to absorb as much solar energy, but the soils absorb that energy, which increases the soil temperature. If the concern is achieving a sufficient number of growing days, a north-south row orientation may give increased sun exposure. However, if the concern is not the length of the growing season, but the coolness of the site, then an east-west orientation may be best because the sun warms up the vineyard soils more effectively.[2]

Decisions regarding vine spacing, trellis systems, row width and length, placement of access roads, and whether or not the vineyard is to be terraced can affect vineyard ambient temperatures, airflow and water drainage, and thereby have positive, negative, or conflicting results on the viability of a *vinifera* vineyard, depending on the site. Due to the widely varying topography of vineyards in the Hudson Valley, this is not the place to discuss this very complicated topic. The books cited in this endnote can provide guidance on the issues raised here.[3]

Selecting *Vinifera* Varieties and Their Clones

I have discussed the importance of finding a site that has: (a) the requisite minimum number of heating degree days to easily ripen a commercial crop; (b) has bud break times after most late spring frosts have occurred and ripens before most early fall frosts occur; (c) has a sufficient growing season to ripen a commercial crop; (d) the proper soil types for vine sustenance and water drainage; and (e) the proper water drainage and airflow to channel water and cold air away from sensitive vines.

Within these more limited parameters, you need to determine if there is a suitable *vinifera* variety that can be grown on this site. This book and others, as well as agricultural bulletins, describe grape varieties' budding times, ripening times, the length of time needed to grow a crop, and other information needed to make an informed choice when selecting a variety.

It is important to note that there can be a wide variation in the growing characteristics of a particular *vinifera* clone. Some take longer to ripen, others bud out earlier or later, and they all have different levels of resistance to fungal diseases. Also, significantly, each clone acts differently in the cellar. Some have lots of color, taste, or nose, which add to a wine's quality; others, while good in the field, may not be suitable for the cellar. The pursuit of clonal selection experimentation which is used to identify the best clone, depending on the circumstances, is still a very new topic in the eastern United States, so there is limited information on this, but for guidance, please review the literature.[4]

Rootstock Considerations

In the eastern United States, French-American hybrids and *labrusca* hybrids grown for wine production or table use often grow on their own roots. This is possible because these varieties have American genes so they can tolerate phylloxera and nematodes, both American insects that live in the soil and feed on grape roots. *Vinifera*, on the other hand, must be grown on American hybrid rootstocks that protect the plant above from phylloxera and nematodes.

Most *vinifera* rootstocks are themselves hybrids of American grape species such as *berlandieri*, *riparia*, *rupestris*, and, to a lesser extent, *cordifolia* and *labrusca*. These hybrids were developed to increase or decrease vigor of the scion (the vine above the graft), impart resistance to phylloxera and nematodes, and impart drought resistance in very dry soils and tolerance to lime soils. The latter two protective features were needed to accommodate the conditions in European vineyards, while the first two attributes helped to address the needs of growing *vinifera* in the United States.

The most commonly used rootstocks in New York are described on the next page to assist growers in determining which rootstock should be used depending on the vineyard site and *vinifera* variety being planted.[5]

Rootstock Hybrids

COUDERC 3309 (3309C)

This rootstock hybrid, bred in 1881 by Georges Couderc (1850–1928), is a cross of *Riparia tomenteuse* × *rupestris*, which initially produced eighteen seeds. Couderc planted them the following year in soils with heightened lime content. All but four plants that survived had indications of chlorosis, an iron deficiency that can adversely affect grape plants that are planted in soils with high lime content.

This rootstock is vigorous in fresh and deep soils, but is not recommended for hot, dry conditions as is found in California. The variety has excellent resistance to phylloxera and medium lime tolerance. It is not recommended for rich, poorly drained soils or droughty soil conditions. The variety both roots and grafts easily for most *vinifera* varieties. This is one of the more common rootstocks used for *vinifera* in the eastern United States.[6]

KOBER 5BB

This was derived from a *berlandieri* × *riparia* grape hybrid that originated around 1886 from a French nurseryman, Euryale Resseguier, who sent 22 pounds of these seeds to Sigmund Teleki (1854–1910) of Hungary, who then planted 40,000 of them. In 1904, Teleki transferred some of his more interesting selections to Franz Kober, an Austrian viticulture inspector, who evaluated them and made further observations.

One of these seeds became Kober 5BB, a vigorous rootstock that produces a large amount of wood for propagation and has a relatively shorter vegetative cycle, which means that it is more suited for northern climates as the plant becomes dormant at the appropriate time in the fall.

Kober 5BB is well-suited to humid clay soils and not extremely dry soil conditions. It has a high tolerance for lime and good resistance to nematodes. The variety roots well, but can present some problems in grafting, especially field grafting due to the high incidence of scion rooting, which leads to the eventual death of the rootstock.[7]

SELECTION OPPENHEIM NO. 4 (SO4)

This rootstock was selected at the viticultural school at Oppenheim, Germany, from Sigmund Teleki's *Berlandieri–Riparia* No. 4, of the same origin as Kober 5BB, noted previously. Like *riparia*, SO4 is a vigorous rootstock that develops rapidly with early fruit set and advanced maturity. While the rootstock imparts productivity to its scion, it may be less productive than 5BB. SO4 is suited to humid and clay soils, and does not work well in very dry soils. It has good resistance to nematodes and good tolerance to lime. This rootstock roots well, field grafts well, and is satisfactory in bench grafts.[8]

MILLARDET ET DE GRASSET 101-14

This is a *riparia* × *rupestris* hybrid developed in 1882 by Professor Millardet of the Bordeaux University, in association with Marquis de Grasset.

This hybrid rootstock is more vigorous than *Riparia Gloire de Montpellier* (see below), but less vigorous than Couderc 3309. It does, however, have a shorter vegetative cycle than 3309C, hence it may be more preferable in situations where early ripening is the goal. The rootstock does well in fresh clay soils and has at best only moderate lime tolerance. Its root system is thin and branching much like its parent *riparia*. It also roots and grafts easily.[9] This rootstock was the seed parent of the Kuhlmann French-American hybrids Maréchal Foch and Léon Millot.

RIPARIA GLOIRE DE MONTPELLIER

Among the many *riparia* brought to Europe at the beginning of the phylloxera crisis, this hybrid rootstock has proven to be popular. The variety has excellent resistance to phylloxera, but lacks vigor and favors quality over quantity in the grapes produced. It also ripens quickly. This variety prefers fresh humid soil and is sensitive to lime. Riparia Gloire roots and grafts extremely well.

Vineyard Owner Constraints

A prospective vineyard owner or manager needs to consider the amount of financial resources and time that he or she plans to commit to growing *vinifera*. If the owner or manager is doing this on a part-time basis or as a second job or hobby, after the needs of the "day job" and the many needs of family life have been met, then growing *vinifera* is probably not the best plan at this stage in a person's life. As mentioned previously, growing *vinifera* in the Hudson Valley can be done, but there is little or no room for error, so if all of the critical steps needed to cultivate *vinifera*—the pruning, tying, spraying, mowing, summer hedging, and placement of nets to reduce bird damage—cannot be done on a timely basis, then do not plant *vinifera*.

Market and Economic Considerations

While some *vinifera* varieties are on average very productive over the years, they may command a low or uncertain price per ton. Other varieties, on the other hand, can command a high price per ton, but have low or inconsistent yields per acre or are expensive to grow. Before growing *vinifera*, it is important to determine if the market will pay the price needed to cover annual production costs and make a profit.

NOTES

1. See the following books: Jeff Cox, *From Vines to Wines: The Complete Guide to Growing Grapes and Making Your Own Wine* (Pownal, Vt.: Storey Communications, 1994); Philip M. Wagner, *A Wine-Grower's Guide* (New York: Alfred A. Knopf, 1985); Tom Plocher and Bob Parke, *Northern Winework: Growing Grapes and Making Wine in Cold Climates* (Hugo, Minn.: Eau Clair Printing Company, Inc., 2001); and Lucie T. Morton, *Winegrowing in Eastern America* (Ithaca, N.Y.: Cornell University Press, 1985).

2. The effect of row orientation and aisle spacing depends on the latitude at which grapes are grown, the slope of the vineyard in question, and the intensity of the sun on vineyards during the growing season. For more information on this topic see: Cox, *From Vines to Wines*, 36–37; Plocher and Parke, *Northern Winework*, 45–46; and Wagner, *Wine-Grower's Guide*, 102.

3. See the following books on vineyard management practices and how they are influenced by how a vineyard is set out: Cox, *From Vines to Wines*; Wagner, *Wine-Grower's Guide*; Plocher and Parke, *Northern Winework*; and Morton, *Winegrowing in Eastern America*.

4. Refer to industry periodicals, which often have articles on the latest clonal selection work being done by wineries and wine associations. Periodicals such as *Vineyard & Winery Management* and *Practical Winery & Vineyard Journal* are among the best, but also look for past articles from the now-defunct *Wine East*, one of my favorite publications on winemaking in the east.

5. These rootstock descriptions come primarily from: Pierre Galet and Lucie T. Morton (translator), *A Practical Ampelography: Grapevine Identification* (Ithaca, N.Y. and London: Cornell University Press, 1979), 214, 209, 208, 215, 207; and Double A Vineyards, *Double A Vineyards Nursery Catalog*, 1989–90 through 2009–10.

6. Galet and Morton, *Practical Ampelography*, 214.

7. Ibid., 209.

8. Ibid., 208.

9. Ibid., 215.

CHAPTER FIVE

The Principles of Winemaking

THIS CHAPTER IS INTENDED to be a primer on winemaking; it is not a substitute for the many excellent books that fully outline the winemaking process and its chemistry. Instead, this chapter presents my perspective as a winemaker in the Hudson Valley for the past thirty-five years. In addition to home winemaking, I also worked at Benmarl Vineyards in the late 1970s and early 1980s, and have made many small batches of commercial wine for the past seven years at Hudson-Chatham Winery in Ghent, in Columbia County.

The endnotes for this chapter cite several good books that detail the highly technical, chemistry-based approach to winemaking, or the more "how-to" approach for the home winemaker.[1] These books offer a complete understanding of the winemaking process, the chemistry that underlies it all, and, most importantly, how to turn fruit into a palatable wine to be enjoyed by family and friends. There are many differences of opinion on the approaches to making wine, thus confirming that it is an art and not a science.

Winemaking in cool climate growing regions is unlike winemaking in warm climates such as California because cool climate grapes tend to have higher acid levels and lower sugar levels than those grown in warmer regions. So, in this chapter I will cover topics such as correcting grapes by adding sugar, or ameliorating the grape juice or must with water to make a good wine. In the end, winemaking can be a fun process that brings family and friends together for a common purpose in a shared endeavor, whether it is during the fall harvest and processing season, the racking of wines in the spring, or ultimately bottling the fruits of these labors the following year. The pursuit of making wine results in the creation of a concrete product that can be shared in the future, and used as a vehicle to relive the memorable events that lead to its production.

Harvested grapes ready for pressing. *The key to making good wine relies on the quality and character of the harvested grapes.*

The Big Picture

Fermentation is the process of converting fruit, such as grapes, into wine. It is the work of one-celled microorganisms called yeast, which are classified as a kind of fungi. Wild yeasts are everywhere and cover everything, including fruit, barrels, winemaking equipment, and buildings where wine is made. These cultured or wild yeasts, in the right environment, consume sugar in the juice of crushed fruit (also called "must"), which allows them to reproduce exponentially through a process of budding whereby a yeast cell multiplies by splitting apart to become two yeast cells.

As yeast cells consume sugar to satisfy their energy needs, they secrete two main waste products: ethanol alcohol (the kind humans can consume) and carbon dioxide. The amount of alcohol and carbon dioxide generated is approximately half and half, with just a slightly higher amount of alcohol being produced at 48.4 percent of the volume of sugar as opposed to 46.6 percent carbon dioxide. The remaining waste product is mostly glycerin, which contributes softness and body to a finished wine.[2] Since yeast cells multiply at an exponential rate, the must bubbles rapidly and violently until all of the sugar is consumed, at which time the bubbling mass subsides. Depending on the temperature, this process can take between five to ten days or as long as a few months. The warmer the temperature, the faster the fermentation process.

Basic Equipment & Supplies Needed to Make Wine

The equipment and supplies described below can be purchased at neighborhood beer- or winemaking supply store. These businesses are also great places to meet other local winemakers and to share information on making wine. At the end of this chapter is a list of just a few of the many beer- and winemaking supply shops in the northeastern part of the United States.[3]

Grape Crusher

The crusher grinds or mashes the grapes, breaking the grape skins and converting the raw fruit into a slurry. This process facilitates the release of juice for winemaking and the fermentation of the must.

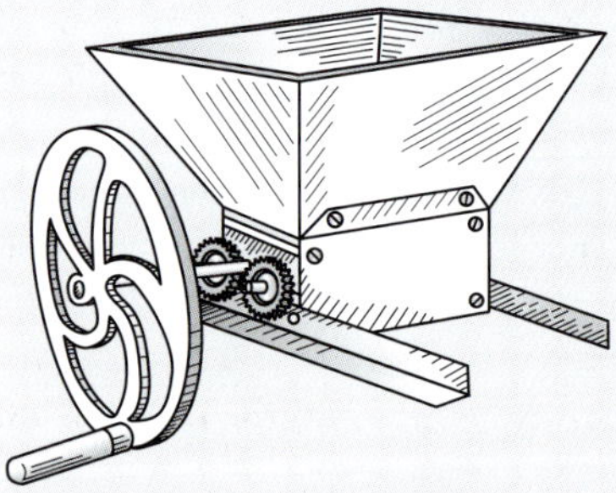

Maceration Vat

The maceration vat, a container that holds the crushed grapes, can be a plastic garbage can, stainless-steel tank, or ceramic pot in which the skins, seeds, juice, and sometimes stems can ferment together. These vats have large openings at the top so that the must can be stirred or punched down at regular intervals. To keep fruit flies and other insects out and carbon dioxide in, the vat is completely covered by a plastic cover or tarp secured to the top.

Wine Press

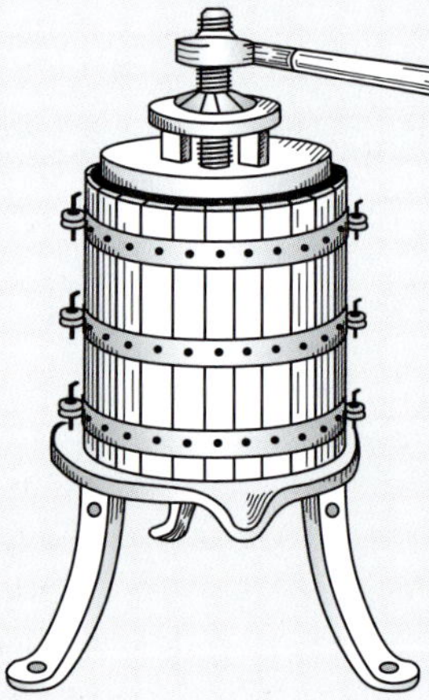

The wine press is a round basket made of wooden slats with narrow gaps between each slat. It sits on a tripod that has a catch basin to collect the pressed liquid. The recently crushed unfermented must (generally of white grapes) or fermented must (generally of red grapes) is placed into the basket to separate the white juice or raw red wine from the skins, seeds, and stems. A plate above the basket is pressed down by a screw-like device to squeeze out the liquid and separate it from the grape solids. These grape solids are called "pumice" or "marc."

Fermenter

This is the container into which the newly pressed raw wine or unfermented white juice is placed. It has a much smaller opening on top for a cork or stopper with a fermentation lock so that the remaining carbon dioxide can continue to escape while preventing insects from entering the container. Fermenters can include glass carboys, stainless-steel beer kegs, or demijohns. Wooden barrels can also be used, but transferring sediment-laden raw wine into such containers makes the barrels difficult to clean later, so nonstick, smooth containers such as glass, plastic, or stainless steel are generally used.

Fermentation Lock

In essence, this is a gas trap with water in it, so that carbon dioxide generated by the fermentation process can be released, but nothing can get in. Fermentation locks need corks or stoppers.

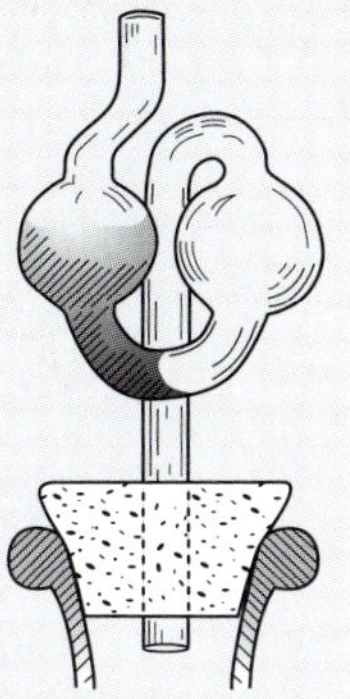

Wooden Barrel or Container

Nearly finished or finished wine is placed in a wooden barrel or wine container for long-term storage and aging. Wooden barrels impart flavors of oak, chestnut, or beech to the wine. Stainless-steel, glass, or plastic containers can be used to store whites and lighter reds until they are ready for blending or bottling.

Hydrometer

This is a weighted glass tube that measures the specific gravity of the must. Specific gravity is the measure of the weight or density of a liquid when compared to the same volume of water. This instrument measures the amount of sugar in the must. The amount of suspended solids (sugar) is measured in units called either the Brix scale or the Balling scale. Hence, must or juice that has a measurement of 24° Brix is a solution that contains 24 percent sugar.

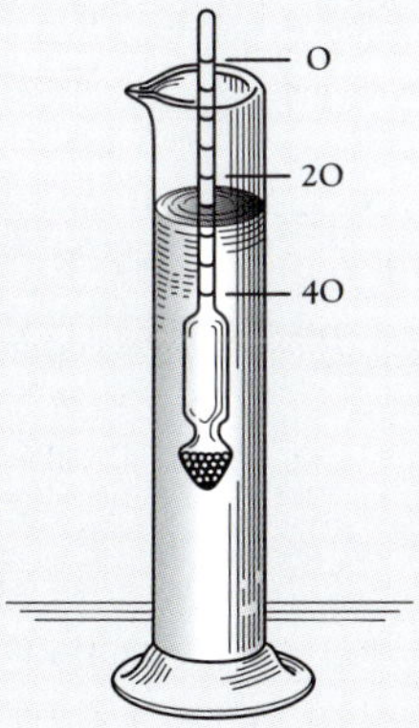

Siphon Tube

This is a flexible, good quality, clear rubber hose or tube used to siphon wine from one container to another during the winemaking process. Siphoning wine is called "racking" the wine.

Wine Pump

To facilitate the racking process, a wine pump can be used to transfer the wine horizontally from one container to another instead of vertically by a gravity-feed siphon.

Sulfur

Sulfur is an antioxidant that reduces oxygen. Adding chemicals such as potassium metabisulfite to wine releases sulfur dioxide, which is essential to protect wine from exposure to oxygen. Sulfur dioxide, an active sulfur compound, acts as an antiseptic, antioxidant, preservative, and defense mechanism against unwanted enzymes and aldehydes. Sulfur comes in the form of potassium metabisulfite, sodium metabisulfite, campden tablets, or sulfur sticks.

Facilities to Make Wine

A suitable place to make and store wine at a cool temperature throughout the year should have access to water for cleaning equipment and bottling wine, adequate lighting, and electrical outlets for power, and yes, a radio for entertainment. Ideally, it should have a slightly sloping concrete floor so that excess water drains away from the work area.

Cleaning Products

Cleaning products are needed to keep storage containers, fermentation locks, presses, pumps, floors, and processing equipment sanitary and free of microbiological organisms and fruit flies. Access to water is important. Also needed are hoses, spray nozzles, long- and short-handled scrub brushes, soda ash, bottle-brushes, and washcloths. When cleaning winemaking equipment, never use chlorine or bleach. If it is not completely rinsed off, it can ruin the flavor of a wine.

Miscellaneous Processing Equipment

Measuring cups or spoons, funnels, buckets, and shovels are needed to help transfer grapes, must, or wine from one container or vat to another. Plastic to cover vats and solid bungs to cap finished wines are also needed.

Bottling Equipment

A corker, corks or other caps, bottles, and labels are needed to bottle the finished wines.

Overview of the Winemaking Process

Harvesting Grapes

Grapes need to be harvested, secured, or purchased. The time of harvest depends on many variables, including the grape variety, the style of wine to be produced, the sugar and acid levels of the grapes, the kind of growing season that occurred (i.e., wet, dry, cold, warm, sunny, overcast, early or late bloom date), the condition of the fruit, and the availability of labor to harvest the crop.

The condition of the grapes at the time of harvest or their potential condition in the near future based on weather forecasts should be considered. In the fall, certain thin-skinned varieties, such as Baco Noir, swell, crack, and fall apart very quickly after heavy rains during the harvest season, so they should be picked before, not after, a heavy rain. For *labrusca* varieties such as Concord, the skins can crack, split and shell and then fall off the cluster as they become overripe. If high winds are forecast, those grapes should be harvested beforehand.

The sugar content of the grapes in the field should be measured with a refractometer. The sugar level of the harvested grapes should be 14°–17° Brix for most *labrusca* grape varieties and between 19°–24° Brix for French-American hybrids and *vinifera* varieties. The grape variety descriptions elsewhere in the book specify the potential sugar levels of each variety.

All of the parameters needed for good grape quality may not align on the same day or week. Determining the optimum time to harvest the crop is a difficult and stressful judgment call. Often there is no choice as to when to harvest grapes. Persistent rain in warm weather can dilute sugar content and increase fungal disease. Lack of sun and warmth can slow the ripening process while birds and wildlife feast on the bounty. Or an early fall frost can bring the season to an abrupt end. The circumstances of each harvest season change each year based on the circumstances of that growing season.

What can be controlled is ensuring that the grapes are picked and stored in clean, stackable containers for transportation. In addition, green and brown stinkbugs should be removed from grapes before processing. These invasive insects, as the name suggests, impart a foul smell and taste to wine when they are crushed with grapes.

Crushing Grapes

Use the grape crusher to crush the grapes and release the juice to facilitate fermentation. The crushed grapes are then placed in a maceration vat (generally for red wines) or directly into the wine press (generally for white wines). To retain the fruity flavors of white grapes, they are pressed out immediately. It is important to crush the fruit as quickly as possible to minimize the breeding of fruit flies and to ensure that the grapes remain sound and free of disease, acetic acid, and other pathogens.

When crushing grapes, do not sheer, tear, or squeeze the stems or crack the grape seeds, which will impart bitter flavors and detract from the quality of the wine. Before fermentation begins, remove any stems that are bright green and not brown or woody.

Correction of Must

This is probably the most important part of the process. Mistakes made here are difficult to correct later. Use the hydrometer to measure the specific gravity—that is, the sugar content of the must. The amount of alcohol projected to be in a wine needs to be based on its sugar content to achieve the style of wine desired. For light fruity (and probably highly acidic) whites, an alcohol level should be approximately 11 percent alcohol or less. If a semisweet or sweet white or red wine is the goal, a higher alcohol level is desired. For a light fruity dry red, the alcohol level should be no higher than 12 percent alcohol. For big complex reds, in the eastern United States, alcohol levels of 13 percent or slightly more should be the goal.

To achieve the desired alcohol level, remember that slightly less than half the sugar in the must is converted to alcohol, so if the initial sugar level of the must is 23° Brix, then the alcohol level will be slightly less than half of that or just slightly above 11 percent alcohol. If the alcohol level desired is 12 percent, then the sugar level of the must should start at just slightly more than 24° Brix.

Grapes grown in the Hudson Valley and in other cool climate regions tend to have slightly more total acid and lower pH levels than is desired to produce a soft, approachable dinner wine. High acid levels, as are commonly found in cool climate grapes, are desired when producing sweet dessert wines. The high acid levels offset the high sugar levels, so the wine's taste is balanced with sufficient fruit to support the body and mouth feel of the wine. Without this balance, the wine would taste like flat, sugary grape juice and not like an interesting, balanced wine.

If the goal is to make a soft, rounder, dry dinner wine, the high acid levels in grape varieties grown in the Hudson Valley and other cool climates need to be reduced. This process is known as ameliorating the must with water. In northern France, as in the eastern United States, water can be added to the must to lower the overall acid levels. Grape varieties such as Baco Noir, Marquette, and Frontenac tend to have, on average, even higher total acidity levels than other grapes grown in the area. The amount of water added depends on the total acidity of the grapes. I generally have had to add between 10 and 20 percent water to reduce acid levels in most reds. For lighter-hued red varieties that do not have excessive acid levels such as Pinot Noir or Cascade, water is not added.

After water has been added to reduce the must's acid levels, the sugar levels will be lower. The hydrometer is used to determine the new sugar levels, and cane sugar can be added to raise the Brix level to what the winemaker wants it to be.

In addition, some winemakers test the must's total acidity and pH levels. Titration kits and portable handheld meters can be purchased for acidity tests. The pH levels should be above 3.2 and below 3.6. If the acid levels are too high or low for the desired wine style, this is the time to correct acid levels by adding water or tartaric or malic acid.

Maceration

After crushing the grapes and correcting the must, the must is then transferred to the maceration vats. Fermenting red must on the skins conveys red pigments, tannins, body, aromas, esters, vanillins, and other flavor or aromatic elements to the wine. Generally, if a red grape is crushed and pressed immediately, the result will be a white or pink wine.

Macerate the must based on the style of wine desired. The longer the maceration time, the bigger the wine will be with more and deeper color, more tannins and complexity. For a white or rosé wine, the skin contact time may be for only five to twelve hours; for a medium-colored and medium-bodied red wine, the time period may be one to five days; for a very big, heavy red wine, the time period may be from seven days to three weeks.

The number of times that the skin cap is punched down also affects the maceration process and skin contact time. As carbon dioxide bubbles up in the fermenting must, it pushes the skins to the top, where they form a hard layer called the cap. To reduce the chance that the cap will harbor bacteria, which will create vinegar and other pathogens, the cap should be punched or pushed down at least once or twice a day. This will reduce the chance of bacteria and increase skin contact, which adds more color, tannins, and other flavor elements to the wine.

The temperature of maceration also significantly affects the aromatics of a wine, its fruitiness, acid profile, and color. How it affects these flavor profiles depends on the variety. For white wines, generally ferment around 55°–70°F so that the fruity aromatics will be preserved or enhanced. For red wines, fermenting at warmer temperatures, but no higher than 85°F, can help to extract more color and tannins from the red skins.

Yeast culture is added at this point in the process. Wild yeasts live everywhere, so if the must is left to its own devices, it will begin to ferment on its own. Before adding a new yeast culture, sulfur should be added to prevent wild yeast from multiplying.

Different yeast cultures have unique attributes that can impart different flavor profiles to a wine, alter the balance and body of a wine, ferment better at different temperatures, ferment at different rates, clean up damaged fruit, ferment in higher alcohol or acid environments, and foam at different levels. Some of the many commercial yeasts available are Premier Cuvée, Pasteur Champagne, Montrachet, Epernay 2, 71-B, K1-V116, Pasteur Red, among others. Each yeast manufacturer gives a description of the fermenting characteristics of their yeasts. Also, one can refer to the numerous studies conducted by Cornell University and the University of California–Davis on the qualities imparted to wine based on the strain of yeast used.

Yeast cultures are granular and come in sealed packets. To start a yeast culture, open the sealed packet and dump the contents into warm—not hot—water that has had some sugar added. Then cover the quart- to half-gallon-size container, depending on the volume of must to be inoculated, with a cloth and place the container in a dark warm area. Stir it a few times and after two hours or so, the liquid will have a foamy head, like a stout. The foaming action indicates that the liquid is fermenting and that the number of yeast cells is rapidly expanding. At this point, the yeast starter can be added to the must, then cover the maceration vat with a plastic tarp.

Adding Sulfur

Sulfur should be added each time the wine is transferred from one storage container to another because free sulfites dissipate over time and during such transfers. The amount of sulfur needed to protect white wines at crush and for subsequent rackings is in the range of 45–55 parts per million (ppm). For reds, the range should be 40–45 ppm at crush and for all future rackings. At bottling, the free sulfur levels should be around 80 ppm for sweet wines, 60–65 ppm for dry white wines, and 50–55 ppm for dry reds.

There are charts and instructions for all sulfur products to help winemakers determine the proper level of free sulfur in wine. For example, Presque Isle Wine Cellars' potassium metabisulfite increases free sulfur by 150 ppm for every gram added to one gallon of wine. To attain appropriate levels of sulfur, which they suggest should be around 40 ppm, approximately one and one-third grams of potassium metabisulfite would be needed to treat five gallons of wine, which is approximately a quarter teaspoon of powder. To achieve slightly higher sulfur levels, more should be added. Other sulfur products have different concentrations of free sulfur, so read the instructions or contact the manufacturer.

Pressing the Must

The length of the maceration process determines when the must should be pressed to separate the new raw wine (if red) or raw juice (if white) from the must. The raw red wine is transferred to the fermenter, where it will finish fermenting until the wine becomes dry—that is, until there is no sugar left in the liquid so that yeast activity stops. Yeast activity will also stop or slow down as the alcohol level rises above 14 percent or as the temperature drops when autumn progresses to winter.

The unfermented white juice is between 16 and 24 percent sugar, with no amount converted to alcohol. Pour this juice into a container with an open top, so that the white juice can be corrected as described in the "Correction of Must" section before putting the juice in the fermenter.

Transferring Pressed Wine to the Fermenter

The next step after pressing the wine is to transfer the new raw partially fermented red wine or the corrected raw unfermented white juice into the fermenter, which can be a glass carboy, stainless-steel tank, beer barrel, or a 55 gallon plastic blue barrel. The red wines can fill most of the container, but leave some room at the top. Insert the fermentation lock.

Unfermented white juice needs a good amount of space in the container for the violent fermentation and bubbling, so fill only seven-eighths of the container. Adding sulfur will inhibit fungal or bacterial growth and to prevent the new wine from browning. Insert the fermentation lock.

First Racking

By mid-November, about one and a half to two months after the new wine was transferred to a fermenter, the fermentation should subside for both red and white wines. If not, that indicates that a sweet dessert wine is being made by a cool fermentation process, or that there is a problem with the wine.

At this point, the container should have clear new wine on top, with lees settled at the bottom of the container. The wine should be racked to separate it from the lees, either by gravity feed or with the assistance of a wine pump, to a new container. Fill this container almost to the top and add sulfur. Taking the wine off the lees, except for barrel-fermented whites, helps to reduce off flavors, the chance of microbiological infections, and the harsh malic and tartaric acids concentrated in the lees. (For further explanation of "lees" see page 51.)

Some winemakers combine different varietal wines into new blends. A varietal wine is made up solely or primarily from one grape variety. I tend to wait until the second or spring racking to blend varietal wines so I will have a better idea of the nose, flavor profile, balance, body, positive attributes and deficiencies, and the overall potential of the wine.

Pre-second Racking Evaluation

Before racking the wine a second time in the late spring of the following year, each batch of varietal wines should be evaluated on two aspects. The first is its chemical stability as indicated by cloudiness, microbiological problems, acid profile and balance, and other attributes as either a varietal wine or as a component of a blended wine. The second is its sensory attributes: nose, bouquet, flavor profile, balance, body, and finish.

This evaluation is important because it determines how a wine will be used, whether as a varietal or in a blend. Also, those wines that have chemistry flaws or that are unstable should be fixed so that they do not adversely affect other sound wines in the blend.

To clear up cloudy wines, refer to the section "Cloudy Wines" toward the end of the chapter.

Second Racking

The second racking can be done after all the wines have been evaluated. This racking, which can take place anywhere from April until June, helps to further separate the remaining lees from the wine. It is also an opportunity aerate the wine to hasten the fermentation of any remaining sugars and to add more sulfur at a rate of 45–55 ppm to preserve white wines and 40–45 ppm for red wines. At this point wines should be blended.

Blending Wines

Varietal wines are combined into blended wines that together have better appearance, bouquet, balance, acid profile, mouth feel, body, and presence. The art of blending takes into consideration the available wines, their potential, and how best to combine them in a manner that improves the overall quality of the blend's component parts. Do not preclude the notion of blending red and white wines to enhance both component parts.

Blending is used to enhance the balance of a wine so that it does not have rough edges or gaps in its flavor profile. Some wines have a nice fruity or complex nose, but are disappointing because they do not have sufficient body, balance, or taste. On the other hand, a varietal wine might have a nice taste and presence, but a muted nose that does not complement the body of the wine.

In the end, the blender is creating a new wine that is more integrated and balanced in all aspects—color, nose, bouquet, body, acidity, fruit taste, and finish—so that it is a more complete wine. Blending wine is a highly personal endeavor and the goal is to increase the sensory enjoyment of the consumer. Throughout the book I have suggested grape varieties that blend well with others and that can either complement both varieties or offset the deficiencies of a particular variety.

Barrel Aging and Care

Oak or chestnut barrel aging improves wine quality. It rounds out both red and white wines and adds complexity to the bouquet. However, it is very important to use only sound barrels that are not infected with bacteria, which will produce acetic acid or brettanomyces. Once a damaging bacterium has infected the wood of a barrel, it will be impossible to remove, so save time, money, and angst by removing that barrel from the winemaking area.

Bottling

After sufficient aging, the wine is ready to bottle. For a fall or Nouveau wine, the aging process is two to three months. For a light spring or summer white, the time period is seven to ten months. For a big complex red, the aging time may be from eleven months to two years.

To bottle wine, a corker, bottles, and corks or another type of cap, is needed. Before bottling, I rack the wine one last time and add the recommended amount of sulfur plus 15–20 percent more for bottling, which means 50–55 ppm for reds, 60–65 ppm for whites, and 80 ppm for sweet wines. Waiting a week or so after the wine is moved can help to ensure that the wine is completely stable for bottling or allow the remaining carbon dioxide in the wine to dissipate.

The easiest way to bottle wine is to use a siphon tube with a bottle nozzle at its end to automatically shut off the flow of wine as one clean bottle after another is filled. Only clean bottles that have been recently rinsed should be used. The books cited in the endnotes of this chapter give detailed instructions on how to bottle wine.

If the wine is not stable, has bubbles, is cloudy, has a haze, or has off flavors, do not bottle it. No amount of sulfur, fining, or filtering will correct a problem wine. Why spend the time, energy, and materials to bottle a faulty wine that will not please the consumer, become cloudy again, or explode? If the wine is faulty, the best thing to do is to wait until the problem has been identified and corrected, or combine the wine into a blend that addresses the underlying problem.

It is also prudent to rack the wine a day or so before bottling it because moving the wine sometimes activates a new spurt of fermentation, produces bubbles, or makes the wine hazy or cloudy. These conditions indicate that the wine is not stable. By waiting a few days, one will have a better idea if the wine is in fact stable and ready to be bottled. If not, bottling should be postponed until it is.

By following all these steps, a quality wine will be produced. The next section briefly outlines some of the things that can go wrong, how to identify the problem, and how mitigate it.

FOUR CARDINAL RULES OF THE WINEMAKING PROCESS

The process of making wine is actually very simple. However, there are a few cardinal rules that must be followed in order to produce a quality wine:

1. Cleanliness is very important because wine is a food product. Use clean, sound fruit, not rotting, fruit fly–ridden fruit. Take protective measures to mitigate against spoilage.

All grape-processing equipment, floors, and storage containers must be kept as clean as possible. Dirty equipment, floors, or storage containers will attract fruit flies, which will introduce acetic acid (also commonly known as vinegar) and other pathogens into the wine. Once contaminants are present, they can spread rapidly to other wines and to the walls and floors of the wine-processing area.

Acetic acid is a by-product of various forms of bacteria that consume either ethanol alcohol or sugar in an aerobic environment—that is, where oxygen is present. In a fermentation process, on the other hand, yeast consumes sugar and produces alcohol in an anaerobic (oxygen-free) environment. The alcohol acts like an antiseptic. As the fermentation continues, the rising level of alcohol protects the must from bacteria, which generate acetic acid and other pathogens.

In addition, the fermenting must creates carbon dioxide, which covers and protects the must from exposure to oxygen. To control acetic acid, keep the environment free of vinegar-producing bacteria and process fruit quickly so that bacteria have no opportunity to propagate.

2. Use your eyes, taste buds, and particularly your nose when making wines. By using your nose to identify off flavors; taste buds to identify wine balance, body, and flavors; and your eyes for off colors and protein hazes, you will be able to quickly identify and correct a problem before it permanently damages what could have been a quality wine.

An adequately trained nose, eyes, and taste buds, which are closely linked to the sense of smell, can identify problems well before any laboratory test. Trust your senses and, if necessary, perform laboratory tests to confirm your initial findings or to quantify the extent of the problem.

3. Keep accurate notes on the sugar levels of the grapes, the harvest dates, the volume of grapes harvested, the pH levels, the yeasts used, and how much, if any, water and sugar is added to your wines. Also note the dates in which wines are pressed, racked, treated (such as the addition of sulfur or other chemicals), bottled, and the number of cases you bottled. If a problem occurs, you can refer to your notes and quickly diagnose and correct the problem. If the final product is to your liking, you will have a record to refer to if you want to replicate these fine results. These notes should also include your tasting notes about young wines before they are bottled.

4. Continually look for indications of problems in the wine such as off smells, flavors, cloudiness, unusual colors, or lack of or too much acidity. It is much easier to correct these problems early rather than after the damage has been done.

Techniques for Making Different Styles of Wine

Before crushing your first grape, know the style of wine that you want to make. Is it to be a Germanic or Alsatian-style Rhine wine, fruity white or red table wine, full-bodied red or white wine, rosé, a seasonal wine, dessert wine, sparkling wine, or something in-between? Once the wine style has been chosen, you will need the right winemaking equipment and the right grape varieties. In the grape variety descriptions of this book, various winemaking techniques have been suggested if appropriate to the grape variety in question for making a specific style of wine.

Briefly, here are a few techniques for making different styles of wine. The books cited in the endnotes of this chapter describe in more detail these different wine styles and winemaking techniques.

Barrel and Sur-Lie Fermentation

To make a buttery, creamy, vanilla-like, leesy, and complex white wine, similar to the heavily oaked California Chardonnays, wines are often partially fermented in wood on the lees. This process combines the technique of barrel fermenting wines with the sur-lie technique—that is, fermenting the wines on the lees.

Unlike the normal procedure of fermenting white wines in which the must is completely fermented in a neutral, stainless-steel fermentation vat and then transferred to a barrel for aging, in the sur-lie process, raw wine that is partially fermented is placed in the barrel to finish fermenting. During fermentation, the suspended solids of the must, also known as "lees" or "mud," settles to the bottom of the barrel. The wine is left in the barrel with a thin layer of fine lees for several months so that the wine can pick up the yeasty flavors of sediment. To obtain even more "leesy" flavors, some winemakers periodically stir the wine to remix the lees with the wine. Sur-lie techniques add a creamy texture to the wine and complex, yeasty flavors.

Fermenting wine in a barrel facilitates the wine's ability to absorb vanillins, esters, and caramel flavors from the wood of the barrel and fosters a malolactic fermentation (the conversion of malic acid to lactic acid), which also adds softness to a wine. Barrel fermentation—used almost exclusively for white wines—is used to achieve buttery, vanilla-like flavors. Suitable grapes for this wine style include Chardonnay, Seyval Blanc, and Vidal.

Carbonic Maceration

This style is employed to make very fruity red wines, and is often used to make fall wines in the Hudson Valley and Nouveau Beaujolais wines in France. Carbonic maceration means allowing grape enzymes, rather than yeast cells, to begin the fermentation process. Uncrushed grape clusters and berries are placed in an oxygen-free environment so that enzymes can convert sugar into alcohol and carbon dioxide. After a week or so, the pressure of the fermenting juices within the berries ruptures the berries and breaks them down. At this point, a yeast starter is added so that a full fermentation can begin.

Often a partial carbonic maceration will be done, which means that whole clusters—about 20–60 percent of the grapes—are put into the fermentation vat. The remainder of the grapes are crushed, corrected to the appropriate sugar level, and then a yeast starter is added so that this portion of the must will generate the carbon dioxide needed to protect the whole clusters, which are fermenting from enzymes and not yeast.

Carbonic maceration is used to make very fruity or seasonal fall and Nouveau wines, which are simple, have lots of fruit, few acids, and little aging potential or complexity. This technique can be used with partially ripe fruit or those varieties that may impart herbaceous flavors that detract from the fruit-flavor profile of the wine.

Cold Soaking

To make a more complex and richer white wine, the skins are left on the must for six to twenty-four hours, and then sulfur is added to inhibit browning of the grapes. This technique, called "cold soaking," adds complexity, body, enhanced mouth feel, esters, and other beneficial flavors to a wine. If this technique is used, only sound, undamaged grapes should be cold soaked. The grape stems should be brown, woody, or ripe and not overly bright green to ensure that bitter flavors are not transferred from the stems to the wine. Taste the grapes first to ensure that the skins will not impart a bitter or herbaceous flavor to the wine. Cold soaks also increase juice yields by up to 10 percent. Grapes such as Chardonnay, Seyval Blanc, Vidal, and Vignoles can benefit from a cold soak.

Settling Juice

If the goal is to make a light, simple, neutral, or fruity white wine, then the juice should be settled. After pressing the white juice, add sulfur to inhibit fermentation and then allow the juice to settle. After a day or so, the solids of the juice will settle on the bottom, and the clear juice on top can be racked into another container to be fermented separately. The remaining lees-laden must can either be fermented separately or added to a red must that is still fermenting. The resultant wine, depending on the grape variety used, can range from a light fruity wine that is pleasant but not complicated to a more fruit-forward wine. Grapes such as Aurora, Seyval Blanc, and Delaware can benefit from settling the juice before adding yeasts.

Malolactic Fermentation

This type of fermentation reduces the sharpness of the malic acid which can be prevalent in many red grapes grown in the Hudson Valley and in other cool climate areas of the United States and Canada. High-acid *riparia* hybrids, such as Baco Noir, can benefit from a malolactic fermentation. In all cool climate areas, red wines tend to have an overabundance of malic acid, which has a sharp, harsh taste on the front of the palate. Malic acid is most often associated with the tart taste of green apples. Too much malic acid can throw off the balance of a wine and make it too angular and disjointed in its flavor and acid profile.

Malolactic fermentation, also called secondary fermentation, is encouraged by adding a malolactic bacteria culture after the wine has finished its primary fermentation. The wine needs to be rather warm at around 70°F, the sulfur levels relatively low, and there should be little or no sugar available for the bacteria to consume. Sometimes in the spring, after a wine begins to warm up, it can spontaneously go through its own malolactic fermentation without any intervention.

This secondary fermentation is different from the primary fermentation. In a primary fermentation, yeast cells consume sugar and secrete alcohol and carbon dioxide. In a secondary fermentation, Leuconostoc bacteria consume malic acid and secrete lactic acid, a milder acid (the same acid found in sour milk), and carbon dioxide. This secondary fermentation occurs over a longer period of time and is much less violent than primary fermentation. It is important that there is no sugar in the wine because these bacteria will consume it with adverse results for the wine.

In addition to reducing harsh malic acids, malolactic fermentation increases the wine's softness and complexity, but it may also reduce some fruit flavors. Malolactic fermentation is not suitable for white wines, in which crispness from higher acidity is desired to balance high concentrations of fruit flavors and/or sugar in the wine.

Dessert Wines

Dessert wines are a specialty in the Hudson Valley and other cool climate growing regions, and a subject in themselves. They include sherries, ports, late-harvest wines, and fortified fruit wines such as cassis. Sweet and semisweet sherries and ports have a few things in common. First, they have sufficiently high levels of total acidity to balance their sweetness. Second, since oxidation of these wines is part of their charm, the oxidation process must be done in a manner that protects these wines, which means fortifying them with additional amounts of alcohol, in the form of brandy, to inhibit bacterial growth.

Sherries are made by filling three-quarters to five-sixths of a wooden barrel with red, white, or blended wines. This large air space encourages oxidation, which browns the wines and gives them a nutty bouquet and taste. To preserve them, brandy is added to raise the alcohol content to 18 percent. Through the Solera method, various batches of sherries are back blended with each other so that the product contains many different grape varieties and vintage years. The Solera method increases the consistency of sherries from year to year because each bottle will contain wines that can be 10 years of age or less.

Ports are made in a similar way, but the only difference is that to preserve their deep ruby-red colors, brandy is added to very young wines that already have bright red and purple hues. The alcohol added preserves the deep red color common among port-like wines. While sherries tend to be back blended with other newer and older sherries, ports tend not to be back blended as much, so there are more vintage ports, while vintage sherries are very rare indeed. In port production, the barrels are filled up higher than they are in sherry production, but not to the top as a small air space is still needed to encourage the wines to mellow and caramelize. Ports, on average, tend to be younger than sherries.

Second-Run Wines

Second-run wines, or as we call them *vino d'aqua*, can be delightful wines that add new tools to the arsenal of a winemaker. Simply put, *vino d'aqua* is taking the pomace (the pulpy residue) after pressing the first run of a wine and rehydrating it to make a second batch of wine. In the Hudson Valley and the eastern part of the United States and Canada, our cool climate grapes are generally high in acid, and locally grown grapes, such as the French-American hybrids, have high concentrations of color, so they are suitable for a second-run wine. In addition, the quality can be good as these wines have a softer balance and tannin structure, are more approachable, and have a higher concentration of attractive glycerins, which brighten a wine, give it more viscosity, and make it more lively.

To make a second-run wine, take the recently pressed pomace and put it back in the maceration vat. Add approximately one-half of the volume of juice or wine from the first-run pressing to the pomace. Correct it so the sugar is between 20°–23° Brix. Then mix up the pomace, sugar, and water so that all the skins are rehydrated. There is no need to add yeast because the skins have all the yeast they need. For a second-run white wine, the must should macerate for no more than one to two days. For second-run red wines, the maceration period can be as specified for the first-run pressing.

After that, all of the other steps are the same. The resulting wines are soft and clear up quickly, so they can be bottled and consumed within nine months of their production. Also, surprisingly, they age very well. Some home winemakers bottle these second-run wines separately or blend them back with their first-run wines to soften them up. For a commercial bonded winery, under federal law, 26 U.S.C. section 5383, no more than 35 percent of the volume of juice (calculated exclusive of pulp) and ameliorating material combined can be added as water, so there are some legal limitations to using these practices in bonded wineries.

Adverse Winemaking-Related Problems

Stuck Fermentations

Most stuck fermentations occur with white juices. If the juice does not begin to ferment within thirty-six hours or so, it means that the fermentation is stuck. At this point, add a new yeast culture, preferably a very vigorous one that is made for stuck fermentations. If this does not work, perhaps too much sulfur was added to the juice. If this is the case, rack the juice to help reduce the free sulfur level, warm it up to at least 60°F, and add a new yeast culture to prompt the must to begin fermenting. Sometimes the wine will partially ferment and then become stuck. Should this happen, warm the must and add a new yeast culture. If the wine remains stuck, and there is another available wine that is still fermenting, blend the two together so that the fermentation can proceed.

Oxidation of Wines

Probably the most common problem is the over-oxidation of wines, particularly white wines. This results from inadequate sulfur levels in the wine to protect it from oxygen exposure. This problem can occur at the press site to after the wine has been bottled. Signs that oxidation is occurring in white wines include a noticeable brown color, a reduced once-crisp acid level, and an appearance and taste similar to a fino sherry or tawny port that has a nutty nose and taste. In fact, sherries and ports are fortified wines in which a controlled oxidation has occurred.

This problem can also occur in red wines, but since red wines are more tolerant of exposure to oxygen and do not need as much sulfur to protect against oxidation, it is not as common a problem. However, if it does occur, the same symptoms are evident; red wines change color from red/purple to brick reds or dirty browns, the wine becomes more mellow, and nutty flavors become more evident.

To curb this problem, ensure that sulfur levels remain high, oxygen exposure is minimized, and wine is not racked more than two or three times. Once the wine begins to brown, add sulfur to inhibit this process. If the oxidation process has gone too far for the wine to be consumed as a table wine, it can be used in the production of a sherry or port.

TROUBLESHOOTING PROBLEMATIC WINE CONDITIONS

ADVERSE BACTERIAL AND YEAST CONTAMINATIONS

Most winemaking problems originate from lack of hygiene, which leads to the spread of microorganisms that produce acetic acid, brettanomyces, or other pathogens; inadequate sulfur or acid levels that left the wine exposed to microorganisms; the failure to top up wines or allowing too much oxygen to mix with wines; and the failure to recognize persistent protein hazes.

Acetic acid and ethyl acetate: This is caused by insufficient sulfur, low acid levels, or exposing the wine to too much oxygen, which creates an environment that allows acetic acid–producing bacteria to thrive. It can also lead to the establishment of ethyl acetate, which can give wine the smell of nail polish remover, geraniums, or petrol. Once this problem worsens to the point when the consumer can taste these microbiological by-products, it is too late to correct this situation. If a faulty wine is blended with other wines, it will infect them.

The best way to avoid this problem is to add adequate amounts of sulfur (40–55 ppm) to the must. In addition, jump-starting the fermentation process is important because fermentation generates carbon dioxide, which protects the wine from oxygen.

Lactic acid bacteria: These bacteria thrive in warm places that have access to oxygen. Unlike the beneficial Leuconostoc bacterium, these wild bacteria consume malic acid and produce off-putting smells and flavors of vinegar, nail polish, geraniums, sour milk, or urine. This problem cannot be fixed or remediated.

Wild film yeasts: Certain wild yeasts form a white or pale slick or powdery film that floats on top of the wine in a storage container or barrel. Wines that have too large an air space in the storage container, particularly those stored in wood barrels or that have low acid or sulfur levels, are susceptible to these colonies of film yeasts. If left untreated, they will oxidize the wine, consume protective alcohol and malic acid, and produce acetic acid and ethyl acetate. Wine stored in cool places, that have alcohol levels over 11 percent, and that are in full containers with adequate sulfur protection should not have this problem. To remedy this, spray an ethanol/sulfur mixture on the top of the floating film and on the fermentation lock. Also add sulfur to the water in the fermentation lock. After several such treatments, the problem should go away.

Brettanomyces: This bacteria, also known as "Brett," tends to show up in wines that have been aged in barrels for well over a year and in which adequate sulfur levels have not been maintained. If all the cellar sanitation-management practices are followed, this problem should not occur. Brettanomyces, in small amounts, can add a beneficial complex "barnyard" bouquet and taste to a wine. In larger doses, the barnyard smell becomes overpowering and is mixed with a mousey to mousey urine smell. Also, as the brettanomyces spreads, the alcohol levels, acids, and fruit are stripped, which leads to a rapid decline in wine quality. If the problem is not too severe, this condition can be ameliorated in a blend. However, it is best to dispose of the barrel as it can infect the rest of the winemaking area.

Enzyme-Based Browning

This occurs when white grapes or their recently pressed juice is exposed to too much air or sun before fermentation. The juice becomes brown due to the overexposure to oxygen. Some varieties, such as Aurora and Delaware, brown very easily when their juice is exposed to air and sunlight. To curb this problem, once these grapes are harvested, press them immediately and put the juice in opaque containers with a lot of sulfur as added protection from the air and sun.

Hydrogen Sulfides

The sulfurous rotten-egg smell, which is generally limited to red wines, occurs when a new raw wine is left on the gross lees for too long after it has been pressed. That is why it is best to rack wines at least two months after they have been pressed in the fall. Once this smell is discovered, rack the wine off the lees immediately, add sulfur at rates between 40 and 50 ppm, and vigorously agitate the wine to aerate it. Aerating the wine helps to release the hydrogen sulfides in the wine.

If not treated early, these hydrogen sulfides convert to mercaptans, which, while less volatile than hydrogen sulfide, become even more embedded in the wine. Mercaptans have a skunk-like smell. After a few more months, these mercaptans will be more oxidized and become disulfides. Disulfides are harder to eliminate from a wine, and smell like sewer gas. There are complicated processes to reverse this, but in the end, the resulting wine quality will not be good.

Cloudy Wines

This can be caused by the presence of pectin, which is found in Jell-O. This can be remedied by adding a pectic enzyme to clarify the wine. If the problem is not caused by pectin, it is caused by a protein. To treat this problem, add fining agents such as bentonite or gelatin for either red or white wines, egg whites for red wines, and casein or isinglass for white wines. After the haze settles, rack and add sulfur to the wine.

A cloudy wine can also be cleared by cooling it down to around 25°–35°F so that the tartaric acids fall to the bottom of the container. This process should take no more than three to six weeks. Then rack the wine to separate it from the tartaric crystals.

Conclusion

The winemaking practices I suggest may be more holistic than those of other winemakers. To correct deficient acid balances in the wine, for example, I rely heavily on cross-blending instead of adding tartaric or citric acid, tannins, or calcium carbonate. As for filtering or adding potassium sorbate, if the wine was properly made, it should be crystal clear and not need preservatives. However, as a commercial winemaker, I insist on running lab tests when the wine is finished to ensure that it is stable and ready for bottling.

In the end, winemaking should be fun, so meet other home winemakers through the local wine- and beer-making supply shop or join a local chapter of the American Wine Society. If a local chapter has not been formed in your area, contact the society and they can help you to establish one.

NOTES

1. See the following books: Philip M. Wagner, *American Wines and Wine-Making* (New York: Alfred A. Knopf, 1969); Philip Wagner, *Grapes Into Wine* (New York: Alfred A. Knopf, 1976); Jeff Cox, *From Vines to Wines* (Pownal, Vt.: Storey Books, 1994); Daniel Pambianchi, *Techniques in Home Winemaking* (Montreal: Véhicule Press, 2002); Lucie T. Morton, *Winegrowing in Eastern America: An Illustrated Guide to Viniculture East of the Rockies.* (Ithaca, N.Y.: Cornell University Press, 1985); Homer Hardwick, *Winemaking at Home* (New York: Wilfred Funk, Inc., 1954); Walter S. Taylor and Richard P. Vine, *Home Winemaker's Handbook* (New York: Harper & Row, 1968); American Wine Society, *The Complete Handbook of Wine-Making* (Ann Arbor: G. W. Kent, Inc., 1993); Philip Jackisch, *Modern Winemaking* (Ithaca, N.Y.: Cornell University Press, 1985); Emile Peynaud and Alan F. Spencer, trans., *Knowing and Making Wine* (New York: John Wiley & Sons Inc., 1984); M. A. Amerine, *The Technology of Winemaking* (Westport, Conn.: AVI Publishing, 1980).

2. Wagner, *American Wines*, 94–95; Wagner, *Grapes Into Wine*, 116–17.

3. Some suppliers in the New York and New England area: *Beer & Wine Hobby*, 155 New Boston St., Woburn, MA 01801, 800-523-5423; *Beer & Winemaking Supplies*, 154 King St., Northampton, MA 01060, 413-586-0150; *DIY Brewing Supply*, 289 East St., Ludlow, MA 01056, 413-459-1459; *Fall Bright Wine Makers Shoppe*, 10110 Hyatt Hill Rd., Dundee, NY 14837, 607-292-3995; *Fulkerson Winery*, 5576 Rt. 14, Dundee, NY 14837, 607-243-7883; *Hennessy Homebrew Emporium*, 470 N. Greenbush Road, Rensselaer, NY 12144, 518-283-7094; *Mayer's Cider Mill*, 699 Five Mile Line Rd., Webster, NY 14580, 585-671-1955; *NFG Home Brew Supplies*, 72 Summer St., Leominster, MA 01453, 978-840-1955; *Pantano's Wine Grapes and Home Brew*, 249 Rt. 32 S, New Paltz, NY 12561, 845-255-5201; *Presque Isle Wine Cellars*, 9440 W. Main Road, North East, PA 16428, 814-725-1314; *Prospero Equipment Corp.*, 123 Castleton St., Pleasantville, NY 10570, 914-769-6252; *E. J. Wren Homebrewer*, 209 Oswego St., Liverpool, NY 13088, 800-724-6875; *Strange Brew*, 416 Boston Post Rd. E, Marlborough, MA 01752, 508-460-5050.

A WINE
WHICH IS
THE PURE JUICE
OF THE GRAPE.
MELLOWED BY AGE.

CHAPTER SIX

Selected American Grape Species Used for Breeding

IN THE UNITED STATES, there are over seventy species of grapes. Grapes belong to the genus *Vitis*, in the flowering plant family designated as *Vitaceae*. *Vitis* has an extremely broad range of habitat and diversity in growth habit, leaf patterns, and berry and cluster size. Fewer than ten of these American grape species have been used in breeding programs in Europe and America to impart superior fungal disease resistance, increase winter and drought hardiness, delay budding times to minimize late spring frost damage, or minimize insect damage for new commercially cultivated grapes for wine and table use. Further, these varieties are still used extensively in commercial grape production in Europe and the Americas as rootstock to protect *vinifera* plants above the ground from phylloxera.

With the exception of a few grape species, such as the muscadines, all seventy-plus species of American grapes are infertile. This means that they can be used for cross-fertilization to obtain hybrid seedlings. This allows grape breeders to use most American wild grape species to increase the diversity of traits in the development of new interspecific hybrids. Hybridization work was conducted extensively in France from the 1860s to 1930s as one way to address the phylloxera epidemic, which destroyed much of France's vineyards and to combat fungal diseases that spread from the United States to continental Europe.[1]

The Pure Juice of the Grape. *Advertisements for Underhill's Croton Point Wines emphasized the quality of the grapes grown on his Hudson Valley property to produce "pure" wines. Circa 1870s.*

The various species of *Vitis* grapes can easily be bred with each other. There is even a lot of genetic variety within each grape species. For example, within the *vinifera* designation are Chardonnay, Grenache, Riesling, Thompson Seedless, and Cabernet Sauvignon. However, they have very different characteristics in growing habitats, growth patterns, taste, and appearance. The same holds true for other species of *Vitis* such as *labrusca*, *riparia*, and all of the others.

To help the reader better understand the genetic make-up of the interspecific hybrid grape varieties covered later in this book, the seven *Vitis* species that have been combined in various ways to produce new varieties for wine production and table use are detailed on the following pages. This chapter also highlights the growing characteristics of the varieties, where they grow in the United States, and how they impart certain traits to its prodigy, both in the field and cellar.

Throughout this book, next to the name of each hybrid will be an indication of the grape species that is part of that variety's genetic makeup. This information should give the reader a better idea of the attributes, both in the field and cellar, of such grape variety based on its genetic heritage. The references used include: U. P. Hedrick, *The Grapes of New York*; Lucie T. Morton, *Winegrowing in Eastern America*; Pierre Galet and Lucie T. Morton, *A Practical Ampelography: Grapevine Identification*[2]; U. P. Hedrick, *A Manual on American Grape Growing*; *Bushberg Catalogue* (1895)[3]; the work of Dr. George Engelmann[4]; Jancis Robinson, *Vines, Grapes & Wines*; and J. R. McGrew et al., *The American Wine Society Presents: Growing Wine Grapes*.[5]

Outlining the seven major grape species used for breeding in this chapter should contribute to a better understanding of these genetic attributes. Further, it should help to explain what these species can impart to the interspecific grape hybrids described throughout the book.

Vitis aestivalis (Michaux) (Aest.)

The French botanist André Michaux (1746–1802) was one of the first to use the name *aestivalis* to describe this grape.[6] This wild grape variety grows in a wide geographical area all across the United States east of the Mississippi River and westward to Missouri and portions of Arkansas. There are several branches or regional variations of this variety based primarily on where it grows: *V. aestivalis* var. *lincecumii* or Southwestern Aestivalis (Post Oak grape); *V. aestivalis* var. *bourquiniana* or Southern Aestivalis; and the *V.* var. *bicolor* or Northern Aestivalis. Some of its most important descendent hybrids include Jaeger 70 (*rupestris* × *lincecumii*), many early Seibel hybrids, Couderc hybrids, Norton, Cynthiana, Delaware, and perhaps Eumelan. With all of these regional subspecies, the core of *aestivalis* grows in the Mid-Atlantic and southern states, with the Norton grape being the most well-known example.

The growth habit is vigorous and climbing. It likes to grow in thickets and openings in the woods and, unlike *riparia*, dislikes streams. Unlike *V. labrusca*, it dislikes deep woods, so it is generally confined to uplands. It does well in lighter and shallower soils unlike *labrusca*, and can endure drought, although not as well as *V. riparia* or *V. rupestris*. The leaves are not injured by sun and it can resist insect damage. Further, it is resistant to downy and powdery mildew, black rot, Pierce's disease, and insect damage. Its hard roots enable it to resist phylloxera.

***Vitis aestivalis* (Mich.)** *Botanical illustration showing leaf, flower, fruit, and seed. Circa 1896.*

Aestivalis tends to bloom just after *labrusca*. Its growing season, on average, is also longer than that of *labrusca* or *riparia* grapes. However, the Northern Aestivalis has about the right growing season for the Hudson Valley. The cluster, small and fairly compact with few branches, has medium-sized round black berries with relatively thin, tough skin, described as being either very pulpy or devoid of pulp. *Aestivalis* berries cling to the cluster after they ripen much better than those of *labrusca*. This is a grape that lends itself to wine production. The grapes are sufficiently large for commercial production, have lots of tannin, but also have enough sugar to balance its fruit acids. Further, it has a very rich coloring that remains stable over time.

The taste of the grape has been described as peculiarly spicy and pleasant, astringent, tart, and acrid due to its high acid. Even with all of this acid, it has a high sugar content so that it is one of the best native American grapes for red wine production. These red wines are rich and highly colored, with a coffee-like character. Norton and Cynthiana are both *aestivalis* grape varieties. Some common names for these grapes are Common Blue grape, Bunch, Pigeon, Rusty, Chicken, Summer, Sour, Little, Swamp, Duck-shot, and Winter grape. The grape does not propagate easily by cuttings, but it can be done.

There are several other varieties of grapes that have been classified as part of the species *aestivalis*.

Vitis aestivalis var. bourquiniana (Bailey): The name Bourquiniana or Southern Aestivalis was given by Thomas Volney Munson (1843–1913), the well-known American-born nurseryman, viticulturalist, and plant breeder, who had a long-established nursery in Denison, Texas.[7] It tends to have thinner leaves and considerably larger fruit than other *aestivalis* grapes. Further, it is generally sweeter and more juicy. This variety tends to be more susceptible to fungal diseases and black rot.[8] Well-known members of this subspecies are Herbemont, Lenoir, Jaquez, and Delaware (Delaware being the only northern grape of importance).

Vitis aestivalis var. bicolor (Le Conte): Also known as Northern Aestivalis (hereinafter called *bicolor*), it makes its home primarily in the northeastern part of the United States from New England to Wisconsin. Bicolor is not confined to streams and riverbanks, and frequently grows on higher land. The *bicolor* vine (left) is vigorous, climbing, and very winter hardy, and the cluster is of medium size, compact, and simple. The berries are small and black.[9]

Vitis aestivalis var. lincecumii (Buckley) Munson: Also known as the Southwestern Aestivalis, this grape is a native of the eastern part of Texas, western Louisiana, Oklahoma, Arkansas, and southern Missouri. It grows on high sandy land, frequently climbing post-oak trees, hence one of its names—the Post Oak grape. Both Hermann Jaeger (1844–1895),[10] the famed Swiss grape grower and nurseryman who settled in Missouri and whose grape hybrids help to address the phylloxera blight in Europe, and T. V. Munson used this grape in their breeding programs. The benefits of this grape are its vigor, fungal disease resistance, hardiness, ability to survive hot, dry summers, and its large cluster and berry size. *Lincecumii's* growth habit is often more bushy than climbing, according to Dr. George Engelmann, who wrote extensive commentaries on grapes for the *Bushberg Catalogue*.[11]

Vitis berlandieri (Planchon) (*Berl.*)

In 1880, the French botanist Jules Emile Planchon (1823–88) described this species of grape under the name *berlandieri*. The description was made from herbarium specimens collected by the Belgian botanist Jean Louis Berlandier (1805–51) in 1834. Berlandier left Europe for America around 1828 and became a druggist in Matamoras, Mexico. He was one of the first botanists to explore Texas and northern Mexico. Unfortunately, in 1851, Berlandier drowned while attempting to cross a stream in Mexico. Many of his papers, drawings, and plant material were preserved in the herbarium at Harvard University.[12]

This variety is found only in the central and southwestern part of Texas and south in Mexico. The vine is moderately vigorous and climbing. The clusters are conical, compact, compound, with many lateral branches, and are medium-large in size. The small round berries have been alternately described as having very little pink acidic juice or black with thin bloom, juicy, and rather tart, but pleasant tasting when fully ripe. Further, it has a long vegetative growth cycle and its fruit ripens late.

In Texas, *berlandieri* grows from the tops of hills to the base of hills near creek beds. It can withstand lime and chalk soils, and its roots are strong, thick, and very resistant to phylloxera. It has been described as being hard to propagate from cuttings. It is a grape of value to hybridizers and those interested in rootstock because of its resistance to phylloxera and high-lime content soils (as is found in Europe). SO4 and 5BB rootstocks are *riparia* × *berlandieri* crosses.

Its common names include Uva Cimarrona, Mountain grape, Little Mountain grape, Sugar grape, Fall grape, Winter grape, Sweet grape, Surett grape, and Spanish grape.

Vitis cinerea (Engelmann) (*Cin.*)

Cinerea closely resembles *aestivalis* and during most of the nineteenth century was considered a variation of it. However, by 1883, Dr. George Engelmann, of St. Louis, Missouri, after much study, gave *cinerea* its own designation. *Cinerea's* habitat has been alternately described as New York State west to Nebraska and Kansas with a northern limit at the 40th parallel and then south to the Gulf of Mexico,[13] or as growing in the Old South, Texas, and the states bordering the Mississippi River up to Iowa.[14] *Cinerea* grows along streams, mostly in limey soils, and is seldom found on very dry land. This species is very late in blooming and is a significant variety because it does well on wet lands. The clusters are conical, loose, medium-large, and winged with small black berries that yield very little acidic juice. Black Spanish and Herbemont are two well-known *cinerea* hybrids. Dr. Engelmann reported that *cinerea* is found in rich soil in the Mississippi River Valley from central Illinois to Louisiana and Texas, especially on bottom lands and along banks of lakes.[15]

Vitis labrusca (Linnaeus) (*Labr.*)

Labrusca, also known as the Fox grape, is probably the best-known grape species in America. It includes grapes such as Concord, Catawba, Niagara, and Delaware. It is the most recognizable in flavor with its grapey Welch's Concord taste as found in countless peanut butter-and-jelly sandwiches. It is also the base of kosher wines and many quality sherries and ports made in the eastern United States.

This species has provided the genetic base for most of the interspecific grape hybrids produced in the United States. The reason for this is that it was found on the east coast of the United States which was first settled by Europeans. Its high productivity, attractive cluster appearance and size, and ease of cultivation in wide areas of the country that were first settled by European colonists and where fruit production was an important part of its economy are probably the reasons for its popularity in cultivation and use in hybridization.

Vitis labrusca. *Botanical illustration showing leaf, flower, fruit, and seed. Circa 1896.*

The variety was mentioned in many early writings of New England by its first settlers. It was first described as a grape species by the Swedish botanist Carl Von Linné (1707–78), better known by his Latin name Carolus Linnaeus. Trained as a physician, Linnaeus was a prominent systematist who developed methods to categorize plant material. Although he named many plants, he was not an explorer who collected plant material, but a botanist who categorized plants collected by other travelers. Linnaeus used the name *labrusca*, an old descriptor for wild grapes in Italy, for this variety of grapes in America.[16]

Labrusca hails from the eastern portion of the United States, from southeastern Canada to Georgia, and is especially abundant in New England and the Mid-Atlantic states, growing as far west as Illinois and Tennessee. There is a northern *labrusca* subspecies in New England and New York that looks similar to Concord. This variety is a blue grape that can grow in the deep woods and climbs trees to reach the sunlight. There is also a southern *labrusca* subspecies that tends to have reddish to amber/rust berries, such as Catawba. However, *labrusca* grapes are unique in the United States for having black, blue, red, and white grape varieties.

The vine is moderately vigorous to vigorous, stocky, and climbing. The leaves are broad, leathery, and thick with a deep green/purple-green hue on the top and light lime green to white on the underside. Its vegetative cycle is very similar to that of *vinifera*. The clusters are of medium to large size, cylindrical, and usually shouldered. The round berries are the largest of all native American grape varieties (about the size of a cherry), with thick slip skins and are generally handsomely colored. However, when ripe, the berries can drop from the cluster and can crack if it rains during harvest time. This shattering or shelling is a serious weakness in some varieties of *labrusca*.

The variety is resistant to fungal diseases, but not as resistant as *aestivalis* or *riparia*. More particularly, *labruscas* have good powdery mildew resistance, but are more susceptible to downy mildew and black rot. Further, the vines do not endure drought as well as *aestivalis* or *riparia*. It has good winter hardiness, but its leaves are subject to sun scald toward the end of a hot August.

Labrusca grapes are also very unique in the respect that there is a layer of very sweet, mild juice under its thick slip skin. Beneath this layer is an acidic pulpy layer that surrounds its seeds. Since *labruscas* grow on the eastern seaboard, they are able to grow in humid conditions in the Hudson Valley. This fairly vigorous variety is generally productive and tolerant of many different soil types. It grows readily from cuttings, and its winter hardiness is between *riparia* (the hardiest) and *aestivalis* (not as winter hardy).

The root system does not penetrate the soil deeply, but does better than *aestivalis* in deep clay soils. It endures excess water in the soil, and needs more water than *aestivalis* or *riparia* to grow successfully. However, while it can grow in wet clay soils, it prefers loose, warm, well-drained soils. Its roots are soft and fleshy for an American grape, so it can be subject to attacks of phylloxera, more so than all other native American grape varieties covered in this chapter. Further, in heavy clay soils, *labrusca* is weakened and more susceptible to phylloxera. Overall, *labrusca* submits well to vineyard culture and is one of the least "wild" cultivars in commercial vineyards.

Overall, the rule is that *labruscas* have relatively high acid and low sugar, so amelioration is needed to balance these components to make a stable wine. Within the *labrusca* species, Concord and Niagara have low sugar content in relation to its acid; Catawba tends to average higher in both acid and sugar; while Delaware has a good balance of high sugars and appropriate acids suitable for wine production.

Labruscas can be fine table grapes due to their large cluster and berry size and have big, powerful flavors. They are not generally viewed as suitable wine grapes because of their relatively low sugar content, high acids, and powerful, foxy flavors.

Many critics maintain that these "foxy" or musky-smelling and -tasting grapes are uncomplicated and overpowering, and not sufficiently sophisticated to produce quality wine. In other words, if you have tasted one *labrusca* wine, you have tasted them all. I have actually found the flavors of *labrusca* grapes to be, while wild and strong, also complex with many flavor nuances. The complicated *labrusca* flavors can be broken down into four categories: (1) foxy, described as like a wet dog or musty; (2) highly perfumed or floral; (3) strawberry and raspberry/cotton candy; and (4) methyl anthranilate or bubble gum.[17]

Labrusca flavors have more diversity than is usually given credit. For example, Niagara is considered to be foxy, while Catawba is not. Concord has a lot of methyl anthranilate or bubble gum flavors, while Catawba does not, though it is highly perfumed. Ives has a more strawberry-like flavor. In the red Cornell hybrids, this big strawberry/raspberry flavor, which comes from *labruscas*, is very pronounced and carries through to Cornell's newer red hybrids.[18]

It is interesting to note that while the other species of grapes have a multitude of names, *labruscas* do not, possibly because they were so prevalent that the names associated with them became the individual cultivars that we know them best by such as Concord, Delaware, or Niagara. With that said, some of its common names are Fox grape, Northern Fox, Northern Muscadine, Swamp, Plum, and Skunk grape.

Vitis riparia (Michaux) (*Rip.*)

This species is one of the most important varieties for both past and future grape-breeding programs due to the vine's winter hardiness, early ripening time, disease resistance, and vigorousness. This variety was also known as *Vitis vulpina* as described by Linnaeus. We use the term *riparia* because botanist André Michaux gave a more comprehensive description and categorization of this important native species.[19] Some of *riparia*'s prodigy include Baco Noir, Maréchal Foch, Frontenac, and Clinton, also known as Worthington. Also, *riparia* is now being extensively used in the University of Minnesota's grape-breeding program. The name *riparia* was given to this variety because it was found along riverbanks, on islands, or in upland ravines. It prefers rich, deep soil and does not do well in droughty conditions.

This grape was clearly identified and described by F. Andre Michaux as early as 1803.[20] *Riparia* is found over most of North America, from southern Canada to the border of the deep South by North Carolina, west to Oklahoma, and up through the Plains states of Nebraska and then back north to Canada. *Riparia* has one of the earliest bud breaks and grape-ripening times of native grapes in North America.

The vine is vigorous to very vigorous with wild growth. It is a climbing variety with long trailing canes that spread laterally. The leaves are broad and large, but thin in thickness, with a pale to medium-green color. The grape propagates very readily to effortlessly from cuttings. The grape has been used extensively in American grape-breeding programs because of its winter hardiness, short growing season, and vigor. Further, it adapts to many soil types, climates, and locations in the eastern United States. Not only is the grape not susceptible to winter injury, but it is tolerant of the summer's heat. Due to its early blooming, it is vulnerable to late spring frosts. However, this is offset by *riparia*'s ability to set a second crop from its secondary buds.

Riparia is not a swamp grape, but it does like water. Further, it likes rich soil and is not nearly as capable of withstanding drought as *rupestris*. It does not like limestone or calcareous marls. However, French wine growers of the early twentieth century maintain that *riparia* is more resistant to excess lime than either *rupestris* or *aestivalis*.[21]

The clusters of *riparia* tend to be small to medium in size and compact, sometimes with long, tapering bunches. However, the berries, which are generally black, are small. The fruit is free of foxiness or any disagreeable "wild" taste. U. P. Hedrick aptly describes the flesh of *riparia* as neither pulpy or solid, and dissolves in the mouth as the seeds separate easily from the skins and flesh.[22]

As for its winemaking qualities, one can say that it is superior to excellent. Even as long ago as 1908, U. P. Hedrick wrote that the grapes, while very high in acid, had a good balance and high sugar content to make solid wines. Further, that without sugar or water, the *riparia* wines must be kept for a very long time in order to mellow and be drinkable. However, the addition of sugar and water makes the *riparia*-based wines more approachable yet full, but able to be consumed within five instead of twenty years. Then as now, the flavor of *riparia* wines is enjoyed by most wine consumers because of its pronounced berry flavor profiles and absence of disagreeable wild flavors. *Riparia* wines can be herbaceous and are very highly colored wines with dark purples predominating. These dark purples indicate that the wines have a lot of acid.[23]

Riparia is very resistant to phylloxera because its roots are small, hard, numerous, and branch freely. The roots feed close to the surface and are not well adapted to force their way through heavy clays or hardpan formations. *Riparia* is less resistant than *aestivalis* to fungal diseases and black rot, but somewhat more resistant than *labrusca*. However, due to its thin, soft leaves, it is more susceptible to insect damage, particularly from Japanese beetles. This variety is sometimes confused with cordifolia because of similarities in growth habit and range where it is found, but it flowers late and ripens later than *riparia*.

Some of its common names have included June, Winter, River, Riverside, and Riverbank grape, and Sweet-Scented grape. One common *riparia* rootstock that is used with *vinifera* is Riparia Gloire de Montpellier (RGM).

Vitis rupestris (Scheele) (*Rup.*)

This grape was first described and named in 1848 by the botanist George Heinrich Aldolf Scheele (1808–64).[24] The range for this grape is most of Texas and to the north and east to include Oklahoma, Louisiana, Arkansas, southern Missouri, as well as Tennessee and Mississippi. One commentator, Lucie Morton, a leading American grape and viticultural expert, has suggested that whatever was said of *riparia* above, the reverse is true for *rupestris*. *Rupestris* grows more like a bush than a vine, with short spindly shoots and a root system that grows down like a taproot for water and minerals, unlike *riparia*'s root system, which grows horizontally. Furthermore, unlike *riparia*, *rupestris* grapes (left) mature late and it dislikes humid soils.[25] Like *riparia*, its roots have good resistance to phylloxera because they are slender and hard, but the leaves can get galls. The most widely used *rupestris* rootstock in planting and breeding is St. George.

The grape variety tends to like gravelly banks along mountain streams or rocky beds of dry water courses. It is considered drought resistant in the Hudson Valley because it originates in the Texas hill country. The variety is variable as to its type of its growth pattern and vigorousness. It does not take well to grafting as well as *riparia* or *aestivalis* because at its graft joint, it produces many suckers. Due to its extreme vigor, long vegetative cycle of approximately 260 days, late maturity, and dislike of humid soils, rootstocks from this species are not recommended for the Hudson Valley.[26]

The clusters are cylindrical and small to very small, with berries about the size and color of black currants and vary from sweet to sour. The grape possesses much pigment and is sprightly, but is not so productive. Further, its foliage is very resistant to rot and mildew, but can be susceptible to anthracnose. With its good disease resistance and winter hardiness, it was used in the breeding programs of Hermann Jaeger, T. V. Munson, the horticulturalist and grape breeder George W. Campbell (1817–98), and Professor Pierre Marie Alexis Millardet (1838–1902) of the University of Strasbourg and the University of Bordeaux. Its common names are July grape, Sand, Sugar, Beach, Bush, Currant, Ingar, Rock, and Mountain grape. Dr. George Engelmann reports that this grape makes a good wine.[27]

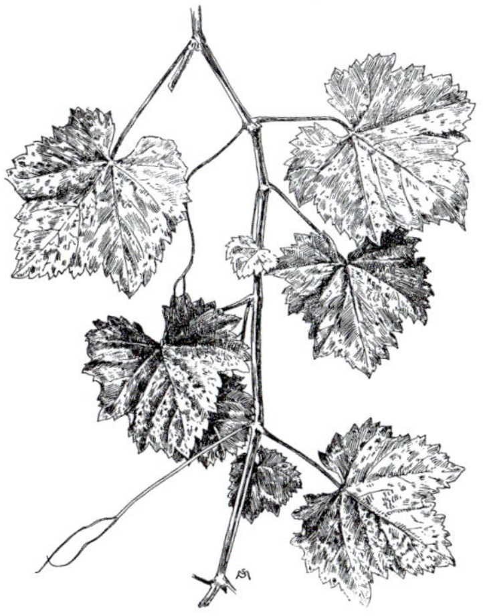

Vitis vinifera. *Early twentieth-century illustration depicting the leaf shape of this grape species that has been cultivated for more than 3,000 years.*

Vitis vinifera (Vin.)

Some would say that we have saved the "best" for last. *Vinifera* varieties came from Eurasia and Europe and can be used to produce either quality wines or table grapes. Further, often they are the parents of many of the interspecific grape hybrids that we use today to produce quality wines in the Northeast such as Baco Noir, Vidal, Vignoles, or Chelois. Varieties such as Chardonnay, Riesling, Pinot Noir, Cabernet Sauvignon, Cabernet Franc, and Aligote are all pure *viniferas*.

Unlike the previous six grape species outlined above, *vinifera* was not "discovered" by a botanist sometime after the eighteenth century. It has been around long before the rise of modern Western civilization. Further, perhaps more than 3,000 years before the birth of Christ, varieties of *vinifera* grapes have been used to produce wine in Ancient Egypt, Greece, and the Roman Empire, hence the term *vinifera*, Latin for "to make wine." As Christianity marched westward throughout Europe, so did the *vinifera* to produce sacramental wine for the celebration of the Roman Catholic mass.

U. P. Hedrick and Dr. George Engelmann both suggest that botanists have never agreed as to whether *vinifera* is a single species, or a combination of two or more species that have been cultivated for so long that it is impossible to ascertain its origin.[28] More than several botanists have been credited with naming this species. What does seem clear is that its original home is somewhere by the Caspian Sea in western Asia. In Europe, *vinifera* grapes mutated into smaller, highly flavored clusters used in wine production, large thick-skinned grapes for table consumption, and seedless varieties used in raisin production. With all of the variety in this grape species and the loss of its natural habitat, it is difficult to describe its attributes and growth habits in the wild. Unlike in North America, where there are over seventy grape species, in Europe *vinifera* is the only significant grape species.

However, some generalizations can be made about *vinifera*. First, its skin is attached closely to the flesh, is flavorful and not bitter, and it can be eaten with the fruit. Second, unlike many North American grape varieties, its flesh is firm, but tender and uniform. Third, its flavor is particularly sprightly, a quality known by some as being "vinous" because it characterizes this species. (The Geneva Research Station often uses the word "vinous" to describe their grapes and wines, so it should be helpful to know where this term comes from.) Vinous means a grape that has the capacity for winemaking applications. Fourth, the berry stays attached to the pedicel and does not shatter or shell from the cluster. Fifth, the roots are relatively soft and spongy, and therefore very susceptible to phylloxera. However, these same roots can penetrate dense clay and hard dry soils better than many North American grapes. Sixth, leaves and fruit of *vinifera* are much more vulnerable to fungal diseases such as black rot, downy mildew, and powdery mildew than most North American grape varieties. Seventh, they tend to thrive in moderate climates that do not get too cold as can be the case in the Hudson Valley. Eighth, the ripening season for *vinifera* is lengthy for proper fruit ripening and cane hardening to ward off winter cold damage.[29]

NOTES

1. Bruce I. Reisch, "Developing Disease Resistant Grapevines," *Wine East*, November–December 1995, 8–15.

2. Pierre Galet and Lucie Morton, *A Practical Ampelography: Grapevine Identification* (Ithaca, N.Y. and London: Cornell University Press, 1979). Pierre Galet's American graduate student, Lucie Morton, translated Galet's work into English.

Pierre Galet (b.1921), an influential French ampelographer and author, was born in Monaco. He has devoted his life to developing a system of scientific description to accurately classify and identify grapevines. Before the advent of DNA testing, this system was the most accurate method available to identify a grapevine. It identifies salient characteristics of a grape variety by looking at: the shape, contours, and angles of its leaf veins; the structure of its leaves, petioles, growing shoots, and shoot tips; and describes grape clusters by analyzing their size, color, seed size and shape, and flavor. His internationally acclaimed work has been very important to growers, academicians, and hybridizers who need to identify the characteristics of the seed and pollen parents of interspecific hybrids and how these characteristics are passed down to future generations. Further, due to the common misnaming of grape varieties in the field and in nurseries, this system helped growers and nurserymen to identify the true variety of their grapes in the vineyard or in the nursery stock.

Galet's first book in this area was *Précis d'Ampelographie Pratique*, published in 1952. This book quickly sold out, and he began his next very comprehensive book in 1955, *Cepages et Vignobles de France* (Varieties and Vineyards of France). This 3,500-page, four-volume set outlining the growing characteristics of such varieties, published between 1956 and 1964, became a much-used reference throughout the world for identifying grape varieties. In 1968, a new edition of *Précis d'Ampelographie Pratique* added notes on the histories and special aptitudes for each variety, in addition to descriptions for grape identification. The third and fourth editions of this book, published in 1971 and 1976 respectively, became the basis for the English translation, with the assistance of Lucie Morton, of the above mentioned *A Practical Ampelography: Grapevine Identification*, published in 1979. In addition, in 1977 and 1982 he wrote the two-volume set, *Maladies et Parasites de la Vigne*, on the various diseases and ailments of grapes, and *Dictionnaire Encyclopédique des Cépages*, a comprehensive catalogue of grape varieties in the world and their international synonyms.

During World War II, Galet hid from the German authorities at the University of Montpellier. During this time, he studied at the University's Department of Viticulture and examined the department's extensive grapevine specimens from all over the world. After the war, Galet accepted a teaching position at the university from 1946 to 1989. He was active at the École Nationale Supérieur Agronomique de Montpellier. He was made an *Officier de l'Ordre du Mérite Agricole*. Galet's work is of particular importance in documenting and classifying French-American hybrids that grow in the United States and their growing characteristics.

3. *The Bushberg Catalogue. A Grape Manual. [An] Illustrated Descriptive Catalogue of American Grape Vines. A Grape Growers Manual by Bush & Son & Meissner, Viticulturalists and Proprietors of Bushberg Vineyards and Grape Nurseries, Jefferson Co., Missouri* (4th ed.) (St. Louis: R. P. Studley, Printers, 1895). This was the nursery catalogue for Bushberg & Son & Meissner Nursery of St. Louis, Missouri, which was for many years during the middle to latter part of the nineteenth century one of the leading nurseries for the sale of grape plants and cuttings in the United States and to France. This catalogue—in a manner similar to the books written by A. J. Downing and Charles Downing, which described fruit and tree fruit varieties in the United States, *The Fruit and Fruit Trees of America* (1st ed., 1845; rev. 2nd ed., 1869) and *The Horticulturalist*—disseminated, for the first time, extensive information on grape varieties that were available and useful for either commercial viticulture, table use, or for landscaping and horticultural purposes. This catalogue was really more of a reference guide that listed hundreds of grape varieties, their growing characteristics, and attributes in the field and cellar. Further, it was used as a reference guide by other nurserymen, horticulturalists, and as a textbook in agricultural schools in the United States. It was also translated into French and Italian.

The proprietor of Bushberg's Nursery was Isadore Bush (1822–98), born in Prague, Bohemia. Due to his political involvement in the revolution of 1848 in central Europe, he left shortly thereafter for America. Upon his arrival, he went to Missouri—then a common destination for Germans emigrating from Europe—because of its large German population. During the American Civil War, he was secretary to General Fremont and held other positions of trust during the war. In 1865, after the war, he established Bushberg & Son & Meissner Nursery, and with the aid of Dr. George Engelmann, an immigrant from Germany also living in St. Louis, wrote the *Bushberg Catalogue and Grape Manual*. This manual was first published in 1869, and then went through three more editions in 1875, 1883, and 1895. In addition, Bush operated the Bush Wine Company. See U. P. Hedrick, *The Grapes of New York. Report of the New York Agricultural Experiment Station for the Year 1907*.

Department of Agriculture, State of New York. Fifteenth Annual Report, Vol. 3 – Part II. (Albany, N.Y.: J. B. Lyon Company, 1908), 119; Thomas Pinney, *A History of Wine in America, From the Beginnings to Prohibition, vol. 1* (Berkeley: University of California Press, 2007), 184–86, 390–94; Sherrie S. McLeRoy and Roy E. Renfro Jr., *Grape Man of Texas: Thomas Volney Munson & the Origins of American Viticulture* (San Francisco: The Wine Appreciation Guild, 2008), 48, 132–33.

4. Dr. George Engelmann (1809–84) was a successful German-American medical doctor and botanist who traveled extensively, described, classified, and wrote about America's flora, especially specimens from the western part of the country and native American grape species. He was born in Frankfurt, Germany, the eldest child of thirteen children born to Julius Bernhardt Engelmann and his wife Julie Antoinette May, nine of whom reached maturity. He was educated at the universities of Heidelberg, Berlin, and Wurtzburg, and received his medical degree from Wurtzburg University. In 1832, he sailed to America, acting as his uncles' agent to purchase land in the United States. Some of his relatives had already settled in Illinois, not far from St. Louis, Missouri. He traveled extensively in Illinois, Missouri, and Arkansas on behalf of his uncles, evaluating the quality of land for potential purchase. Meanwhile, Engelmann also surveyed the flora of the region.

In 1835, he established a medical practice in St. Louis, which he built up sufficiently so that by 1840, he was able to return to Germany to marry his cousin, Dora Horstmann. Dr. Engelmann continued his very successful medical practice and raised a son, Dr. George Engelmann Jr., who became an obstetrician, in St. Louis. By 1869, Dr. Engelmann Sr. no longer kept a medical office. However, during the entire time of his medical practice, he continued his work as a botanist by evaluating, describing, and classifying native plant material, including grapes.

By at least 1860, he began to focus on plant diseases affecting grapes and identifying local grape species. The knowledge that he obtained in this area became critical as the phylloxera epidemic devastated the French wine industry in the 1870s. Engelmann was able to help identify disease-resistant American grape varieties that could be resistant to phylloxera in France. The third and fourth editions of the *Bushberg Catalogue* (1883 and 1895) included his articles, which described American grape species and the growing characteristics of some of the many grape plants offered for sale in that catalogue.

Dr. Engelmann helped to found the St. Louis Academy of Sciences, and several species of grapes and desert plants are named in his honor. In addition, his botanical collection was given to the Missouri Botanical Garden, which led to the establishment of the Henry Shaw School of Botany as a department at Washington University, where the Engelmann professorship of botany was named in his honor. See Charles A. White, "Memoir of George Engelmann, 1809–1884," (Washington, D.C., Judd & Detweiler, Printers, 1896), read before the National Academy of Sciences, April 1896; Hedrick, *Grapes of New York*, 131–32; Pinney, *History of Wine*, 183–86; McLeRoy and Renfro, *Grape Man*, 132–33.

5. Hedrick, *Grapes of New York*, 95–156; U. P. Hedrick, *Manual of American Grape-Growing* (New York: Macmillan, 1924), 311–29; Lucie T. Morton, *Winegrowing in Eastern America: An Illustrated Guide to Viniculture East of the Rockies* (Ithaca, N.Y.: Cornell University Press, 1985), 21–23, 61–73; Galet and Morton, *Practical Ampelography*, 140–53; and *Bushberg Catalogue* (1895), 12–30; Jancis Robinson, *Vines, Grapes, and Wines* (New York: Knopf, 1986), 8–9; J. R. McGrew, J. Loenholdt, T. Zabadal, A. Hunt, and H. Amberg, *The American Wine Society Presents: Growing Wine Grapes* (Ann Arbor: G. W. Kent, Inc., 1993), 59–71; Reisch, "Developing Disease Resistant Grapevines," 8–15.

6. Hedrick, *Grapes of New York*, 139.

7. Thomas Volney Munson (1843–1913) was a preeminent American nurseryman, viticulturalist, and grape breeder, who wrote the landmark book *Foundations of American Grape Culture* (New York: Orange Judd Co., 1909). This grape-breeding reference book was credited by Elmer Swenson (1913–2004), the preeminent grape breeder of his own cool climate hybrids and many of those related to the University of Minnesota's breeding program, as the book that started him on his own grape-breeding path. Munson was born near Astoria, Illinois, in 1843, the son of William Munson (1808–90) and Maria Linley Munson (1810–90). He graduated from the University of Kentucky at Lexington in 1870 and worked in Kentucky as a nurseryman. After graduating from college, he married and then moved near Lincoln, Nebraska, to continue his business as a nurseryman. In 1873, he became interested in the improvement of American grape varieties and set out on his own systematic grape-breeding program to develop new varieties that could thrive on the Plains and in the southwestern part of the United States. His experiments failed in Nebraska due to harsh climatic conditions and locust damage, so he moved to Denison, Texas, in 1876, where his two brothers lived.

At Denison, located in north Texas on the Red River across from Oklahoma, Munson made his viticultural mark. In 1877, he established a nursery business and set out to breed American grape varieties for quality wine production and table use, grapes that could be grown locally where varieties that had originated from the East Coast of the United States had failed. His nursery business, T. V. Munson & Son, is where he indulged his passion for collecting, describing, categorizing, and hybridizing native American grape varieties. While the climate was less harsh at Denison than in Nebraska, the winters were still severe and the summers hot and dry, so his breeding program reflected the need to grow grapes in these conditions.

In his breeding program, he used many local grape varieties. In addition, he documented and categorized the botany of native grape species growing in the area and their growth characteristics. Through his breeding program, he released approximately 300 new interspecific hybrids, probably more than any other American. Some of his named varieties include Bell, Brilliant, America, Captivator, Carman, Delicatessen, Headlight, and Rommel. Throughout his career, he wrote bulletins on grape culture for the U.S. Department of Agriculture and the Texas Experiment Station. Munson had three ambitious goals for his grape-breeding program: (1) to create a series of table grapes ranging from early to late season ripening dates so that the entire season could be filled with a succession of grapes ready for harvest; (2) to create grapes that could live in the harsh Texas and American Southwest climate, which included developing cold- and heat-hardy grapes that had strong resistance to fungal and Pierce's disease; (3) to develop wine grapes that could be grown in different parts of America.

Munson's breeding program was complex and long-lived. He incorporated many different grape species and interspecific hybrids into his newer hybrids. After surveying his work and the species he used, it looks like that he brought in *labrusca* and *vinifera* varieties to impart either large table grape size, *labrusca* table grape flavors, or higher potential wine quality. For fungal disease resistance and cold and heat hardiness, he used local native grape species such as *V. aestivalis* var. *lincecumii*, *V. champini*, *V. candicans* (Mustang grape), *V. rupestris*, *V. riparia*, and *V. berlandieri*. He incorporated superior grape qualities from both seed and pollen parents, and alternated superior quality grapes with those with better cultural characteristics from both the seed and pollen parent side. By the middle of his program, both the seed and pollen parents had heritages of superior quality grapes of *labrusca* and *vinifera* combined with local varieties that had disease, cold, heat, and drought resistance.

His greatest impact on viticulture came from his work with European grape growers to identify suitable rootstocks that would be resistant to the phylloxera epidemic devastating European vineyards. In honor of his work, the French government named him as a *Chevalier du Mérite Agricole* in 1888. He died in Denison in 1913. After Munson's death, many of his hybrids were lost, but recent efforts have been made to establish new vineyards with Munson hybrids so that these varieties can be propagated and sold. Currently, the Munson Memorial Vineyard in Denison provides this service. In addition, the T. V. Munson Viticulture and Enology Center serves as a repository for his documents and those items related to his work. For an excellent biography of his life and work, see McLeRoy and Renfro, *Grape Man*. Also see Hedrick, *Grapes of New York*, 122; Pinney, *History of Wine*, vol. 1, 408–10, 445–46. See also charts and grape variety genealogy by Carl Camper at the Chateau Stripmine Experimental Vineyard, www.chateaustripmine.info/Breeders.htm

8. Hedrick, *Manual of American Grape-Growing*, 321–22.

9. Ibid., 322–23.

10. Hermann Jaeger (1844–95) was a trained viticulturalist from Switzerland who came to Missouri sometime between 1865 and 1867. Along with his brother John, Jaeger planted a vineyard at Monark Springs, near Neosho, which is in the most southwestern part of Missouri, near the border of Kansas and Oklahoma. The brothers became grape growers and winemakers. To advance the success of his own and neighboring vineyards, Jaeger searched the area for native grape species such as *aestivalis* var. *lincecumii*, *cordifolia*, and *rupestris*. He used them as the basis of the approximately 100 selections that he either identified in the wild and cultivated or bred with other local varieties. From the literature, the objectives of Jeager's breeding program was to secure more vigorous, productive, and disease-resistant varieties. However, unlike other breeders, he seems to have used local *cordifolia*, *rupestris*, and *aestivalis* var. *lincecumii* varieties to the exclusion of all others, including *labrusca* or *vinifera*. Also, unlike other breeders, he simply propagated varieties that were already in the wild (*aestivalis* var. *lincecumii*) or made very simple hybrids between wild *cordifolia* and *rupestris*. Some of these hybrids were ultimately used in France as rootstocks and by French direct producer hybrid grape breeders Albert Seibel (1844–1936) and Georges Couderc (1850–1928) in their hybridization work. The most prominent hybrid used by these French hybridizers was Jaeger 70, a simple *lincecumii-rupestris* hybrid.

During Jaeger's short life, he wrote about his work for scientific and horticultural journals. He communicated extensively with viticulturalists in France and with local viticulturalists, winemakers, scientists, and nurserymen such as Isadore Bush of St. Louis; George Husmann of Hermann, Missouri (who was a winemaker, nurseryman, and the first professor of horticulture at the University of Missouri at Columbia); T. V. Munson; Charles V. Riley (Missouri's first state entomologist (1843–1895); and Dr. George Engelmann concerning the development of new hybrids for direct production and those to be used as disease-resistant rootstocks. This local concentration of scientific, viticultural, and entomological knowledge and practical experience in America was rivaled only by the viticultural and hybridization work occurring around the same time in the Hudson Valley. While the goals were different, with grape breeders in the Hudson Valley concentrating on developing new table and wine grape varieties based on *labrusca* and *vinifera* and pruning systems, the close proximity of talent created a synergy that rapidly advanced viticultural practices and scientific knowledge for the entire United States grape industry.

Jaeger's nursery business alone shipped approximately seventeen rail boxcars of hardy phylloxera-resistant rootstocks and cuttings to France. In addition, the nurseries of Hussman, Bush, and Munson, shipped millions of grape plants and cuttings to France. The greatest shipment of cuttings from the Missouri area to France occurred between 1873 and 1876. For his efforts, in 1893, Hermann Jaeger was made a *Chevalier de la Légion d'honneur*, as was T. V. Munson in 1888.

Hermann Jaeger died at the young age of 51. The legend is that he took his own life after facing financial and legal troubles. In the 1890s, well before the national imposition of Prohibition, Neosho enacted a local law to ban the sale of spirits in that municipality. In an effort to circumvent the new law, Jaeger's winery sold German cookies and pastries, along with a glass of wine to accompany the dessert. Jaeger was indicted, and with his legal troubles, was determined to sell his Neosho vineyard and relocate to the friendlier nearby town of Joplin, Missouri, to establish a new fruit farm and vineyard. However, on May 16, 1895, Jaeger left the Joplin farm and his family, telling his wife that he was traveling to Neosho to deal with legal issues. Some days later, the family received a postmarked letter from Kansas City, written in German, which said that "When you read these lines I will be alive no more." The letter suggested that Jaeger planned to end his life in a manner where his body would not be found. The letter was signed "Your unlucky Hermann." See Mark Parker, "Legend of Hermann Jaeger," April 2009, www.MissouriRuralist.com. For a more in-depth look at the life and work of Jaeger and his collaboration with T. V. Munson, see McLeRoy and Renfro, *Grape Man*. Also see Pinney, *History of Wine*, vol. 1, 393–94, 409. See Chateau Stripmine, www.chateaustripmine.info/Breeders.htm, and Hedrick, *Grapes of New York*, generally.

It is interesting that the Hudson Valley grape industry was also going through the same difficult economic times and contraction due to the long-term national economic recession, which lasted from 1873 to 1896, with the acute Panics of 1873, 1890, and 1893 causing additional stress for grape breeders and nurserymen. The Hudson Valley hybridizer Andrew Jackson Caywood and his son Walter J. Caywood of Marlboro, New York, who were facing acute financial problems, died in the early part of 1889, followed by A. J. Caywood's wife, Deborah Cornell Caywood the following year. Further, James H. Ricketts, the Hudson Valley hybridizer from Newburgh, New York, may have lost his home and vineyard to foreclosure in 1877 after the Panic of 1873. Could it be that Jaeger's financial difficulties related to the national economic downturn caused by the Panic of 1893 or may have been an additional reason for Jaeger to have possibly taken his own life, in addition to the violation of local Prohibition laws? For a more indepth look at the life and work of H. Jaeger and his collaboration with T. V. Munson, read McLeRoy and Renfro, *Grape Man*.

11. *Bushberg Catalogue* (1895), 9–30.

12. Hedrick, *Grapes of New York*, 130–31.

13. Ibid., 132.

14. Galet and Morton, *Practical Ampelography*, 145–46.

15. Hedrick, *Grapes of New York*, 131–32; *Bushberg Catalogue* (1895), 15.

16. Hedrick, *Grapes of New York*, 149–51.

17. Morton, *Winegrowing in Eastern America*, 65.

18. Ibid., 65.

19. Hedrick, *Grapes of New York*, 118.

20. Ibid., 118.

21. Ibid., 120.

22. Ibid.

23. Ibid.

24. Ibid., 114.

25. Morton, *Winegrowing in Eastern America*, 73.

26. Ibid.

27. *Bushberg Catalogue* (1895), 17.

28. Hedrick, *Grapes of New York*, 154; *Bushberg Catalogue* (1895), 17.

29. Hedrick, *Grapes of New York*, 155–56.

THE NEW
White Grape,
NIAGARA.
This new purely native White Grape is a cross between Concord and Cassady. Hardy; as strong grower as Concord; earlier and better in quality. Bunches large uniform; very compact. Enormously productive, a four-year old vine producing 140 clusters, weighing from 8 to 16 ounces each. Possessing qualities as a market grape. Excelling any known variety, and decidedly the most valuable for the vineyardist, and is being extensively planted. Selling at 15 to 25 cents per pound, when best California grapes brought 10 cents. The most desirable white grape ever produced. THE GRAPE FOR THE MILLION. Every vine sold at retail has the seal of Niagara White Grape Co. attached.
COPYRIGHTED
BY MENSING & STECHER, ROCH. N.Y.

CHAPTER SEVEN

Labrusca Hybrids

TO THIS DAY, THE VAST majority of grapes grown in commercial and family vineyards east of the Mississippi River are *vitis labrusca* hybrids. The names are all too familiar—varieties such as Concord, Niagara, Delaware, and Catawba are all *labrusca* hybrids. They have been very successful commercially because of their winter hardiness; their ability to grow in a wide variety of soil types; their utility for different purposes such as juice, jellies, table consumption, or wine; and their high productivity and ease of growth in the field.

Even though these *labrusca* hybrids have been extensively used in wine production in the East, most of them are not really suited for table wines as far as the general public is concerned. However, *labrusca* still has a solid following either because of tradition or because of their ability to make very fruity, country wines that can easily be consumed. If *labrusca*-based wines are made properly, they can be fun wines that have a lot of wild fruit flavors reminiscent of fruit-based wines, unlike the more "serious" *vinifera*-based wines. Further, *labrusca*s have been successfully used in the production of quality sherries, ports, and kosher wines for more than 100 years. In fact, in the eastern United States, *labrusca* grapes have been the basis of a large and successful sparkling wine industry since the 1860s. The Pleasant Valley Wine Company of Hammondsport, New York, and the Brotherhood Wine Company of Washingtonville were noted for their production of quality "champagnes" and sparkling wines, which won many awards both in the United States and in Europe in the nineteenth and twentieth centuries.[1]

Niagara grape. *Heralded as the "White Grape for the Millions," this labrusca hybrid was developed in New York in 1868, and for many years dominated the eastern American white grape market.*

For the production of dry table wines, *labruscas* generally have too little sugar, are too acidic, and they generally possess a wild and very unsubtle fruity taste that, as noted earlier, has been described as "foxy." Before I describe the six *labrusca* grape varieties, it is best to tackle this commonly used term "foxy." It is, for some, a derogatory term used to describe all things *labrusca*. However, the term is too broad and inaccurate. It does not reflect the many nuanced and complex flavors that *labrusca* wines possess. There are five different elements of the *labrusca* flavor: (1) strawberry or cotton candy; (2) the naturally occurring presence of the chemical compound methyl anthranilate, sometimes equated with bubblegum smells and flavors; (3) heavy perfumey and floral flavors, and smells of guava and pineapples; (4) musty, musky, and wet-dog flavors; and (5) wild grapey/fruity that is hard to describe, but we all know as belonging to native American *labrusca* grapes.

Each *labrusca* variety has some or all of these aromatic and flavor elements. For example, Niagara is very perfumey in taste and smell. Concord has a strong methyl anthranilate smell, while Catawba has very little of it, but is very perfumey with ripe strawberry flavors. Ives, on the other hand, has intense strawberry or cotton candy in its nose and taste, with lots of heavy grapiness. One overarching characteristic is that they all have a very intense, forward, and pronounced grapey/fruity nose and taste. Also, *labruscas* can have a somewhat bitter finish. That is why many are finished semisweet to offset this high-acid middle and finish. Grapes such as Delaware, Diamond, Dutchess, and Iona are considered to be quality *labrusca* grapes for white wine production because they do not have these strong pronounced characteristics and possess other complicated nuanced flavors that make these wines interesting.[2]

The major *labrusca* hybrids are described on the following pages. They were either purposely bred or are chance seedlings that were found by horticulturalists or hobbyists in their backyard gardens or along the road throughout the East. These new hybrid *labruscas* were isolated, studied, and then propagated for commercial cultivation.

In Chapter Eight on the Hudson Valley hybridizers there are descriptions of many more hybrids with *labrusca* genes that were developed and grown in the Hudson Valley.

CATAWBA

PARENTAGE
labrusca, vinifera

HARVEST DATE
Very late

A-	B	C	A	B

Catawba was one of the first widely grown commercial wine grape varieties in America. As of 2006, there were 1,291 acres of Catawba grown in New York.[3] The red, not blue, grape was first introduced by John Adlum of Washington, D.C., about 1823. Adlum obtained cuttings from Mrs. Scholl of Clarksburgh, Montgomery County, Maryland, in the spring of 1819. The vine had been planted by Mrs. Scholl's husband, who died before Catawba's origins could be clearly ascertained. Mr. Scholl had always called it Catawba.

Adlum, reportedly, also found the same variety, Catawba, on a trellis on land owned by Mr. J. Johnston, near Fredericktown, Maryland, and at another farm in Lycoming County, Pennsylvania, where it had been named Muncy. It is thought to have originated in North Carolina. In U. P. Hedrick's *The Grapes of New York* and Jancis Robinson's *Wine Grapes* there are other interesting accounts of where Catawba may have come from, but in the end, the derivation of the grape is unknown.[4]

Louis Ravaz and Professor Pierre Marie Alexis Millardet thought Catawba was a *labrusca-aestivalis* hybrid, while in the late 1890s to 1910, T. V. Munson and Hedrick thought it was a *labrusca-vinifera* hybrid.[5] Regardless, the *labrusca* clearly dominates. In the 1920s, Hedrick later altered his opinion and believed the variety was probably a purebred *labrusca*. Further, all indications are that the grape originated on the banks of the Catawba River in North Carolina around 1800.[6] Since the publication of the first edition of this book, DNA analysis has confirmed that Catawba's pollen parent is the *vinifera* variety Sémillon, and seed parent is a wild American red or black species. Further, that Catawba is the pollen parent of Concord.[7]

For most of the nineteenth century, Catawba was a dual-purpose grape, used mostly in the production of sparkling wines and quality still table wines. The vine is productive, of medium to standard vigorousness, and winter hardy, but not as hardy as Concord. However, it ripens a bit too late for cultivation in the Hudson Valley, except for the warmest and longest-growing sites. At best, it ripens no earlier than the third or fourth week of October. In sum, it is a very late ripening grape that needs a long growing season. Its late ripening may indicate that this grape comes from the South, but was not very good for cultivation there due to its lack of resistance to fungal disease.

The vine blooms by mid-season. The clusters are medium-large to large, moderately compact, rather long and broad, cylindrical, and tapering. It is generally single-shouldered and somewhat compact. The slip-skin berries are medium-large, oval to roundish, and of a dull purple-red. In the late nineteenth and up to the mid-twentieth century, Catawba was primarily a wine and juice grape, but had an attractive appearance, so it could be sold as a table grape.

The variety does better in sand, gravel, or friable loam clay soils as long as there is good drainage and rich organic matter for fruit cultivation. One of the reasons for the rapid expansion in the cultivation of Catawba was its adaptability to many different soil types and environments.

Its fruit and foliage are susceptible, however, to black rot and downy mildew. It has minor resistance to powdery mildew and good resistance to botrytis, but it is less resistant to fungus diseases than Concord. Catawba is not sensitive to sulfur, so these treatments can be used.

This high-quality *labrusca-vinifera* is still a relatively foxy, musky, spicy, and fruity variety that is fragrant and moderately grapey with firm acid, but it is not overpowering like Concord. The juice has high sugar content for a *labrusca* and rich, soft flavor. Now with more modern winemaking techniques, light press of the must, and cold fermentation, the juice is very suitable for the production of sparkling wines and fun, soft pink country wines. The wines can be made in a variety of styles, so the wine descriptors change, depending on the type being made. Catawba wines can be made white, but are generally a light pink color like the famous pink Catawba, otherwise known as "Pink Cats." The wines have a hefty acid profile, but are not as acidic as Concord, and can be made approachable with some sugar and amelioration.

The flavor of semi-sweet Catawba is grapey, musky, and spicy, but with a soft grapiness, unlike Concord, which is much harsher. It is similar in flavor profile to Iona, but again, is a bit harsher. The wine is really a fruit wine. It is perfumey and has a nose and taste of strawberries, cherry bubblegum, and tutti-frutti. There is an underlying presence of spice. Even with sugar, Catawba has a firm acid structure with a long, clean finish. The wines are clean and enjoyable like fruit wines with simple flavors of watermelon, and an even balance.

Catawba wines are not offensive as is often suggested by wine critics. They are simple, fun wines that have a place in the Hudson Valley, whether it is on the dining room table or the picnic table. As Catawba needs a long growing season, local growers should not overcommit to this variety, but for historical interest, each vineyard with a suitably warm site should have a few Catawba vine planted.

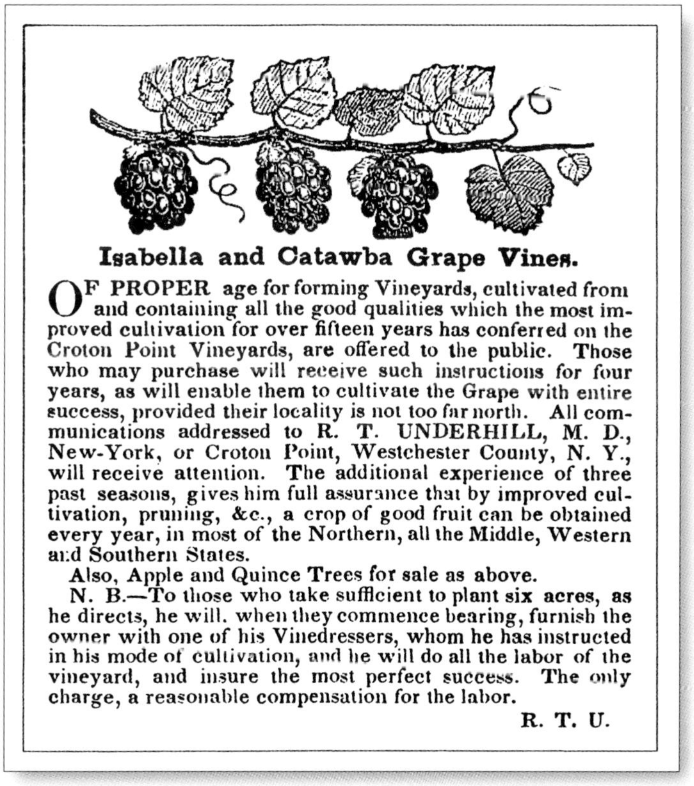

Isabella and Catawba Grape Vines.

OF PROPER age for forming Vineyards, cultivated from and containing all the good qualities which the most improved cultivation for over fifteen years has conferred on the Croton Point Vineyards, are offered to the public. Those who may purchase will receive such instructions for four years, as will enable them to cultivate the Grape with entire success, provided their locality is not too far north. All communications addressed to R. T. UNDERHILL, M. D., New-York, or Croton Point, Westchester County, N. Y., will receive attention. The additional experience of three past seasons, gives him full assurance that by improved cultivation, pruning, &c., a crop of good fruit can be obtained every year, in most of the Northern, all the Middle, Western and Southern States.

Also, Apple and Quince Trees for sale as above.

N. B.—To those who take sufficient to plant six acres, as he directs, he will. when they commence bearing, furnish the owner with one of his Vinedressers, whom he has instructed in his mode of cultivation, and he will do all the labor of the vineyard, and insure the most perfect success. The only charge, a reasonable compensation for the labor.

R. T. U.

Isabella and Catawba Grape Vines. *Early advertisement for two of the most popular grape varieties of the time, offered by Richard T. Underhill of Croton Point, Westchester County. Circa 1860s.*

CONCORD

It is difficult to overestimate the influence that Concord has had on the grape and wine industry in the eastern United States. It is the most widely planted red grape variety east of the Mississippi River, with approximately 20,200 acres of the grape growing in New York State alone.[8] Unlike most other grape-growing regions, the Hudson Valley grew many different kinds of *labrusca* grape varieties for the table, juice, jellies, and wine, however a larger percentage was always Concord.

PARENTAGE
labrusca, vinifera

HARVEST DATE
Mid-season to early late season

A+ A B A+ B-

Unlike Delaware or Catawba, it is clear where the Concord grape came from—Concord, Massachusetts. It was once thought to be a purebred *labrusca*, the seed of a wild grape that was planted in the fall of 1843 by Ephraim W. Bull (1806–1895).[9] It was later transplanted from beside a field fence to the garden, where another grape, Catawba and a wild vine, was open to cross-pollination. One of these seedlings was named Concord and exhibited at the Massachusetts Horticultural Society in the fall of 1852. The new grape was introduced in the spring of 1854, by C. F. Hovey and Company of Boston. The grape spread rapidly throughout most fruit-growing regions to become a leading grape variety grown in the eastern United States.

It was once difficult to ascertain the genetic makeup of Concord other than to surmise that it was 100 percent *labrusca*, but others maintained that it had at least some *vinifera* in its heritage.[10] Because of Concord's prowess in the field, breeders have used this grape extensively to impart better qualities related to its growing abilities, productivity, and ease of cultivation. U. P. Hedrick maintains that Concord, when bred and cross-bred, has prodigy that are often white and are usually of better quality than the parent; perhaps the white color is correlated to an enhanced quality of taste. From this we can surmise that one of Concord's parents was a white or light pink grape of high quality.[11] The quality white grape Dutchess is a Concord hybrid of White Concord bred with mixed pollen of Delaware and Walter. Also, Niagara, Diamond, and Jefferson are some of the more important examples of the hundreds of Concord hybrids that exist. Since the publication of the first edition of this book, DNA analysis has confirmed that Concord's pollen parent is the white *vinifera* variety Sémillion.[12]

In its first year of introduction in 1854, it was placed on the grape list of the *American Pomological Society Fruit Catalogue* as a variety that would "promise well." In 1858, it was placed on the regular list and widely planted thereafter.

Due to the economic, social, and political climate before and during the American Civil War, including religious movements such as the Great Awakening, which also pushed for the elimination of slavery and the promotion of temperance, or at least the curtailed consumption of alcoholic beverages, the demand for a grape such as Concord could not have been better placed. In 1865, it won the Greeley Award, which was sponsored by the American Institute of the City of New York, named after its donor Horace Greeley, the founder and editor of the *New York Tribune*. Greeley coined the expression that Concord was "the grape for the millions" because it could be so cheaply produced that no other grape could compete economically against it.[13]

Concord's western expansion was also swift. It was firmly established as one of the most widely planted grapes in the nation's newest and rapidly expanding grape-growing areas of western New York and Missouri just before and during the Civil War.[14]

There are several reasons why Concord is so popular with growers and so widely cultivated in the eastern United States. First, it is a vigorous to very vigorous variety that is very productive in many different soil types and in different climates. Second, it is a fruitful variety that consistently bears large crops each year when compared to most other varieties. Third, it is very winter hardy and relatively resistant to fungal diseases and insect damage with the maintenance of a proper spray program. Fourth, it ripens by mid-season to late mid-season throughout the northern United States and Canada. Fifth, it has multiple uses so it can be sold for table use, juice, jellies, or wine.

The introduction of Concord significantly affected grape culture in the United States because few other varieties could compete against it on the basis of yield and cost inputs. Its cost to produce—and hence its cost to the consumer—was much less than any other grape variety grown. It was very cheap to purchase, yet gave an ample return to the grower. Few other grape varieties could economically compete with it, unless that variety had a specialized use for table grapes, still wine, sherries, ports, or sparkling wines.

While Concord competes very well against other varieties in terms of price, there are many other much better-tasting *labrusca* varieties that look better in the basket and produce superior wines. To be more specific, the quality of Concord is not high; it lacks richness and delicacy of flavor and does not make very good wines when compared to other *labrusca* varieties such as Bacchus, Dutchess, Iona, or Jefferson. However, as a table grape, the cluster is of good size and attractive, and the berry is large.

As noted earlier, Concord is very winter hardy and not particular about the soils it grows in. The variety buds out relatively late so it avoids late spring frost damage. Of the fungal diseases, Concord is most susceptible to black rot and somewhat susceptible to powdery mildew, but is resistant to botrytis and downy mildew. Further, it has some sensitivity to sulfur. Concord is more or less resistant to insect damage.

Concord ripens mid-season to early late season around late September to early October, along with varieties such as the French-American red hybrid Chelois and a bit earlier than Chambourcin. Its somewhat compact or crowded clusters are medium-large to large. The cluster tapers and is conical, with a single shoulder, sometimes two. Concord's round blue berries have a heavy blue bloom and are large to very large. The slip-skin berries separate easily from the pulpy flesh. The skins are tough and bitter, can crack during heavy rains during harvest time, and can shell from the stem. In ideal weather, Concord grapes can hold onto the vine, which helps to extend its two- to three-week harvest season.

While most Concords are utilized for juice, jellies, and table consumption, a large quantity is made into still, sparkling, and dessert wines. The still wines are unique and favored by those who grew up familiar with the Concord taste.

The wines can be bright red, scarlet, or very dark in color with high acid. Many vintners add water to reduce the intensity of the color and reduce the acid levels. Also, Concord wines are generally finished semisweet to offset their high acids. For a wine grape, Concord has no more than 16° Brix of sugar, so sugar must be added to make a stable table wine.

If it is used to produce a dry wine, the flavors are strong, pungent, and very aromatic. The wines are big and flavorful without a lot of tannin, or perhaps they have so much flavor that the tannin structure is overshadowed. Detractors suggest that they are harsh, coarse, and not sophisticated in any way. It is difficult to describe a Concord wine because we are all so familiar with them that it is like describing another

Concord grapes crate label. *For many decades of the early twentieth century, especially during Prohibition, most grapes from the Hudson Valley were labrusca varieties such as Concord, sold as table grapes throughout the Northeast. Circa 1930s.*

ubiquitous American food item, the hamburger. However, as the United States Supreme Court Justice Stewart Potter stated in his concurring opinion when trying to define pornography in *Jacobellis v. Ohio*, "I know it when I see it,"[15] the same can be paraphrased for Concord, "I know it when I taste it." However, this statement does a disservice to the many interesting flavor elements that Concord wines possess.

The wines are very aromatic, grapey, fruity, and pungent, but musty and musky at the same time. They can have a nose of big coarse violets and lilacs, strawberries and lemons, as well as tutti-frutti and boysenberries. While there is much more flavor in these wines, their body is much thinner than a French-American hybrid or *vinifera* wine. The wines can be bright, very perfumey, and wild. They should be compared with other fruit wines instead of other grape table wines.

Concord is also extensively used in sweet wine blends. As Concord taste is very assertive and dominant, just a little Concord wine added to a blend can give it that Concord taste. Its powerful flavors, however, can easily cover up the faults of other wines in the blend.

The best use of Concord is in the production of sherries and ports. The old method, used until recently by several wineries including Widmer Wine Cellars, was to store wooden barrels full of Concord and other *labrusca* wines on the winery's roof to help them oxidize. In the summer the wines baked, in the winter they froze, and this helped to oxidize the wines and give them that nutty taste and tawny to brown color.[16] However, the more common way to make sherries now is to heat and bubble oxygen through them to hasten the oxidation process. Also, flavoring additives can be incorporated to give these wines that nutty aged sherry taste. Concords are also used extensively in making sweet kosher wines, such as those made at Royal Kedem Winery in Milton.

Overall, Concord is what it is. It is a grape that is commonly grown in New York and the Hudson Valley and used by winemakers to make semisweet wines and dessert sherries and ports. There is now a large enough supply of this grape, so it is suggested that growers consider other varieties if they are expanding their vineyards. With that said, it is important to keep Concord in the Hudson Valley winemaking heritage.

DELAWARE

Delaware was considered the premium American white wine grape from its discovery in 1849 until the arrival of French-American hybrids in the 1950s and *viniferas* today. It still has many positive attributes in the cellar and should be considered for production as a varietal wine or used in blends. In the early twentieth century, Delaware was one of the most widely planted grapes in the Hudson Valley. In 2006, there were 265 acres of Delaware in New York.[17]

PARENTAGE
aestivalis, labrusca, vinifera, or bourquiniana, labrusca, vinifera

HARVEST DATE
Early mid-season to mid-season

A B C B+ A

It is not completely clear how Delaware was bred or where it originated. It is believed that the grape we now call Delaware was first noticed by Abram Thompson, the editor of the *Delaware Gazette* in Delaware, Ohio, in the summer of 1849. Mr. Thompson saw the beautiful red fruit in town, which was produced by Mr. Warford and Mr. Heath near the banks of the Scioto River, just a few miles from the town. Mr. Warford brought this grape from New Jersey more than twenty years earlier. It was known in this local Ohio community as Heath or Powell.[18]

Mr. Thompson sent the fruit to the Hudson Valley horticulturalist A. J. Downing, of Newburgh, who named it Delaware after the town in Ohio. The Delaware vines secured by Mr. Warford could be traced back to the garden of Paul H. Provost, a Swiss immigrant then residing in Frenchtown, Kingwood Township, Hunterdon County, New

Jersey. Provost had brought with him many European grapes, which he cultivated in his garden when he immigrated to the United States.[19]

It was difficult to ascertain exactly where Delaware came from, but according to one account, it was obtained from Mr. Provost's brother, who resided in Italy. The other account is that it was brought to Provost's garden by a German, who had been in the country for a short while and spent time with Hare Powell of Philadelphia. It is not certain if the German secured the vines from his native homeland or from Powell.[20] The *Bushberg Catalogue* states that Delaware was a seedling found in Provost's garden, an accidental cross of the grapes he brought from Europe with other local grapes.[21]

There are also different theories as to the heritage of Delaware. The French botanist Millardet considers Delaware to be a hybrid of *vinifera*, *labrusca*, *cinerea*, and *aestivalis*, while T. V. Munson of Texas maintained that it is a *labrusca-bourquiniana* hybrid, with the possibility that there is some *vinifera* in its background.[22] While modern-day ampelographers Pierre Galet and Lucie Morton believe it to be a *labrusca-aestivalis-vinifera* hybrid,[23] U. P. Hedrick designated this as a *labrusca-bourquiniana-vinifera* hybrid.[24] In the end, I would put my money on Galet and Morton because the variety is somewhat susceptible to phylloxera, so it probably has some *vinifera* in its genes. Further, one of its offspring is Dutchess, which also looks like it has some *vinifera* in its gene pool.

The *Bushberg Catalogue*, like Galet and Morton, (but unlike T. V. Munson) suggests that Delaware is an accidental *aestivalis-labrusca-vinifera* hybrid and not a *bourquiniana-labrusca* hybrid. This is because it is more closely allied to *aestivalis* in its leaf and wood, and is more winter hardy. Further, its cuttings are more difficult to propagate, unlike a vigorous *bourquiniana* hybrid (Herbemont) and it is a less vigorous grower. Delaware, which seems to have very few black grapes, more often has white or pink seedlings. It seems that Delaware wants to pass down its gene for producing either red or white grapes, which may give some indication as to who its ancestors were.[25]

Many of these early *labrusca* hybrids, like Delaware, were first identified in a backyard garden and then became popular because of the quality of their fruit or prowess in cultivation and production. They were not purposefully bred to produce fruit, either for the table or for wine. However, these new *labrusca* hybrids also seemed to appear in the gardens of amateur horticulturalists or recent European émigrés who had brought Old World *vinifera* grapes with them, which were then planted near other local grapes, so some of the early *labrusca* hybrids may indeed have some *vinifera* heritage.

Regardless of Delaware's origin or genetic heritage, it is still a fun grape that is tasty as a table grape and is balanced for sugar, acid, and flavor to make a quality white wine. Further, it oxidizes rapidly so it can be made into quality sherries. A testament to how highly it was regarded in the nineteenth century is that scores of hybridizers, including local Hudson Valley hybridizers such as A. J. Caywood, James H. Ricketts, and Dr. William A. M. Culbert used Delaware as a major part of their breeding program to create bigger and better "Delawares" that were more productive or disease resistant.

For a grape that is of high quality in the cellar, Delaware is also not bad in the field. The grape can be finicky but is generally tolerant of different soils, although it tends to prefer deep, rich (but not too rich) fertile loam or slightly clay loam soils that are well drained and warm. Its roots are slender, of medium toughness with a rather soft phloem, and are not inclined to branch out much.[26]

The small- to medium-sized vine has moderate to average vigor. Delaware is winter hardy to very

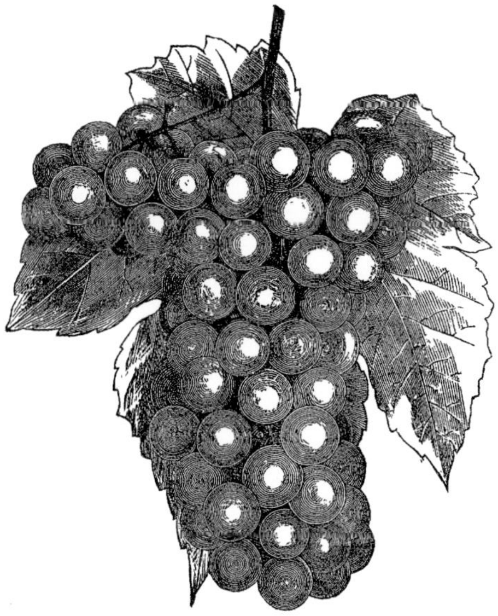

Delaware grape. *Nineteenth-century depiction of this dark pink grape of mixed genetic heritage, praised for both quality wine and table use.*

winter hardy, but less so than Concord. It is moderate to above average in productiveness if in the right soils. To increase production for Delaware, it is best to decrease the vine spacing and prune to long canes with spurs. The grape is initially a slow grower and does not propagate freely from cuttings. Further, the vine seems to have a compact growth habit. The grape can be grown anywhere in the Hudson Valley.

Its bud break is mid-season. But, like many *labruscas*, it does not produce a large secondary crop if hit by a late spring frost. It is, on average, more susceptible to fungal diseases than Concord, except for black rot, but it is significantly more susceptible to botrytis because of its relatively thin skins, susceptible to downy mildew, and somewhat resistant to black rot and powdery mildew. The berries can split if heavy rains occur just before harvest time. It is not sensitive to sulfur treatments. Overall, Delaware is more resistant to fungal diseases than most French-American hybrids, so an adequate spray program will ensure a good crop.

Delaware ripens early mid-season to mid-season, with high sugars. Further, it has a great acid balance to make a table wine or juice, unlike some other high-acid *labruscas*. Delaware can hang on the vine for a long period of time, increasing the concentration of sugar, and even if attacked by *Botrytis cinerea*, the grapes hang together and can produce truly exceptional dessert or ice wines.

The one-shouldered clusters are very pretty, compact, small-medium in size, and have a slender, blunt, cylindrical-conical shape. The deep red to pink berries are small-medium to medium in size, and the clusters can have a beautiful deep red/pink color when ripe, or a pink/greenish color when not ripe.

For the home vineyard or garden, Delaware is a highly desirable grape because of its small compact growth habit, handsome red fruit, and lustrous, delicately formed, and pointy dark olive green leaves. Further, it is great as a table grape.

Delaware's well-kept secret is its versatility in the cellar. It can make an interesting fruity/floral dry white table wine with multiple layers of taste. Or, it can be made into a delightful semidry aperitif wine or a substantial late-harvest dessert wine. Further, it can be used as a blender to add fruit, substance, and body to other lighter white French hybrid wines or to reduce their acid. It can also be used as a base for a complex and nutty sherry since it has the predisposition to oxidize and turn to a dark golden color. Further, it is commonly used as a base for sparkling wines. The sugars in the must can easily reach 22° Brix and as high as 27° Brix or more.

For a *labrusca hybrid*, the flavor of Delaware is flowery, fruity, and clean, with some light spice, yet mild and subtle. The wine can be light green in color or a light golden color. It has a delicious slightly nuanced foxy aroma and somewhat musky flavor that is soft, but with good body for a white. To keep a Delaware light in color and flavor, it is very important to reduce the amount of oxygen exposure during the winemaking process and keep sulfur levels high.

Unlike many *labrusca* hybrids, Delaware wines can be delicate, very interesting, and complex without the harsh foxy *labrusca* nose and taste. From my own tasting notes, dry Delaware wines can be pale green to a slight yellow in color and have interesting and consistent flavors of white peaches, guava, honey, and almonds with a musky element in both the nose and body. The wines are grapey, but aromatic, floral, with flavors of tropical fruits, ripe bananas, and almonds. The almonds come out in the taste, acid profile, and finish. For semisweet or late harvest, Delaware can have a big nose and flavor profile of ripe guava, honey, apricots, orange rinds, ripe bananas, muscat, and almonds. Further, the wines have a dank, musky feel, which integrates well with Delaware's fruit. If made in a certain style, they are reminiscent of Vignoles grapes with lots of floral tropical fruits, honey, orange rinds, pineapples, and integrated acids. These semi-sweet and late-harvest wines tend to be syrupy, but the acid maintains the structure.

Delaware grapes are grown in the eastern United States, as well as in Brazil and Japan as a table grape.[27] While Delaware acreage in New York is declining, it should be considered by Hudson Valley growers because of its solid attributes in the field and versatility in the cellar. Historically, Delaware was widely grown in the Hudson Valley, and it should be widely grown here again.

DIAMOND

Diamond or Moore's Diamond is a quality white *labrusca* variety that is used to produce soft *labrusca*-tasting white table wines. Along with Delaware, Dutchess, and Iona, Diamond is a quality white *labrusca* grape variety for wine. This white variety is gaining more popularity. In 2006, there was 93 acres of Diamond grown in New York, and that acreage is increasing.[28] It is being made as varietal wines and used as a component in light *labrusca*-flavored wines.

PARENTAGE
labrusca, vinifera

HARVEST DATE
Early mid-season

A+ A C+ A A

Jacob Moore[29] (1835–1908) of Brighton, New York, bred Diamond around 1870 from a Concord seed fertilized by Iona,[30] a local Hudson Valley grape that was introduced by Dr. Charles W. Grant in 1864. Moore started in the nursery business early in his life and began to conduct his own hybridization program after 1860 to help improve the quality of grapes he could offer in his nursery business.[31]

Diamond was introduced around 1885, and gained attention relatively slowly, not appearing in the *American Pomological Society Fruit Catalogue* until 1897.[32] Diamond is one of the parents of the grape variety Ontario, which was used extensively in the breeding program at the New York Agricultural Experiment Station at Geneva, New York, to breed Cayuga White, Horizon, Melody, and many other new white grape varieties for wine production.

The grape is an early mid-season-ripening grape that ripens about one week to ten days before Concord. It is very winter hardy and productive, and is a vigorous grower for a white variety. While Niagara is more productive, the quality of Diamond fruit for wine is far superior. U. P. Hedrick maintained that Diamond is a diluted hybrid between *labrusca* and *vinifera*. Further, the touch of the *vinifera* grape gave Diamond a rich flavor and the refreshing sprightliness of its American heritage.[33] Regardless of these comments, there seems to be little evidence of *vinifera* in Diamond's wine characteristics.

The plant resembles its seed parent, Concord, but it is not as vigorous. However, it grows as well in the same soils and climate as Concord does, so it has a wide range of habitat. It buds out by mid-season. The disease resistance is as good as Concord, but it is more susceptible to black rot. However, it is more disease resistant than Niagara. It has an upright growth habit with good, but not excessive, vigor. It is best trained to cordons with spurs.

The pale green clusters can be somewhat variable in size—medium to short, rather broad, somewhat blunt, and cylindrical. Further, they are single-shouldered and compact. The berries are of medium size and green with a tinge of yellow on riper berries, but less yellow than Niagara.

In wine quality, when compared to Niagara, Diamond does not have the same clawing perfumey fruit qualities, nor does it have the musty, musky, bitter "wet-dog" qualities either. Diamond's color is light pale green to yellow. Further, it oxidizes or browns as rapidly as Delaware or Aurora.

The wine is clean, balanced, spicy, and rich, but mild and lightly perfumey for a *labrusca*. It can be an assertive white wine with flinty overtones and some spice with fewer of the hard and somewhat astringent *labrusca* flavors. Diamond is very perfumey, but not as much as Niagara. It is floral with clean notes of guavas and mead. The overall taste is of honey/mead, banana, some pineapple, and melons. It is floral like muscat wines, but in a nice, not overpowering way and has the *labrusca* notes and taste. The wines are approachable and complex, but with a steely middle, even if finished off-dry.

Diamond tastes like its pollen parent Iona, not like the harsh, big, angular flavors of its seed parent Concord. In sum, Diamond is a softer, less grapey, and mellower version of Iona wine with lighter yellow color than Iona's more-amethyst colors.

This grape should be considered for further expansion in the Hudson Valley to produce a unique white *labrusca* wine.

ELVIRA

Elvira is not widely planted in the Hudson Valley, but it has many beneficial characteristics that could be of use to local growers on cold marginal sites. This white grape was bred by Jacob Rommel (1837–1916) of Morrison, Missouri. Rommel was a nurseryman and grape grower who hybridized many commercially successful grape varieties that were designed specifically for wine production and suited to the growing conditions and environment of the Midwest and Prairie states. His father, who lived in Hermann, Missouri, also had a nursery business and was interested in grape growing and winemaking.[34]

PARENTAGE
riparia, labrusca

HARVEST DATE
Mid-season to late mid-season

A+	A	A	A+	B

Elvira is from a seedling of Taylor that was possibly pollinated by Martha in 1862.[35] Taylor is a white *riparia* × *labrusca* chance seedling found in the early 1800s in the Cumberland Mountains near the Kentucky and Tennessee border. It came to the attention of Judge John Taylor of Jericho, Henry County, Kentucky, who was an amateur horticulturalist. He disseminated the newly named grape Taylor to grape growers for testing in the 1840s. It was considered a wine grape.[36] The possible pollinator, Martha, is a white grape that is a seedling of Concord sent by E. W. Bull to Samuel Miller of Calmdale, Lebanon County, Pennsylvania, in the early 1860s.[37]

Elvira was first introduced and disseminated by the *Bushberg Catalogue* during the 1874–75 season.[38] Elvira is used for the production of wine in the South and Midwest, but is becoming a more commonly grown variety in the Finger Lakes due to its ease of cultivation and high fungal disease resistance.

This is a great grape in the field, especially for poor sites in the Hudson Valley. Its *riparia* heritage has helped to create a grape variety of great productivity, very high fungal disease resistance, very high vigor, and high winter hardiness. The grape ripens about the same time as Concord. It is a stocky grower that has stout, long canes with well-developed laterals.

The variety buds out early, so is vulnerable to late spring frost damage. However, due to its *riparia* high vigor and ability to produce a secondary crop, its early budding can be somewhat mitigated. Like Baco Noir, another *riparia* variety, even with its high productivity, its *riparia* vigor gives the plant sufficient energy to ripen its vast crop. In the late 1980s to early 1990s the grape was recommended by Canandaigua Wine Company to its growers in the Finger Lakes because of its high yield, good disease resistance, winter hardiness, and production of bulk relatively neutral wines. In 2006, it covered 587 acres in New York.[39]

Its clusters are short, cylindrical, single-shouldered, and compact to very compact. The berries, of a dull greenish-yellow with a red-brown tinge to them, are of medium size, but with thin skins that can split close to harvest time if the season is wet.

Elvira ripens mid-season to late mid-season along with Concord and Chelois. The grapes have thin skins, which can lead them to shatter when ripe, so they need to be closely monitored to avoid loss of juice and oxidation of Elvira's wines. The flavors of this white grape are *labrusca*-like, but more neutral and muted, so it can be used as a base for either still or dessert wines.

This grape should be considered for those in the Hudson Valley who have all but the most impossible sites. Also, for breeders, it imparts to its offspring good growing characteristics and produces grapes that have more neutral flavors suitable for wine production.

NIAGARA

PARENTAGE
labrusca, vinifera

HARVEST DATE
Mid-season

A	A–	B	A+	B–

Niagara is the most widely planted white grape east of the Mississippi River.[40] Niagara is a cross made by Claudius L. Hoag and B. Wheaton Clark of Lockport, Niagara County, New York. Its breeders state that it came from a seed of Concord fertilized by Cassady that was planted in 1868 and first fruited in 1872. It was introduced for sale in 1882 by Hoag and Clark's Niagara Grape Company, and by 1885, had been listed in the *American Pomological Society Fruit Catalogue*.[41] The grape is widely grown north of the Mason-Dixon Line, in Canada, and east of the Mississippi River. It is also extensively grown in Brazil and to some extent in New Zealand and Japan.[42]

One of its parents, Cassady (*labrusca, vinifera*) is a chance white seedling from H. P. Cassady of Philadelphia that first fruited in 1852. Cassady is reported to be of medium vigor, productive, with a medium-sized cluster. The berry is round, compact, and greenish-white with an occasional faint salmon tint. Cassady ripens late to very late with Catawba.[43]

Niagara is not as vigorous or winter hardy as Concord, but is still as vigorous and winter hardy as other white *labrusca* varieties and more winter hardy than the hardiest white French-American hybrids such as Vignoles or Vidal. It is, however, very productive, almost as productive as Concord. Niagara can be grown in many different soils in a manner similar to Concord. Niagara is more than moderately susceptible to powdery and downy mildew, is slightly more susceptible to all fungus diseases than Concord, and particularly susceptible to black rot. It is not sensitive to sulfur treatments.

The vine buds out by early mid-season to mid-season. The grape ripens by mid-season along with the French-American hybrid Baco Noir. Further, it can hold onto the vine during non-wet seasons, which helps to extend its harvest season. Its large to very large clusters and berries are both well formed and bigger than Concord. When ripe, the berries are pale yellow.

Its berries do not crack as easily as Concord, but shell freely when ripe. Further, it is a more handsome cluster and is very suitable for sale as a table grape. Similar to the economics of growing Concord, Niagara completely dominated the eastern American white grape market because of its high productivity. This has led to a very low price per ton in the market, but due its high productivity, growers can still make a profit. During the late nineteenth and early twentieth centuries, this dynamic allowed Niagara to supersede all other competing white grapes in the market.

The wine, whether finished dry or sweet, has a very highly perfumey fresh grape taste reminiscent of the vineyard at harvest time. This massive fruit-bowl taste can be really clawing and very foxy. The foxiness is a very pronounced "wet-dog" musty, musky flavor with a somewhat bitter finish. No matter what technique is used to vinify Niagara, there will still be the overwhelming sensation of eating fresh grapes. This fresh-grape taste is intense, rich, and musty, with elements of petroleum that can finish with a somewhat bitter petroleum/kerosene taste. The complex and intense tastes include lots of guava, ripe bananas, ripe pineapples—a regular mixed fruit bowl of flavors. Because of its highly powerful and pungent fruit, Niagara needs sweetness to support these huge flavors.

This grape is widely available in New York. Due to its wide availability, limited winemaking capabilities, and better substitutes for table grape varieties, I do not recommended that growers in the Hudson Valley plant this variety.

NOTES

1. Lucie T. Morton, *Winegrowing in Eastern America: An Illustrated Guide to Viniculture East of the Rockies*. (Ithaca, N.Y.: Cornell University Press. 1985), 31–33.

2. Ibid., 65; this is also my own opinion on these wines.

3. Jancis Robinson, Julia Harding, and Jose Vouillamoz. *Wine Grapes: A Complete Guide to 1,368 Vine Varieties, Including Their Origins and Flavor.* (New York: HarperCollins, 2012), 206.

4. U. P. Hedrick, *The Grapes of New York* (Albany, N.Y.: Department of Agriculture, State of New York, 1908), 205–7. In *Wine Grapes*, Robinson maintains that it is clear where Catawba comes from, but she is not convinced of that fact. See Robinson et al., *Wine Grapes*, 205.

5. Pierre Galet and Lucie T. Morton, *A Practical Ampelography: Grapevine Identification* (1st American ed.) (Ithaca, N.Y. and London: Cornell University Press, 1979), 159.

6. U. P. Hedrick, *Manual of American Grape-Growing* (New York: The Macmillan Company, 1919), 8.

7. F. Huber, F. Röckel, F. Schwander, E. Maul, R. Eibach, P. Cousins, and R. Töpfer, "A view into American grapevine history: *Vitis vinifera* cv. 'Sémillon' is an ancestor of 'Catawba' and 'Concord,'" *Vitis – Journal of Grapevine Research*, vol. 55, no. 2 (2016), Geilweilerhof: Julius Kühn-Institut, Institute for Grapevine Breeding, 53-56.

8. Jancis Robinson et al., *Wine Grapes*, 265.

9. Ephraim W. Bull (1806–1895) was born in Boston, Massachusetts. Bull, by trade, made and sold gold leaf. He was an avid horticulturalist who specialized in growing and breeding new grape varieties. In addition to propagating the seedling Concord, he cultivated many other grape seedlings. However, none of these seedlings attained Concord's magnitude of success. Bull made very little money for himself on his discovery of Concord because other growers quickly propagated and sold the vine without restraint. His last days were not prosperous. The inscription on his tombstone in the Sleepy Hollow Cemetery in Concord reads "He sowed, others reaped." For more biographical details on Ephraim Bull's life and work, see Chapter Sixteen, pages 270–273. See also Thomas Pinney, *A History of Wine in America: From the Beginnings to Prohibition, vol. 1* (Berkeley: University of California Press, 2007), 211–14; Hedrick, *Grapes of New York*, 221.

10. Robinson et al., *Wine Grapes*, 264.

11. Hedrick, *Grapes of New York*, 221.

12. F. Huber, et al, "*Vitis vinifera* cv. 'Sémillon' is an ancestor of 'Catawba' and 'Concord,'" 53-56.

13. Ibid., 220.

14. Pinney, *History of Wine*, 212–13.

15. Jacobellis v. Ohio, 378 U.S. 184, 197; 84 SCT 1676, 1683 (1964).

16. I saw this process firsthand when I toured the Widmer Winery with Jim Finkle, the grower-relations representative for the Canandaigua Wine Company that owned the facility. I visited the winery, located in Naples, New York, in the spring of 1988. On the flat roof of the winery were hundreds of barrels of wine that were being aged for sherry production. These wines would then be fractionally blended by the Solera method to produce consistently excellent quality sherries each year, the best of them being sold as Empire Cream Sherry.

17. Robinson et al., *Wine Grapes*, 295.

18. Hedrick, *Grapes of New York*, 232–33.

19. *The Bushberg Catalogue* (Jefferson City, Mo.: Bush and Son & Meissner, 1895), 113; Hedrick, *Manual of American Grape-Growing*, 358–59; Galet and Morton, *Practical Ampelography*, 160. Jancis Robinson is more confident about the origins of Delaware, but I am not as confident, see Robinson et al., *Wine Grapes*, 294.

20. Hedrick, *Grapes of New York*, 233.

21. *Bushberg Catalogue*, 113–14.

22. Hedrick, *Grapes of New York*, 233–34.

23. Galet and Morton, *Practical Ampelography*, 160.

24. Hedrick, *Grapes of New York*, 231.

25. *Bushberg Catalogue*, 113–14.

26. Ibid., 113; Hedrick, *Manual of American Grape-Growing*, 358–59.

27. Jancis Robinson, *Vines, Grapes, and Wines* (New York: Alfred A. Knopf, 1986), 262; Robinson et al., *Wine Grapes*, 295.

28. Robinson et al., *Wine Grapes*, 658.

29. Jacob Moore (1835–1908) was born in Brighton, New York, in 1835. He was a careful and exacting grape breeder who began hybridizing grapes around 1860 in Attica, New York. He cross-bred *labrusca* and *vinifera* varieties with each other with the goal of melding the good characteristics of the Old World and New World grapes into new superior varieties. It does not seem that his breeding program was extensive, but the varieties that he released or sold to commercial introducers were truly quality grapes. He also seemed to test them thoroughly before releasing them to the public. In breeding his hybrids, he alternated between having the *vinifera* and *labrusca* variety as the seed or pollen parent. His named varieties are: Diana Hamburg, bred in 1860 and exhibited at the New York Agricultural Society in 1864; Brighton, bred around 1867 and exhibited at the New York Horticultural Society in 1872; Diamond, bred around 1870, but introduced much later between 1882 and 1885; and Geneva, bred in 1874 and introduced thereafter. Of his varieties, both Brighton and Diamond are still in commercial cultivation and sold by nurseries.

For many years, Moore pushed the U.S. Congress to pass a patent law to protect the intellectual property behind newly developed fruit varieties, but to no avail. This seems to have embittered him as he had devoted significant time and capital to their development, but his varieties could be commercially sold for a profit by other nurserymen, and he was not entitled to receive royalties for their development. He died in comparative poverty in January 1908. See Hedrick, *Grapes of New York*, 191–93, 236–38, 241–42, 274–75; T. V. Munson, "Beginning and Progress in the Development of American Grapes," in *Transactions of the Kansas State Horticultural Society, The Proceedings of the Forty-Second and Forty-Third Annual Meetings, Vol. XXX* (December 1908 and 1909), Walter Wellhouse, ed., (Topeka, Kans.: State Printing Office, 1910), 195–204. See charts and grape variety genealogy by Carl Camper at the Chateau Stripmine Experimental Vineyard, www.chateaustripmine.info/Breeders.htm.

30. Hedrick, *Grapes of New York*, 237; Hedrick, *Manual of American Grape-Growing*, 359–60.

31. Hedrick, *Grapes of New York*, 192.

32. Ibid., 237.

33. Ibid. U. P. Hedrick reaffirmed his belief that Diamond's heritage included *vinifera* eleven years after the publication of *The Grapes of New York*. Hedrick, *Manual of American Grape-Growing*, 359.

34. Jacob Rommel Jr. (1837–1916) was born in Philadelphia, Pennsylvania, in 1837. In 1838, the family moved to Hermann, Missouri, where his father, Jacob Rommel Sr., was engaged in the nursery business. We are not sure if Rommel Sr. became interested in grape growing and winemaking while in Hermann, or if he moved here because this was an excellent area for grape growing and winemaking. In 1860, Rommel Jr. moved to Morrison, Missouri, where he went into business with the Sobbe (or possibly the Shobe) family to grow nursery stock and grapes. Like other grape growers and nurserymen in Missouri, they were dissatisfied with the available eastern grape varieties because they were generally for table grapes and not suitable for wine production, nor were the growing conditions in Missouri suitable. Rommel Jr. set out to create new varieties that could be used for winemaking in the Missouri area. In his breeding program, of which thirteen hybrids are recorded, the seed parent was generally either Taylor, Elvira, Delaware, or a wild *riparia*. The pollen parent was generally open pollinated. Rommel's hybrids had extreme vigor and productiveness. All of his hybrids were named, the most important being Elvira, Black Delaware, Amber, and Etta. Rommel was married to Katharina Buehrlen Rommel and buried in the Clary Cemetery in Morrison, Osage County, Missouri. See Hedrick, *Grapes of New York*, 259–61, 265–66, 351–53; Chateau Stripmine, www.chateaustripmine.info/Breeders.htm; Munson, "*Beginning and Progress*," 195–204.

35. Hedrick, *Grapes of New York*, 260.

36. Ibid., 408–9.

37. Ibid., 341–42.

38. *Bushberg Catalogue*, 124–25.

39. Robinson et al., *Wine Grapes*, 327.

40. There were 3,468 acres of Niagara in New York in 2006 and 3,520 acres grown in Michigan; See Robinson et al., *Wine Grapes*, 727.

41. Hedrick, *Grapes of New York*, 361; Hedrick, *Manual of American Grape-Growing*, 418–19. Claudius Lamb (C. L.) Hoag and Benjamin Wheaton (B. W.) Clark were two grape farmers from Niagara County, New York, who crossbred the Concord and Cassady grapes sometime in the 1860s. Their innovation paid off four years later with their first fruiting, on a very vigorous vine with qualities similar to Concord, though in a white grape. Hoag and Clark ultimately formed the Niagara Grape Company in 1879, and opened several vineyards across New York. Planting and sales of the grape continued to grow, and in 1889, the Niagara Grape Market was established. Gradually, the market became oversaturated with Niagara grapes and prices of the fruit plummeted. By 1915, the Niagara Grape Market Company had folded. See J. J. Thomas, *The Niagara Grape* (Lockport, N.Y.: The Niagara White Grape Co., 1888).

42. Robinson, *Vines, Grapes, and Wines*, 265; Robinson et al., *Wine Grapes*, 728.

43. Hedrick, *Grapes of New York*, 445.

IT WILL BE

to the advantage of those intending

To Plant Vineyards

to get our Prices and Terms for the

ULSTER PROLIFIC

AND

Po'keepsie Red Grapes.

Inquire as to their hardiness and quality.

A. J. CAYWOOD & SON, Marlboro, N. Y.

RICKETTS' NEW SEEDLING GRAPE THE LADY WASHINGTON; the Best Hardy White Grape ever offered to the Public. Also my new EXOTIC SEEDLING GRAPE, WELCOME; the Best Grape in existence. Send for new Circular and Price List of Vines and Wood, with Stamp.

JAMES H. RICKETTS, Newburgh, N. Y.

Small Fruits. Grape Vines

E. P. ROE offers one of the largest and finest stocks of plants ever grown in the country, at very low rates. All the leading new and standard varieties. Ill. Descriptive Catalogue free. Address **E. P. ROE**

Cornwall-on-Hudson, N. Y.

Grape Vines for Vineyards

CONCORD CUTTINGS.

Rows four feet apart, stocky; as large as vines often are at two years. When well grown this age is preferred by vineyardists.

$8 per 100, $75 per 1000.

30,000 of other desirable varieties, the leading kinds predominating at usual rates, including the celebrated

MONTGOMERY,

The largest hardy white grape in cultivation, clusters often weighing two pounds each; a prodigious grower and bearer.

Oct. 2t $1 00 each. $75 per 100.

FERRIS & CAYWOOD, Poughkeepsie, N. Y.

SING SING

GRAPE VINES.

All my business is to grow

Grape Vines

and sell only what I grow.

The Grape Vines grown under the firm of J. F. DELIOT & RYDER, were all grown by J. F. DELIOT; it must be understood that I **warrant true to name** only the

Grape Vines

Sold by J. F. DELIOT.

Send for Catalogue to

J. F. DELIOT,
Sing Sing, N. Y.

A NEW WHITE GRAPE.

The Croton is a White Grape, and is acknowledged by all who have tasted it to surpass in quality and beauty any white Grape yet introduced, that will succeed in the open air, and is, in delicacy of flesh and flavor, fully equal to

The Finest Foreign Varieties.

It has held its foliage well in all parts of the country, and in many places better than any other variety. Will undoubtedly prove to be a most valuable market grape.

THE SENASQUA

is a black grape; the vine a healthy, vigorous grower, with every appearance of a pure native, but its fruit more closely resembles the fleshy foreign grapes than any variety that has as yet been introduced. It is considered by some of our best pomologists as the finest hardy grape they have tasted.

For fine cut of Croton Grape, and further description, history of origin, reports of success in various localities, list of premiums awarded, etc., etc., send for circular.

PRICE LIST:

	Each.	Per Doz.
Croton, extra, 1 year	$3 00	$30 00
Senasqua, extra, 2 years	3 00	30 00

The above vines are grown from well ripened wood, and are all of very large size, perfectly healthy, and free from insects or disease.

Will be sent by mail, post-paid, at above rates.

STEPHEN W. UNDERHILL,
Croton Landing P. O.,
Croton Point, New York.

The Iona Nurseries.

We have a full and general assortment of all the leading varieties of **BLACKBERRIES, RASPBERRIES, STRAWBERRIES, CURRANTS, etc.,** which have been grown with requisite skill and care for the production of best plants.

All are invited to investigate the merits of the **Eumelan** grape, and the quality of our stock.

Address **HASBROUCK & BUSHNELL,**
(Successors to C. W. GRANT,)
Iona, near Peekskill, Westchester Co., N. Y.

CHAPTER EIGHT

The Hudson Valley Hybridizers

THE HUDSON VALLEY, in the second half of the nineteenth century, was a cradle of horticultural activity and learning. Many professional and amateur horticulturalists and grape breeders developed new interspecific grape varieties and collaborated with other local and nationally recognized grape breeders, horticulturalists, and writers. At that time, there were many experienced fruit growers and nurserymen who lived nearby who could test these new locally developed varieties in the field and provide comments on such varieties, and ultimately, propagate them for commercial production.

The goal of these hybridizers was to increase the quality of table grapes grown in the Hudson Valley for the fresh-fruit market in New York City and, for some, to increase the quality of grapes grown locally for wine production. The ancillary breeding goals were to increase productivity and disease resistance, but by looking at their breeding methodology, it is clear that the primary goal was to incorporate the enhanced quality characteristics of European *vinifera* varieties into the *labrusca* varieties that were commonly being grown in the Valley. As a result, seed parents of *labrusca* varieties were mixed with pollen parents of *vinifera* varieties, or seed parents of *vinifera* varieties were mixed with pollen parents of *labrusca* varieties. Some hybridizers, such as James H. Ricketts, extensively used *riparia* varieties to impart disease resistance or to increase the ability of a new interspecific hybrid to be used in wine production. In sum, the Hudson Valley hybridizers were trying to convert locally grown grapes into European *vinifera* varieties for table use or wine production.

Hudson Valley grape growers. *Selection of advertisements from prominent mid nineteenth-century Hudson Valley viticulturalists and hybridizers.*

The Hudson Valley hybridizers' efforts contrasts with grape hybridizers of Missouri, Texas, the American Midwest and the Plains states, such as Thomas Volney Munson, and recent European immigrants Isador Bush, Hermann Jaeger, or Jacob Rommel. These hybridizers were attempting to breed cold-, heat-, and drought-tolerance, and insect- and disease-resistance into East Coast *labrusca* table and European *vinifera* grapes so that they could grow wine grapes in their newly adopted home. Their goal was to create new types of wine grapes that could produce local wines in their new harsh climate, while in the Hudson Valley the hybridizers' goal was to grow better quality table grapes with European flavor for the New York City fresh market. To accomplish their goals, the Midwestern hybridizers, along with botanists such as George Engelmann of St. Louis, Missouri, collected, identified, classified, and used locally growing native grape species that they found in the wild, such as *V. aestivalis*, *V. berlandieri*, *V. candicans*, *V. cinerea*, *V. cordifolia*, *V. riparia*, and *V. rupestris* to impart the disease- and temperature-tolerant characteristics that the varieties would need to survive. To impart the desirable table grape characteristics of large berry and cluster size and big pleasant fruit flavors, they bred in *labrusca* and *vinifera* varieties from the East Coast and Europe. To impart desirable winemaking characteristics and flavors, the Midwestern hybridizers also bred in *labrusca* and *vinifera* varieties.

The Golden Age

During the Hudson Valley's "Golden Age" of grape breeding, many of the breeders as detailed on the following pages were enthusiastic amateurs, and had national reputations for their work and writings on horticultural topics. In addition, these breeders were active members of national and regional horticultural organizations such as the American Horticultural Society, the American Pomological Society, the New England Horticultural Society, the New York Horticultural Society, and others. Further, they actively participated at these meetings, exhibited their new grape varieties and other fruits, and wrote articles to be presented before these groups. Nationally recognized writers on horticultural issues who lived in the Hudson Valley—such as Andrew Jackson Downing, his older brother Charles Downing, and Edward Payson Roe—often discussed their work with local nurserymen such as William B. Brown, Andrew Jackson Caywood, Dr. William A. M. Culbert, Dr. Charles W. Grant, James H. Ricketts, J. G. Burrows, and several generations of the Barnes, Clark, and Underhill families. The close proximity of so many grape breeders and horticultural experts helped to advance the individual work of each breeder.[1] In fact, the Downings, the Ricketts, Dr. Culbert, and Dr. Grant all lived in Newburgh, within four or five blocks of each other, while Andrew Jackson Caywood lived just five miles north in Marlboro.

In addition to the more prominent grape breeders such as Caywood, Ricketts, Grant, and Underhill, the horticulturalists—such as Edward Payson Roe, William D. Barnes, and J. G. Burrows—also dabbled in grape breeding, and actively participated in meetings of the American Society of Pomologists and local agricultural societies such as the Newburgh Bay Horticultural Society and the Orange County Agricultural Society. At such meetings they also presented their latest new interspecific hybrids, papers on breeding and propagating grapes, and conducted informal discussions on these subjects.

Hudson Valley Horticulture and the Downing Brothers

One of the most nationally known horticulturalists to come from the Hudson Valley was Andrew Jackson Downing (1815–52) of Newburgh, a famous landscape designer and rural architect, who published the book *The Fruits and Fruit Trees of America* (1845).[2] In addition, he published the *Treatise and Practice of Landscape Gardening Adapted to North America, with a View to the Improvement of Country Residences, with Remarks on Rural Architecture* (1841)[3] and *Cottage Residences* (1842).[4]

A. J. Downing

In 1846, A. J. Downing became the editor of *The Horticulturalist* and *Journal of Rural Art and Rural Taste.* He contributed an essay for that journal every month until his untimely death in 1852 at the age of thirty-seven. This journal set the national standard for the selection of plant material, landscaping, rural architecture, and the cultivation of fruit in the United States from the 1850s until 1875. In 1851, he was commissioned to lay out and plant the public grounds of the Capitol, the White House, and the Smithsonian buildings in Washington, D.C. He drowned in the Hudson River in July 1852 near Yonkers, New York, after the *Henry Clay*, the Hudson River steamer he was aboard, caught fire while racing the steamer *Armenia.*

While A. J. Downing was the more famous brother, it was his older brother, Charles Downing (1802–85), who had a much more lasting influence on fruit and grape culture in America. The Downing brothers were born in Newburgh to Samuel and Eunice Downing. The couple moved to Orange County from Lexington and Cambridge, Massachusetts, upon their marriage. Samuel Downing intended to pursue his trade of making carriages and had a shop first in Montgomery, Orange County, and then in the village of Newburgh at the southeast corner of Broad and Liberty Streets. But due to ill health, he abandoned that trade and became a nurseryman.

Samuel Downing died in 1822 when Charles was twenty years old and Andrew was seven. Samuel had established a successful nursery business and taught Charles the business. Upon his father's death, Charles took over the management of the nursery and

Charles Downing

supported his mother and siblings until they reached adulthood, when the firm then became known as C. & A. J. Downing Nursery, but it remained under that management scheme for only a few years.

In 1837, Charles moved one mile north to the outskirts of Newburgh to the suburb of Balmville on what is now Downing Avenue, while Andrew stayed at the family homestead at Broad and Liberty streets in Newburgh. Charles continued the nursery business for the next thirty years and became one of the foremost horticulturalists and pomologists in the United States. He studied the forms, varieties, and qualities of different fruits, which enabled him to improve many varieties of fruits and originate others. Further, he was constantly exchanging information with other nurserymen and fruit growers across the United States and in Europe about new fruit varieties and growing techniques.

Charles seems to have delighted more in the cultivation, study, and growth of those plants that his younger brother simply wrote about. However, later in his life, Charles became a regular contributor to *The Horticulturalist* and other periodicals. He twice revised and corrected *The Fruits and Fruit Trees of America*, which was originally written by his more famous younger brother, and added two new appendices that listed new fruits, making the final book almost twice the size of the first edition. Upon retiring from his nursery business, he lived at his townhouse at the southeast corner of Chambers and South Streets in Newburgh. Charles continued to write about horticulture and advise people throughout the country on horticultural issues for many years.[5]

Heirloom Preservation

The lives and work of the major Hudson Valley grape breeders in the nineteenth century and the grapes that they created are outlined in this chapter. Unfortunately, many of these cited grape varieties are no longer widely grown or not available at all. However, some of these varieties are preserved at the Plant Genetic Resources Unit (a component of the National Plant Germplasm System), located on the Geneva, New York, campus of Cornell University, so they are still available for propagation. At my farm, Cedar Cliff, we are propagating many of these heirloom varieties to establish new vineyards of these grapes so that we can provide cuttings to other growers. Brotherhood, America's Oldest Winery in Washingtonville is assisting our efforts by establishing their own heirloom vineyard at their tasting facility to showcase some of these long-forgotten American hybrid grapes.

Author's Methodology

The descriptions for the grapes on the following pages are based on my personal observations and those of other growers, as well as a compilation and evaluation of literature, including sources such as: U. P. Hedrick's *The Grapes of New York* and *Manual of American Grape-Growing*;[6] the *Bushberg Catalogue*;[7] the additional writings of the Geneva Experiment Station; and the writings of nurserymen, horticulturalists, and ampelographers such as the late Philip Wagner, Lucie Morton, T. V. Munson, and other sources.[8]

I made value judgments on what was reported on these varieties at the time and compared it to information about other grapes grown in the nineteenth century that are still commonly grown today, such as Concord, Delaware, Niagara, and Hartford as a point of reference. Much time and effort was devoted to comparing and integrating this information to give accurate descriptions on how the plants grow, the conditions that they prefer, and the resultant fruit and their use at either the table or in the cellar. Further, for those varieties that had recorded must readings for sugar content, such as Eumelan, Bacchus, Delaware, or Dutchess, I assumed that such varieties were used, at least to some extent, for wine production, so they were designated as wine grapes as opposed to table grapes.

I also evaluated the activities and methodology of each hybridizer's recorded breeding work in its entirety, specifically as it related to the grapes used in the breeding program and the resultant interspecific hybrids created, to determine the objectives of each hybridizer and if the final product met the hybridizer's goals. Some wanted to create superior table grapes, others wine grapes, and still others dual-purpose grapes that were suitable for both wine and table consumption. Often the commentators disagreed on the positive and negative attributes of a variety or on more basic information about the quality of the variety both in the field and for table or wine use.

Andrew Jackson Caywood

(1819–1889)

Andrew Jackson Caywood was a veteran horticulturalist and nationally known writer who was a member and exhibitor of fruit at the American Pomological Society, the Western New York Horticultural Society, and local agricultural societies. He worked extensively with Charles Downing in his fieldwork and contributed considerably to Downing's written works on the subject of grapes. In addition, he was one of the founding members of the Poughkeepsie Scientific Society, which ultimately became the Vassar Brothers Institute.

The youngest of eight children, he was born in Orange County, near the hamlet of Modena, in Ulster County, on August 20, 1819. His father, who was a farmer, died when Andrew was quite young. Caywood continued farming until he was about twenty-two or twenty-three years old, at which point he attended the Amenia Seminary, located in Amenia, Dutchess County. He attended this Methodist Church–directed institute of higher learning for two years, making special studies of botany and geology. During his early years, he was also a mason and a contractor.

When he was twenty-five years old, A. J. Caywood became interested in grape and fruit cultivation. This interest may have been inspired by William T. Cornell, the brother of his future wife Deborah Cornell, who had established a vineyard in Ulster County sometime around 1845, which was about the same time Caywood was courting Deborah.

A. J. Caywood and Deborah Cornell (1823–90) were married in 1848.[9] They had five children: Anna Cornell, Frank, twins Harriet Eliza and William Cornell, and the youngest, Walter J. Caywood. Around 1850, Caywood bought a small farm in Modena where he started his own fruit farm, nursery business, and grape-breeding program. He was one of the first in Ulster County to plant a vineyard and peaches for market purposes, prompting other farmers to grow fruit in Ulster County. Today, Ulster County continues to be the largest fruit-growing county in the Hudson Valley and one of the largest in New York State.[10]

In 1861, Caywood moved to Poughkeepsie where it is believed that his grape-breeding activities continued. In 1865, the grape variety Walter was brought to the attention of the public at various state and regional agricultural exhibitions. The new variety was named for his youngest son and listed in the *American Pomological Society Fruit Catalogue* in 1871.

In 1877, Caywood moved to an old 84-acre farm on the west side of Old Post Road, just south of the hamlet of Marlboro that was on the north side of the Old Mill House Creek (also known as Jews Creek), near Buckley's Bridge. He relocated to Marlboro from Poughkeepsie to either expand or start his own nursery business in connection with fruit propagation, first under the name Ferries & Caywood, and later under the name Caywood & Son when his son Walter joined the business. The nursery business was also known as Riverview Nursery. The property had one large residence on it near Old Post Road and several other large buildings that had been used for manufacturing enterprises related to the water power provided by the creek.[11]

Many of A. J. Caywood's selections were introduced after 1878, following his acquisition of the Marlboro property. Since he had also lived in Poughkeepsie for a time while he was hybridizing grapes and had named two of his first selections after the area, Dutchess and Poughkeepsie Red, it is surmised that the Modena or Poughkeepsie property is where he did most of his hybridization work. During his lifetime, Caywood developed many varieties of grapes, some of which were named and sold to the public.

Breeding Methodology

A. J. Caywood was a very dedicated fruit breeder who wanted to improve existing varieties, and devoted much of his time and money to this endeavor. In his long-lived program, he sought to combine the hardy constitution of the native grapevine with the increased size and superior flavors from cultivated grapes.

Caywood's breeding program appears to have differed from the breeding practices of other contemporary Hudson Valley grape breeders of his day in that he produced second- rather than first-generation hybrids. Every year he would cross his selected plants to produce seeds, and then plant and cultivate thousands of the resultant grape, raspberry, and blackberry seedlings. After cultivating and observing these seedlings, he would slowly thin them out after watching them for three to six years. After this observation period, he would select and retain the most promising hybrids and discard the rest.[12]

As best as can be determined from the records available, the named grape varieties bred by Caywood, in order by their introduction date, were Walter (first seed planted in 1850 and introduced to the public in 1871); Modena (introduced in 1867); Poughkeepsie Red (planted in the early 1860s and introduced in 1880); his most famous grape, Dutchess (planted in 1868 and introduced in 1880); Nectar, also known for a while as Black Delaware (first planted in 1883 and introduced in 1888); and subsequently Ulster (introduced in 1885); and Little Blue (introduced in 1888). Other named Caywood grape hybrids that were developed at some undetermined date included Metternich, Florence, Hudson, Mabel, White Concord, and White Ulster. He also developed the Marlboro Raspberry and Minnewaska Blackberry.

In reviewing Caywood's breeding program from the sketchy records available and the genetic makeup of his hybrids, he seems to have been fond of using quality *labrusca* grape varieties of the time such as Delaware and Iona, and his own quality hybrids such as Dutchess and Walter to produce new offspring. For resistance and production, he seems to have favored the use of White Concord or Concord seedlings as one of the parents of his new breeds. While interested in developing new grape varieties that could be used as table grapes, he seems to have been equally interested in developing grape varieties that could be used in the production of quality wines. Consequently, Caywood's hybrids tended to be more susceptible to winter injury and fungal diseases than the work of other hybridizers of his day. In 1957, when Mark and Dene Miller purchased the Wardell property (the upper portion of the old Caywood property), the vineyards still had primarily Delaware plants and some old Caywood hybrids that were Delaware-based.[13]

Some literature maintains that Caywood used *vinifera* to improve the quality of his new cultivars, but this is not the case. Actually, he seems to have used a superior white Concord seedling he either obtained or grew as the base of his many quality hybrids. Dutchess is one example held out as a variety that has significant *vinifera* heritage. However, in tasting Dutchess wines today, it is clear that the muted White Concord flavors are present. While Dutchess has a complex flavor, it is clearly a *labrusca* hybrid of some sort. The only *vinifera* in its heritage come from its other parent, Delaware.

Later Years

A. J. Caywood's later years were clouded by financial troubles and failing health. "He became dissatisfied, somewhat quarrelsome and indifferent about his affairs, to which poor health probably contributed."[14] He died at his home in Marlboro on January 13, 1889. Ironically, his only son Walter, who had an interest in his breeding work and nursery business, died shortly after his father's death on March 29, 1889, just before his twenty-eighth birthday. Then on the following day, April 1, 1889, Walter's wife Ruth gave birth to their son. A. J. Caywood's wife, Deborah Cornell Caywood, died shortly thereafter on October 14, 1890. If the nursery business was not doing well, as reported in the *Bushberg Catalogue*, the deaths of its founder in 1889, his son two months later, and his wife a year after that seems to have sealed the fate of this nursery business, as these occurrences reduced the family's ability to propagate and sell Caywood's newer selections.

Caywood Legacy

There is a more upbeat summation to the life, work, and death of A. J. Caywood that should be explored. In *The Grapes of New York* and the *Bushberg Catalogue*, the biographical sketches of the nation's best grape breeders tended to cast their ultimate demise in terms of failed projects that first led to financial ruin and then death, and in one case even suicide. The biographical sketches of breeders such as A. J. Caywood, Ephraim W. Bull, Jacob Moore, James H. Ricketts, Dr. Charles W. Grant, and Hermann Jaeger all had the theme of failed projects, bitterness, and ultimate financial ruin, poverty, and then death. However, this may not have been the case, according to the obituaries published for Caywood. In one, the obituary writer spoke to him in September 1888, four months before Caywood's death. In their conversation, Caywood told him he had a large number of seedling vines in his trial grounds that he was testing, and that it was now one of the greatest pleasures of his old age to go out among these vines to examine them and learn of their qualities. Further, the obituary praised Caywood's work over his lifetime and said that his death would be a loss to all who knew him.[15] Another obituary stated that Caywood had over 100 new varieties that were bearing fruit, and that he could give the parentage and origin of each variety merely by examining its leaf and cane.[16] This does not seem to be the kind of bitter end to a life devoted to horticulture as was suggested by *The Grapes of New York* or the *Bushberg Catalogue*. It portrays Caywood as someone who, while recognizing his old age, still truly enjoyed what he did and was looking to the future, not the past.

A final editorial comment on the emotional disposition of horticulturalists. We seem to be, all things considered, a generally happy lot, who are devoted to the work that we love, interested in the work of our peers and enjoy discussing our work with our colleagues. So the themes contained in the biographies of *The Grapes of New York* and *Bushberg's Catalogue* of failed projects, financial ruin, bitter lives, and finally death, does not seem to fit into a characterization that we would commonly referred to as horticulturalists.

Caywood's Descendants

The lives of the five Caywood children were very much intertwined. In 1880, Harriet married the neighboring farmer's son, Lawrence E. Wardell, in Marlboro. After Harriet died in 1885, Wardell married his sister-in-law Anna Cornell Caywood in 1886. He and Anna had one child, William L. Wardell (1887–1970), who would farm the land until it was sold to Mark and Dene Miller in 1957. Harriet's twin, William Cornell Caywood (1855–1901), married Florence Ormsby in 1891 and had two children, Edith and Marion, who would live and work on the Caywood property until 1983. William C. Caywood died in 1901 long after the nursery business had ceased to exist. His wife continued to operate the farm and run a boarding house to support her two children.

Florence then married Frederick Schramm, an illustrator and commercial artist from Queens, New York. The couple had three more children: Margaret, Elizabeth, and Frederick. Florence Caywood Schramm managed the lower farm (then known as Shady Brook Lodge), operated the boarding house, raised a family, and, with her sister-in-law Anna Cornell Caywood Wardell, subdivided the Caywood estate to distribute it to the various Caywood descendants. Florence Schramm, her daughter Edith Meckes, and Anna Cornell Caywood Wardell's work influenced how the property would be managed. The upper, or western, portion of the greater Caywood property, farmed by Lawrence E. Wardell, his wife Anna, and their son William Lawrence would eventually become Benmarl Vineyards,[17] which still operates today under the name Benmarl Winery. They grew wine and table grapes such as Delaware, Dutchess, Bacchus, and Worden, and had a small packing house, which still exists today. Another, now defunct winery, Cottage Vineyards, operated by Allan MacKinnon, was located on the lower or eastern part of the Caywood property near the boarding house in the 1970s and early 1980s.

A. J. Caywood's youngest son, Walter J., married Ruth A. West in 1884 and had Bessie C., who died after only a few months, and Walter C. Caywood, who was born in 1889. This family may have lived either in Poughkeepsie at 98 North Clinton Street or on the farm in Marlboro.[18]

Marion Cornell Caywood at Shady Brook Farm. *Marlboro, New York. Circa 1930s.*

DUTCHESS

Dutchess, sometimes referred to as Duchess, was named after Dutchess County, New York, the place of its origin. Dutchess is a white grape that came from a seed of a white Concord seedling pollinated by mixed pollen of Delaware and Walter planted in 1868. The white-fruited seed parent was an offspring of Concord, pollinated by Montgomery.[19] This grape was hybridized in 1868, probably in Poughkeepsie, and introduced in 1880. It was listed in the *American Pomological Society Fruit Catalogue* in 1881.

PARENTAGE
vinifera, labrusca, bourquiniana? aestivalis?

HARVEST DATE
Late

C	B	C	B	A

Dutchess, along with Delaware, Diamond, and Iona, was one of the top premium native white wine grapes developed in the United States in the nineteenth century. According to Pennsylvania grower Doug Moorehead, this is one of the very few native white wine grape varieties to improve with age, up to ten years. It does well as a dry wine with oak, but is generally finished sweet. This grape is still grown on a limited basis in New York and Pennsylvania, as well as Brazil, with 27 acres recorded as being grown in New York in 1996.[20]

This high-quality grape is grown on a vine that has been characterized as weak to vigorous, depending on the soil conditions. The vine should not be closely pruned as it is susceptible to winter damage. However, this susceptibility to winter injury might be attributed to the fact that Dutchess is more particular about the soils that it is grown on. Either way, it is accurately rated as being somewhat susceptible to wintery injury, similar to varieties such as Riesling or Verdelet. Dutchess has been called an uncertain bearer that is susceptible to mildews, particularly powdery mildew in some locations. The vine is self-fertile, so a few vines can be planted in the home garden.

The canes are of intermediate length and number. The vine's blooming date is late. The fruit ripens no earlier than late midseason, but generally by late season. The clusters, when well grown, are medium to large and form rather long, slender, tapering, and cylindrical compact bunches. The round berries are of medium size with a pale yellowish-green color that can range from golden yellow to light amber. The skin is freckled with small black dots. Dutchess can have many small shot berries when grown on vigorous sites. Further, it may not ripen evenly and, even when ripe, the grape bunch is rather hard and does not soften up like most other varieties as they ripen. It has very good storage qualities, though, when shipped as fresh fruit.

Dutchess is best grown on well-drained, moderately fertile soil. Culturally, the grape is capricious to grow, so it should not be grown in rich or heavy soils. Cluster thinning is recommended to enhance crop quality, to advance its ability to ripen fruit, and to facilitate the canes to harden off before the onset of winter.

It seems that this variety was sometimes used in the nineteenth century as a table grape because its long storage capabilities were noted, and its few small seeds separated easily from the pulp. However, it was prized as a wine grape because it was one of the few grown in New York that imparted few or no *labrusca* flavors. While the use of this grape has been generally replaced by the many available French hybrid grapes in the production of neutral white wines, Dutchess should still be cultivated because of its own unique winemaking attributes.

The wine is subtle, delicate, and soft, yet complex. Its mild acid profile contrasts with the many high-acid white wines being produced in the Hudson Valley today. It is clean, reminiscent of pears and melons, and not grapey at all. The wine, which has been described as smoky, has almost no *labrusca* aroma and none of its high acids. However, with age, the wines can acquire a petroleum nose and taste if not aged in oak.

In the early to mid-twentieth century, U. P. Hedrick and others at the Geneva Experiment Station maintained that Dutchess was a variety for the "amateur" and not for commercial vineyards.[21] The Geneva commentators tended to recommend the varieties that they had developed in the 1910s to 1930s and not those that had been developed by hybridizers in the Hudson Valley, such as Bacchus, Dutchess, Iona, Jefferson, or Empire State. However, Dutchess and Iona are still some of the most widely planted grape varieties developed in the Hudson Valley. Other grape breeders thought highly enough about Dutchess that they incorporated it in their own breeding programs. Albert Seibel of France chose Dutchess as one of the parents of Seibel 3037 and Seibel 3074 to improve their wine quality. Dr. J. Stayman of Kansas used this grape as one of the parents of his hybrids White Imperial, White Beauty, and Black's Imperial.

This heirloom grape should be seriously considered for cultivation by commercial growers. Its low acid profile makes it very useful in white blends of Seyval Blanc or Vidal Blanc that have high total acidity or which are harsh in the mouth. Further, Dutchess can add complexity to many white blends.

NECTAR

This was initially named by its originator as Black Delaware, but became known to the public as Nectar about 1880. It was placed on the list of the *American Pomological Society Fruit Catalogue* in 1899. Nectar is said to have been crossed in 1883 and came from a seed of Concord fertilized by Delaware.

PARENTAGE
labrusca, bourquiniana, vinifera?

HARVEST DATE
Mid-season

B B B B B

The vines are vigorous, small-leaved, bear medium-sized black berries on clusters that look like Delaware, which are of good but not high quality for table use. The plants, while vigorous, are not always hardy, usually producing light crops, and are very susceptible to mildew. U. P. Hedrick commented that the vines are nearly worthless because of their susceptibility to mildew, and their resemblance to Delaware is not apparent.[22]

However, the *Bushberg Catalogue* description was more kind, stating that the quality was good, the berries do not crack or fall off the stems and are vinous (that is, good for wine), and have a pure, rich flavor. Also, the vine was a fairly good grower, hardy, productive, and not disposed to mildew.[23] The blooming time is mid-season and it ripens between Delaware and Concord. The clusters and jet-black berries are of medium size, somewhat larger than Delaware.

The clusters are intermediate in size and length, and are irregularly cylindrical to tapering. The grapes are a dull, dark purplish-black and do not hold up well for shipping. This does not seem to be a grape worthy of propagation, but it is noteworthy that A. J. Caywood, who tended to favor Delaware and Concord as parents of his hybrids (the former for quality in wine production, and the latter for disease resistance and productivity), chose to cross these two varieties.

POUGHKEEPSIE

Poughkeepsie, also known as Poughkeepsie Red, is from a seed of Iona fertilized by mixed pollen from Delaware and Walter. The plant was bred in the 1860s, and was known only in the Hudson Valley until it was introduced to the public, presumably by the American Pomological Society, around 1880. It was never popular and was reported to be nearly obsolete by 1910.

PARENTAGE
bourquiniana, labrusca, vinifera

HARVEST
Early mid-season

B+ B C B B

U. P. Hedrick stated that "there is no doubt as to its quality, both as a table grape and for wine; in this respect it is considered by many as equal to the best of our American varieties and quite the equal of some of the finer European sorts. But the vine characters are practically all poor and the variety is thus

Poughkeepsie. *Although prominently featured in the national fruit catalogs of the day, this variety never became popular.*

effectively debarred from common cultivation."[24]

The vine and fruit greatly resemble Delaware, but Poughkeepsie does not equal Delaware's vine or fruit characteristics. Further, it is more susceptible to winter injury than Delaware, of intermediate vigor, and less productive. The bloom time is late, but the grape ripens early, even a bit earlier than Delaware. Poughkeepsie's cluster is attractive and marketable as a table grape. The cluster is of only medium size or smaller, tapering to cylindrical and very compact. The berries are medium to small and roundish. The *Bushberg Catalogue* of 1895 prominently highlights this grape and states that the Poughkeepsie Red's cluster is about one-third larger than Delaware, rather darker red, and the flesh melts like Iona. Further, it is claimed to be valuable as a wine grape, and ripens very early like Hartford,[25] a black grape that typically ripens one to two weeks before Concord.

This grape has not been tested much beyond the Hudson Valley. At the Michigan Experimental Station, South Haven, Poughkeepsie was deemed a failure. However, it may have a place in the home garden due to its historical curiosity and because of its beauty and high quality for table consumption. The variety does have potential for seeded raisin production.[26]

ULSTER

Ulster's parents are reputed to be Catawba pollinated by a wild *aestivalis*. The vine and fruit are reported as showing unmistakable traces of *labrusca* and *vinifera* in its heritage, but the *aestivalis* characteristics are not so apparent. Because of its large berry size and beauty, Ulster was for a while touted as an up-and-coming variety. However, it sets too much fruit, which leads to small clusters. Further, this overproduction stresses the vine to the point that it cannot fully recover to produce fruit the following season.

PARENTAGE
aestivalis, labrusca, vinifera

HARVEST DATE
Late mid-season to late

A A C A B+

Ulster was presented to the American Pomological Society meeting in 1883. It was introduced to the public in 1885, after the introduction of Walter, Modena, and Dutchess, toward the end of A. J. Caywood's life. The grape was added to the list of recommended grape varieties by the American Pomological Society in 1899, ten years after his death.

The vine is moderately hardy to hardy, a weak grower, but productive in yield. Hence, it overbears for the vigorousness of the vine. The glossy leaves are small and thick with an upper surface of light-green and an underside that is grayish-white. It is sometimes susceptible to fungal diseases. From U. P. Hedrick's description of the grape over the years, it can be inferred that he was not impressed with this variety.[27]

However, my experience with Ulster at my farm at Cedar Cliff is that the vine grows well and is disease resistant, but may be shy in its production. This observation was confirmed by J. G. Burrow of Fishkill in 1894, when he reported that the grape grew well and was profitable to cultivate, but needed good culture and a fair amount of fertilizer. In addition, L. R. Taft, the horticulturalist of the Michigan Experiment Station, reported in 1894 that Ulster was high in quality and very productive, but slightly lacked vigor.[28]

When grown optimally, the quality of the fruit is very good, much like Catawba. The fruit ripens late season. Its moderately compact clusters are of middling size, cylindrical to slightly tapering with one large shoulder. The round berries are medium to medium-large. The color of the berry varies from a dark dull red like Catawba to a bluish hue. The skin is thick and tough with pale green flesh. Cluster thinning is required to obtain a balanced crop.

The flowers are self-fertile and are reported to bud and bloom early. The grapes ripen late mid-season to late. The skin is thick, tough, adherent, and astringent tasting. The fruit is similar to Catawba and has been further described as faintly aromatic, slightly foxy, juicy, and tender.

This grape should be considered for future cultivation in the Hudson Valley as an heirloom variety.

WALTER

This red/pink grape is one of A. J. Caywood's first hybrids. Walter has been characterized as having *vinifera*, *labrusca*, and *bourquiniana* parentage by U. P. Hedrick,[29] and having *aestivalis* and *labrusca* parentage by the *Bushberg Catalogue*.[30] The plant comes from a seed of Delaware pollinated by Diana around 1850. It was listed in the *American Pomological Society Fruit Catalogue* in 1871, but was disseminated to growers much earlier than 1871.

PARENTAGE
vinifera, labrusca, bourquiniana or aestivalis, labrusca

HARVEST DATE
Early to mid-season

B B– C A A

The bunch and berry resemble Delaware, while the flavor is similar to Diana, which is tender in texture and has an agreeable spicy flavor. In *The Grapes of New York*, it was said that "the variety has been cultivated for nearly half a century but is seemingly less and less grown, a fact to be regretted, for there are few American grapes of more exquisite flavor and aroma and more dainty appearance."[31] The fruit is attractive like Delaware, and ripens evenly unlike its parent Diana. The grape is highly susceptible to fungal diseases, and if planted at unsuitable sites, fungus can attack its leaves, shoots, and fruit to the point that it can lose all of its leaves by mid-season.

Walter is fastidious as to the soils it can thrive on. Generally, it grows in soils that are best suited for Delaware. The vine is moderately vigorous and a good bearer on the right soils, but is only moderately hardy to winter damage. While the cluster has been called "perfect," its berry size is small to medium and the bunch size is small. It flowers by mid-season, so is generally not susceptible to late frosts.[32] The grape ripens early to mid-season in a manner similar to Delaware, with sugars measuring at 23° Brix. It stores well for table use, but was intended primarily as a wine grape.

Walter was rarely grown even as late as the 1920s, but it is included here because it was one of A. J. Caywood's first hybrids. In addition to U. P. Hedrick's glowing description of Walter's fruit, it seems that Caywood was also proud of this variety, so much so that he distributed it widely to area growers. Further, Caywood seems to have thought highly enough about Walter to name it after his own son, Walter J. Caywood, and to use it as one of the building blocks in his breeding program. It is one of the parents of the varieties Poughkeepsie Red and Mabel, which Caywood also bred.

MINOR GRAPE VARIETIES OF CAYWOOD

CAYWOOD NO. 1

labrusca, vinifera, bourquiniana

This is a red grape obtained from a red seedling of Poughkeepsie fertilized by Iona. It was not named and as of the late 1890s was most likely lost.

CAYWOOD NO. 50

labrusca, vinifera

Developed in 1888, this grape is described as vigorous, healthy, productive with a medium cluster, compact, and often shouldered. The black berry, with abundant bloom, is large, roundish, and it shatters. It is reported to be vinous and ripens a little before Worden.

FLORENCE

labrusca, vinifera, bourquiniana?

Little is known about when this variety, a seedling of Niagara pollinated by Dutchess, originated. The white fruit ripens slightly earlier than Niagara and does not keep well. The clusters are above medium in size and rather long to medium and loose. Berries are medium to large, roundish, and green, often with a tinge of yellow, covered with a thin gray bloom. The flavor is said to be aromatic and agreeably sweet.

HUDSON

labrusca, riparia, vinifera

This is a seedling of Rebecca crossed with Clinton around 1870. The goal was to combine the fine quality of the tender seed parent Rebecca with the vigor of Clinton, which is a *V. riparia* wine grape of great strength. Hudson was a white grape that was neither very luscious nor very hardy or productive. Caywood refrained from introducing this grape, but sent it out for testing to several people, including J. W. Prentiss, who introduced the grape Prentiss. This grape seems to be very similar or identical to Prentiss.

LITTLE BLUE

labrusca, vinifera, aestivalis

Developed about 1888, it is vigorous, medium in productiveness, and has a cluster size that is shouldered. The berry is medium, oblong, and black, and ripens after Concord. The flowers are partially fertile with upright stamens.

MABEL

labrusca, bourquiniana, vinifera[33] *or aestivalis-hybrid*[34]

Originated from a seed of Walter, Mabel's vine is a moderately vigorous grower, not always hardy, inclined to mildew, but productive to moderately productive. It is a mediocre grape of medium qualities with regard to the size of the berry and cluster. It is a large, round black/purple grape with a blue bloom. It ripens before Concord, but is not a good keeper.

METTERNICH

riparia, labrusca, vinifera, bourquiniana

It has been alternatively reported as being a seedling of Poughkeepsie crossed with Clinton[35] and as a seedling of Clinton fertilized by Poughkeepsie.[36] The cluster is small to medium and compact; the medium-sized berry is light green to medium-dark red and translucent. The grape had been reported to ripen either early or late, and is moderately productive. It is interesting to note that the reports in *The Grapes of New York* and the *Bushberg Catalogue* completely contradict each other.

MODENA

labrusca

Little is known about the genealogy of this grape, but it may be a Concord seedling developed about 1867. It is vigorous and hardy, with a medium-sized bunch and berry that is roundish, black, and similar to Concord in flavor and ripening time. It is perhaps the same as Mingo, and is reported to make a very dark-colored wine.[37]

WHITE CONCORD

According to the *Bushberg Catalogue*, Walter Caywood said that White Concord, crossed with Delaware, was the parent of Dutchess. This is a very important grape variety that seems to have been used extensively in Caywood's breeding program. Perhaps there was more than one White Concord seedling, for it is not mentioned in *The Grapes of New York*, except for noting that it is one of the parents of Dutchess.[38]

WHITE ULSTER

labrusca, vinifera

An "amateur" variety from a seedling of Caywood's Ulster Prolific (an Ulster seedling) crossed with White Concord.[39]

Dr. William A. M. Culbert

(1822–1890)

Dr. William A. M. Culbert came from a wealthy New York City family. His father, John Culbert, settled in New York soon after the Revolutionary War, and was engaged in mercantile pursuits for almost fifty years. William Culbert was educated at elite private schools, and graduated from the Academic Department of the University of the City of New York with a bachelor of arts degree in 1841, and received a master's degree three years later. In 1846, Dr. Culbert received his doctor of medicine degree from the same university. Early in his professional career, he espoused the principles of homeopathy and remained a strong proponent of homeopathic medicine for the rest of his life.

Shortly after his graduation, he was encouraged by his friend, A. Gerald Hull, MD, a physician from New York City, to move to Newburgh. Dr. Hull had left the practice of medicine earlier, and settled in Newburgh to recover his health, which had been impaired due to the stress of his busy practice. Dr. Culbert moved to Newburgh in November 1847, and established a practice that grew rapidly and was very lucrative. Dr. Culbert treated many of the leading families in the area, and in 1852, he married Henrietta Powell, the daughter of Robert and Louisa A. Powell, a prominent Newburgh family. The couple had one son, Francis Ramsdell Culbert.[40] Around 1855, the Culberts moved into a large, custom-built home on a corner lot at 120 Grand Street, the main tree-lined street where many of Newburgh's leading families had stately homes. Grand Street ran parallel to the Hudson River, so there are spectacular views of the river and Hudson Highlands.

The house was commissioned and designed by the local landscape architects A. J. Downing and Calvert Vaux, who were becoming nationally known for their cottage and villa-style of construction. The two-and-a-half story brick structure is trimmed with sandstone and has a curved mansard roof. Dr. Culbert's medical office was on the lower floor and the upper floor was the residence. This home eventually became the Newburgh City Club. (Tragically, the building was ravaged by fire in December 1981, and remains in a state of disrepair, however, it is on the National Register of Historic Home, so it is protected until it can one day be restored.)

After many years of practicing medicine, Dr. Culbert's health eventually began to fail, so in 1870 and 1871 he traveled to Europe on trips that lasted several months. It is reported that he visited medical institutions in the countries that he passed through.

Unlike many leading businessmen of Newburgh, though, Dr. Culbert did not actively engage in the city's political or social life. Instead, for twenty-five years, his sole recreational pursuit was horticulture. He was an enthusiastic cultivator and propagator of new seedling grapes at his vineyard, which was located north of the city, probably in Balmville. There was scarcely a day in which he did not visit his vineyard. Horticulture was a relaxing pursuit away from his very active medical practice, and he was awarded several prizes for grapes at the Orange County Fair.[41]

Newburgh City Club, formerly the Culbert House. *The Culbert's custom-built residence, a rare 1851–52 design by the short-lived partnership of Calvert Vaux and A. J. Downing, remains standing on Grand Street in Newburgh's Historic District.*

In 1890, he died at his home in Newburgh of Bright's disease and from long-term complications from a broken bone in his leg that limited his physical activities.[42] In his obituary, published in *The New York Times*,[43] he was referred to as "one of the most prominent homeopathic physicians" in New York State, but there was no mention of his work as a grape breeder. All of his hybrids seemed to originate after 1870, when it appears that he began to reduce his medical practice and travel to Europe.

None of Dr. Culbert's hybrids were ever widely planted. However, it is important to catalog his work to understand his approach to hybridization and to note his communications with other local hybridizers. Culbert's breeding program consisted of using high-quality *labrusca* hybrids, such as Delaware or Iona, as the seed parent crossed with the pollen of *vinifera* varieties such as Muscat Hamburg. James H. Ricketts, his neighbor and colleague who lived nearby on City Terrace in Newburgh, also used this same breeding technique. In surveying Ricketts's program, the more successful breeds were *labrusca* × *labrusca* or *labrusca* × *riparia* hybrids.

MINOR GRAPE VARIETIES OF CULBERT

CULBERT SEEDLING

labrusca, vinifera

The grape is a cross of Iona and Muscat Hamburg. The bunch and purplish-black berry are large.

DELAWARE SEEDLING

vinifera, bourquiniana, labrusca

Known as Delaware seedling No. 4, this is a seedling of Delaware crossed with General Marmora (a *vinifera* grape often used for breeding). It is a more vigorous grower than Delaware, hardy, and prolific.

EARLY DAWN

labrusca, vinifera, aestivalis

This grape originated sometime around 1870 and is reported to come from a seed of Israella fertilized by Black Hamburg or perhaps Muscat Hamburg. It is a black *labrusca-vinifera* hybrid of good quality and appearance, but lacks good vigor and productiveness. The grape was described as tender, juicy, sweet, slightly vinous, rich, and of very good quality. The vine is not very hardy. The fruit ripens early—about two weeks earlier than Concord or a week before Hartford—and keeps and stores well. Clusters are medium to large, nearly cylindrical, and irregular in outline. Berries are medium in size, roundish, with an attractive purplish-black color. This grape was never widely grown in the Hudson Valley.

GOLDEN BERRY

vinifera, labrusca

Also known as Culbert No. 5, this white seedling of Hartford and General Marmora is reported to be a hardy and free bearer. It was exhibited as a new fruit at the American Pomological Society in 1877.

NEWBURGH MUSCAT

labrusca, vinifera

Also known as Culbert's No. 3, White Moline, and White Muscat of Newburgh, this seedling of Hartford crossed with Iona is a vigorous and hardy variety that is reportedly not productive. Clusters are medium in size, loose, short, and often single-shouldered. The berries are medium, round, and pale yellowish-green. The grapes shatter badly. The flesh is soft and tender with a mild musky/muscat or rather toned-down foxiness in its aroma and taste. It was exhibited in 1877 at the American Pomological Society.

PURPLE BLOOM

labrusca, hybrid

This is a seedling of Hartford crossed with General Marmora. The vine is hardy, vigorous, and prolific. Its bunches are large and showy, and the berry size is fair. One commentator suggested that it was well adapted for a good market. It was exhibited in 1877, but was not disseminated.

SILVER-DAWN

labrusca, vinifera, aestivalis

A seedling of Israella fertilized by Muscat Hamburg and a brother of Early Dawn, it is reported to be a fine white grape of the best quality, vine-hardy, and vigorous.

Dr. Charles William Grant

(1810–1881)

Dr. Charles William Grant was born in Litchfield, Connecticut, and attended the Female Academy in 1828. He later moved to New York City and practiced medicine in New York City and Peekskill. Eventually he became dissatisfied with medicine and switched to dentistry. In 1840, he settled in Newburgh, where he built up a very large dental practice that lasted from 1840 to 1858. In 1843, he married Jane Forsyth Beveridge (1825–56) and they had three children: Anna Margaret (1847–1923), John Beveridge (1850–1912), and Grace Jane (1855–60).[44] In 1849, he purchased a part interest in Iona Island, also known as Salisbury Island, from his father-in-law, John Beveridge, who had purchased the island sometime before that date.[45]

Iona Island is located near the west shore of the Hudson River on the border of Orange and Rockland counties. The island is surrounded by wetlands called Salisbury and Ring Meadows, which are two large tidal marshes, and the mouth of the Doodletown Bight and Doodletown Brook. The bedrock of the island is made of very erosion-resistant Precambrian gneiss rock, which forms rocky knobs or cliffs that rise approximately 100 feet above the Hudson River. The island is approximately 118 acres and is surrounded to the west by about 200 acres of tidal meadows. This marsh is on the former site of an ancient horseshoe bend of the Hudson River that was bypassed by the last glacier that came through and straightened the river channel to some extent.

Of particular interest to my family, and perhaps to our readers, the Casscles family resided just south of Iona Island along the west shore of the Hudson River well before the Civil War. A map of the Iona Island area in 1854 shows the dock of J. Cassels on the Hudson River.[46] It was there and in nearby brickyards that my ancestors manufactured brick for shipment to New York City. Our family cemetery is still there along Route 9-W and so is the Episcopal church, the Chapel of St. John the Divine, in Tomkins Cove, which my family helped to establish in 1871.[47] This church continues to hold services each week, and some members of the congregation are related to the Casscles family.

Dr. Grant was an active horticulturalist who associated with other prominent men—including Charles

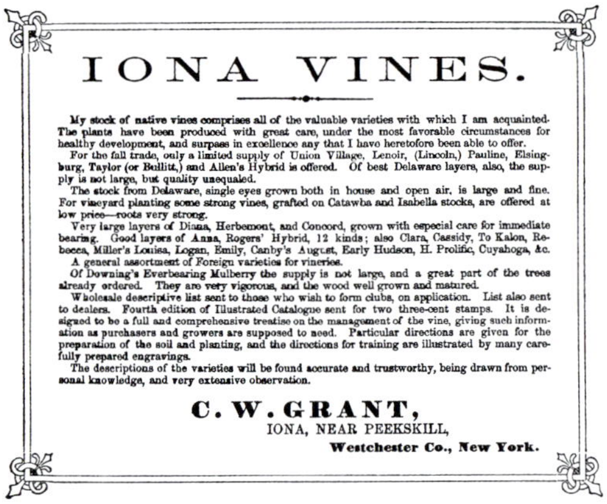

IONA VINES.

My stock of native vines comprises all of the valuable varieties with which I am acquainted. The plants have been produced with great care, under the most favorable circumstances for healthy development, and surpass in excellence any that I have heretofore been able to offer.

For the fall trade, only a limited supply of Union Village, Lenoir, (Lincoln,) Pauline, Elsingburg, Taylor (or Bullitt,) and Allen's Hybrid is offered. Of best Delaware layers, also, the supply is not large, but quality unequaled.

The stock from Delaware, single eyes grown both in house and open air, is large and fine. For vineyard planting some strong vines, grafted on Catawba and Isabella stocks, are offered at low price—roots very strong.

Very large layers of Diana, Herbemont, and Concord, grown with especial care for immediate bearing. Good layers of Anna, Rogers' Hybrid, 12 kinds; also Clara, Cassidy, To Kalon, Rebecca, Miller's Louisa, Logan, Emily, Canby's August, Early Hudson, H. Prolific, Cuyahoga, &c.

A general assortment of Foreign varieties for vineries.

Of Downing's Everbearing Mulberry the supply is not large, and a great part of the trees already ordered. They are very vigorous, and the wood well grown and matured.

Wholesale descriptive list sent to those who wish to form clubs, on application. List also sent to dealers. Fourth edition of Illustrated Catalogue sent for two three-cent stamps. It is designed to be a full and comprehensive treatise on the management of the vine, giving such information as purchasers and growers are supposed to need. Particular directions are given for the preparation of the soil and planting, and the directions for training are illustrated by many carefully prepared engravings.

The descriptions of the varieties will be found accurate and trustworthy, being drawn from personal knowledge, and very extensive observation.

C. W. GRANT,
IONA, NEAR PEEKSKILL,
Westchester Co., New York.

Iona vines. *Advertisement by C. W. Grant, another mid-Hudson Valley physician-turned-viticulturalist, who pursued horticulture for profit and enjoyment.*

Downing, Horace Greeley, Henry Ward Beecher, W. C. Bryant, and Donald G. Mitchell—who had similar interests in horticulture and other rural pursuits.[48] After the death of his young wife in 1856, at the age of 31, Dr. Grant, at the age of forty-six, gave up his dental practice and moved to a spacious mansion 15 miles south of Newburgh to Iona Island on the Hudson River in 1858. In 1859, he married Isabella (also known as Israella) Hasbrouck Proudfit. For twelve years, he lived on Iona Island, where he cultivated his vineyard and orchard, operated several greenhouses, identified and bred new grape varieties, and conducted a grape nursery business.[49] His catalogue, the *Illustrated Catalogue of Vines—With Explanatory Remarks and Indications of Cultivation*, is available at several regional libraries.[50]

Iona Island is a very beautiful island located in a European-like fiord setting. Dr. Grant was an active grape breeder who worked especially hard to improve the quality of commercial grape varieties. His grape and orchard business was particularly active during the Civil War period, but his commercial nursery did not seem to be very profitable. However, Dr. Grant's name is linked to the introduction of the Iona, Israella, Eumelan, and Anna grape varieties.

Some commentators have suggested that while Dr. Grant's grape varieties were widely advertised, the descriptions were much more glowing than their actual performance in the field.[51] By 1868 or 1869, after failing to make a commercial success of his enterprise, he retired from the active pursuits of horticulture. During the last part of his life, he lived in the home of his brother John in Litchfield, where he died in 1881. After selling Iona Island, the property first became an amusement park and then was used as a U.S. Navy munitions depot from 1899 to 1951.

In 1965, the island was transferred to the Palisades Interstate Park Commission to be used for public recreation and as a bird sanctuary. Some of the U.S. Navy's brick military munitions buildings remain on the island. These buildings are still used today to store building supplies and other materials for the Park Commission. There is a marker commemorating the island as a site of a large resort hotel and commercial vineyard during the Civil War era.

EUMELAN

This grape is a chance seedling that grew in the yard of Mr. Thorne of Fishkill Landing (which is now part of the City of Beacon), around 1847. Around 1860, this variety found its way to Dr. Grant, who introduced the vine to the public in 1867 (see page 120). Some grape classifiers maintain that Eumelan is a seedling of Isabella.[52] The *Bushberg Catalogue* designates this variety as an *aestivalis*,[53] while T. V. Munson states that the grape has nothing in it other than *labrusca* and *vinifera*.[54] U. P. Hedrick believes that the *labrusca* is evidenced by the texture of the fruit and seeds; *vinifera* possibly by its appearance and tendency to mildew; and *aestivalis* or *bicolor* due to its spicy taste, difficulty to propagate from cuttings, a bluish bloom on its shoots and canes, and pigment beneath the skin.[55]

PARENTAGE
labrusca, vinifera, aestivalis

HARVEST DATE
Early to mid-season

A+ A− B A+ B+

Hedrick wrote in 1908 that "about forty years ago, the general opinion among leading viticulturalists of the time [was] that it was one of the best black grapes that had been brought to the notice of grape-growers." He added that it has a "great a number of valuable good qualities and as few objectionable ones as almost any other of our black grapes, yet the variety is now but little grown." From this description, it is clear that Hedrick liked this grape, had studied it extensively, and believed that it made a very good red wine.[56] I grow this grape and have found it to be a solid variety in the field, but have not yet had sufficient time to fully evaluate its wine quality, although first impressions are good.

The vine is vigorous, very winter hardy, and very productive. Its clusters and berries are well formed and of good size. Eumelan is suitable for more northern climates such as Michigan, Minnesota, and upstate New York above the Mohawk River. Further, the round, medium-sized grapes are a very attractive black to blue in color with a fine bloom. The cluster is loose, large, and rather long and slender, slightly tapering to cylindrical with one shoulder. While an earlier description of the cluster by Hedrick was favorable, in his 1919 book, *Manual of American Grape Growing*, he classified the grapes as being good only for amateurs.[57] The flavor of the grape has no foxiness and is "rich, sweet, and vinous," which makes a very good and rich wine. The sugars can reach 24° Brix with low acid. Eumelan ripens early, but since it holds on the vine well and does not crack or shell, it can be harvested mid-season to late mid-season. Some of its disadvantages are its susceptibility to fungal

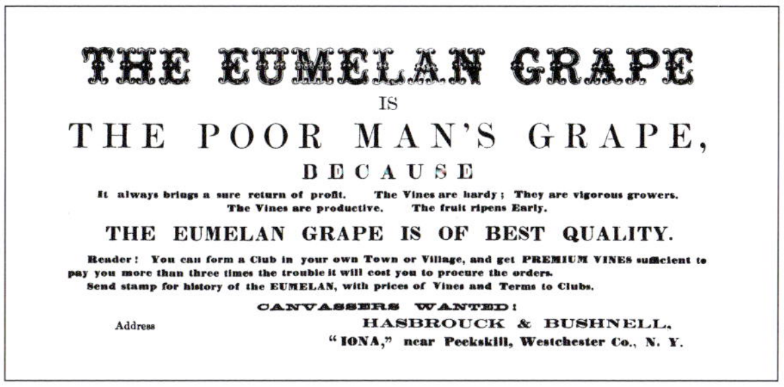

Eumelan grape. *A Hudson Valley–developed wine grape that was used in other French and American wine grape breeding programs.*

diseases, sterile female flowers, and difficulty to propagate, which may be one reason the grape was not widely planted by commercial growers. The flowers bloom late, so a compatible grape pollinator with late flowers needs to be planted for the vine to fruit.

Most horticultural commentators during the last quarter of the nineteenth century and Lucie Morton in her writings in the late twentieth century seemed to like Eumelan and reported on the grape extensively, lamenting that it was not more widely planted. My own observations of growing Eumelan at my farm at Cedar Cliff are that it is vigorous, winter hardy, and very productive. Further, it can produce relatively complex flinty red wines with good body and fruit in the nose and taste for an heirloom grape variety. Eumelan was used in other grape-breeding programs, so it must have been deemed to be a grape variety that had positive attributes that breeders wanted to impart on even newer grape breeds. T. V. Munson bred a seedling of Eumelan with Delaware to create Eumedel.

D. S. Marvin, a grape breeder from Watertown, New York, who worked closely with T. V. Munson, used Eumelan extensively in his work when he was breeding grapes in the 1870s and 1880s. Eumelan was most likely used because of its winter hardiness in the cold Adirondack climate and because of its early ripening ability. Some of Marvin's named varieties that used Eumelan include: Centennial, a cross of a *labrusca* grape and a Eumelan seedling, first fruited in 1875; Laura, a cross of Eumelan with Delaware, originated in 1880; and Corporal, a cross of Eumelan and Worden, was reported in the American Pomological Society's *Report on New Fruits* in 1881.[58]

Due to my own favorable observations, the significant body of work that was reported on Eumelan, and the interest that other breeders had in this grape variety, Eumelan should be reconsidered for new plantings in the Hudson Valley as an heirloom variety, as long as a suitable pollinator grape is planted with it since it is not self-fertile. Further, it should be considered in future breeding work done in New York, particularly in developing more cold-hardy varieties that are capable of producing quality wine.

IONA

Iona is one of the most important grape varieties developed in the Hudson Valley and was widely grown in the 1880s until the 1920s. It was a wine grape used for the production of sparkling wines and quality white/blush wines in the latter part of the nineteenth century. The grape is still grown to some extent in the eastern United States and interest in this variety is increasing. Since it was regarded in the late nineteenth century as a quality grape, it was used extensively in many grape-breeding programs at that time and into the twentieth century. It is a parent of many subsequently developed hybrids such as Jefferson and Moore's Diamond (Diamond), which then begat varieties such as Alden, and many Geneva Experiment Station hybrids such as Buffalo, Cayuga White, Horizon, Melody, Schuyler, and Steuben.

PARENTAGE
labrusca, vinifera

HARVEST DATE
Late mid-season to late

A A C A A

The variety originated on Iona Island, in Rockland County. Dr. Grant maintained that Iona is from a seed of Diana planted in 1855, which fruited four years later. A. J. Caywood, however, says that Grant informed him that it was found growing as a chance seedling under a Catawba plant. Since Diana is a seedling of Catawba, it would be difficult to differentiate the characteristics of the parents to determine Iona's real parents.[59] During the Civil War, the variety was awarded the Horace Greeley prize for new grape varieties of merit that may be adapted to the general cultivation in the United States and which had superior vine characteristics and produced quality grapes. Unfortunately, Iona did not live up to its reputation, so Grant returned the prize.

Iona was introduced in 1864 and placed in the *American Pomological Society Fruit Catalogue* in 1867. However, the grape may not have been received as well as the developer had hoped because he exaggerated the grape's ability to flourish in the field.

U. P. Hedrick maintained that it was not a popular grape with commercial growers in the late nineteenth century because it required more care than growers could provide. Further, Hedrick said that Iona was particular about where it grows, wanting a warm soil that is deep, dry, sandy, or gravelly clays and did not do well in damp, rich black soils, or on poor sand or gravelly soils. The vine is not particularly winter hardy for a *labrusca* so it should be planted on a sheltered site. It is not vigorous or particularly productive, so it requires close pruning to avoid overcropping. Iona is susceptible to fungal diseases and somewhat cold tender for a *labrusca* grape.[60]

I have found this grape easy to grow, with moderate fungal disease resistance and ability to provide an ample crop even in heavy soils.

The vine blossoms by late season. Its fruit ripens late mid-season to late, later than Concord. The berries have a unique pale to dark pink/red hue with some amethyst and watermelon shades that are not uniform. The cluster is medium-large to large, with rather loose large berries of varying size that can ripen unevenly. It is a straggly cluster that is not suitable as a table grape due to its appearance. The cluster is double-shouldered and nearly cylindrical. Its sugars range between 18° and 22° Brix.

Despite the negative cultivation aspects of Iona, it does have a delightful and soft *labrusca* taste that carries through in the wines made from this grape. It is of similar quality to Delaware, but much softer, rounder, and approachable, with elements of guava and watermelon and none of the mustiness of Delaware. It is a very perfumey and welcoming grape that makes a refreshing white wine. This grape seems to be good in the field and is a fine heirloom grape that should be considered for wine production in the Hudson Valley.

MINOR GRAPE VARIETIES OF GRANT

ANNA

labrusca, vinifera

This seedling of Catawba was obtained from Eli Hasbrouck of Newburgh. (The Hasbrouck's would become the in-laws of Dr. Grant's daughter.) The seedling fruited in 1851 and was later introduced by Dr. Grant. It resembles Catawba in appearance of vine, but is unhealthy and feeble; the bunches are medium, loose; the berries are medium in size, pale amber in color, and ripen late with Catawba.

ISRAELLA

labrusca, vinifera

This was grown from a seed of Isabella in 1855. In 1859 or 1860, Peter B. Mead, editor of *The Horticulturalist*, selected this plant from several thousand seedlings and named it in honor of Dr. Grant's second wife. Israella was listed in the *American Pomological Society Fruit Catalogue* in 1867, but was dropped in 1881.[61] The grape was widely tried, but discarded due to its poor quality and unattractive appearance as a table grape. The grape was never commonly grown, and was rarely found at all after 1900. The vine is of medium vigor, generally hardy, but not productive. The fruit ripens a little later than Concord, and its sometimes unattractive clusters are above average in size, strongly tapering, single-shouldered, and compact. Berries are small to medium in size, roundish to oval, and black to purple-black in color. The skin is thick, and the flavor is not as good as Concord, fair at best.

MARION

riparia, labrusca

This variety was promoted by Samuel Miller of Pennsylvania who obtained it from Dr. Grant. It was noted in the 1880s for its dark red wine. The cluster is medium and compact, the berry larger than Clinton. It blooms early, but ripens early-late. Marion is a very vigorous vine, probably due to its *V. riparia* heritage, that is productive and resistant to fungal diseases. It has strawberry flavors and does not have a heavy foxy taste. Further, it is a *teinturier* grape that can produce very dark red colored wines that are used to enhance the color of lightly colored red wines. It was reputedly grown in Bordeaux, France, in the 1880s as a coloring grape and for the production of neutral wine.[62]

James H. Ricketts

(1818/1830–1915)

James H. Ricketts was born in Oldbridge, Middlesex County, Massachusetts, in 1830.[63] However, Ricketts's obituary, published in Washington, D. C.'s *Washington Evening Star* on December 1, 1915, states that he was born in 1818 in Old Bruige, New Jersey (probably Old Bridge, New Jersey), and died at the ripe old age of ninety-seven.[64] While James was a child, the Ricketts family moved to Greencastle, Indiana, where DePauw University is now located. As a young man, Ricketts learned bookbinding and later practiced his trade in New York City, Newburgh, New York, and Washington, D.C.

In 1857, he established a bookbinding business in Newburgh. After 1870, the City of Newburgh City Directory notes that his home address was first at Miller Street, then in 1883 he moved to City Terrace (near Third Street), near the southeastern corner of what is now Downing Park, and then back to S. Miller Street in 1889 just before he left Newburgh in 1890 for Washington, D.C. His bookbinding business was located first at 48 Second Street, and then after 1880, the business moved to 102 Water Street on the northwest corner of Water and Second streets in Newburgh.[65] His house and business was no further than nine blocks from the nursery business of Charles Downing, and six blocks from the residence of Dr. William A. M. Culbert on Grand Street. He did bookbinding work for the West Point Military Academy Library and the register books for the use of visitors at Washington's Headquarters in Newburgh.

James H. Ricketts was married to Ruth M. Ricketts, who died at the age of seventy-one on February 10, 1904. The couple had been married approximately fifty-four years. The Ricketts had five children, sons Eugene K. and Edmond L., and daughter Josephine Painter, who survived their father's death, but his other two children, Virginia A. Ricketts and William M. Ricketts, died before he did.

In Newburgh, Ricketts became interested in raising fruit, and devoted as much time as he could spare from his bookbinding business to pursue his new avocation. In 1861, Ricketts began his work on grape improvement, reading all the books he could on this subject to learn how to breed grapes. In 1862, he built a glass greenhouse to cultivate *vinifera* grape varieties to be crossed outside of the greenhouse with native grape varieties. He also seems to have had an extensive friendship and working relationship with Charles Downing and other local horticulturalists.

A Prolific Breeder

The first grape variety Ricketts introduced was Raritan, a seedling of Delaware crossed with Concord. Of the Hudson Valley hybridizers, Ricketts was more inclined to use *vinifera* varieties as one of the parents of his hybrids. His first hybrid of *vinifera* × native grape was Downing, which was named after his friend, neighbor, and fellow horticulturalist Charles Downing.

While in Newburgh, Ricketts produced hundreds of seedlings. For ten to twelve years, he exhibited them at various fairs, horticultural society meetings, and other places where their "magnificent appearance and fine flavor" attracted favorable attention and won him many medals and prizes.[66] Overall, Ricketts's varieties are known for their large clusters and berries and for their high quality. Because he bred *vinifera* into his new varieties, while the quality was excellent, some of his vines were susceptible to mildew and other *vinifera* weaknesses such as cold tenderness.

During this time, Ricketts became a nationally known hybridizer who was noted for breeding a large number of grape varieties. In the mid-1870s Charles Downing, Patrick Barry, and John J. Thomas had appointed a special committee to examine the more than 100 new varieties that Ricketts had allegedly produced and claimed to have bred. The special committee came back with a glowing report on these new varieties: "We may state, in a general way, that so great has been his success that his collection of nearly a hundred new grapes, in bearing, can scarcely fail to excite little less than astonishment at the results." His experiments were performed on less than an acre of land, the soil of which was a medium loam, with some manure, and thoroughly cultivated. The hills surrounding the vineyard provided some shelter from the cold west winds. The vines were all laid down in the winter, but not covered so that they could produce fruit again in the spring.[67]

Ricketts grape varieties were generally evaluated in his day as beautiful and of good size. Growers of the day maintained that his selections were susceptible to mildew and winter injury, and hence did not grow well for them. This is why his hybrids were not widely commercially grown. However, my own observations of his grape varieties Bacchus, Downing, Empire State, and Jefferson are much more favorable. They all have good fungal disease resistance and are winter hardy and very productive overall.

Ricketts continued to produce these prize-winning grape clusters for public showings at forums such as the American Pomological Society. Perhaps he was an even more skilled cultivator of grapes than a breeder and could have greatly advanced the practice of viticulture if he had passed on those cultivation skills or techniques to other growers. However, the secret of his cultural success was never discovered. For his efforts, the American Pomological Society repeatedly awarded him its Wilder Silver Medal and hundreds of premiums from horticultural societies from across the country.

In reviewing how Ricketts's named varieties were bred, a few things stand out. First, the seed parent was either a resilient disease-resistant *labrusca* variety such as Concord or a *riparia* like Clinton. Second, to impart quality to his hybrids, he tended to use either *vinifera* (Black Hamburg), *vinifera* hybrids, or quality *labrusca* varieties such as Iona. Third, he used Clinton as the seed parent and pollen parent interchangeably to see what results he could get by crossing Clinton with *labruscas* and *viniferas*.

Ricketts, as well as many other grape hybridizers of the day used Black Hamburg and Clinton extensively in their breeding programs. Black Hamburg was used to breed in quality. Clinton, a *riparia*, was used because it had high fungal disease resistance and winter hardiness, but it also made a solid wine. (Descriptions of these grapes are on page 102.)

Later Years

In Ricketts's rather long obituary in the Washington Evening Star it describes him as a ninety-seven-year-old veteran bookbinder and grape culturalist, whose gray beard, thirty-eight inches long and white hair to his shoulders distinguished him on Washington's streets. He experimented over many years at his extensive vineyards in Newburgh, introducing during that time more than seventy new grape varieties, one of which was an unspecified grape that became well known as a wine grape. Further, his specialty was hybridizing different species, crossing them while in bloom. His work has been extensively reviewed in publications issued by the U.S. Department of Agriculture.

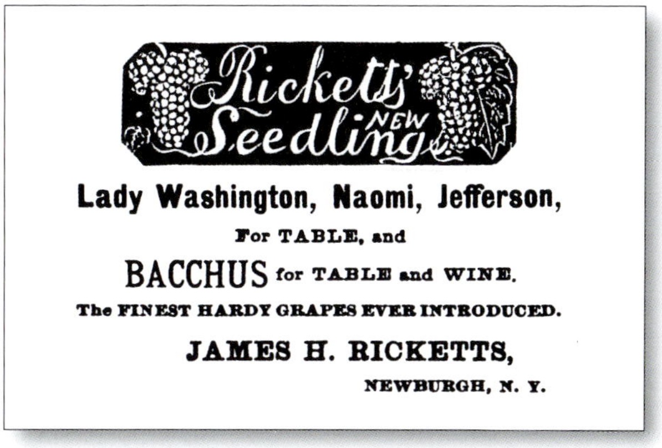

Ricketts' Seedlings. *Advertisement for popular table and wine grapes from James Ricketts, Newburgh, New York. Circa 1878.*

The biography provided by U. P. Hedrick in *The Grapes of New York*, however, conflicts with James Ricketts's obituary. Hedrick maintains that, like many other American grape breeders, Ricketts fell into financial difficulties and in 1877 lost his vineyard and home to foreclosure. Eleven years later, in 1888, he moved to Washington, D.C., to work at his bookbinding trade and began improving grapes and growing a number of new varieties that may or may not have been shown to the public.[68] The obituary maintains that Ricketts moved to Washington, D.C., in 1880 to accept a position in the Government Printing Office, which he retained until 1910. (The information in obituaries in the late nineteenth century, however, was often inaccurate.) In the end, Ricketts probably left Newburgh for Washington, D.C., sometime between 1888 and 1890, and he seems to have had a successful and happy life there.

While in Washington, D.C., Mr. and Mrs. Ricketts lived at 628 or perhaps 638 G Street, SE.[69] Behind his home on G Street, he established a 50 by 100 foot vineyard and vegetable garden, where he devoted the last years of his life to horticulture. He seems to have been in good health until he retired one evening after reading the evening papers and died.

BLACK HAMBURG

PARENTAGE
vinifera

HARVEST DATE
Mid-season

C C B A+ A

Black Hamburg was used extensively by Ricketts and other American breeders in the nineteenth century and by German breeders in the early twentieth century. It is a pure *vinifera* variety that from the 1830s to early 1900s was grown in greenhouses in the United States. Such greenhouses were often lean-to glass houses attached to a building, such as a barn, and generally between 15 and 20 feet wide and from 50 to 100 feet long, either heated or unheated. Heat was supplied by a smoke flue from a wood stove that ran the length of the structure.[70] Even in unheated greenhouses, the environment was sheltered and warmer than in the vineyards, so the survival rate of grapes like Bacchus was higher than for grapes grown outside.

Black Hamburg's virtues as a table grape are its large cluster and berry size, well-formed and uniform clusters, high sugar, low acid, fleshy texture, and non-foxy taste. The pulp and skin are tender and the pulp separates from its seeds. Further, it is easy to cultivate, highly productive, and winter hardy for a *vinifera*. In Britain and the Benelux countries, the variety is known as Black Hamburg and used as a table grape. In Württemberg, it is called Trollinger and is the primary red wine grape used in that significant red wine-growing region of Germany. In Italy, it is known as Schiava Grossa or Vernatsch in the Tyrol region of Italy. Hence, it is a dual-purpose grape that can produce wine and be used as a table grape.

CLINTON

PARENTAGE
riparia, labrusca

HARVEST DATE
Mid-season

A+ A A A B

It is important to note that while Clinton is not a seedling of Ricketts, it is included here because he used it extensively in his grape breeding program. Clinton existed before 1835 and is probably a chance seedling that may have come from the garden of Mr. Peebles above Waterford on the Hudson River. It then came to the attention of L. B. Langworthy, who introduced it as Clinton in the Rochester area. In 1852, the Ohio Pomological Society determined that Clinton was identical to the grape Worthington, which was an old variety known to the nurserymen Adlum and Prince and originated around Annapolis, Maryland.[71]

Clinton is a *V. riparia* hybrid that may have some *labrusca* characteristics, but in the end it looks and grows like other *riparia* grapes such as Baco Noir. It is known for its wild vigorous growth, very good winter hardiness, disease resistance, fruitfulness, high productivity, and early budding and coloring. It ripens by mid-season. As with other *riparia*, its acid and sugars are both high.

The wines are deeply colored. Further, Clinton's high acids and muted *labrusca*-tasting wines do mellow with age. The sugars are between 22° and 23° Brix.[72]

The best known of Ricketts's Clinton-based hybrid varieties are Advance, Bacchus, and Empire State.

BACCHUS

Bacchus is one of the few important red wine grapes developed in the Hudson Valley in the third quarter of the nineteenth century. It was not a dual-use grape suitable for both the table and wine production, however, it was one of the most desirable grapes bred for making a dark red wine with a dark brownish-red color and great body. Others used Bacchus to produce sparkling wines. The sugars are between 22° and 26° Brix.

PARENTAGE
riparia, *labrusca*

HARVEST DATE
Late

A+ A A A+ B

Bacchus is a seedling of Clinton, first exhibited at the American Pomological Society in 1879. It resembles Clinton in vine and leaf characteristics, but it surpasses Clinton in fruit quality and productivity. While not as popular as Concord or Delaware, Bacchus was a commonly cultivated second-tier wine grape grown in New York and the East Coast.

As one of a Clinton group of *riparia* hybrids, Bacchus is vigorous to very vigorous, healthy, productive, winter hardy, and free from fungal diseases and insect attacks. The grape adapts to a wide variety of soils, but does not thrive in droughty soils or soils that contain much lime. It blooms and flowers very early to early, like many other *riparia* grapes of which Clinton is a member. However, it ripens late in the season, but can still be generally grown in the Hudson Valley. The clusters are small to medium, below average in length, slender, often have a single shoulder, and compact. The thin-skinned berries are slightly smaller than medium size, black in color, and glossy in texture. Basically, it has the cluster and berry size of a grape that is between Maréchal Foch and Baco Noir.

Its virtues include resistance to cold, resistance to phylloxera, freedom from fungal diseases and insects, ease of propagation, high productivity, and capability of bearing grafts. It is a grape variety that should be of interest to plant breeders due to its many virtues in the field. This grape may have had a bright future for wine production in the early twentieth century, but its introduction was too close to the beginning of Prohibition when commercial wine production was outlawed in the United States. This may have curtailed its popularity because it clearly had few applications other than as a wine grape.

The wines that Bacchus produces are big fruity and flinty wines with muted *labrusca*-type flavors that have some potential for aging. Barrel aging does help the quality of this wine. The future of this grape should be reevaluated for the Hudson Valley.

DOWNING

Downing was one of Ricketts's first hybrids (known as Ricketts' No. 1), which he developed sometime around 1865. It was first disseminated to local growers around 1870. As with many of Ricketts's varieties, it was a grape of high quality, very handsome in appearance, and excellent keeper in storage. Further, the vine is moderately vigorous and healthy, but must not be allowed to overbear.

PARENTAGE
vinifera, *aestivalis*, *labrusca*

HARVEST DATE
Late

B B C B B+

U. P. Hedrick states that the parentage was variously given as Isabella fertilized by Muscat Hamburg, Croton fertilized by Black Hamburg, and Israella fertilized by Muscat Hamburg. J. G. Burrows of Fishkill, who was associated with Ricketts and his work, maintains that Downing is Israella fertilized by Muscat Hamburg. However, Ricketts himself stated after some review that Downing is from a seed of Concord fertilized by Muscat Hamburg. Hedrick noted that if this is true, it is difficult to account for its apparent *aestivalis* characteristics.[73] The variety, introduced to the public in 1883, was named after his friend, neighbor, and colleague Charles Downing, a nationally prominent pomologist who lived just a few blocks away in Newburgh. J. H. Ricketts must have held this variety in high esteem because he named it after such a close friend and well-known figure.

As with many of Ricketts's hybrids, Downing's weaknesses include having moderate susceptibility to fungal diseases in unfavorable seasons, late-ripening fruit, and being somewhat tender to winter damage. However, fungal diseases, even then, could be controlled with Bordeaux mixture, so with current modern fungicides, this grape may merit further evaluation in the future. The variety was also noted as not being very productive especially in areas where winter kill is prevalent.

Downing vines bloom by mid-season. The fruit ripens somewhat late, after Concord, and keeps in storage until the spring. The attractive clusters are large, long, rather slender, cylindrical to slightly tapering, uniform, well filled, and sometimes loosely shouldered. The blue berries are large to medium-large, distinctly oval, dark purplish-black, glossy, and covered with a light-blue bloom. The grape shows some *vinifera* qualities in fruit texture and taste. Further, like *vinifera*, its seeds and skins are not bitter like many other American hybrids. The flesh is described as being green/yellow, tender, vinous, and mild of very good quality.

Downing is an heirloom variety that should be considered for cultivation in the Hudson Valley.

EMPIRE STATE

Empire State was first fruited in 1879. At the turn of the twentieth century, it competed with Niagara, Diamond, and Pocklington as one of the more popular white grapes in commercial vineyards. It was first shown at the American Pomological Society Exhibition in Boston in 1881. J. H. Ricketts maintains that it came from a seed of Hartford pollinated by Clinton. However, viticulturalists have suggested that this grape does not show the characteristics of either of the reputed parents. George W. Endicott, an experienced horticulturalist, stated in the *Bushberg Catalogue* that it is probably a hybrid of Clinton and a white European grape.[74]

PARENTAGE
riparia, labrusca, vinifera?

HARVEST DATE
Mid-season to late

A A B A+ A

Empire State was introduced around 1884 and was listed in the *American Pomological Society Fruit Catalogue* in 1889. Shorty after, the entire stock of Empire State was purchased by nurseryman George A. Stone of Rochester for the hefty price of $4,000—a record price for an American grape variety at the time.

U. P. Hedrick reported Empire State was comparable to Niagara, Diamond, and Pocklington in its vigorous growth habit, freedom from fungal diseases and insect damage, and productivity. However, it is also less winter hardy and its clusters are not as attractive in appearance. Further, in some localities, the clusters can be small when compared to Niagara, Diamond, and Pocklington, so severe pruning or cluster thinning is recommended to keep the cluster size large. [75]

The grape quality is reported to be very good, much better than Niagara, somewhat better than Pocklington, and nearly as good as Diamond. Its flavor profile is more of Muscats rather than the foxiness of a *labrusca*. All commentators noted the high quality of the grape's taste. The grape is also fine for the table because of its long storage capacity and flavor retention. It was particularly in demand for wine production, especially for sparkling wine production at the turn of the last century. This indicates that it was a quality wine grape like Delaware, Dutchess, and Iona.

It is a moderately vigorous vine, usually healthy, but in some locations it can be only somewhat hardy when compared to other white *labrusca* varieties. The blooming time is by mid-season and the flowers are fertile. The grape ripens mid-season to late, a few days earlier than Niagara. The clusters are variously reported as large to less than medium in size, long, and rather slender and cylindrical to slightly tapering. It is frequently single-shouldered and of medium compactness. The berries are variable in size (averaging medium size), round, with a pretty color of pale yellowish-green to light green.

Growers such as George W. Campbell of Ohio and A. J. Caywood of New York both suggested that Empire State would not fulfill earlier expectations of the grape.[76] Further, the variety, even though generally healthy and tolerably hardy, seems to lack vigor and the root power needed to carry and mature a commercially profitable crop. However, based on Empire State's popularity with commercial growers, the large number of published evaluations that cited its quality, and my own preliminary observations on its solid growing characteristics and fungal resistance, I believe this to be a good grape overall.

Due to its quality for both the table and wine production and its solid growing capabilities, this grape should be reexamined for further cultivation in the Hudson Valley as an heirloom variety.

JEFFERSON

Jefferson is from a seed of Concord pollinated by Iona. It is one of the best known of Ricketts's hybrids and was one of the most widely distributed red grape varieties in the early 1900s. The plant first fruited in 1874 and was introduced to the public in 1880. It was quickly listed in the *American Pomological Society Fruit Catalogue* in 1881 and won the coveted Wilder Silver medal and many other awards when the grape was grown and shown by Ricketts.

PARENTAGE
labrusca, vinifera

HARVEST DATE
Late

B+ A B A A

Jefferson resembles the vigor, productivity, health, and disease resistance of Concord, but is not Concord's equal. In color and quality, the fruit resembles Iona. It is not, however, an Iona fruit on a Concord vine. The fruit is prettier than Iona and almost of equal quality. The vine blooms very late. Further, it produces fruit late, approximately two weeks after Concord or just before Catawba. The vine may not be nearly as winter hardy as Concord because its fruit ripens later in the season and its wood may not have time to properly harden off.

The very attractive red clusters are large, cylindrical to tapering, well formed, and well filled to compact. The berries are medium-large in size and uniform in color, which makes them an attractive light to dark red grape with a moderate amount of lilac bloom. The flavor has been described as juicy with a rich vinous flavor and a delicate aroma that persists even after the berries have dried to raisins. Some have described the flavor as being strawberry-like.

The grape was widely distributed and well known by growers in the East in the early 1900s because it is not particular about its soils, except that it does not like lime soils. Many commentators believed Jefferson to be a quality red grape both in the field and in the cellar, but it was used primarily for the table because of its attractive appearance and very large clusters, and the fact that the grape ships and keeps well.

Based on my own observations at Cedar Cliff and the grape's disease resistance, winter hardiness, and productivity with high-quality *labrusca* flavors, Jefferson should be reexamined as an heirloom table and wine grape for cultivation in the Hudson Valley.

MINOR GRAPE VARIETIES OF RICKETTS

ADELAIDE

labrusca, vinifera

This variety is a cross of Concord and Muscat Hamburg. The vine is of medium size. The oval black berry has purple-red flesh and a sweet, but sprightly flavor.

ADVANCE

riparia, labrusca, vinifera

First exhibited in New York City in 1870 and introduced to the public in 1872, this variety generated much interest because of its *riparia*, *labrusca*, and *vinifera* heritage. It is a cross of Clinton and Duke of Magenta, a grape resembling Black Hamburg or which is actually Black Hamburg. The quality of fruit is an improvement over Clinton, but the *vinifera* characteristics of lack of winter hardiness and susceptibility to fungal disease also carried through to Advance. The vine is vigorous and productive. The fruit ripens very early. Clusters have been described as above medium in size and compact to large and long. The berries are medium to large, oval, dark purplish-black, and covered with a heavy blue bloom. The fruit quality is good and similar to Clinton with a sweet spicy flavor.

ALMA

riparia, labrusca, vinifera?

This is possibly a seedling of Bacchus fertilized with a Ricketts hybrid that crossed a hardy native variety and the Purple Constantia from the Cape of Good Hope. Ricketts described this as a pleasant dessert grape that makes a splendid wine. The wine has been described as having a rose and wintergreen flavor that is delicately integrated.[77] This black grape with blue bloom ripens soon after Hartford. It is vigorous and healthy, and the cluster is medium in size, compact, and seldom shouldered. Its Brix levels were recorded from 23° to 25°, so it must have been used primarily as a wine grape.

ARIADNE

vinifera, riparia, labrusca

This Clinton seedling crossed with a *vinifera* was brought to public notice in 1870. The vine is reported to be moderately vigorous, healthy, and productive. It needs cluster thinning to reduce its overproductive crop. The clusters, which resemble Clinton, are small to medium and compact, but are of much better quality. The berry is black and small to large. It is a wine grape that reputedly could produce a light wine or a very dark and rich red wine with good body and a fine flavor. It can also have a sherry flavor.

ELDORADO

labrusca, vinifera

In 1870, Eldorado came from a seed of Concord fertilized by Allen's Hybrid (Golden Chasselas × Isabella). It was introduced to the public around 1881 and as of 1910, it was still being offered for sale by a few nurserymen. It is a very early white grape that ripens around the same time as Moore Early. It is delicately flavored, with a very distinct aroma and taste that is reminiscent of pineapples that underlies its fine slightly foxy flavor. Along with its sister, Lady Washington, it is a secondary hybrid of *labrusca* and *vinifera*. It inherited most of the good qualities of Concord, but Eldorado does not inherit Concord's ability to set fruit well. It needs cross-pollination and does best in a mixed vineyard. The clusters are small to moderately large and straggling, so it is not recommended for commercial production. The vine is generally a strong grower, generally moderately winter hardy, and an uncertain bearer. The clusters do not set well in fruit and are variable in size. The berries are large to medium, roundish, of a yellowish-green to golden yellow.

EXCELSIOR

vinifera, labrusca

This variety comes from a seed of Iona fertilized by pollen of some unknown *vinifera*. The variety was introduced in 1882. The vine is described as being moderately vigorous, precocious that needs the best of care, not always hardy, and medium to productive. The clusters are handsome, unusually large, long and broad, and moderately compact. It needs cluster thinning. The slightly oval berries are medium to large in size and of a pale to dark red color covered by a lilac bloom. Its flavor has been described as similar to Black Hamburg in many ways. The grape taste is vinous, with a rich

Minor grape varieties, cont'd.

aromatic Muscat flavor. The skins and seeds are not objectionable in taste. The vine is inclined to overbear, so cluster thinning is required. The grapes ripen late to very late at about the same time as Catawba or later. Ricketts prized this grape highly and there was some interest from growers until the 1910s.

GAZELLE

This is a neglected variety of unknown parentage. Grower Sam Miller says the grape is splendid, the cluster is large, the berry is

about the same size as Herbemont, and the color is nearly white and almost translucent. The vine is vigorous, healthy, and somewhat productive.[78]

GOLDEN GEM

labrusca, vinifera, bourquiniana

A seedling of Delaware and Iona that was first exhibited at the American Pomological Society meeting in 1881, this is a solid table grape of a rich golden color. The vine is moderately vigorous, the cluster is small and sometimes shouldered, and the berry is small. The fruit is described as ripening very early, even before Hartford. It can be grown wherever Delaware or Iona can be grown. The flavor is rich with a fine rosé taste.

HIGHLAND

vinifera, labrusca

Originating around 1865 from a seed of Concord fertilized by Jura Muscat, Highland was sold to Messrs. Asher Hance & Sons, who distributed it via their nursery business. It was somewhat popular in varietal and amateur vineyards, but did not become a popular commercial variety. For a black grape, it has good appearance, presence, and quality. The clusters are somewhat compact, unusually large, and weigh up to two pounds. The clusters are rather long, broad, and tapering with beautiful and large dull bluish to purplish-black berries. The berries are large to very large. The grape's flavor is that of its parent, Jura Muscat, which keeps and ships well.

The vine is relatively vigorous and can resemble Concord in vine and foliage. Highland is not very hardy and tends to overbear with usually two clusters per shoot. Cluster thinning is required to produce quality fruit. The fruit ripens very late, even after Catawba. The variety is healthy, and it is unknown whether or not it is susceptible to fungal diseases. At the turn of the century, Highland had been on trial for approximately thirty years, yet it was not widely distributed. However, because of its appearance, quality of fruit flavors, large cluster size, and seeming resistance to fungus, this variety should be reconsidered in the Hudson Valley as an heirloom variety. Further, due to its significant *vinifera* heritage, Highland should be considered for further breeding to produce more advanced hybrids.

IMPERIAL

vinifera, labrusca

Perhaps hybridized around 1870, Imperial is a seedling of Iona fertilized by Sarbelle Muscat. Its most valuable characteristic is its reputed hardiness and vigor, which has been compared to Concord. This is unique for a hybrid with so much *vinifera* heritage. The appearance and quality of this white grape is very good, and if the vine characteristics were better and there were not so many other good white grapes in the early 1900s, Imperial might have been more generally cultivated. The fruit ripens late and the clusters are large, symmetrical, and rather compact. The berries are large, greenish-white, and covered with considerable bloom. The seeds are small and not numerous. The flavor is vinous, sprightly, and soft with muted flavors of Iona–Muscat, and of good quality.

LADY DUNLAP

labrusca, vinifera

This seedling was developed about 1875 and first exhibited in 1881. The cluster and berry are medium and compact, with an amber color and vinous taste. The fruit quality is good.

LADY WASHINGTON

labrusca, vinifera

This variety comes from a seed of Concord fertilized by Allen's Hybrid (Golden Chasselas × Isabella). It was introduced in 1878 and listed in the *American Pomological Society Fruit Catalogue* in 1881. The vine characteristics are mostly of *labrusca*, but the fruit shows some *vinifera* traits.

U. P. Hedrick noted that with so many other quality American hybrid green grapes available, Lady Washington, while good, was not exceptional.[79] The grape has a good appearance for a table grape, but is somewhat short in quality and its vine characteristics are solid, but not stellar. The vine blooms late, is vigorous to luxuriant in growth, and hardy for a

vinifera grape. Further, it is healthy and not highly susceptible to fungal diseases. The variety is somewhat selective as to its soils and locations where it will grow. It seems to do well in Arkansas, in the Midwest, Missouri, and Illinois.

The flowers are self-fertile and open by mid-season to late mid-season. The fruit ripens after mid-season to late season after Concord. The pretty clusters are large to medium in size, irregularly cylindrical, single or double-shouldered, loose to medium compact, and slender to medium. The berries vary in size, are roundish to oblate, and a dark green that changes to a pale or yellowish amber, glossy color, with a delicate rosy tint when the fruit is exposed to the sun. The fruit keeps and ships well.

The flavor is of better quality than Concord. With its heavy *vinifera* heritage and vine characteristics, Lady Washington could be useful to grape breeders. W. H. Lightfoot of Springfield, Illinois, used Lady Washington in his breeding program and developed the grape Alice Lee.[80]

NAOMI

vinifera, riparia, labrusca

This grape is from a seed of Clinton fertilized with Muscat Hamburg that was first exhibited before the American Pomological Society in 1879. This grape was, according to Ricketts, one of his more magnificent table grapes. This white grape is of a pale green color that occasionally has a reddish-yellow tinge in the sun. The compact cluster is large to above average in size. Berries are of medium size and roundish to oval. The vine is very vigorous, hardy, and very productive. The flavor is good, crisp, and sprightly with a trace of Muscat flavor. However, it ripens late—generally too late for the Hudson Valley—and is very susceptible to fungal diseases.

NEWBURGH

labrusca, vinifera

This is a seedling of Concord crossed with Trentham Black. One comment, published in *Our Native Grape* by C. Mitzky in 1893, stated that the bunch and berry are very large, the cluster is heavily shouldered, and the berries are black with a bluish-gray bloom. It is juicy with a peculiar flavor, very vigorous, and a fine amateur grape.[81]

PEABODY

riparia, labrusca, vinifera

This seedling of Clinton, supposedly developed in 1870 and introduced in 1882 is described as of good quality, vigorous in growth, and winter hardy. Its growth habit is that of *riparia*. The cluster is medium to large, long, and quite compact. The berry size and shape are of Iona or smaller, and the color is black with blue bloom. The vine is a fair grower of medium production and blooms mid-season. It appears to do better in the northern states or Canada.

At the Experimental Station grounds in Ottawa, Ontario, it ripened about the same time as Worden. Overall, the grape ripens early. The cluster is too small for table use, but it keeps well, and its value for wine seems to have not been determined even though its parentage would lend it to wine production. The taste is vinous and spicy. The general appearance of Peabody resembles Advance, but it is not as strong a grower, nor as prolific or hardy. The fruit is distinctly different from Clinton, which demonstrates that the variety is not a purebred seedling.[82]

PIZARRO

vinifera, riparia, labrusca

This is a Clinton seedling crossed with an unknown *vinifera*. The cluster is long and loose, the berry is medium in size, oblong, and black. It seems that the Clinton hybrids carry forward *riparia's* traits of winter hardiness and good winemaking ability. Ricketts believed this variety made a quality light red summer wine of great richness. The vine is of medium vigor, productive, and ripens at mid-season.

Minor grape varieties, cont'd.

PLANET

labrusca, vinifera

A cross of Concord and Black Muscat of Alexandria, it is described as healthy and productive. The large cluster is loose, and the berry size is a combination of large and small berries, with small berries having no seeds. It tastes of slight Muscat.

PUTNAM

labrusca, bourquiniana, vinifera

Ricketts' Delaware seedling No. 2, is a cross between Delaware and Concord introduced before 1871. Like Delaware, it ripens very early. Brix readings of between 19° and 20° have been recorded, so this must have been considered a wine grape. The vine resembles Concord more than Delaware. Clusters are medium in size and moderately compact with medium-sized, oval black berries.

QUASSAIC

vinifera, riparia, labrusca

This cross of Clinton and Muscat Hamburg was made in 1870. It is named after the Quassaick Creek, which travels south from Marlboro and Plattekill and then east where is flows into the Hudson River at Newburgh. Quassaic has a large and moderately compact cluster; berries are medium-large to large, oval, and black with a blue bloom. It ripens soon after Concord, is slightly vinous and aromatic, and is generally vigorous and productive.

RARITAN

labrusca, bourquiniana, vinifera

This variety is Ricketts' Delaware seedling No. 1 crossed with Concord. The vine is moderately vigorous, and should be grafted on Clinton to increase vigor. It is somewhat winter hardy; the cluster and berry are reminiscent of Delaware, but the color is black. The flesh is juicy and vinous, and the grape ripens about the same time as Delaware. Its Brix levels are high and can reach between 24° and 25°. It was probably used as a wine grape because its sugar readings were noted in early literature.

SECRETARY

vinifera, riparia, labrusca

This grape is one of Ricketts's earliest hybrids from a seed of Clinton fertilized by Muscat Hamburg. It was planted in 1867, exhibited at the American Pomological Society in 1871, and sold around 1875 to Stephen W. Underhill of Croton Point. Underhill then introduced the variety a few years later. The grape was considered to be the finest new grape at the Massachusetts Horticultural Exhibition in 1872, and Charles Downing pronounced it to be one of Ricketts's best grapes for quality.[83] The vine and grape are highly susceptible to fungal diseases infecting the leaves, young wood, and fruit. When disease-free, the fruit is reported to be of good quality, with well-formed large clusters of medium-sized purplish-black berries. The vine and foliage somewhat resemble Clinton, one of its parents, but it is not nearly as hardy, vigorous, or productive. Its fungal problems made this variety of interest only to amateur growers.

The grape ripens shortly after Concord, and keeps and ships well. Its sugar levels have been recorded at 22° Brix, so it must have been viewed as a wine grape with table grape applications.

UNDINE

labrusca, riparia

A cross of Concord and Clinton, Undine's vine is vigorous with healthy foliage. The cluster and berry size is of Concord, the color is pale green, turning yellowish-white as it ripens. The quality is good.

WAVERLY

One of Ricketts's first efforts around 1870, this is a seedling of Clinton and one of the Muscats. The vine is reported to be vigorous to very vigorous, hardy,

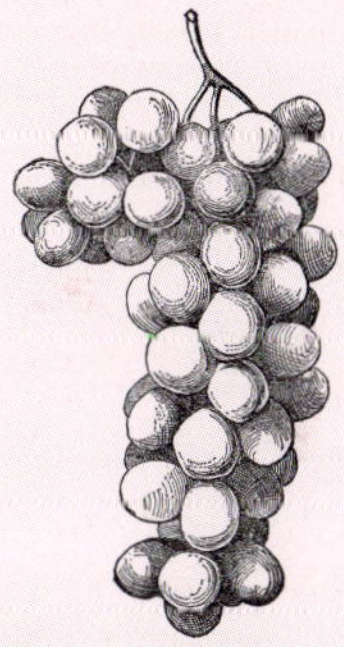

healthy, and productive. The cluster is medium-long and compact, and the berry is medium to large, oval, and black with a thin blue bloom. The vine is so productive that cluster thinning is required. It ripens with the grape Brighton.

WELCOME

vinifera

An exotic grape that is a cross of Pope's Hamburg and Canon Hall-Muscat. The grape is a warm-weather grape that can be grown only under glass or in places such as California or Georgia.

The Underhill Family

The Underhills are one of the most important families in the history of the Hudson Valley grape and wine industry. Three generations of Underhills working at Croton Point could be considered the first dynasty of American viticulture.[84] As with the successful French grape hybridizer Albert Seibel, the Underhills worked as grape growers, hybridizers, and winemakers for more than sixty years, making successful advances in the field of viticulture. The first of the Underhills to become involved with grape culture was Robert Underhill, a Quaker born in Yorktown Heights.

Robert Underhill (1761–1829)

As was commonly the case with many people during this period, Robert Underhill was engaged in several business enterprises at the same time. At one time he was part owner and operator, along with his brothers Joshua and Abraham, of a flour or grist mill at the head of the navigable part of the Croton River (above the site of the Van Cortlandt Manor historic site near the Hudson River). He later sold his interest in this business, and in 1804, at the age of forty-three, moved to a property at Croton Point that he had previously purchased. (Croton Point is now a park operated by Westchester County and the home of the Hudson River Sloop Clearwater's strawberry festival in the spring, and pumpkin festival in the fall.)

Around 1810, Underhill became interested in viticulture in addition to his other agricultural and business pursuits. He was encouraged by André Parmentier to plant European grapes to produce wine at Croton Point sometime between 1825 and 1827. Parmentier, a wealthy and educated Belgian, moved to Brooklyn, New York, at the beginning of the French Revolution. In Brooklyn he planted a large garden that contained many types of European grapes. This garden became a commercial nursery from which he distributed European grapes. As with most other vineyards containing European varieties, the vineyard was ultimately not successful. However, the seed to plant grapes for wine production did flourish.[85]

In 1827, Robert Underhill and his two sons, Richard T. Underhill (1802–71) and William Alexander Underhill (1804–73), planted a small vineyard of Catawba and Isabella on the southern part of Croton Point that was surrounded on both sides by the Hudson River. These vineyards grew in size until they covered 27 acres by 1843 and ultimately expanded to approximately 75 acres.[86] At first these grapes were sold primarily in New York City and nearby towns.

In 1829, Robert Underhill died, and his land holdings of approximately 250 acres on Croton Point were divided between his two sons, with William A. Underhill receiving approximately 165 acres and his brother, Dr. Richard T. Underhill, the remaining 85 acres.

William A. Underhill (1804–1873)

In addition to cultivating grapes and other crops, William A. Underhill concentrated on operating a brickyard on the northern portion of Croton Point, which helped to make the Village of Croton-on-Hudson flourish. Bricks with his initials W. A. U. or with the self-promoting and very un-Quaker-like cryptogram "IXL," a claim that he excelled at brick making, can still be found there along the Hudson River at low tide.[87]

Coincidentally at about the same time, my ancestors also lived on Croton Point and operated a sawmill on the Croton River in the 1770s to 1800.[88] Sometime after 1800, my branch of the Casscles family moved across the Hudson River to Caldwell's Landing (now known as Jones Point) and to Tompkins Cove to manufacture bricks in Rockland County.

Underhill bricks manufactured on the northern portion of Croton Point were used to build barns and two wine vaults that still exist today at Croton Point Park. At the time, the barns were used to store fruit and horse carriages. The two vaulted brick wine cellars were built into the hillside to store wines made from the Underhill's grapes.

One lasting impact of William A. Underhill on the local grape industry is that his wife, Abby Underhill, was a sister-in-law of William T. Cornell, who subsequently established a vineyard of Isabella grapes in 1845 near Clintondale, Ulster County. Further,

Mr. Cornell's sister, Deborah Cornell, ultimately became the wife of the Hudson Valley grape breeder A. J. Caywood.

Dr. Richard T. Underhill (1802–1871)

The elder son, Dr. Richard T. Underhill, was so enthusiastic about grape culture that he discontinued his medical practice in New York City to move to the southern tip of Croton Point to grow grapes. He was one of the first American advocates of the "Grape Cure," a diet of fresh grapes, which was then popular in Europe. This cure was believed to prevent dyspepsia, liver ailments, and a long list of other maladies.[89] In 1846, he built an Italianate villa for his residence and called it Interwasser. Richard T. Underhill, who owned the southern part of the Point, began to expand his apple orchards, rose gardens, and vineyards and to establish his winemaking capacity, but never pursued the brick-making business on his property.

After his father's death in 1829, Dr. Underhill aggressively pursued planting vineyards for wine production, apples, and roses. He established Croton Point Vineyards, one of the first wineries in the Hudson Valley. Although he did not begin to commercially sell the wines he produced until 1851, he had made wines previously for private use. Wine was commercially made on Croton Point from 1859 until 1871. Dr. Underhill aimed to produce a characteristic American wine, and his preference was to produce a "perfectly natural wine"—the pure juice of the grape—which was advertised as "neither drugged, liquored nor watered." Their wines won many awards and were sold in New York City from an establishment called the Pure Wine and Grape Depot. After Dr. Underhill's death in 1871, the last of his rare vintages of wine were offered for sale by Thurber & Company of New York City in 1872. With the death of Dr. Underhill, the winery ceased to produce new wines.[90]

Dr. Underhill also had a national reputation as one of the first great hybridizers in the Hudson Valley. While the development of grape varieties such as Croton, Black Eagle, Black Defiance, and Senasqua have long been attributed to Dr. Underhill's nephew, Stephen W. Underhill, it may have been Dr. Underhill who developed the varieties while Stephen Underhill introduced them to the public after 1865.

Like Dr. William A. M. Culbert and Dr. Charles W. Grant of nearby Newburgh—both doctors who hybridized grapes in their retirement or near retirement—it seems likely that Dr. Underhill may also have bred grapes as a hobby upon retiring from the medical profession or as a way to relax while he was still practicing. While most of the Underhill varieties were first hybridized and seeded around 1863 or slightly earlier, Dr. Underhill's activities may have piqued the interest of his nephew Stephen W. Underhill to start a grape-breeding program of his own. At the time Stephen was no more than twenty-six years old and still helping his father, William A. Underhill, with his farm and extensive brick-making operations.

Stephen W. Underhill (1837–1925)

The son of William A. Underhill, Stephen Underhill became familiar with the grape-growing operations of his father and uncle, and around 1860 became interested in hybridizing to develop new grape varieties. Most of Stephen's work (and presumably with the assistance of his uncle Richard) was done between 1860 and 1870. As noted in the grape descriptions on the following pages, it was said that his varieties generally possessed too many *vinifera* weaknesses for profitable commercial cultivation at the time. With the advances in viticultural practices today, these varieties may need to be reevaluated for commercial production. Regardless of who developed the Underhill varieties, it was Stephen Underhill who introduced and disseminated them to the general public.

Upon the death of Dr. Underhill in 1871 and his father in 1873, Stephen seems to have lost interest in grape cultivation and breeding and devoted his time almost exclusively to brick making. This shift away from grape breeding may have been because of a lack of genuine interest in the activity, or because of lack of time while helping to run his late father's large brick-making operations.

All of the grape varieties following are attributed to Stephen W. Underhill of Croton Point. However, as mentioned earlier, Dr. Richard T. Underhill, his uncle, may have had some hand in their development.

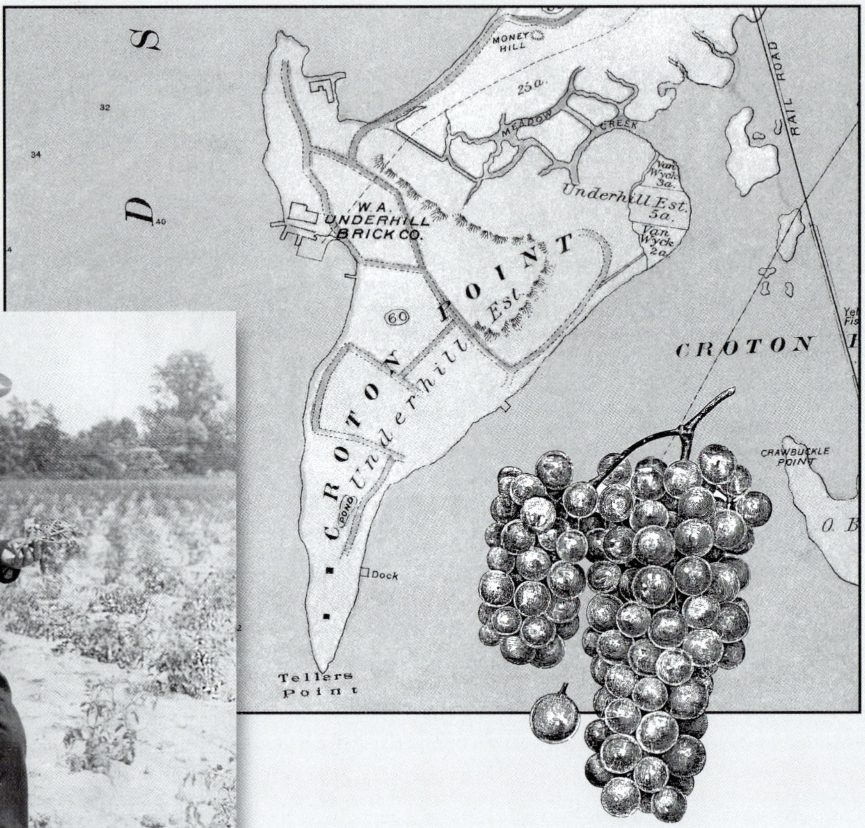

Stephen W. Underhill at the family farm at Croton Point. *Map of Croton Point on the Hudson River, where the Underhills grew grapes, such as Croton (above), made wine, and had a large brick-making business. Circa 1860s.*

End of an Era at Croton Point

Dr. Underhill never married, and thus had no direct descendants to whom to leave his farm and winery, so at his death, there was no family member willing to take over his operation. The farm was slowly converted to other public recreational activities. In 1912, the Point Pleasant Park was opened on the southern tip of the Point where visitors could picnic on the Interwasser mansion grounds.

In 1915, the Underhill brickyard closed because the land was depleted of clay for brick making. The Village of Croton-on-Hudson's economy changed significantly after the brickyard closed. As laborers left the village, there was an influx of artists and writers. During World War I, the journalist and socialist Jack Reed married Louise Bryant and moved to Croton-on-Hudson. During the 1920s, a group of writers and artists from Greenwich Village in New York City started purchasing old farm buildings on Mount Airy Road. Many worked for political monthly magazines of the left, such as *The Masses*. After 1923, the Croton Point Park Amusement Park was established on the north shore near the bathhouses. In 1924, the Westchester Park Commission purchased approximately 500 acres of Croton Point to establish the Croton Point Park.

CROTON

Croton, named after Croton Point, is probably Underhill's most well-known grape variety. It comes from a seed of Delaware pollinated by Chasselas de Fontainbleau. The seed was planted in the spring of 1863 and the vine fruited in 1865. It was first exhibited at the Massachusetts Horticultural Society meeting in 1868 and listed in the *American Pomological Society Fruit Catalogue* in 1871, but was dropped in 1883, chiefly due to its susceptibility to fungal diseases.

PARENTAGE
vinifera, labrusca, bourquiniana

HARVEST DATE
Early mid-season to mid-season

B A– C B A

This very high-quality white grape has been described as a feast for both the eyes and palate. In 1868 and subsequent years, it won prizes at the New York, Pennsylvania, and Massachusetts Horticultural Society meetings.[91]

However, according to U. P. Hedrick, it is difficult to grow as it adapts to few soils (except for sandy soils or rich, silty soils as found at Croton Point). Further, Hedrick has described it as a weak grower unless it is grown on ideal soils, in which case it can be a vigorous grower. Like Riesling, it is moderately susceptible to winter injury. Further, it is susceptible to fungal diseases. While described as an unfruitful variety, it can still be grown and the delicate, superior *vinifera*/Delaware flavors may make its cultivation worthwhile.[92]

Unlike the comments offered by Hedrick, growers on the West Coast have praised its virtues, particularly its taste and cluster appearance and have not made negative comments about its ability to grow. Croton has a refreshingly spritely mild Delaware-*vinifera* flavor with less acidity and about the same sugar levels as Delaware.

I have recently started growing Croton at my farm at Cedar Cliff. My preliminary findings are that it is a moderately vigorous variety when grown in heavy soils. Further, it is not difficult to grow, has good fungal disease resistance, and the cluster is long and pretty. The vine blooms late and is self-fertilizing. The fruit ripens by early mid-season to mid-season, but keeps well on the vine or in storage. Its uniform clusters are described as very pretty with a prominent shoulder, and are large, very long, slender, cylindrical, and moderately compact. The medium-sized berries are somewhat irregular in shape, round-elongated, and a translucent yellowish-green in color. The skin is thin but tough, and it is very juicy with a pleasant *vinifera*-like clean flavor.

It is interesting to note that Croton was also an important grape used for breeding. It is one of the parents of Medora (*aestivalis*-hybrid) which is a cross seedling of Lenoir × Croton bred in Texas at Onderdonk's experimental vineyard.[93]

With the advancements in viticulture practices, this variety should be reexamined for future cultivation in the Hudson Valley as an heirloom variety. Further, due to its appearance and flavors, it should be considered for use in future breeding programs.

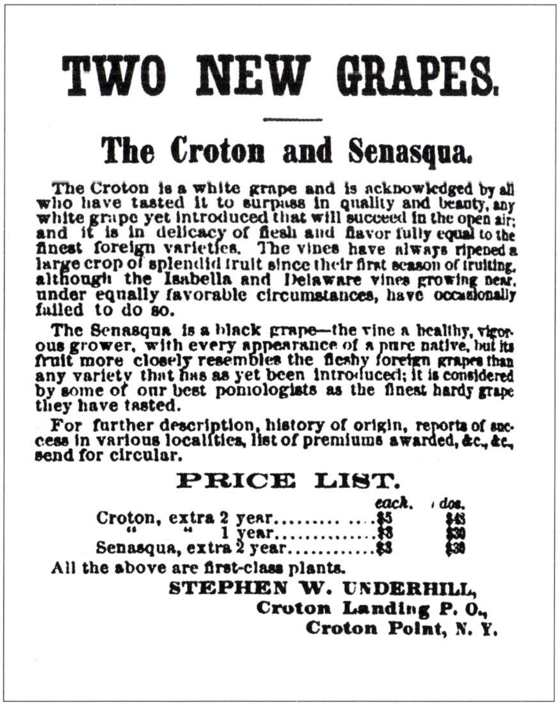

TWO NEW GRAPES.

The Croton and Senasqua.

The Croton is a white grape and is acknowledged by all who have tasted it to surpass in quality and beauty, any white grape yet introduced that will succeed in the open air; and it is in delicacy of flesh and flavor fully equal to the finest foreign varieties. The vines have always ripened a large crop of splendid fruit since their first season of fruiting, although the Isabella and Delaware vines growing near, under equally favorable circumstances, have occasionally failed to do so.

The Senasqua is a black grape—the vine a healthy, vigorous grower, with every appearance of a pure native, but its fruit more closely resembles the fleshy foreign grapes than any variety that has as yet been introduced; it is considered by some of our best pomologists as the finest hardy grape they have tasted.

For further description, history of origin, reports of success in various localities, list of premiums awarded, &c., &c., send for circular.

PRICE LIST.

	each.	doz.
Croton, extra 2 year	$5	$48
" " 1 year	$3	$30
Senasqua, extra 2 year	$3	$30

All the above are first-class plants.

STEPHEN W. UNDERHILL,
Croton Landing P. O.,
Croton Point, N. Y.

Two New Grapes. Stephen W. Underhill introduced Croton, a white grape hailed for its beauty and taste, and Senasqua, a black grape resembling "fleshy foreign grapes." Circa 1870s.

BLACK EAGLE

Black Eagle (Underhill's No. 8–12) is the full brother of Black Defiance, which it resembles, but is superior because it ripens mid-season to late mid-season at about the same time as Concord. Like Black Defiance, it first fruited in 1866 and was sent out for testing, but was introduced without Underhill's consent. The quality of the fruit is described as very good to the best, but the vine can lack vigor, is moderately tender, and not productive. Further, it is self-sterile and the fruit is susceptible to black rot and other fungal diseases.[94]

PARENTAGE
labrusca, vinifera

HARVEST DATE
Mid-season to late season

B A B B B+

The attractive clusters are large, long, tapering, single- or double-shouldered, compact, and reported to weigh nearly two pounds. The berries are oval and black with a glossy thick bloom.

Black Eagle buds out late and its leaves look very much like a *vinifera* leaf. This vine is very erect, with an upright growth habit. Further, it can have vigorous growth like other *vinifera* varieties. The flowers open mid-season and it ripens by mid-season. Black Eagle cannot set fruit by itself, but if it grows near other comparable grape varieties, it can be very productive.

Despite its faults and even though it once failed as a commercial variety, my preliminary observations at Cedar Cliff, where it is grown on not ideally-suited heavy soils, are that Black Eagle has good vigor and cold resistance in the field, and is moderately resistant to fungal diseases.

Black Eagle is a promising grape which is why it is included here as a possible heirloom variety that should be considered in the future.

MINOR GRAPE VARIETIES OF UNDERHILL

BLACK DEFIANCE

labrusca, vinifera

This is reputed to be a *labrusca-vinifera* hybrid that first fruited in 1866. U. P. Hedrick maintains that Black Defiance came from a seed of Concord fertilized by Black Prince (*labrusca, vinifera*).[95] The *Bushberg Catalogue* states that it is a cross of Black St. Peters and Concord (Underhill's No. 8–8).[96]

A telling fact about this grape is that it was introduced without the originator's consent, but it is listed as a major grape in *The Grapes of New York*, in which it is described as ripening about three weeks after Concord.

It was looked upon with favor in France during the phylloxera epidemic where it was used as a direct producer to some extent before grafting *vinifera* vines became a more common practice. For a brief period it was a somewhat popular late table grape, but was superseded by thriftier varieties.[97]

The fruit is distinguished by its large size in both berry and cluster, lustrous black color, and handsome bloom. Not much is known about its growth habit, disease susceptibility, or winter hardiness, but it was viewed favorably in France in the second half of the nineteenth century. It has been described as of medium growth, large clusters with an oblong large base, and frequently branched. The fruit ripens late to very late.

IRVING

vinifera, labrusca

Also known as Underhill's No. 8–20, the parentage of this white grape is disputed. The *Bushberg Catalogue* maintains that this is from a Concord seed crossed with

Minor grape varieties, cont'd.

pollen from White Frontignan that first fruited in 1866.[98] However, *The Grapes of New York* maintains that Irving is a seedling of Catawba fertilized by Chasselas de Fontainbleau in 1868. It is described by U. P. Hedrick as having vigorous, short-jointed canes and large, long clusters with shoulders that are not well filled. Hedrick did not prominently highlight this grape in *The Grapes of New York*.[99]

The berry is considerably larger than Concord and a yellowish-green to yellow/white in color with a thick bloom that has a slight tinge of pink when very ripe. The flavor is vinous and very good. The vine is a moderately vigorous grower with large thick foliage that rots and mildews badly. The fruit ripens late to very late, just before Catawba. The *Bushberg Catalogue* considers the Irving as more deserving of dissemination than Croton.[100] The name is probably derived from the author Washington Irving, who lived in nearby Sleepy Hollow, New York. This variety should be looked at more closely as an heirloom variety for cultivation and wine. This could be a diamond in the rough.

SENASQUA

labrusca, vinifera

This *labrusca-vinifera* hybrid is named Senasqua, one of the Indian names for Croton Point, the place of its origin. The variety comes from a seed of Concord pollinated by Black Prince. The seed was planted in 1863, bore its first fruit in 1865, and the variety was introduced in 1870.

The vine is described as lacking somewhat in vigor, winter hardiness, productivity, and health. Further, the variety is somewhat susceptible to attacks of mildew. The vine and leaves show little trace of *vinifera* heritage, but the fruit does. The grapes are of good quality, and when well grown, the variety is about average compared to other *labrusca-vinifera* hybrids. The berries, however, have a tendency to crack, which is aggravated because the clusters are very compact. The skin adheres to the flesh like a *vinifera* variety and is not a slip-skin variety. The cluster and berry vary from medium to large in size, and the cluster can be stout and irregularly tapering. The berry color is reddish-black to black with a blue bloom.

The bud break on Senasqua is late, so it is seldom hurt by late spring frosts. The *Bushberg Catalogue* maintains that the vine is vigorous and productive in rich soils and moderately hardy. Clay soil is not the best for this grape; it requires a light, deep soil. The grape ripens just a few days after Concord. The *Bushberg Catalogue* suggests that Senasqua is not as desirable as Black Defiance or Black Eagle.[101]

The grape's flavor and consistency is *vinifera*-like, with brisk and vinous flavors. Further, in the Drôme and Lot-et-Garonne districts of France, the grape was cultivated somewhat in the mid-1880s because it was one of the better American hybrids.[102] Even by the early 1900s, the grape, while widely tested, was no longer grown commercially, and even Stephen Underhill did not recommend it for commercial cultivation, but for amateur gardeners.

Other Hudson Valley Growers and Grape Breeders

What makes the Hudson Valley the birthplace of American viticulture is the concentration of grape hybridizers, horticulturalists, farmers, nationally renowned horticultural writers, and those who specialize in developing new pruning and cultivation practices. The Hudson Valley also has close proximity and ease of transportation to New York City, one of the largest markets in North America. This access to New York City markets made it easy to sell many different kinds of grape varieties to its specialized markets, and one reason why Hudson Valley growers up until the 1920s had a much wider variety of grapes under cultivation than quite possibly any other grape-growing district in the country.

Following are some of the additional grape growers, horticulturalists, and the grape varieties they developed that helped to make the Hudson Valley a very special place to be during the second half of the nineteenth century. This large concentration of talent created a synergy that facilitated the work of those who worked in this industry. It is clear that they communicated with each other often and extensively in their own horticultural work.

The histories and activities of the families profiled below are described in great detail to give a glimpse of their lifestyles and economic well-being. Without prosperous and dedicated growers such as the Barnes and Clark families and other local growers in the Middle Hope area such as John W. Bingham, W. J. Fowler, V. J. Kohl, and other dedicated growers throughout the Valley, there would have been no testing, field research, or consultation on the value of locally developed new varieties before they were introduced to the general public.

The Barnes Family

The Barnes family of Middle Hope, Orange County, featured prominently in the grape-breeding activities of the Hudson Valley. Reviewing the history of the Barnes family may help to illustrate the workings of a prominent fruit-growing family and their ongoing relationship with local grape breeders and their own efforts at breeding new grape varieties.[103]

Nathaniel Barnes (1782–1879)

Nathaniel Barnes was born in Litchfield, Connecticut. His father, Isaac Barnes, lost his property through a defective land title and was forced to move his young family to Schoharie County, New York, and then Otsego County, New York. At the age of twenty-one, Nathaniel Barnes set out on his own. He was employed by contractors to build portions of turnpike roads in Orange and Ulster counties which was a popular occupation in the early 1800s. Nathaniel, over time, expanded his contracting business and helped to build turnpikes such as the Farmers', Huckleberry, Snake Hill, Newburgh, and Plattekill turnpikes in Orange and Ulster counties.

In 1822, at the age of forty, Nathaniel Barnes purchased his first property in Middle Hope, Orange County, on what is now part of the Cedar Hill Cemetery Association. In 1828, Nathaniel Barnes married Effie Dusenberre (1796–1880), the daughter of Dr. William Dusenberre (Dusinberre) of Plattekill, Ulster County. They had four children, William D., Nathaniel Jr., Daniel D., and Mary E.

It seems that Nathaniel Barnes came from modest means. However, through his own industry, attempts to receive at least a rudimentary education, and his marriage to Effie, the daughter of a doctor, he was able to secure the capital needed to purchase adjoining small farms in Middle Hope, and to expand his farm operation. This farm, about four miles north of Newburgh and near Roseton, was located along what is now U.S. Route 9-W, was gradually expanded to over 300 acres of very choice fruit lands when Nathaniel Barnes died in 1879. At the time of his death, he was among the first families who cultivated fruit in Orange County.

All of Nathaniel's sons—William D., Nathaniel Jr., and Daniel—continued to commercially cultivate fruit in Middle Hope and actively participate in local community, charitable, and political affairs.

William D. Barnes (1828–1904)

William D. Barnes—also sometimes spelled Barns—was the most prominent horticulturalist of the three brothers. He graduated from Columbia College and for a time before his marriage was a schoolmaster with an excellent reputation. At the age of twenty-five, he took over his father's farm operations at the old family homestead, which still exists today behind the Overlook Farm Market on Route 9-W. (His father was injured in 1852 and left completely blind.) William D. and his wife, Elizabeth A. Carpenter, married in 1860 and had five children.

During William D. Barnes's lifetime, he wrote about grape growing in the Hudson Valley and pursued horticultural knowledge, both on an individual basis and in collaboration with nearby hybridizers A. J. Caywood of Marlboro, as well as Charles Downing and James H. Ricketts of Newburgh. He developed his own grape hybrids, including a named variety called Paradox. He was a longtime officer and member of the Orange County Agricultural Society, superintendent of the fruit department at the Orange County annual fair, and an exhibitor and member of the Newburgh Bay Horticultural Society, the Eastern New York Horticultural Society, and the Western New York Horticultural Society. In 1893, he was appointed by Governor Flower as a member of the board for the New York State Experiment Station in Geneva. William D. was also active as a commissioner of the local almshouse and children's home.

For more than twenty-five years, William D. was an elder of the Presbyterian Church in Marlboro, and active in politics in the Democratic Party.[103] At the time of his death in 1904, his eldest son, Edwin W., continued to farm the old homestead under the firm name of Edwin W. Barnes & Son. Over 70 acres were devoted to fruit culture, from the earliest strawberry to the latest apples.

Nathaniel Barnes Jr. and Daniel D. Barnes

Nathaniel Barnes Jr. also farmed a large adjoining fruit farm in Middle Hope. He and his wife Martha Waring married in 1854 and had four sons and one daughter. It seems that his children worked in other business enterprises such as dry goods and the wholesale produce business in New York City. Mr. Barnes was actively involved in politics, serving as the town of Newburgh's tax assessor and as the first supervisor of the reformulated boundaries of the town of Newburgh, which had excluded what was then the Village of Newburgh (later the City of Newburgh in 1865).

The third son, Daniel D. Barnes, also a fruit farmer, operated a farm that adjoined his father's old homestead. He married Hester D. Carpenter, daughter of Captain L. S. Carpenter of Marlboro, New York. They had three daughters and one son, Nathaniel C. Daniel D. distinguished himself more as a breeder of fast trotting horses and showed some interest in public affairs. As of 1908, his son, Nathaniel C. Barnes, born in 1863, succeeded his father to operate the farm when Daniel D. retired from active farming and served as the Town of Newburgh's justice of the peace.[104]

The only grape attributed to the Barnes family is Paradox. They were more interested in doing field testing of already developed varieties.

PARADOX

labrusca, vinifera

A seedling of Hartford by Iona raised by W. D. Barnes of Middle Hope, Orange County. Paradox, also known as Seedling No. 502, was propagated by James H. Ricketts, but was given to Barnes to grow and observe. Mr. Barnes was a large commercial grower who had a long friendship with grape breeders A. J. Caywood and J. H. Ricketts. The vine is alternately described as of variable vigor to not always hardy; a precocious bearer to a vigorous and productive plant, with clusters medium to above average in size and compact, and the black or dark purple berry medium in size with blue bloom. It ripens with Concord or a bit earlier, and is described as juicy, sweet, vinous, and with a good-quality, pure flavor.

The Clark Family

Edson H. Clark (1813–1885)

Edson H. Clark was born at East Hampton, Massachusetts. Clark's father was a master millwright from an old New England family. Edson spent a portion of his childhood in Bloomingburgh, Sullivan County. He went to Newburgh to learn about and work in the stone-cutting trade until 1839, when he was twenty-two years old, after which he became associated with Oakley & Davis in some unknown trade. In 1843, he joined a newly formed foundry and machine shop, Stanton, Clark & Co., which was located in Newburgh at the foot of South Street on the Hudson River. The firm was dissolved in 1851, at which time he built a new ironworks on Washington Street, Newburgh, under the firm name Clark & Kimball from which he retired in 1875.

In addition to being a member of the Village of Newburgh Board of Trustees in 1854 and 1865 and the representative of the Third Ward in the first City of Newburgh Common Council, Clark was interested in pomology. After retiring in 1875, he studied pomology with his nearby friend Charles Downing. Edson Clark was an authority on raising fruit and the nomenclature of fruits, and actively participated in the American Pomological Society, the Newburgh Bay Horticultural Society, the Orange County Agricultural Society, and fairs sponsored by such organizations. For many years, he and Charles Downing were coworkers in this field, and in their later years spent much time together studying this science. In his garden was a tree that once bore approximately 200 distinct varieties of apples. Married to Ruth Ann Clark of Newburgh (not a relative), they had four children: Leander Clark Jr., Ruth Ann, Mrs. Elizabeth Upright, and Mrs. Albert Coutant of Chicago. Clark died in 1885 at the age of seventy-two at his residence on Liberty Street, Newburgh.

Leander Clark Jr. (1837–1906)

The son of Edson H. Clark, Leander Clark is another individual from Newburgh who, in addition to working in other industrial fields, also cultivated grapes in Orange County.

Leander Clark Jr. was born in Beattiesburgh, Sullivan County. The family moved to Newburgh when he was six months old. He graduated from Newburgh Academy in 1853 and then learned the trade of iron-founder at his father's foundry. Clark Jr. became the corresponding secretary and bookkeeper for Dr. Charles W. Grant, the eminent horticulturalist at Iona Island, Rockland County. Clark remained at the island until the start of the Civil War in 1861. While in military service, he was a paymaster's steward and then was promoted to a paymaster's clerk before being honorably discharged in August 1864. From 1867 to the spring of 1869, he was superintendent of the Newburgh water works. After resigning from that position, he became involved in the manufacture of brick and in fruit growing in the town of Newburgh until 1888.

Since 1858, Clark Jr. had taken a great interest in horticulture and agriculture. He was a member of the Newburgh Bay Horticultural Society, which later became part of the Orange County Agricultural Society. He was a good progressive grower who had been instructed for many years by Charles Downing. In 1888, Clark became a resident of the city of Newburgh once again after he retired from active business. After retirement, he resided at 287 Liberty Street, where he looked after his large real estate holdings and was a director of the Quassaick National Bank.[105]

Edward Payson Roe
(1838–1888)

Edward Payson Roe of Cornwall-on-Hudson, was a nationally recognized novelist, horticulturalist, and Presbyterian clergyman. In addition, he was a prolific writer on small fruit cultivation for the industry and writer of popular pieces for the general public on fruit cultivation for magazines such as *Scribner's* and *Harper's*. Rev. Roe was born in Moodna, Orange County. After extensive travels in the United States, he moved back to Cornwall-on-Hudson to write and evaluate fruits and vegetables. He tested and evaluated many varieties of fruits cataloged and offered by nurserymen and seedmen at his home in Cornwall along the Hudson River.

He wrote *Success with Small Fruits*,[106] which outlined issues related to the cultivation and propagation of small fruits such as strawberries, currants, raspberries, blackberries, and gooseberries. It is clear from his writings that he traded advice on horticultural issues with other local hybridizers and horticulturalists such as Charles Downing and A. J. Caywood, and with Marshall P. Wilder of Boston, president of the Massachusetts Horticultural Society (1840–48), and founder and president of the American Pomological Society (1848–86). Rev. Roe also wrote *The Home Acre*, a home gardening book geared to rural homesteaders who needed a concise how-to manual on raising fruits and vegetables for the home.

The only grape attributed to Edward Payson Roe was Storm King.

STORM KING
labrusca

A "sport" of Concord, Storm King resembles its parent in all respects, except that the berries are about twice as large and less foxy in taste. The grape is named after Storm King Mountain, which is near the village of Cornwall-on-Hudson where Roe lived.

MINOR GRAPE VARIETIES BY OTHER HUDSON VALLEY GRAPE BREEDERS

Below is a list of grape varieties developed by the Hudson Valley's lesser known grape breeders, horticulturalists, and amateur hobbyists. The name of the hybridizer or person who identified the grape variety is included if the name is available.

ALICE
labrusca, aestivalis? vinifera?

One of several grapes named Alice, this variety was a "chance seedling" of unknown parentage, found growing near an old stone wall by Ward D. Gunn of Clintondale, Ulster County. It was transplanted into his vineyard in the spring of 1884, where the vine was propagated. After being tested for ten years, the Alice grape was introduced by Gunn in 1895 and attracted great attention, winning great praise from the American Institute and the NY Agricultural Experiment Station, as well as from the Massachusetts Horticultural Society, who awarded it a first-class certificate of merit. It was then introduced to the market by Fred E. Young of Rochester, New York for sale in his nursery business. The Alice grape was considered a good "winter-keeping" grape, valuable for its long-keeping qualities, even without cold storage, and as the skins are thick and tough, without astringency, it was also valued for its shipping potential.

Minor grape varieties, cont'd.

Alice is a *labrusca* grape with a few characteristics that indicate *aestivalis* and *vinifera* heritage. The vine is average in most ways from vigor to production. Fruit ripens with Concord or slightly earlier. It has a dull pale red color that is lighter than Catawba. It is of good quality like Diana or Catawba.[107]

ANNA

labrusca, vinifera

A seedling of Catawba from Eli Hasbrouck of Newburgh. The vine fruited in 1851 or 1852 and was later introduced for sale by Dr. Grant of Iona Island. It resembles Catawba in vine appearance, but is unhealthy and feeble; the bunches are medium in size and loose; the berries are medium in size, pale amber in color, meaty, vinous, and ripen with Catawba.

BROWN

labrusca, vinifera?

A grape identified by William B. Brown of Newburgh, that was found in his yard about 1884 or 1885. Charles Downing examined the vine and determined that it was a seedling of Isabella. The vine was awarded first prize at the New York State Fair in 1892 and a diploma and honorable mention at the World's Columbian Exposition in 1893. The variety was distributed in 1907 to subscribers of the *Rural New Yorker* as Brown's Early. In spite of its many awards, its quality has been described at New York's Geneva Agricultural Experiment Station as only good and the berries shatter badly. As of 1908, the grape had little or no commercial grower following. The vine is medium in vigor, hardy, and very productive. The fruit is reported as being large and keeps well. However, in the same description it states that the clusters are small to medium, slender, and cylindrical or tapering. It adds that the berries are of medium size, oval, black in color with a thick bloom, and the taste is slightly foxy. The berries drop as soon as the grape ripens. The vine blooms by mid-season.[108]

BURROWS NO. 42C

labrusca, vinifera

A seedling of Concord crossed with Jefferson, which New York's Geneva Agricultural Experiment Station received from J. G. Burrows of Fishkill in 1888. It can occasionally be unproductive, with a medium-sized, compact, and handsome bunch; the medium-sized or larger berry is dark red with a lilac bloom, juicy, sweet, tender, slightly vinous, has a fine flavor, and ripens with Concord.

DENNISTON

labrusca

A native grape found on an island in the Hudson River below Albany by Isaac Denniston about 1823. It is described as very vigorous and hardy, with a large yellowish-red berry and a slight musky flavor.

DORINDA

labrusca

Reputedly a seedling of Rebecca found in Hudson, New York, around 1858. The cluster is medium in size, the berry oval, greenish-white, sweet, and sprightly with scarcely any pulp.

EARLY AUGUST

labrusca

Also known as Burton's Early August. A native seedling from the United Society of Shakers, Lebanon, New York, it is described as large, early, and foxy.

EUMELAN

labrusca, vinifera, aestavalis

This important grape, as described in the "Dr. Charles W. Grant" section earlier in this chapter, is a chance seedling found in 1847 in the yard of Mr. Thorne in the Village of Fishkill Landing, which is now part of the city of Beacon, New York. It is supposedly a seedling of Isabella. The grape variety, by some reports, "fell into the hands" of Dr. Grant of Iona Island in 1860, while others suggest that Dr. Grant purchased the original vines from Mr. Thorne in 1866. It was Dr. Grant who introduced the grape as Eumelan in 1867.

This is a quality grape that is good for winemaking. It is a significant grape because breeders used it to create new grape varieties such as Centennial, Corporal, Eumedel, and Laura. The flavor profiles and growing characteristics of Eumelan can be found on pages 97–98.

JUMBO

labrusca

Grown by Mrs. Reuben Rose of Marlboro, New York, this was probably a Concord seedling. It is vigorous, hardy, and productive,

Minor grape varieties, cont'd.

and produces large clusters that weigh one pound. The black berries are very large, almost the size of small blue plums. The fruit ripens with Concord, sometimes earlier, and is described as a good market grape. The *Bushberg Catalogue* describes Jumbo as a favorite market grape in New York, presumably the New York City market.

NORTHERN MUSCADINE

labrusca

This variety originated at New Lebanon, Columbia County, New York. D. J. Hawkins and Philemon Stewart of the United Society of Shakers brought this grape to the public's attention in 1852. It was listed in the *American Pomological Society Fruit Catalogue* in 1862 and dropped in 1871. The aroma and taste of this grape are very foxy. The qualities of the vine are good, being of medium vigor to very vigorous, productive, and of moderate hardiness for a *labrusca*. However, its cloyingly foxy taste has limited its popularity among consumers and hence growers. The fruit ripens with Worden or about two weeks before Catawba and does not keep well. Clusters are medium in size, and the berries are medium to large, round to oval in shape, and dark amber to dull brownish-red, covered with a thin gray bloom. The grapes drop as soon as they ripen.

REBECCA

labrusca, vinifera

An accidental white seedling found in the garden of E. M. Peake at Hudson, New York, it fruited in 1852 for the first time. The vine was introduced to the public four or five years later by W. Brooksbank of Hudson, New York. It received a silver medal in 1856 from the Massachusetts Horticultural Society. In 1856, the American Pomological Society placed it with Concord and Delaware as one of the coveted "new varieties which promise well." It was listed in the *American Pomological Society Fruit Catalogue* in 1862 and removed in 1891.

A. J. Caywood claimed that he had originated the variety and named it Hudson. Caywood, upon the advice of other grape experts, shipped the grape to about sixty growers, including J. W. Prentiss, to obtain their opinions on it. Those who examined the fruit of Hudson and Rebecca deemed them to be very similar, if not identical. The fruit is of exceptional quality, very juicy, has a pleasing aroma, and a slightly foxy taste. However, the vine lacks hardiness and vigor, is susceptible to fungal diseases, and is productive only under the best conditions. The fruit ripens with Concord or later, and keeps and ships well. The medium to small cylindrical clusters are well formed. The berries are intermediate in size, oval, and described as green with a yellowish tinge that verges to amber or alternatively as yellow-white and semi-transparent. The taste is vinous and slightly foxy. Must average is 17° Brix. Due to its high quality, this grape should be reexamined for cultivation in the Hudson Valley.

RELIANCE

vinifera, bourquiniana, labrusca

Exhibited by J. G. Burrows of Fishkill, New York, at the American Pomological Society meeting in 1881. It may be a cross of Delaware and Iona. The grape is described as vigorous, hardy, and very productive; the bunch resembles Delaware in size, but is not so compact; the medium-sized berry is light red, tender, juicy, sweet, and ripens with Delaware.

SCHOONEMUNK OR SKUNNYMUNK

labrusca

A native seedling found by W. A. Woodward of Mortonville, Orange County, New York (southwest of Newburgh, New York), around 1860. It is named after nearby Schunemunk Mountain. Based on the location of the mountain, Mortonville may in fact be Mountainville because Mountainville is even closer to Schunemunk than Mortonville, which was the Revolutionary Army headquarters for General Henry Knox. This grape is reported to be hardy, and productive; the fruit is equal in size and flavor to Concord, though it ripens earlier.

THOMPSON'S SEEDLINGS

David Thompson of Green Island, New York (near Troy, New York), raised a large number of seedlings in the late 1860s or early 1870s that had *vinifera* parentage. These varieties were not successful commercially since *vinifera* grape varieties could not then be cultivated in an open vineyard. Among his named seedlings were David Thompson, General Grant, L. H. Tupper, Nathan C. Ely, A. B. Crandall, Bonticue, Early August, William Tell, Lavina, Elenor, and Jas. M. Ketchum.

As discussed earlier in this chapter, the goal of the Hudson Valley hybridizers was primarily to enhance the quality of locally grown *V. labrusca* and *labrusca* hydrids by incorporating European *vinifera* characteristics to increase the quality of their table grapes and, as an ancillary goal, some developed grapes that were suitable for wine production. Unlike the grape hybridizers of the Midwest and Plains states, such as T. V. Munson, who had complex grape-breeding programs that lasted for two or more decades, most Hudson Valley hybridizers, with the exception of James H. Ricketts and A. J. Caywood, created breeds that were not complex and did not incorporate many different American grape species into newer, more complex hybrids. Further, except for the work of Caywood and Ricketts, most of their breeding programs generally lasted less than a decade.

The next two chapters will look at the work of French grape breeders during the same time period starting in the 1860s and ending in the 1950s. The goal of these French hybridizers was similar to that of the Midwest and Plains states hybridizers and very different from the Hudson Valley hybridizers. The French hybridizers' goal was to impart enhanced fungal disease and insect resistance into the American grape species that the French were then using for wine production.

NOTES

1. Edward Payson Roe, *Success with Small Fruits* (New York: Dodd, Mead & Company, 1880). Originally serialized in *Scribner's Magazine*, this book details the extensive amount of communication exchanged among grape hybridizers and growers in the Hudson Valley. Mr. Roe had his home in Cornwall-on-Hudson, New York.

Roe's interaction with local hybridists and horticulturalists are indicated in his Preface and Dedication:

"Dedicated to Mr. Charles Downing, a neighbor, friend, and horticulturist from whom I shall esteem it a privilege to learn in coming years, as I have in the past.... I shall, moreover, always cherish a grateful memory of the aid received from my brother, the Rev. A. C. Roe, and from Mr. W. H. Gibson, whose intimate knowledge of nature enabled him to give so correctly the characteristics of the fruits he portrayed. I am greatly indebted to the instruction received at various times from those venerable fathers and authorities on all questions relating to Eden-like pursuits—Mr. Chas. Downing of Newburgh, and Hon. Marshall P. Wilder of Boston, Mr. J. J. Thomas, Dr. Geo. Thurber; to such valuable works as those of A. S. Fuller, A. J. Downing, P. Barry, J. M. Merrick, Jr., and some English authors; to the live horticultural journals in the East, West and South, and, last but not least, to many plain, practical fruit-growers, who are as well informed and sensible as they are modest in expressing their opinions."

2. A. J. Downing, *The Fruits and Fruit Trees of America: Or the Culture, Propagation and Management in the Garden and Orchard of Fruit Trees Generally; with Descriptions of All the Finest Varieties of Fruit, Native and Foreign in this Country* (New York: Wiley and Putnam, 1845).

3. A. J. Downing, *A Treatise on the Theory and Practice of Landscape Gardening, Adapted to North America: With a View to the Improvement of Country Residences* (New York: Wiley and Putnam, 1841).

4. A. J. Downing, *Cottage Residences: or a Series of Designs for Rural Cottages and Cottage Villas, and Their Gardens and Grounds, Adapted to North America* (New York: Wiley and Putnam, 1842).

5. John J. Nutt, *Newburgh: Her Institutions, Industries, and Leading Citizens* (Newburgh, N.Y.: Ritchie & Hull, 1891), 196, 197.

6. U. P. Hedrick, *The Grapes of New York. Report of the New York Agricultural Experiment Station for the Year 1907. Department of Agriculture, State of New York. Fifteenth Annual Report, Vol. 3 – Part II.* (Albany, N.Y.: Lyon Company, 1908); U. P. Hedrick, *Manual of American Grape- Growing* (New York: The Macmillan Company, 1919).

7. *Bushberg Catalogue: A Grape Manual* (Bushberg, Jefferson County, Mo.: Bush & Son & Meissner, 1895).

8. See Walter Wellhouse, ed., *Transactions of the Kansas State Horticultural Society, The Proceedings of the Forty-second and Forty-third Annual Meetings, Topeka, December*

1908 and December 1909, vol. XXX (Topeka, Kans.: State Printing Office, 1910).

9. Caywood's obituary states that he was married at the age of seventeen, which would place the marriage around 1836. However, his family records indicate otherwise, unless his marriage to Deborah Cornell was his second marriage in 1848. See Bailey, *Annals of Horticulture*.

10. See the obituaries of A. J. Caywood, *Vick's Monthly Magazine*, February, 1889 (Rochester, N.Y.: James Vick Seedsman); L. H. Bailey, *Annals of Horticulture in North America for the Year 1889: A Witness of Passing Events and a Record of Progress* (New York: Rural Publishing Company, 1890); and *Popular Gardening and Fruit Growing*, March 1889.

11. Map of Section 16, portion of Orange, Ulster, and Dutchess counties, by Watson & Company, 1891.

12. See obituaries of A. J. Caywood, *Vick's Monthly Magazine*; Bailey, *Annals of Horticulture*.

13. Mark Miller, *Wine—A Gentleman's Game, the Adventures of an Amateur Winemaker Turned Professional*. (New York: Harper & Row, 1984), 23–25.

14. *Bushberg Catalogue*, 100.

15. See generally, the obituaries of A. J. Caywood, and specifically *Newburgh Daily Journal*, January 15, 1889, 3; *Vick's Monthly Magazine*, February, 1889.

16. Obituary of Andrew J. Caywood, *Newburgh Daily Journal*, January 15, 1889, 3.

17. Benmarl Vineyards was purchased by Mark and Dene Miller and their family in 1957, and sold in 2006 to Victor Spaccarelli Jr. and his family, who continue to operate the business under the name Benmarl Winery.

18. Probate papers of Deborah Caywood, July 8, 1891, Marlboro Free Library, Historical Section.

19. Jancis Robinson maintains that Dutchess, also called White Concord, is a seedling of Niagara, pollinated with a mix of pollen from Delaware and Walter. See Jancis Robinson, Julia Harding, and Jose Vouillamoz, *Wine Grapes: A Complete Guide to 1,368 Vine Varieties, Including Their Origins and Flavor*. (New York: HarperCollins, 2012), 318. However, this is probably not correct since Dutchess was bred in 1868, the same year when the variety Niagara was bred and which Niagara fruited for the first time in 1872. Niagara was then introduced in 1882. Further, Caywood extensively used his own variation of a White Concord and rarely used Niagara in his breeding program. As Niagara was a widely planted grape shortly after it was introduced in 1882, there should have been little confusion between White Concord and Niagara. See *Bushberg Catalogue*, 107.

20. Robinson et al., *Wine Grapes*, 318.

21. Hedrick, *Grapes of New York*, 246.

22. Ibid., 358.

23. *Bushberg Catalogue*, 160.

24. Hedrick, *Grapes of New York*, 381; also see Hedrick, *Manual of American Grape-Growing*, 427–28.

25. *Bushberg Catalogue*, 170.

26. Ibid.

27. See Hedrick, *Grapes of New York*, 414; Hedrick, *Manual of American Grape-Growing*, 441.

28. *Bushberg Catalogue*, 183.

29. Hedrick, *Grapes of New York*, 419; Hedrick, *Manual of American Grape-Growing*, 443.

30. *Bushberg Catalogue*, 185.

31. Hedrick, *Grapes of New York*, 420.

32. Ibid., 419; Hedrick, *Manual of American Grape-Growing*, 443.

33. Hedrick, *Grapes of New York*, 483.

34. *Bushberg Catalogue*, 151.

35. Ibid., 155.

36. Hedrick, *Grapes of New York*, 487.

37. *Bushberg Catalogue*, 156.

38. Ibid., 107.

39. Ibid.

40. Francis R. Kowsky, *Country Park & City: The Architecture and Life of Calvert Vaux* (New York: Oxford University Press, 1998), 36.

41. Obituary of Dr. William A. M. Culbert, *Newburgh Daily Journal*.

42. Dr. William A. M. Culbert's biography (with his photograph) is a summary of information contained in Edward Manning Ruttenber and L. H. Clark, *A History of Orange County, New York, with Illustrations and Biographical Sketches of Its Many Pioneer and Prominent Men* (Philadelphia: Everts & Peck, 1881), 184–85; John J. Nutt, *Newburgh*, 86–87; obituary of Dr. William A. M. Culbert, *Newburgh Daily Journal*, November 10, 1890, 1.

43. Obituary of William A. M. Culbert, *The New York Times*, November 11, 1890.

44. Emilie Kracen, Litchfield Historical Society, "Grant Family Paper, 1817–1869," www.litchfieldhistorical society.org.

45. Hedrick, *Grapes of New York*, 304.

46. Elizabeth "Perk" Stalter, *Doodletown: Hiking Through History in a Vanished Hamlet on the Hudson* (Bear Mountain, N.Y.: Palisades Interstate Park Commission Press, 1996), Appendix, Map Section.

47. My family members who helped to establish Chapel of St. John the Divine were Charles M. Casscles, warden, and Joseph Casscles, vestryman, along with William Springstead, vestryman, an ancestor of the singer and songwriter Bruce Springsteen. See Rev. David Cole, DD, *History of Rockland County, New York with Biographical Sketches of Its Prominent Men* (New York: J. B. Beers & Company, 1884), 332; Joanne Potanovic, *The Chapel of St. John the Divine, Our History* (Tompkins Cove, N.Y.: The Chapel of St. John the Divine, updated January 2012).

48. "Iona Island and the Fruit Growers Convention of 1864," *New York History*, 1967, 332–51; and Hedrick, *Grapes of New York*, 304.

49. Kracen, "Grant Family Paper."

50. Libraries include the New York State Library in Albany, the New York Public Library, the New York Historical Society Library, The LuEsther T. Mertz Library at the Bronx Botanical Gardens, the Chicago Botanic Garden, the University of Minnesota, etc. Every few years, Grant issued his catalogue *Illustrated Catalogue of Vines, etc., etc.: With Explanatory Remarks, and Indications for Cultivation*, with different printer/publishers (John A. Gray, printer and stereotyper, corner of Frankfort and Jacob streets,), and engravings by Henry Holton, engraver. For some reason, Mrs. C. W. Grant's name is also attached to the catalogue. Grant also seems to have published an earlier version in 1859 (already in its fourth edition) with the similar title of *Descriptive Catalogue of Vines, etc., etc.: with Explanatory Remarks, and Indications for Cultivation.* This was published by C. M. Saxton in 1859, and is at the New York State Library in Albany (Manuscripts and Special Collections, call no: 634.8 G76 1859).

51. Hedrick, *Grapes of New York*, 304.

52. Ibid., 267. Hedrick seems to have backed off on his claim that Eumelan was a seedling of Isabella and remained silent on the matter in his *Manual of American Grape-Growing*, 371–72.

53. *Bushberg Catalogue*, 126.

54. Hedrick, *Grapes of New York*, 267.

55. Ibid.

56. Ibid., 266.

57. Hedrick, *Manual of American Grape-Growing*, 372.

58. *Bushberg Catalogue*, 101, 108, 147.

59. Hedrick, *Grapes of New York*, 304.

60. Ibid., 302–5.

61. Ibid., 312.

62. *Bushberg Catalogue*, 152.

63. Hedrick, *Grapes of New York*, 318.

64. Obituary: "James H. Ricketts Dies in Ninety-Seventh Year," *Washington, D.C. Evening Star*, December 1, 1915, 9.

65. *Newburgh City Directory* (Newburgh, N.Y.: L. P. Waite & Company, 1870–89).

66. Hedrick, *Grapes of New York*, 318–19.

67. Commissioner of the United States Department of Agriculture, *Report of the Commissioner of Agriculture for the Year 1875* (Washington, D.C.: Government Printing Office, 1876), 385–86.

68. Hedrick, *Grapes of New York*, 318–19.

69. There is a conflict in the Rickettses' address in Washington, D.C., based on James's address in his obituary, and the address cited in the obituary of his wife, Ruth M. Ricketts, who died on February 10, 1904. See "James H. Ricketts Dies ...," 9; obituary: "Ruth M. Ricketts," *Washington, D.C. Evening Star*, February 10, 1904.

70. U. P. Hedrick, *A History of Agriculture in the State of New York* (Albany, N.Y.: Printed for the New York State Agriculture Society, 1933), 393.

71. Jancis Robinson relates the story of Clinton's origin, which was also recounted in Hedrick's *The Grapes of New York*, that it was first planted in 1821 by Hugh White, then a student at Hamilton College on the grounds of Prof. Noyes's home on College Hill in the Town of Clinton, which is east of Syracuse, New York. The seedling came from hundreds of seeds planted in Mr. White's father's garden in Whitesboro in 1819. Noyes named it after the Town of Clinton. This account appeared in 1863 in the *Elmira Advertiser*. This story was not refuted, perhaps because the introducer in Rochester had died. Robinson gives only this account of the origin of Clinton, but there other theories as well. See Robinson et al., *Wine Grapes*, 252; Hedrick, *Grapes of New York*, 214–15.

72. Hedrick, *Grapes of New York*, 213–16; Hedrick, *Manual of American Grape-Growing*, 350.

73. Hedrick, *Grapes of New York*, 242–43.

74. *Bushberg Catalogue*, 125.

75. Hedrick, *Grapes of New York*, 262; Hedrick, *Manual of American Grape-Growing*, 370.

76. *Bushberg Catalogue*, 125.

77. Ibid., 84.

78. Ibid., 128.

79. Hedrick, *Grapes of New York*, 327.

80. *Bushberg Catalogue*, 84.

81. Hedrick, *Grapes of New York*, 493.

82. *Bushberg Catalogue*, 168.

83. Ibid., 179.

84. Thomas Pinney, *A History of Wine in America* (Berkeley: University of California Press, 2007), 194.

85. Ibid., 191–94.

86. Ibid., 192.

87. Sarah Gibbs Underhill, "A Brief History of Croton-on-Hudson, Tales from Croton Point," www.brickcollecting.com.

88. James Albert Casscles, with his three sons James Albert, Samuel, and Charles, operated the sawmill.

89. Leon D. Adams, *The Wines of America* (Boston: Houghton Mifflin Co., 1973), 121–22.

90. Ibid., 192.

91. *Bushberg Catalogue*, 109.

92. Hedrick, *Manual of American Grape-Growing*, 355–56.

93. *Bushberg Catalogue*, 154.

94. Hedrick, *Grapes of New York*, 185–86; Hedrick, *Manual of American Grape-Growing*, 338.

95. Hedrick, *Grapes of New York*, 184.

96. *Bushberg Catalogue*, 90.

97. Ibid.

98. Ibid., 141.

99. Hedrick, *Grapes of New York*, 473. The variety was eliminated completely from Hedrick's *Manual of American Grape-Growing*.

100. *Bushberg Catalogue*, 141.

101. Ibid., 179.

102. Ibid., 180.

103. Obituary: "Wm. D. Barns," *Newburgh Sunday Telegram*, October 23, 1904.

104. Information on the Barnes family biographies was obtained from Russel Headley, ed., *The History of Orange County* (New York: Van Deusen & Son, 1908), 342, 345, 639–40, 779–80; *Portrait and Biographical Record of Orange County, NY* (New York and Chicago: Chapman Publishing Company, 1895), 513–15; Ruttenber and Clark, *History of Orange County*, 368; and conversations with Jackson Baldwin in the winter of 2009, the subsequent owner of a portion of the Barnses' farms since the 1920s.

I am very familiar with the old homestead of the Nathaniel Barnes family as it is my paternal grandmother's family fruit farm in Middle Hope, where I grew up and where I still grow grapes. It is no more than 200 yards from a portion of the Barnes family's lands. Further, our family friends, Jackson and Marian Baldwin, purchased their farm from the spinster Barnes daughters in the 1920s. When the Baldwins purchased the Barnes home, a very old center hall colonial home dating from the late 1790s, most of the vineyards and apple trees were in disarray. It seems that the economic conditions of the first quarter of the twentieth century were not kind to the Hudson Valley fruit industry. Older Delaware, Niagara, and Concord vineyards were all that remained. Nancy Baldwin, the daughter of Jack and Marian, still lives there and cultivates herbs. In front of her home is the farmstand that the Baldwins operated until the 1970s, but it is now owned by Jim Lyons and called Overlook Farm Stand.

105. The biographical information on the Clark family was obtained from Nutt, *Newburgh*, 202–3.

106. Roe's *Success with Small Fruits* was originally serialized in *Scribner's Magazine* in the late 1880s.

107. Hedrick, *Grapes of New York*, 164. The New York Experimental Station in Geneva had first noticed the grape in 1892, and announced Alice as a new grape in their 1896 annual report. Although the 1895 *Bushberg Catalogue* correctly noted the Alice grape (as Alice II), and its qualities, they erroneously listed Gunn as "Ward B. Greene". *Bushberg Catalogue*, 84.

108. Hedrick, *Grapes of New York*, 195.

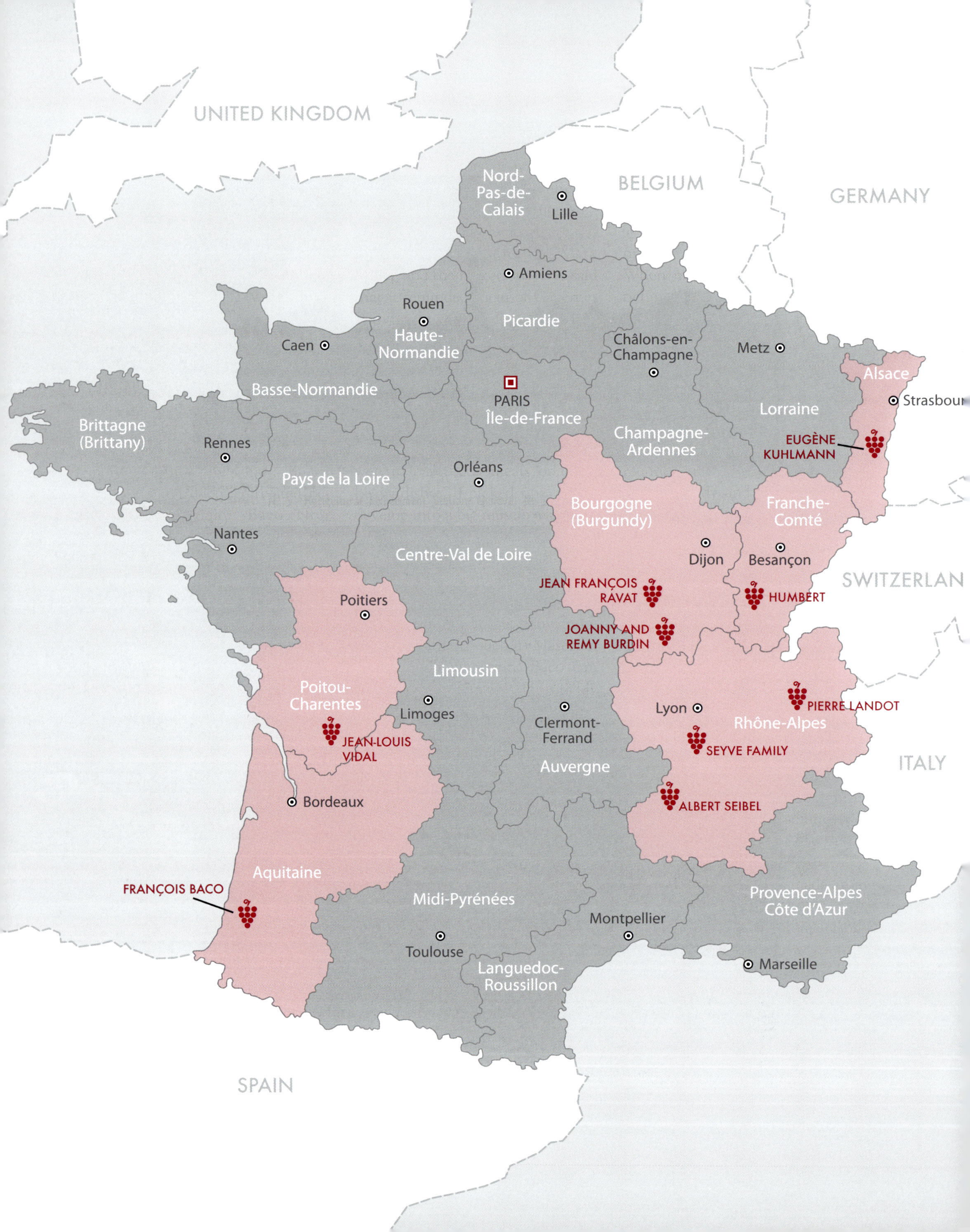
UNITED KINGDOM
BELGIUM
GERMANY
Nord-Pas-de-Calais
Lille
Amiens
Picardie
Rouen
Haute-Normandie
Caen
Basse-Normandie
Châlons-en-Champagne
Metz
Alsace
Strasbourg
PARIS
Île-de-France
Lorraine
Brittagne (Brittany)
Rennes
Champagne-Ardennes
EUGÈNE KUHLMANN
Pays de la Loire
Orléans
Bourgogne (Burgundy)
Franche-Comté
Nantes
Centre-Val de Loire
Dijon
Besançon
SWITZERLAN
JEAN FRANÇOIS RAVAT
HUMBERT
Poitiers
JOANNY AND REMY BURDIN
Limousin
Poitou-Charentes
Limoges
Lyon
PIERRE LANDOT
Clermont-Ferrand
Rhône-Alpes
JEAN-LOUIS VIDAL
SEYVE FAMILY
ITALY
Auvergne
Bordeaux
ALBERT SEIBEL
Aquitaine
FRANÇOIS BACO
Midi-Pyrénées
Provence-Alpes Côte d'Azur
Montpellier
Toulouse
Marseille
Languedoc-Roussillon
SPAIN

CHAPTER NINE

Early French Hybridizers 1875–1925

BY THE MIDDLE of the nineteenth century, the importation of American grapevines to Europe for horticultural study and scientific use led to the inadvertent introduction and spread of the North American root-sucking insect, phylloxera, and fungal diseases such as powdery mildew, downy mildew, and black rot to France and the rest of Europe. Their rapid spread decimated European vineyards after 1868. Within twenty years, virtually every vine in France was injured or killed. In reaction to this economic, political, and social crisis, two schools of thought—which I call the Viniferists and the Hybridists—developed in France to address this critically important problem.

The goal of the Viniferists—professors and accredited horticulturalists who based their academic work at leading French national universities—was to continue growing their traditional *vinifera* varieties such as Chardonnay, Pinot Noir, Cabernet Sauvignon, and Sauvignon Blanc. Their benefactors were the wealthy wine producers based in Bordeaux and Burgundy, who wanted their traditional grape varieties.

Hybridization programs in France. *Beginning in the latter part of the nineteenth century, breeding programs were established in wine regions across France, including Alsace, Jura, Rhône, Burgundy, and Bordeaux.*

Viniferists addressed this multifaceted problem by first developing rootstocks to protect *vinifera* vines from phylloxera, which attacked soft *vinifera* roots and, second, developing sprays to protect the fruit and leaves from newly introduced American fungal diseases. The phylloxera-resistant rootstocks had to be derived from American grape species roots that were either immune or highly resistant to this insect. The second problem of protecting vines from American fungal diseases was addressed in 1885 by the botanist Professor Pierre-Marie-Alexis Millardet of Bordeaux University, who developed the fungicide called "Bordeaux Mixture," which is a combination of powdered lime and copper sulfate.

The Hybridists, who were primarily amateur horticulturalists and nurserymen, lived in the backwater areas of France where the economy relied heavily on the production of bulk wine for everyday consumption. This group looked to developing new "direct-producer" grape varieties that could produce bulk wines inexpensively. Direct-producer grape varieties had the benefit of not needing expensive rootstock grafting or extensive fungicide spray applications, both of which increased material costs and required additional vineyard labor and new fungicide-spraying equipment.

Collectively, the Hybridists had to address five concerns as they developed their new hybrid grape varieties. They wanted interspecific grape hybrid direct producers with the following characteristics:

1. The varieties had to have enhanced fungal disease resistance to downy mildew, powdery mildew, and black rot.

2. These grapes needed to have enhanced wine quality compared to the disease-resistant American grape varieties, such as Noah, Clinton, Othello, Herbemont, Lenoir, or Isabella, which were then temporarily being grown in France to produce wine because local *vinifera* vineyards were dying or disease-ridden. These American varieties had solid phylloxera and fungal disease resistance, but they did not make palatable wine.

3. The varieties needed enhanced resistance to high-alkaline soils. Such soils are common in many French vineyards, but American grapes did not grow well in such soils, while *vinifera* grapes did.

4. The hybrids needed to have delayed bud break to avoid damage from late spring frosts or which could produce a good secondary crop if a late spring frost occurred.

5. They had to be capable of producing high yields.[1]

All of the early Hybridists were born between 1845 and 1865, and started their hybridization work between 1875 and 1885 in reaction to the American invasion of insects and fungal diseases that were destroying French vineyards. In sum, they were all young men in their late twenties to early thirties who were confronting a serious economic crisis that directly affected their local communities and who resided in areas outside of where classic noble grape variety production occurred.

These pioneers included Albert Seibel, Bertille Seyve Sr. and his sons Bertille Seyve Jr. and Joannes Seyve, François Baco, and Eugène Kuhlmann, detailed on the following pages. I did not include the hybridizers Georges Couderc, Pierre Castel, or Ferdinand Gaillard because few of their hybrids are, or could be, grown in the Hudson Valley.

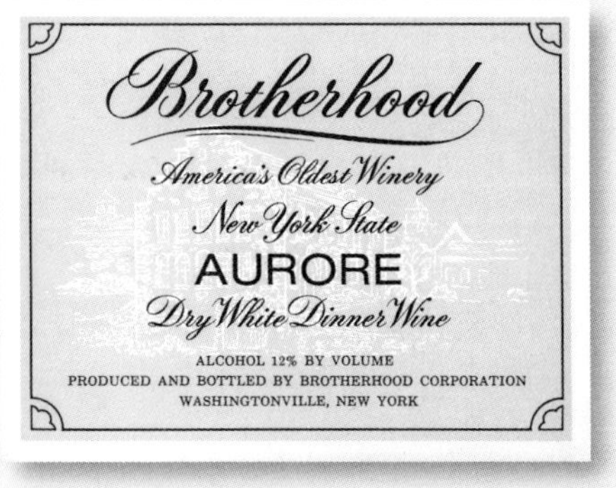

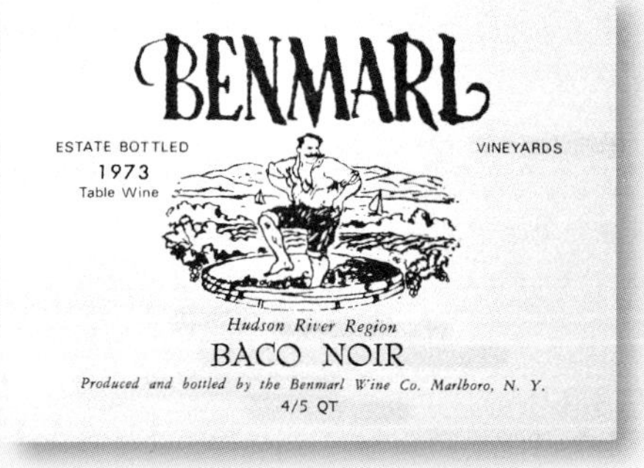

French hybrid wines. *Above: Philip M. Wagner, former editor of the Baltimore Sun and founder of Boordy Vineyards, was the first to introduce French-American hybrid grapes to the United States. Left: Wine labels from Hudson Valley vineyards featuring French-American hybrids. Circa 1970s.*

Albert Seibel
(1844–1936)

Albert Seibel was the most prominent and prolific hybridizer of French-American hybrid direct-producer grapes in France. At one point, his hybrids in France covered more ground than the acreage of all other hybridizers combined. Shortly after World War II, it was estimated that over 500 different commercially grown Seibel varieties covered more than approximately 370,000 acres in France. This is an area about half the size of the entire Catskill Mountain State Park Preserve. The more commonly grown varieties included Chancellor, Rayon d'Or, Aurore (also called Aurora), Chelois, Verdelet, Plantet, and Seinoir.

Even as late as the 1960 French vineyard registration, Seibel hybrids accounted for approximately one-third of all French-American hybrids grown in France. At that time, French-American hybrids made up approximately one-third of all vineyard land in France. The percentage devoted to Seibel hybrids and French-American grape varieties in general has gone down substantially since then to under 10 percent. In the 1960 vineyard registration cited above, the most popular Seibel varieties were Chancellor (99,190 acres), Plantet (68,944 acres), Roi des Noirs (30,790 acres), Rayon d'Or (17,224 acres), Rosette (9,094 acres), Flot Rouge (8,698 acres), S.1 (7,883 acres), Subereux (4,065 acres), Gloire de Seibel (3,188 acres), Colobel (3,133 acres), Seinoir (2,842 acres), S.5437 (unnamed) (2,820 acres), Aramon du Gard (2,718 acres), Duc Petit (2,545 acres), Pate Noir (2,286 acres), and Chelois (2,236 acres).[2]

Seibel had a very ambitious breeding program that incorporated many different sources of genetic material. Further, he had a successful commercial nursery business to disseminate his new hybrids. In Seibel's first nursery stock offerings, he provided a two- to three-line description for each variety that included genetic origin, disease resistance, fertility, and harvest time.[3]

His work is still relevant today because of the large amount of acreage of his grapes that are cultivated in France and the United States. His work is also very important because his hybrids were the basis of the work of future French hybridizers such as Joanny Burdin, Bertille Sevye Jr., Joannes Seyve (Seyve-Villard), J. F. Ravat, and Jean-Louis Vidal, and current grape-breeding programs in Canada, Germany, Hungary, and New York.

Early Hybridization

Albert Seibel was born in 1844 in the commune of Aubenas, Ardèche department, in the Rhône-Alpes region of France, which is south of Lyon. He died there in 1936. Dr. Seibel was a physician who was also a viticulturist. He began hybridizing in 1874 at the age of thirty. Part of his success was due to his long and productive life. His first hybrids had mediocre results, so Seibel gave up hybridizing for a period of time, starting up again in 1886 when his neighbor, Eugène Contassot (a local pastry maker), offered him seeds from his hybrids in which the female plant was Jaeger 70 (a *V. lincecumii* and *V. rupestris* hybrid produced by Hermann Jaeger of Neosho, Missouri). Similar seeds were also offered to Georges Couderc. These seeds gave Albert Seibel his first plants, which he numbered 1, 2, 3, etc., until his final crosses tallied at over 19,975.[4] The names of some of his more significant grape hybrids grown in the Hudson Valley include Chancellor, Rayon d'Or, Aurore, Chelois, De Chaunac and Verdelet. Nearly all grape hybridizers today use at least some of the progeny that he developed, and it is interesting to note that Seibel used the work of Hudson Valley hybridizers such as A. J. Caywood. In 1895, Seibel founded a school to teach grafting methods.

Methodology

Seibel was by far the most prolific and successful grape hybridizer who produced for sale hundreds of different varieties of grapes that were of greatly varying characteristics both viticulturally and in terms of quality. He developed both red and white varieties that can be grown in most of the continental United States and which have very different winemaking characteristics and flavor profiles.

La Maison Seibel. *French advertisement promoting the nursery of the late hybridizer Albert Seibel, under the direction of his nephew Louis Seibel. Circa 1938.*

It is very difficult to summarize Albert Seibel's hybridization program as it is so extensive that it deserves a book of its own. Of his approximately 20,000 recorded crosses, named varieties include the aforementioned Chancellor, Rayon d'Or, Aurore, Chelois, Verdelet, Plantet, and Seinoir, as well as Pate Noir, Rosette, Flot d'Or, Bienvenu, Rougeon, Subereux, Colobel, De Chaunac, Rubilande, Cascade, and Bellandais. Most of these varieties have been used by other hybridizers to advance the propagation of even newer grape varieties. Seibel hybrids such as Plantet, Subereux, Rayon d'Or, Chancellor, and Seinoir were those most often used in the work of Seibel's contemporaries and subsequent hybridizers.

The genetic material that Seibel used to develop his hybrids included a wide variety of plant material: American hybrids such as Munson, Ganzin 1, Jaeger, and Othello; various *vinifera* varieties, such as Aramon, Piquepoul, Alicante, and Sauvignon Blanc; wild American grape species such as *berlandieri* and *rupestris;* interspecific hybrids such as Noah, Herbemont, Vivarais, Gaillard 2, and Baynard; Hudson Valley–developed hybrid grapes such as Dutchess; and many strains of his own Seibel hybrids. It is interesting that one *vinifera* variety, Aramon, found its way into Seibel's most popular hybrids. From the late 1800s to the mid-twentieth century, Aramon was one of the most widely planted grapes in France. Further, one of Aramon's parents, Gouais Blanc, is also one of the parents of Chardonnay and Riesling as determined by DNA testing.

It seems that the base of many quality *vinifera* whites such as Chardonnay and Riesling are also in the genetic makeup of Seibel hybrids such as Rayon d'Or, Subereux, and Plantet, which then became the parents of higher quality Seibel hybrids such as Verdelet and Colobel. It is important to note that Subereux and Plantet, which have Aramon in their genetic makeup, became the parents of the more noted later Seibel hybrids. Hence, it seems that just a few grape varieties are the ones that impart quality to later generations of interspecific hybrids. Again, an entire book could be written about this subject.

After reviewing Seibel's breeding program and the varieties that he crossed to develop new ones, he seems to have tried to maximize genetic diversity by crossing many American species and American hybrids with locally grown productive *vinifera* varieties (either directly or by hybrids whose parents were grapes such as Ardèche, Cinsault, Raisaine, and Alicante-Bouschet). In his early work he hybridized American varieties directly with European varieties and used the breeding material of Georges Couderc, Victor Ganzin, Ferdinand Gaillard, and Thomas V. Munson. As his long career progressed, he crossed his own hybrids with each other.

By 1894, Seibel had reached his 2000 series of hybrids. The 4000 and 5000 series used less foreign material, and by the 6000 series, hybrids other than Seibel hybrids were rarely used. Over time, he introduced new *vinifera* grapes to improve the gene pool for the production of quality wine and to increase resistance to high-alkaline soils. As he crossed different varieties, he would relentlessly select offspring with desirable growth habits, disease resistance, and winemaking potential as parents for his next series of crosses.[5]

Because French-American hybrids such as those developed by Seibel required less than one-third of the sprays to produce a ton of grapes, these hybrids accounted for up to 55 percent of all French wines made during and immediately after World War II. However, after the war, great quantities of cheap *vin ordinaire* were brought into France from its colony of Algeria and other southern European countries such as Spain, Italy, and Yugoslavia.[6]

To compete, the French government adopted the strategy of emphasizing quality and not quantity of wine produced, and began to curtail and then ban the production of wines made from the older hybrids. In 1934, direct producers with much American *Vitis* parentage such as Isabella, Clinton, Noah, Othello, Jacquez (Black Spanish), and Herbemont were officially banned from further plantings. While these grapes had superior disease resistance, their winemaking capabilities were rejected. After the end of World War II, as there was no longer a need to produce large quantities of wine with few production inputs, the French government, in 1955, began a new program to discourage later-bred hybrids from being planted and sold for wine production. Grape varieties were divided into three categories: "recommended," "authorized," and "tolerated." In 1970, the French government and the European Common Market, using Regulation 2005/70, placed what we now know as French hybrids into the last two categories. Further, the grapes designated as "tolerated" were further downgraded to "temporarily authorized," which placed even more restrictions on hybrid varieties.[7]

Under these and later regulations, only "authorized" varieties (in order of importance) could be grown, such as: Villard Noir (S.V.18-315), Plantet (S.5455), Couderc Noir (C.7120), Villard Blanc (S.V.12-375), Baco Blanc (Baco 22A), Baco Noir (Baco 1), Garonnet (S.V.18-283), Seinoir (S.8745), Rayon D'Or (S.4986), Oberlin Noir (O.595), Varousset (S.V.23-657), Chambourcin (J.S.26.205), Seyval Blanc (S.V.5-276), Colobel (S.8357), Chelois (S.10878), Ravat Blanc (R.6), Rubilande (S.11.803), Landal (L.244), Roucaneuf (S.V.12-309), Léon Millot (K.194–2), Maréchal Foch (K.188-2), Valerien (S.V.23-410), Bellandais (S.14.596), Ambror (S.10.173), and Florental (B.7705). The chill of government regulation was felt in the field as nurserymen reduced the number of direct producers, or French hybrids, sold from 68 million or 27.6 percent of all grape cuttings sold in France in 1958, to 8.6 million or 2.4 percent of all nursery cuttings in 1975. Further, while in 1958 a total of 993,724 acres (about 31 percent) of the total vineyard area in France was devoted to French hybrid grapes, that amount declined to 739,399 acres (23.5 percent) in 1968, and to less than 494,000 acres of French hybrids (about 20 percent) of total vineyard acreage by 1979.[8]

With this change in the winemaking environment, René Seibel (the grand-nephew of Albert), who was running Seibel Nurseries, found it more difficult to make a living. Sales of his hybrid cuttings fell from a postwar high of approximately 70 million cuttings per year to less than 5 million. Further, in 1953, the city of Aubenas expropriated Seibel's nursery farm for the development of factories and warehouses.[9]

While Seibel hybrids have become less of a factor in the French wine industry, they still remain very important to the cool climate wine industry in the United States, England, Canada, and New Zealand. Their disease resistance and cold hardiness are valuable building blocks for future wine grape hybrids.

On the following pages are six grape varieties developed by Albert Seibel that are grown in the Hudson Valley.

AURORA (S.5279)

This white hybrid, also known as Aurore, which was among the final third of Seibel's hybrids, is composed of S.788 × S.29, two of his earliest crosses. S.788 is a simple white hybrid of Sicilien × Clairette Dorée Ganzin and S.29, which is a black grape cross of a Munson hybrid × an undisclosed *vinifera*. In 2001, there were 738 acres of Aurora planted in New York which were reduced to 621 acres by 2006, but then increased to 659 acres in 2011.[10] At one point, there were 712 acres of Aurora in the northeast and southwestern part of France.[11]

PARENTAGE
lincecumii, *rupestris*, *vinifera*

HARVEST DATE
Very early to early

A A- B- A+ B

Aurora, sometimes called Aurore,[12] is one of the earliest ripening white French-American hybrids. It was initially bred as a table grape that could be made into wine, but it did not hold up well for table grape applications and its grapes fell off the stems. Nevertheless, it is a reliable producer of mildly fruity wines that have some flint in its body, but it is coarse. Aurora has an early bud break, so it is susceptible to late spring frosts. It is winter hardy, but not as hardy as Seyval Blanc. Aurora is susceptible to black rot, powdery mildew, and bunch rot, but less so to downy mildew. It has no sulfur sensitivity. The vine is vigorous and productive to very productive. It has large to very large clusters that are long, meaty, and somewhat loose. It was named after Aurora, the Roman goddess of the dawn,[13] and as its name suggests, the berry color is white-pink to grayish-yellow to amber when very ripe. As noted, when ripe, the berries tend to fall off the cluster.

Since Aurora ripens so early, bird damage and fruit rot can be a problem. The growth habit is drooping rather than lateral. Cordon or cane pruning is recommended depending on the fertility of the site. The grapes tend to break down when fully ripe, producing more wild native American grape flavors. The wines are balanced for acid and sugar.[14]

The wine is light, pale, fresh, fragrant, and delicate of medium body. It is often used for blending to reduce the high acid of other whites. It oxidizes rapidly, so it should not be racked often and sulfur levels should be kept high. The wines can be neutral and flabby, but can have some elements of Alsatian wines. They tend not to be of very high quality, oxidize early and are not long lived. Therefore, it is best to use Aurora in blends or alone as a young wine to be consumed early. With its neutral flavors and mild acids, Aurora is a good blender with other more highly flavored and acidic whites.

The wine descriptors for Aurora are that it is floral, has lots of melon, some light mead, and peach/apricot, but in the end it has mostly metallic neutral flavors. It can also be coarse, watery, and not very unique if not made well. Aurora is most like a simple Muscadet or Moselle.

Aurora is a suitable grape for the production of neutral wines, but is not highly recommended for further expansion of acreage in the Hudson Valley. It is a very good grape for areas with a colder and shorter—but not longer—growing season in the Hudson Valley.

CASCADE (S.13053)

Cascade was one of the last named grape varieties to be hybridized by Seibel that has been commercially available since 1938.[15] The variety was named Cascade in 1970.[16] It has a complex genetic heritage and was Seibel's later attempt to develop a quality grape for both the field and cellar. It was never really grown in France because by the time it was developed in the 1920s, the pendulum was already beginning to swing away from direct-producing hybrids and toward more established classic *vinifera* grapes such as Chardonnay and Cabernet Sauvignon.

PARENTAGE
aestivalis, cinerea, labrusca, lincecumii, riparia, rupestris, vinifera

HARVEST DATE
Very early to early

A+ A B+ A A

Cascade is a hybrid of S.7042 × Gloire de Seibel (S.5409). In the United States, it is believed to either have a virus infection in most of its propagating wood or be very sensitive to soil-borne viruses, so it is rarely planted anymore and not sold by many nurseries. In 1975, there were 183 acres of Cascade planted in New York, but many of those vineyards have been pulled up.[17] This is unfortunate since I like the quality of its light-bodied, strawberry-like colored and flavored wine.

Cascade has medium-sized long, loose clusters that have small- to medium-sized black berries. It is productive and very winter hardy in the Hudson Valley. It is a vigorous variety with lots of vegetative growth and a trailing growth habit. It should be spur pruned due to its high vegetative growth. Cascade is a reliable producer that is generally resistant to fungal diseases, but is somewhat susceptible to downy mildew. The variety ripens very early to early, so it needs netting to minimize bird damage.[18]

Because of its low acidity, it is ideal for blending with many other commonly grown high-acid French-American hybrids such as Baco Noir, Chancellor, and De Chaunac. Using it as a blender can soften and improve a wine by making it more marketable for sale earlier in the year. It also makes an attractive soft rosé wine.

Cascade is included in this book because I like this variety despite its difficulties with viral disease. Its wines are fun and interesting. Cascade is suitable for the colder, shorter growing areas of the Hudson Valley and foothills of the Catskills and Berkshire Mountains.

CHANCELLOR (S.7053)

Before 1939, Chancellor rivaled and then surpassed Plantet (another Seibel red hybrid, S.5455) as the most widely planted red French-American hybrid grape in France. Even as late as 1958, there were almost 100,000 acres of this variety alone in France.[19] The grape variety was named Chancellor in 1970.[20] In 2001, there were 49 acres of Chancellor being grown in New York, comparable to 46 acres as of 2006 and 42 acres as of 2011.[21] Chancellor is a hybrid of S.5163 (S.2510 × Gaillard 2) and S.880. One of Chancellor's parents, S.5163, is the same parent for Chelois and De Chaunac. Chancellor is the pollen parent of Chambourcin.

PARENTAGE
cinerea, labrusca, lincecumii, riparia, rupestris, vinifera

HARVEST DATE
Late mid-season

A B– B+ A+ B+

The vine is vigorous to very vigorous (but not like the wild growth of Baco Noir), and the variety is highly productive. The cylindrical cluster is medium-large to large and well filled, but not too compact. The blue berry is medium to large in size and round or slightly oval. Chancellor's bud break is early, which makes it vulnerable to late spring frosts; however, it has a good secondary crop if frost hits. In fact, one reason Chancellor was so popular in France, in addition to its high productivity, was because of its ability to bear a large secondary grape crop after a late spring frost. The grape ripens late mid-season to late, after De Chaunac, but before Chambourcin.

While very productive, cluster thinning is recommended to reduce its crop so that superior wines can be produced. Chancellor is a heavy producer, with every sucker carrying flowers. Therefore, it must be drastically pruned, especially in the early years, otherwise it can overbear itself to death. For vines older than five years old, cluster thinning is still very much needed to balance the crop and maintain vine viability for the next season.

The vine is hardy to very hardy to winter damage in the Hudson Valley, more so than Baco Noir or Chambourcin, but less than Maréchal Foch or Concord, so it is safe to grow throughout the Valley. Its growth habit is upright, which helps to expose its fruit to the sun. This is very important because Chancellor is highly susceptible to powdery mildew and slightly less so to downy mildew. The vine is however, more resistant to black rot and botrytis. Further, fungicide treatments must be limited because it is very sensitive to sulfur. It is very important to maintain a carefully monitored spray program starting at the beginning of the year to veraison to keep Chancellor's mildew problems in check. In the interests of suppressing its large crop and to spread out its leaf growth along the trellis, spur pruning is recommended. Overall, Chancellor is easy to grow, not particular to soils, productive, and cooperative.[22]

For a French-American hybrid, Chancellor produces a deeply colored, rich red wine. It can be full-bodied and complex, especially if aged in wood for a while to make a Claret-style wine. Chancellors are well balanced, but they can be so well balanced that they can become bland or neutral. Older Chancellors have been compared to a good old Bordeaux or Rhône, but not to Burgundies. While big in color and taste, they do not have the mouth feel, full body, or tannin structure that can generally carry these deeply colored and flavorful wines to their logical conclusion, like Merlot or Cabernet Sauvignon.

Chancellors are distinctly varietal. They have a lot of blackberry and cooked mulberry fruit flavors, but they can also have a gamey, greasy, or buttery quality that can be compared to Hudson River catfish. Like those big, meaty fish (which I like), Chancellors have the same taste from beginning to end, but can be musty, oily, or greasy, like gin. In the end, they do not have a lot of finesse unless they are manipulated to make them more interesting wines. In France, Chancellor was grown to produce simple, well-balanced, ordinary wines that have clean fruit and a piney herbaceous accent.

Even though Chancellors can be simple neutral wines that are no more than ordinary, some extraordinary Chancellors have been made that are complex and interesting. The nose, while muted and somewhat musty, can have fruit flavors of blackberries, blackberry jam, blueberries, chocolate, cooked prunes, violets, and black olives. Despite its medium body and tannin structure with a buttery texture,

Chancellor is rich for a French-American hybrid. On the palate, its middle has structure provided by flavors of earth, cigar box, black olives, leather, soft tobacco, with a buttery or oily quality that provides its body. Further, it can have many cooked flavors of black olives, molasses, cooked mulberries, and chocolate. The finish of Chancellors is generally sweet with a hint of cloves or cinnamon.

If a comparison had to be made to help describe Chancellor to those not familiar with the wine, the comparison would be to Merlot. In this case though, the Merlot would be a richly colored and well-balanced wine, but soft, fat, unassuming, and somewhat weak in tannin structure with a consistent taste throughout the wine.

Chancellors are good for blending with other reds. In particular, adding Chancellor to Cabernet Sauvignon can help to make the Cabernet more approachable and soften its fruit and acid profile. Chancellors, generally, have not been made into rosé or blush wines, but based on the balance they provide would probably make solid dessert ports.

While there should be a place for Chancellors in the Hudson Valley, this may not be the best first selection for a red grape variety. However, if the primary consideration is the coldness of the site and spring frosts, Chancellor does provide a good second crop for production purposes. It also has good resistance to winter injury. It is not a grape for the shortest growing season areas of the Hudson Valley due to its relatively late ripening date.

CHELOIS (S.10878)

Chelois is one of my most favorite Seibel red hybrids for the top-quality wine that it makes. It is also a solid grape in the field. Chelois is a cross of S.5163 × S.5593 and shares the same parent (S.5163) with the Seibel hybrids Chancellor and De Chaunac. The *vinifera* heritage of Chelois includes Dattier, Aramon, Alicante Bouschet, Black Hamburg, Grenache, Piquepoul, and several other unknown *vinifera* grape varieties. Chelois has about 50 percent *vinifera* genes. Some have suggested that Chelois is a cross of Bienvenu (S.2859) × Roi des Noirs (S.4643).[23] The variety S.10878 was commonly called Chelois in France.[24]

PARENTAGE
aestivalis, cinerea, labrusca, riparia, rupestris, vinifera

HARVEST DATE
Late mid-season to early late season

A A B- B A

In France, there were 2,236 acres of this variety in 1958, and 1,609 acres in 1968. In New York, there were 156 acres of Chelois in 1975[25] and, in 2001 there were less than 50 acres of this grape.[26]

Chelois is a vigorous vine that is moderately productive. To ripen properly, this red variety must be picked no earlier than late mid-season to early late season. It ripens about ten days after Baco Noir. Its bud break is late, which reduces the risk of spring frost damage, and if it is injured, it has a small secondary crop.

The variety is hardy to very hardy to winter damage in the Hudson Valley, but less so than Maréchal Foch or Baco Noir. If there is winter damage, the damage tends to manifest in dead cordons, also called dead arm, or what appears to be live buds that die later in the spring. But there is generally enough wood, even in severe winters, to generate sufficient fruitful buds to obtain a full crop. To ensure success, Chelois should be planted on better sites to increase production, but, more importantly, to bolster its already superior winemaking abilities.

The vine is healthy, vigorous, and large with a slightly upward to lateral growth habit that is somewhat bushy. Chelois tends to grow with an open canopy, which increases sunlight to the grapes and facilitates spray applications. The variety should either be short cane or spur pruned. The very compact cluster is medium in size and cylindrical with a round medium-sized blue-black berry. The vine likes deep, well-drained, and rocky soils and does not do well in droughty growing conditions. Further, cluster thinning is needed to increase wine quality and to advance the date in which Chelois can be picked. Chelois has a broad climatic tolerance, so it can be planted throughout the Hudson Valley.[27]

The grape is not generally susceptible to either black rot or downy mildew, but is susceptible to powdery mildew. Further, due to its very compact cluster, it can get bunch rot and its berries can split during harvest if a heavy late rain comes in before harvest. To protect against bunch rot, it is best to spray heavily just before the growing berries of the cluster close in so that the spray gets behind the berry. The grape and vine are not sulfur sensitive. The variety, or the cuttings available in the United States have virus problems (perhaps tomato ringspot virus), which can lead to the early decline of Chelois plants. One way to avoid this virus or systemic condition is to have Chelois grafted. My own experience is that a small percentage of Chelois vines (about 10 percent) in a vineyard block will decline after ten or fifteen years or so, but those vines can be replaced by layering.

The wine quality of Chelois is excellent. Philip Wagner maintained that a good deal of Chelois was grown on the "wrong side of the road" in French Burgundy.[28] Chelois has soft mature fruit, a medium-bodied tannin structure, and an approachable acid profile that ages very nicely for twenty years. Walter S. Taylor wrote that "a wonderful balance is found in this variety. Lightly flavored both in bouquet and taste, its color is nicely rubied and its tartness neutral."[29]

Chelois is a complicated wine so it has many descriptors. Some commentators suggest it is Burgundian in character[30] and others maintain that it is more like a Bordeaux as it ages. Some, myself included, maintain that it is something different and has a unique quality all its own. If forced to compare it to another wine, it would be similar to a dark Pinot Noir with elements of a light Cabernet Franc.

The color is a medium red, with no purple/blue hues like many other French hybrids. The color turns brown sooner than Baco Noir and becomes brick red with hints of orange within seven to twelve years. The nose of this medium-bodied red wine is aromatic, complex, and layered with elements of dried fruits, smoky wood, cedar box, black cherry, raspberries, and other berries, strawberry jam, and a nice spiciness reminiscent of anise and eucalyptus. Some darker Chelois will have in the nose, taste, and finish the flavors of tobacco, burnt toast, and barbequed meat that meld with dark black cherry and black raspberry flavors. Chelois has a firm but approachable acid/tannin profile that can stand on its own, or it can be blended with other reds such as Cascade, Maréchal Foch, Baco Noir, Chambourcin, or Burdin red hybrids to add complexity.

What distinguishes Chelois from other French-American hybrid wines is its soft spice and black pepper which permeates the nose and taste. This interesting and layered wine has the same nose as its taste and then its finish. As the wine gets older, the fruit flavors remain, but the structure becomes more noticeable with earth and wet-brick flavors, and the body, while still velvety, has a grainy complex presence that gives structure to the entire wine. The wines age well and needs oak to enhance its quality immeasurably.

Since Chelois tends to be medium bodied with a soft, but firm tannin structure it is also good blended with high-acid wines such as Baco Noir, or with heavy-bodied wines such as Chancellor, Cabernet Franc, or Cabernet Sauvignon to make hefty long-lived wines. Chelois also blends well with lighter bodied wines, such as Cascade or Burdin red hybrids, to soften them. Further, because of its medium body and forward berry nose, it has been successfully made into quality rosé and Nouveau wines.

More Chelois should be planted in the Hudson Valley due to its strengths in the field, its varietal wine qualities, and for its great blending potential.

SAISON 1937-1938

Producteurs Directs Nouveaux de Seibel

Adresser la correspondance à M. SEIBEL à Aubenas (Ardèche) Tél. 225 - Télég. Seibel-Aubenas ou à son bureau à Montboucher-s-Jabron (Drôme) Téléphone 3 Télégr. Seibel - Montboucher

Hybridations poursuivies depuis 1884 par Albert SEIBEL puis continuées par ses neveu et petit-neveu Louis et René SEIBEL à AUBENAS (Ardèche) C. P. SEIBEL - Lyon 109-36

NOTICE

Nos Producteurs Directs Nouveaux — Plants Nouveaux ou P. N. pour employer la terminologie actuelle — ont en bon sol une résistance suffisante au phyloxéra, et certaines de nos variétés ont même une résistance égale à celle des meilleurs porte-greffes.

Ces Plants Nouveaux sont infiniment plus résistants aux maladies cryptogamiques que les vieilles vignes françaises (viniferas). Certaines variétés ont une résistance absolue, tandis que d'autres placées en milieux humides se contentent pour sauver leur récolte d'un ou deux sulfatages, alors qu'il en faut normalement dix à douze pour les viniferas.

Nos Producteurs Directs ne sont pas ou peu atteints par la cochylis et l'œudemis.

Enfin beaucoup de variétés de nos Producteurs Directs repoussent à fruit après un accident : gelée ou grêle.

Il y a là un ensemble d'avantages importants en faveur de la culture de vignes plantées avec des Plants Nouveaux plutôt qu'avec des Viniferas, surtout si l'on compare les frais et les résultats que donnent ces cultures.

C'est ainsi qu'en 1936, année de mildiou, toutes les récoltes de Plants Nouveaux ont été sauvées avec deux et au maximum trois sulfatages, tandis que des Viniferas traités quinze et dix-huit fois n'ont donné qu'une demi récolte, quand ils l'ont donnée.

Dans la même année, les gelées des nuits des 13 et 20 avril ont, dans certaines régions, détruit entièrement la récolte des Viniferas, tandis que les vignes qui dans les mêmes régions étaient plantées avec des Producteurs Directs de SEIBEL entre lesquels : 5.474, 6.905, 7.053, 8.745 etc... ont malgré ces gelées donné une récolte normale.

Cette année on nous signale que certaines régions du Midi ont eu leur récolte d'Aramon anéantie par l'œudémis, tandis qu'à côté nos Producteurs Directs restent à peu près indemnes.

VINS. — La qualité d'un vin est fonction de trois facteurs : l'emplacement du vignoble, la nature du cépage, les procédés de vinification.

Le raisin arrivant à maturité s'imprègne de tout ce qui l'entoure : les sucs de la terre, la saveur des plantes avoisinantes, l'odeur des arbres des alentours, tout contribue à donner au vin que produit ce raisin le goût propre à chaque région, à chaque cru. Et c'est pourquoi nos Grands Crus ne pourront jamais être détronés et que l'on ne fera jamais du vin de bourgogne qu'en Bourgogne et du champagne qu'en Champagne.

Chaque région a ses cépages de vieille vigne française — Viniferas — à côté desquels viennent peu à peu prendre place les Plants Nouveaux, à mesure que l'on s'aperçoit que leur culture est infiniment plus économique et que le vin qu'ils produisent est tout aussi bon, si ce n'est parfois meilleur, que celui des Viniferas.

Il n'est en effet plus question de discuter la qualité des vins des Plants Nouveaux, il y a quelques années tant critiqués. Dans certaines régions ces vins font prime sur le marché et sont payés plus cher par le Commerce que les vins de viniferas, et les Concours de Dégustation que tous les ans organise le « CENTRE D'ETUDES VITICOLES DE MACON » prouve bien la qualité des vins que l'on peut obtenir avec nos Producteurs Directs Nouveaux.

Seibel catalog of French-American hybrids. *Most French hybridizers, such as Albert Seibel, had a nursery business to propagate and sell their planting material. This catalog featured 53 grape varieties, including Rayon d'Or, Verdelet, Chancellor, and Chelois. Circa 1937.*

DE CHAUNAC (S.9549)

PARENTAGE
labrusca, lincecumii, riparia, rupestris, vinifera

HARVEST DATE
Late mid-season

A A B A+ B+

De Chaunac is a very productive and reliable red French-American hybrid grape. It is noted as a cross of S.5163 × S.793 in the New York State Agricultural Experiment Station's *report, Vineyard and Cellar Notes,*[31] but Galet and Morton maintain that the origin of De Chaunac is unknown, but that it is probably related to S.5163.[32] S.5163 is also the parent of Chancellor and Chelois.

The grape was named after the Canadian enologist Adhemar F. de Chaunac (1896–1972) of Brights Winery in Niagara Falls, Ontario, Canada. De Chaunac was responsible for Brights Winery's research program. As early as 1946, he imported at least thirty-five French-American hybrids, including varieties that we now call Chelois, Foch, and Chancellor, among others.

S.9549 was first named Cameo by the Finger Lakes Wine Growers Association in 1970, but the name was changed in 1972 because it was determined to be an infringement of a proprietary name already being used and did not reflect the red color of the variety. In 1972, the joint American-Canadian committee, the Great Lakes Grape Nomenclature Committee, agreed to rename S.9549 as De Chaunac in honor of Adhemar de Chaunac.[33] Ironically de Chaunac's favorite variety was S.10878, which had been officially named Chelois by a French governmental body a decade earlier.[34]

Adhemar F. de Chaunac. *Renowned enologist and researcher at Brights Winery, Niagara Falls, Canada.*

The grape De Chaunac, while grown on a small scale in France, was from 1975 to 1990 one of the top red wine grape varieties in acreage grown in New York and the Niagara Peninsula in Canada. In 1975, there were 899 acres of De Chaunac in New York, and by 2001, that acreage had decreased to 165 acres, declining further to 101 acres by 2006, and to 97 acres in 2011.[35]

The variety ripens by late mid-season along with Chancellor, and is vigorous and very productive. It is also considered to be hardy to very hardy to winter damage, not as hardy as Maréchal Foch, but hardier than Baco Noir or Chambourcin. Further, it is not generally vulnerable to false springs so has little susceptibility to late spring frost damage. Its growth pattern is generally lateral so it needs shoot positioning if the benefits of a high cordon training system are used. The cylindrical cluster is semi-loose and medium-large to large in size, and sometimes winged. With a loose cluster, the variety has few problems with bunch rot. The blue-black berry is small to medium in size and round.[36]

To improve wine quality and to maintain a constant yield year to year, it must be cluster thinned. Without cluster thinning, the variety will overcrop. After achieving ripeness, this variety will hold on the vine sufficiently well so that other varieties that do not hold well can be harvested first. Also, the De Chaunac has relatively thick skin, so it will not split due to harvest rains. For some unknown reason, it does not seem to sustain as much bird damage as other red French-American hybrids.

De Chaunac has particularly good disease resistance to black rot, botrytis, and downy mildew, but is susceptible to powdery mildew. The grape has a slight sensitivity to sulfur so this treatment should be limited. It has also been reported that the variety is susceptible to soil-borne viruses. Philip Wagner recommends that the vine should be cane pruned.[37]

To reduce De Chaunac's tendency to produce grassy wines, the crop load should be reduced and the clusters should get maximum exposure to sunlight as the vine grows throughout the season. De Chaunac tends to have a bushy canopy when grown on fertile ground, so placing these grapes on less fertile sites can help to reduce fruit shading and hence herbaceous flavors. However, the vine cannot be encouraged to bear too small of a crop because this will facilitate denser and more bushy growth that shades fruit even further.

The wine produced from De Chaunac is generally fair to average. It is often inky in color, but of medium body that has a distinct leather, grassy, and cinnamon flavor. Truth be told, De Chaunac is a challenge in the cellar for several reasons. It generally produces medium-bodied wines that do not match their very dark color and pronounced nose. It can also be made in a dry, fruity rosé style. However, the wines brown easily, so rosés produced from De Chaunac should be consumed early to avoid browning. Since De Chaunac oxidizes and turns brown more rapidly than other French-American hybrids, these wines should always be topped up to avoid exposure to air.

De Chaunac's tendency to brown early is a positive attribute when used alone or in blends to make dessert wines such as cream sherries or tawny ports. De Chaunac added to these dessert wine blends hastens the browning process. Further, it gives these wines the quality of being older and mustier than their age would otherwise indicate. A Concord/De Chaunac blend is highly successful in making a cream sherry or tawny port.

The grape, noted for its attributes in the field, led to massive plantings of this variety in the late 1960s and early 1970s at the suggestion of the Taylor Wine Company and the Geneva Experimental Station. Unfortunately, by the time these deeply colored wines appeared on the market, the wine consumers had shifted their interest to light fruity white wines. This market reality forced many winemakers in the East to make De Chaunac into alternative wines such as ports, sparkling wines, rosé, Nouveau, and blush wines. Wagner Vineyards of the Finger Lakes is one of the more enterprising producers of alternative De Chaunac wines. Acreage of this grape variety has been gradually reduced so that its smaller supply now is beginning to match its demand within the industry.

Since De Chaunac has been widely planted and a high volume of wine has been produced with it, many talented winemakers have been able to produce interesting wines that are not in the standard De Chaunac mold which can be a medium-bodied inky red wine that is grassy and one dimensional with an underlying rubber tire in the finish. To reduce these pronounced rubber tire and leather flavors, it is recommended that the fermentation temperature should be kept low and the skin time shortened to three to five days. Also, using some whole fruit in fermentation can help to mitigate De Chaunac's tendency to make inky flat wines with muted fruit flavors.

Many of the techniques suggested above can be used to bring out the fruit flavors of other red hybrid wines, as well, however, using these techniques with De Chaunac is critical to overcome the fruit's tendency to produce only average wines.

If made properly, De Chaunac can have the fruit of soft red cherry, cooked mulberries, prunes, and blackberries. De Chaunac's dark flavors can be made into wines that are appealingly musty with an earthy and mushroom-like flavor profile and cigar-box nose, middle, and finish. Further, they can have complex fruit flavors and an underlying structure that has elements of cooked prunes, mint, charcoal, and burnt toast that can remind one of a Sandeman's Porto-lite. As stated previously, De Chaunac can easily become a flat wine that tastes like old rubber tires, burnt sugar, burnt toast, with pronounced cinnamon and green bean flavors and a metallic finish. Its tendency to produce pronounced grassy flavors, however, can be converted to more appealing mellow mint flavors.

De Chaunac can use wood aging to great advantage to smooth out its rough edges and to meld the wine into a unified taste profile that can be interesting, and hold up from five to seven years. Unfortunately, the fruit flavors of De Chaunac can be muted when young and may not grow and develop as the wine gets older. As it ages and its acids mellow, the fruit flavors and nose may not develop as one would expect: it just simply falls apart. Rarely do De Chaunac wines age long enough for their harsh acids to soften and meld with the fruit flavors to become an enjoyable and integrated wine. Therefore, one solid use of De Chaunac is to add no more than seven percent of it to other red blends to make the blended red wine appear and taste as if it is much older than it really is.

Overall, this is not the first grape variety that I would recommend for cultivation in the Hudson Valley and other cold weather climates. However, there is a large supply of it in the region, so utilizing the suggestions above can help to make De Chaunac into a simple, but approachable table wine or a better than average dessert wine.

VERDELET (S.9110)

This white grape is one of my favorite varieties. The wines are excellent with a floral, perfumey nose and a solid acid background that makes it a good wine for still or sparkling wine production. It is one of Seibel's later crosses of S.5455 (Plantet Noir) × S.4938 (a white grape). The heritage of Verdelet is complex and includes the Hudson Valley–developed grape Dutchess; *vinifera* grapes such as Aramon, Grenache, Alicante-Bouschet, and Chasselas Musque; and the disease-resistant Jaeger 70. For its high quality, it is surprising that only one-third of its heritage is *vinifera*. The heavily productive grape variety Plantet (S.5455), one of Verdelet's parents, is also the same parent of Burdin 6055, another highly productive quality red variety. The grape was named in 1970.[38]

PARENTAGE
aestivalis, cinerea, labrusca, lincecumii, riparia, rupestris, vinifera

HARVEST DATE
Late mid-season to late

B	B	C	A	A+

Verdelet is a very productive variety of only medium vigor, but has average production if managed well with pruning and cluster thinning. Its bud break is neither early nor late. The variety matures by late mid-season to late, but hangs well, so can remain on the vine much later in the season. Letting the grape hang on the vine benefits its wine quality as its sugar content increases and its very high acid profile diminishes. The clusters are large to very large, loose to well filled, and very pretty, with large tear-shaped berries that ripen to a beautiful pink-yellow to light golden color. The skin is resilient but not tough and adheres to the flesh, which allows the grape to remain on the vine after it ripens. The variety can be used as a table grape as its skin is not bitter. Due to its large clusters and the medium vigor of the vine, Verdelet should not be allowed to overcrop. It has a tendency to overproduce, so spur pruning and cluster thinning are recommended. The berries can split if heavy rains coincide with harvest time.[39]

The variety is very sensitive to downy mildew and has been considered by some to be as tender to winter injury as Riesling. However, at Cedar Cliff, I have found that it is hardier than Riesling to winter injury. Verdelet is susceptible to phylloxera, both on its leaves and roots, so it should be grafted onto a phylloxera-resistant rootstock, but it can still be successfully grown commercially on its own rootstock. In France, in 1958, there were 146 acres of this variety.[40] Philip Wagner maintained that the grape was tender and was susceptible to fungal diseases, but I have found that this is not the case.[41] With all that said, the variety should be planted on warm, well-drained sites that are out of the wind. While too precious not to harvest for the cellar, its beautiful, very large clusters of tasty, firm-fleshed grapes are also very suitable as table grapes.

The wine quality is excellent with a delicate, floral, and perfumey nose that has a flinty finish. Verdelet wines tend to be very pale in color with some hints of green. Its acid can be very high, so it is suitable for sparkling wine production. Due to its high acid, the wine lends itself to being made as a semidry wine to offset the high acid.

The flavor profile of Verdelet is also very floral and perfumey with flavors of green apples, lemons, melons, lots of white peach, bananas, and light apricots. As the wines age, they can take on flavors of orange rinds and hazelnuts. The wine has lots of structure, which tends to make it flinty, clean, steely, metallic, and with grapefruit flavor and an acid profile that sometimes has almond pits. The complex yet simple wines made from Verdelet are hard to compare to other wines, but it is most similar to Chenin Blanc.

Lucie Morton summed it up best by suggesting that Verdelet is a "delicate vine with delicate white clusters that make a delicate wine."[42] For those who are adventurous in the Hudson Valley, Verdelet is highly recommended as its quality of wine will not disappoint and it has a secondary use in the table grape market.

The Seyve Family

The Seyve family—the father, Bertille Seyve Sr. (1864–1939), and his sons, Bertille Seyve Jr. (1895–1959) and Joannes Seyve (1900–66)—individually and collectively made significant contributions to the European and American wine industry by their development of many useful French-American hybrid grapes that were extensively planted.

Bertille Seyve Sr. was from the Bougé-Chambalud commune, Isère department, in the Rhône-Alpes region of France, which is south of Lyon. He began hybridizing grapes in 1895. Over his lifetime, Bertille Sr. introduced more than 200 varieties for sale to other growers.

Joannes eventually took over Bertille Sr.'s nursery and farm, while Bertille Jr. set off to establish his own business. In 1919, Bertille Jr. married the daughter of another French hybridizer, Victor Villard, and they started their own nursery business in that same year. To avoid confusion with Joannes Seyve's nursery, Bertille Jr.'s hybrids went under the name Seyve-Villard. Approximately 100 Seyve-Villard hybrids were introduced for sale to growers on the market.[43]

This section will look at several Seyve family's hybridization programs to improve the grape variety stock, and some of their many hybrids that are suitable for growth in the Hudson Valley.

Bertille Seyve Sr.

(1864–1939)

Bertille Seyve Sr., like his two sons, was part of a generation of grape hybridizers who propagated most crosses between their own interspecific hybrids or used other breeders' complex interspecific hybrids. The approaches taken by Bertille Sr. and Albert Seibel (1844–1936) were very similar in that they both first began hybridizing to increase disease resistance and improve crop yield and then shifted over time to breeding for increased production or wine grape quality.

Both Bertille Sr. and Seibel started breeding grapes in the last quarter of the nineteenth century with similar plant stock in their hybridization program to increase disease resistance (i.e., Munson, S.14 (Munson × *vinifera*), Noah (*labrusca* × *riparia*), a seed of Taylor, and Ganzin 1). However, Bertille Sr. used Couderc hybrids 4401 and 132-11, while Seibel did not use Couderc hybrids at all. Seibel probably hybridized more than ten times the number of plant selections than Bertille Sr., so Seibel's program of incorporating different strains of disease resistance, frost resistance, and increases in yield was far more comprehensive. However, Bertille Sr.'s program was also very extensive. Both Seyve and Seibel, about one-third to one-half through their respective breeding programs chronologically, began to cross the hybrids that they had developed with each other to create grape varieties of fairly complex genealogy.

More specifically, Bertille Sr.'s grape-breeding program relied heavily on Ganzin 1, S.14 (Munson × *vinifera*), Noah (*labrusca* × *riparia*), and Vivarais (S.2003), and unlike Seibel, on varieties such as Couderc hybrids (a neighbor of Seibel's), which were hybridized to resist phylloxera and lime in the soil. To a lesser extent, he used Othello, Gaillard, and La Cost. Interestingly, he did not rely nearly at all on 100 percent *vinifera* varieties, but brought in *vinifera* via Ganzin 1 (Aramon × Rupestris Ganzin) and S.14 (Munson × *vinifera*).[44]

Both of Bertille Sr.'s sons relied very heavily on the work of Seibel, and only sparingly used their father's work in their own hybridization programs.

Of the more than 150 hybrids that Bertille Sr. bred, only three varieties were named, all of which had military-themed names: Le Colonel (BS 2667), Le Commandant (BS 2862), and Le General (BS 5563).

LE COLONEL (BS 2667)

Le Colonel is a cross of Couderc Noir ((C.7120)(Jaeger 70 × *vinifera*)) × BS 872 (S.85 × Gaillard 2). While not a grape commonly grown at all, it shows promise in the Hudson Valley. It has large, somewhat compact clusters with medium-sized black berries. It is moderately winter hardy with solid disease resistance. The vine is of a standard size, but it is a vigorous grower with moderately high yields. The grape buds out very late, thus avoiding late spring frosts. Its growth habit is to have extensive foliage on medium-length canes that grow laterally, but in a way that is airy so that the grapes receive sun and air ventilation, and spray penetration is good.

PARENTAGE
labrusca, lincecumii, riparia, rupestris, vinifera

HARVEST DATE
Late mid-season to late

B A B A- A

The wines are serious and interesting, yet fun. The fruit is not forward, but is complicated, with flavors of cherry, cranberry juice, raspberries, light strawberries, and beach plums. The body of the wine is serious, which is steely and flinty, but bright. It is aromatic, approachable, and somewhat floral, and presented in layers. Even though the color is a light red, it has a big mouth feel with soft developed tannins. The wines are *vinifera*-like, and not hybrid-like, even though just a little more than one-quarter of its genetic makeup is *vinifera*.

This is a grape that shows promise both in the field and in the cellar in the Hudson Valley. More of Seyve Sr.'s hybrids should be reevaluated to find "diamonds in the rough."

BERTILLE SEYVE JR.

(1895–1959)

All of Bertille Seyve Jr.'s hybrids are under the name Seyve-Villard (S.V.). This name was used for his nursery business and grape varieties to avoid confusion with his father's already existing nursery business, which was ultimately taken over by his brother, Joannes Seyve.[45] Bertille Jr.'s wife was the daughter of the grape hybridizer Victor Villard.

Of the two brothers, Bertille Jr. had the more extensive breeding program and created several widely planted grape varieties in France, England, and the eastern United States, such as Seyval Blanc, Roucaneuf, Villard Blanc, and Villard Noir. Perhaps one reason for his more prolific breeding program was that Joannes, his younger brother, continued to operate his father's farm and nursery, and was presumably still evaluating his father's older hybrids. In addition, Bertille Jr., more so than Joannes, tended to name more of his selections. Some of these named varieties include Seyval Noir (S.V.5-247 (S.5656 × S.4986)); Roucaneuf (S.V.12-309 (S.6468 × S.6905)); LaRouge (S.V.12-327 (S.6468 × S.6905)); Garonnet (S.V.18-283 (Chancellor × S.6905)); Perle Noire (S.V.20-347 (Panse × S.V.12-358)); Dattier de St. Vallier (S.V.20-365 (Panse × S.V.12-358)); Pierrelle (S.V.20-366 (Panse × Villard Blanc)); Muscat St. Vallier (S.V.20-473 (S.V.12-129 × Panse)); Valerien (S.V.23-410); and Varousset (S.V.23-657 (S.4668 × S.6905)).

While Bertille Jr.'s breeding program was almost as extensive as his father's, Bertille Jr., relied primarily on crossing interspecific hybrids such as S.5124 (Bienvenu × Aramon du Gard), Subereux (S.6905), and S.6468 (S.4614 × S.3011), and his own creations such as Seyval Blanc (S.V.5-247). He did not, like later generations of hybridizers such as J. F. Ravat or Jean-Louis Vidal, directly introduce "superior" *vinifera* varieties, such as Gamay and Chardonnay, to increase wine quality, and he rarely used his father's hybridization work. Toward the middle of Bertille Jr.'s breeding program, he began to rely much more heavily on interspecific hybrids that he had developed under the Sevye-Villard name.[46] Of his work, Philip Wagner stated that Bertille Seyve Jr. was one of the more successful producers of hybrids for hot climates and produced some hardy hybrids adaptable to short-season conditions.[47]

SEYVAL BLANC (S.V. 5-276)

Seyval Blanc is one of Bertille Jr.'s first early successes in hybridizing. This white wine grape was developed in 1921.[48] Galet and Morton maintain that Seyval Blanc is a cross of S.5656 × S.4986 (Rayon d'Or),[49] while the Geneva *Vineyard and Cellar Notes* maintains that the cross may have been S.4995 × S.4986.[50] It is agreed, though, that Rayon d'Or (S.4986) is one of its parents and shares this parent with Vidal Blanc. This grape, after its popularity began to grow, became known in France as Seyval Blanc,[51] the name being the contraction of Seyve and Vallier.[52] Seyval Blanc has been used to breed Cayuga White, Chardonel, La Crosse, Melody, and St. Pepin.

PARENTAGE
lincecumii, rupestris, vinifera

HARVEST DATE
Mid-season to late mid-season

A A- C A A

The grape is adaptable to different regions and climates and is grown throughout the eastern United States, northern France, and England.[53] It is a versatile grape that can be made into many different wine styles, such as very fruity, semi-dry Germanic whites; Sancerre-like wines or Sauvignon Blanc–style wines that are dry with crisp fruit, grassy, or herbal notes; Chablis-style Burgundies made with Chardonnay; and California-style, fat and buttery Chardonnays. Many consider Seyval Blanc to be one of the signature white grapes of the Hudson Valley and has been likened to the Valley's Chardonnay.

Seyval Blanc buds out fairly early, but has a sizable secondary crop if a late spring frost hits. It is moderately winter hardy, but less so than Baco Noir, Maréchal Foch, or Delaware. It is susceptible to black rot, powdery mildew, and botrytis at maturity, and but more resistant to downy mildew. On average it is more susceptible to fungal diseases than other white hybrids such as Vidal or Vignoles. Further, it has no sulfur sensitivity.

The cylindrical to slightly tapering conical clusters are medium-large to large and compact to semi-compact and even semi-loose, depending on the clone and the soil that it is grown on, with berries of medium size. The compactness of the clusters can lead to berry splitting at harvest time, particularly if it rains before harvest. This berry splitting leads to bunch rot. Seyval grows on a standard-sized vine of medium vigor, and it ripens mid-season to late mid-season. Due to its susceptibility to fungal diseases and cluster compactness, it is important to use canopy management techniques to increase sunlight and ventilation. Also, before the cluster closes in to become compact, a fungicide spray application should be made to reduce fungal diseases and the grape berry moth that harbors within the middle of the enclosed cluster.

The vine is of medium vigor, but is reliable and highly productive. It should be cluster thinned to produce fruit for quality wine production. It can be somewhat capricious as to soils, which can affect its productivity and winter hardiness. It does not do well in droughty or shallow soils. Without cluster thinning, the vines tend to be thin and weakened, thereby making it susceptible to severe winter injury. The vine has an upright growth habit and should be spur pruned to reduce the risk of overcropping and to balance its crop load to the quantity of grapes that the vine can fully ripen. Seyval can be subject to poor fruit set if the spurs are pruned too short and overcropping if pruned too long.[54]

The plant had a limited following in France in the 1950s and 1960s. There were 274 acres of Seyval Blanc in France in 2008, down from 393 acres in 1998 and 3,235 acres in 1958.[55] It is, however, adaptable to different growing regions and climates and is popular from Pennsylvania north to Canada and in England. In 2001, there were 331 acres of Seyval Blanc grown in New York, which increased to 373 acres by 2006.[56] Seyval is also one of the parents of the University of Minnesota hybrid St. Pepin and the Geneva Experimental Station's Chardonel.

Most wineries in the Hudson Valley produce at least some wine that has Seyval Blanc in it. When Sevyal is fermented at a cool, but not cold, temperature and aged in neutral containers such as glass or stainless steel, the result is a high-acid, Germanic-style white wine. The wine's fruit flavors include elements of green apples, pineapples, and citrusy notes of lemon, grapefruit, and pear. It is a bright wine that is clean and crisp—even metallic. To offset the high acid, it is often finished as a semidry wine.

If the wine is made from ripe to very ripe grapes, fermented cool, and then aged in oak barrels for a short time, the wine can taste more like a French Sancerre or Muscadet, or wines made from Sauvignon Blanc or Semillon. These Loire-like whites are more complex, with an herbaceous flint-like body and softer fruit flavors of pears, melons, peaches, and apples. The wines are rounder and more subtle than the Germanic-style Seyvals. Also, Seyvals lend themselves to be made as sur lie–style wines.

Seyval, if picked ripe—but not too ripe—and aged in oak for six months or more, can result in a wine that is a slightly thinner version of a Burgundian-style Chardonnay. The flint and steely finish is still there, but it is more austere with fruit flavors of apples, cider, peaches, and pears with a complex yeasty and toasty vanilla finish. These wines age well—I have had twenty-five-year-old Seyvals in this style and they remain fresh and interesting as any old Burgundian Chablis made from Chardonnay of the same age.

The grape can also be made as a California-style Chardonnay that is fat, buttery, and full of wood and vanilla flavors. If very ripe grapes are fermented in oak barrels and left on the lees while aging, this results in a lighter version of a California-style Chardonnay. These wines tend to have big wild flavors of melons, bananas, honey, ripe peaches, orange blossoms, almonds, and hazelnuts. The body is round, warm, buttery, and complex with lots of vanilla throughout, with no hint of herbaciousness. Barrel fermentation and malolactic fermentation can enhance the body of Seyval.

Seyval Blanc, like Baco Noir, should remain a signature grape variety for the Hudson Valley. Its planting should increase, as it is well suited to grow in the cool climates of the Hudson Valley, and it is also very versatile in the cellar.

VILLARD BLANC (S.V. 12–375)

Villard Blanc is a white wine grape variety that can double as a nice table grape. It is a very vigorous and productive variety that ripens late to very late. Villard Blanc[57] was at one time one of the most popular French-American hybrids grown in the Midi region of southern France. It is a cross of S.6468 (commonly used by Bertilles Seyve Jr.) × Subereux (S.6905) that was introduced in 1924.[58]

PARENTAGE
berlandieri, lincecumii, rupestris, vinifera

HARVEST DATE
Late to very late

B A A A B

The vine is at best moderately winter hardy, in line with grapes such as Chambourcin or Riesling, but likes places with winters that are less severe than those that often occur in the Hudson Valley. The grape is generally resistant to fungal diseases, except for powdery mildew, to which it is susceptible.

The widely branched conical clusters are very large and loose with oval berries that are deep golden yellow. Villard Blanc is not a grape for short-season areas, but may do well in warmer places in the Hudson Valley as long as there is adequate air circulation in the summer. The grape is remarkably vigorous and a heavy producer of compound clusters. Further, it does best in areas of low humidity and with a long growing season since it ripens late to very late. It has a spreading growth habit, which helps to reduce fungal diseases.[59]

Spur pruning and rigorous cluster thinning is required to help reduce its heavy crop load. It has a slight sulfur sensitivity. A testament to its productivity and ease of cultivation in the right environment is that even though the French government, by law, has designated it a "discouraged" variety for cultivation in the south of France, growers continue to grow this variety extensively. In 1968, there were 52,873 acres of this grape grown in France, which made it the third most widely planted white grape variety in the country. In 1979 its acreage was reduced to 14,500 acres, and in 1999 there were conflicting reports of between 1,825 to 2,790 acres remaining, with 862 acres of it being grown in 2008.[60]

The wine quality is only average and has a bitter finish. Further, the grape is uncooperative in the press and does not clarify well due to its high iron content. If grown on rich soils, it can develop a protein haze during fermentation. For quality wine production, it is best to grow it on shallow, mild soils with lime content.[61] In France, it is said to provide good fruity, everyday table wine that should be consumed young.[62] Walter S. Taylor suggested that it produces wine that is neutral and soft, with good body.[63] It is wine that is good for blending. Philip Wagner thought it to be a heavy producer, but only a fair wine that was neutral in flavor.[64]

My own evaluation of this wine is more positive. It can be a subtle wine that has elements of green apple, mint, pears, melons, grapefruit, orange rinds, and orange blossoms. Further, it can be a flowery and perfumey wine that has citrus hints with elements of fresh-cut herbs. I have had Villard Blancs that were clean and which had a body that consists of earthy pineapple, flint, and nuanced smoky flavors. It is "very French" with a long clean finish.

In the end, Villard Blanc can be made into a relatively soft wine that can be used as a base wine to round off the rough edges of a Seyval Blanc or Vignoles, for example. Villard Blanc is included here because it should be considered for cultivation in the Hudson Valley as a high yielding variety utilized as a blender to enhance other locally produced wines, or as a base wine to which more flavorful wines could be added to create a final blend.

S.V. 18-307

This red grape is a cross, as reported by the Geneva *Vineyard and Cellar Notes*, "supposedly" of Chancellor (S.7053) × S.V.12-375 (Villard Blanc).[65] However, Carl Camper of Chateau Stripmine Experimental Vineyard and other researchers maintain that the variety is a cross of Chancellor and Subereux (S.6905).[66] This is not a commonly grown grape in the Hudson Valley, but it has been used extensively in Cornell's grape-breeding program, so it is worth describing this varietal grape and wine in more detail. It is a parent of Corot Noir, NY 64.533.2, and NY 70.816.5. This grape has been extensively field-tested at my farm at Cedar Cliff, so the observations are not from ancillary sources.

PARENTAGE
bourquiniana, cinerea, labrusca, lincecumii, riparia, rupestris, vinifera

HARVEST DATE
Mid-season to late mid-season

A B B A B+

Bud break of this variety is not early, nor is it late. The grape has medium to medium-large, semi-loose clusters of black grapes. Its berries are small to medium in size. The vine is winter hardy to very winter hardy, but has only average resistance to fungal diseases, especially downy and powdery mildew, when compared to other French-American hybrids.

The vine size is standard to large and its growth is vigorous with a thick canopy similar to a dense, vigorous Maréchal Foch. The growth habit of this vine is bushy, with short, medium, and long canes. In addition, the new canes have lateral growth, which makes the canopy even denser. The growth is somewhat upright to lateral. One reason, perhaps, for its somewhat below-average disease resistance is its thick canopy, so alternative pruning and training systems may help to reduce disease in this variety. The grapes ripen by mid-season to late mid-season and it is a reliable but, depending on the site, average to heavy producer. In warmer climates such as Arkansas, S.V.18-307 can produce wines with very soft tannins and no herbaceous flavors, but very little bouquet.

The red wine is deeply colored to inky. It is full, has a heavy body, and good balance and tannin structure, but can be very herbaceous and weedy. The wine is, at best, quite ordinary and similar to an inexpensive California Cabernet Sauvignon. Also, the finish can be bitter. The wine can be neutral and used as a blender to round out the tannin structure or deepen the color of another wine. It can be firm, but not heavy. In my test wine trials, while the wine is integrated, it is dank with some weeds, smoke, and herbal flavors. Its fruit is of cooked mulberries, blackberries, heavy plums, mint, coffee, and black pepper. It has a somewhat complex body of earth/charcoal that is gamey, and can have some soft Rhône wine elements, as well.

Viticulturally, S.V.18-307 is suitable for the Hudson Valley as its yield potential is average to above average, though this is a wine that should not be made and sold on its own. However, it may have a place as a consistent producer of heavy red wines that can be used to give substance to other red wines or for the production of ports.

VILLARD NOIR (S.V. 18-315)

Galet, Morton, and Jancis Robinson report that Villard Noir is a cross of Chancellor (S.7053) × Subereux (S.6905).[67] However, the *Geneva Vineyard and Cellar Notes* maintain that Villard Noir has the same parents as S.V.18-307, which in the Geneva *Vineyard and Cellar Notes* is a cross of Chancellor (S.7053) × Villard Blanc (S.V.12-375).[68] However, the *New York State Fruit Testing Catalog*, published later on in the 1990s, stated that it was a Chancellor × Subereux hybrid.[69]

PARENTAGE
bourquiniana, cinerea, labrusca, lincecumii, riparia, rupestris, vinifera

HARVEST DATE
Late to medium-late season

A A C A B

The grape was released around 1930 and became commonly known as Villard Noir in France.[70] Villard Noir has replaced Chancellor (which since 1977 has not been designated as an authorized grape variety in France) as the most planted red hybrid direct producer in France. Villard Noir is grown primarily in the south France for the production of *vin ordinaire*. In 1968, there were 75,058 acres grown in France, which declined to less than 20,000 acres by 1979 and to 1,485 acres in 1999.[71]

Villard Noir is fruitful and a productive vine of medium vigor. Since it is very productive, growers need to cluster thin and watch the vine's vigor to balance its growth and production level to maintain its winter hardiness. To increase plant vigor, grafting is suggested, particularly on heavy soils.

Its bud break is by mid-season, but it produces a good secondary crop if hit by a late spring frost. The vine is as winter hardy as Chambourcin or Baco Noir, but can be more so. It is very resistant to downy mildew, but is sensitive to powdery mildew and botrytis. To maintain vine health over the years, the plant should be spur pruned to suppress production.

The grape ripens late to medium-late season. It has blue berries on medium- to large-sized moderately loose clusters. Due to its requirement for a long growing season and to improve wine quality, again, cluster thinning is necessary.

Villard Noir makes a sound but neutral wine of deep red color and high alcohol. However, the wine can be herbaceous and astringent. Morton reports that the wine can be light-bodied, fruity, and low in tannin. The wines can be suitable for a soft, lunchtime rosé or light red. Further, it is easy to grow and productive, but is best suited to warm climates not found in the Hudson Valley.[72] Both Philip Wagner and Walter S. Taylor suggest that it is a heavy-bodied wine with lots of color and a good blender.[73]

Villard Noir probably becomes a bigger, heavier wine with lots of color when grown in warmer climates. It is still widely grown in southern France, but can be found in Loire, Hermitage, and Bordeaux.

To reduce its sometimes astringent edge, it can be finished with some sweetness. My own tastings of Villard Noir remind me of a thinner, more extrinsic Chancellor (one of its parents). However, it can have complex flavors of blackberries, black pepper, eucalyptus, prunes, tobacco, cedar, and cigar box.

This grape has been included here because it produces neutral, sometimes thin wines that can then be blended with other locally grown hybrid wines that have too prominent grapey/berry flavors and high acid levels. This grape is not widely planted in the Hudson Valley today, but should be reexamined in the future due to its high productivity.

Seyval Blanc. *Of all the hybrids developed by the Seyve family, Seyval Blanc has been one of the most popular grape varieties for winemakers in the Hudson Valley for several decades.*

JOANNES SEYVE

(1900–1966)

Joannes Seyve took over his father's nursery business, while his brother, Bertille Seyve Jr. joined forces with Victor Villard (his father-in-law) to establish a different nursery. While Joannes's own grape-breeding program was much less extensive than Bertille Jr.'s, he is still credited with more than fifty crosses propagated by his nursery.

Like his brother, Joannes extensively used S.6468 (S.4614 × S.3011) as the seed parent, but relied on Chancellor as one of his primary pollen parents. In addition, he used Seyve-Villard crosses of Seyval Noir (S.V.5-247) and S.V.12-417, and crosses that had Subereux (S.6905) as one of its parents.[74] By far, Joannes's most widely known grape—and the only one named—is Chambourcin (J.S.26-205). Overall, like his older brother, Joannes used Seibel hybrids for his breeding work. He crossed Seibel hybrids to the exclusion of all others, except for the few crosses done with his father's B.S.4825 (Le Colonel (BS 2667) × Subereux (S.6905)).

Philip Wagner, a contemporary of Joannes Seyve, cites that Joannes was, like Galibert, chiefly interested in producing hybrids for the warm and humid Mediterranean climate of the Midi area of France.[75] In addition to Chambourcin, Joannes hybridized other grapes, including a white grape called J.S.23.416 (B.S.4825 (Le Colonel × Subereux) × Chancellor). This fruit, incidentally, has very large clusters of amber to pink berries that are oval and good for table use or wine, the wine being described as pleasant in nose and body in a German style. J.S.23-416 is also one of the parents of Traminette.

In addition to J.S.23-416, there are experimental plantings of white grape J.S.12.428 (a mutation of S.6468), as well as J.S.26-627 (white), J.S.26-674 (white), and J.S.13-756 (red), all currently being evaluated at my farm at Cedar Cliff.

CHAMBOURCIN (J.S. 26-205)

Chambourcin is a very versatile grape that can make big Rhône or northern Italian–type reds, Anjou-type rosés, Nouveau or fall wines, and soft enjoyable table reds. It has been commercially available only since 1963 and could now be one of the most highly propagated direct hybrid producers in France.[76] Morton and the Geneva Vineyard and Cellar Notes state that the parentage of this variety is uncertain. Other references suggest that Chambourcin may be a cross of seed parent S.V.12-417 (S.6468 x Subereux (S.6905)) x Chancellor.[77] However, recent DNA testing indicates that its true parents are J.S.11369 x S.5455 (Plantet).[78] The name Chambourcin is derived from the municipality in which it was developed by Joannes Seyve, Bougé Chambalud, only the name is reversed.[79]

PARENTAGE
berlandieri, cinerea, labrusca, lincecumii, riparia, rupestris, vinifera

HARVEST DATE
Late to very late

A– A C A A+

This grape is popular in the Loire Valley, Touraine, Nantes, Muscadet, and Savoie, France. It is one of the most widely planted French-American hybrids grown in France, though it is on a downward trend. In the late 1970s, there were 8,310 acres, 2,343 acres in 2006, and 1,977 acres in 2008. In the late 2000s, in the United States, it was a popular grape in Michigan (42 acres), Missouri (154 acres), Pennsylvania (146 acres), Virginia (96 acres), and is grown in most other Mid-Atlantic states and Australia.[80]

Chambourcin is popular, perhaps, because like Cabernet Franc, the other popular local *vinifera* grape grown in the Loire Valley, Chambourcin can make quality big, rich reds as well as light fruity rosé wines. In the Hudson Valley, it can make big reds of great aging potential or steely rosé wines. In fact, many Chambourcins are reminiscent of Cabernet Franc.

Chambourcins grown in colder locations can have a thin austere quality, so the site chosen in the Hudson Valley needs to be warm enough to produce varietal-quality wines. The Hudson Valley is about Chambourcin's northern limit in the United States, not due to winter hardiness problems, but because it needs a long growing season to ripen properly. In too short growing seasons, the wines can be too acidic and austere to be enjoyed. In 2001, there were 41 acres of Chambourcin in New York, but acreage has for some inexplicable reason been reported as declining to 31 acres as of 2006, but was back up to 40 acres in 2011.[81]

Chambourcin has a late bud break, but will still produce a secondary crop should a very late spring frost hit the vineyard. The vine is at best moderately vigorous, has a standard vine size, and produces large to very large clusters of moderately loose bunches of big blue grapes. The vine is a very consistent producer and a productive variety that needs cluster thinning to sustain the vine and to produce quality wines. The vine is sensitive to lime soils and should not be planted in droughty soils. It does much better in deep, well-drained soils.

Chambourcin has a reputation for being at best only moderately winter hardy. At my farm in Cedar Cliff, it is only slightly less winter hardy than Baco Noir and about as hardy as Chelois, and rarely sustains winter injury. The grape is susceptible to black rot and less so to powdery mildew and even less for downy mildew. It is resistant to botrytis and most bunch rots due to its thick skin, loose-forming clusters, and late harvest date. However, spray applications should not be neglected, particularly for black rot and downy mildew. It is sensitive to sulfur treatments.

Due to its airy leaf canopy and loose clusters, the plant is receptive to fungal spray applications and dries out quickly because its cane/leaf canopy allows for good air and sunlight circulation. It also dries out quickly after summer rains.

The grape ripens late to very late, depending on the harvest season, but its thick skins make it resistant to botrytis and frost damage. Chambourcin has a drooping growth habit, therefore, the vine should be high cordon spur pruned, with three to five buds on each spur.

The wine is superior to most other red French-American hybrids when it is made into a big aromatic red because it has a solid tannin structure and a rich complex flavor profile. Chambourcin is not overly grapey, berry/jammy, herbaceous, or have hints of gun powder or dank rubber tire like some other red hybrids. The wine blends well with *vinifera* wines such as Cabernet Sauvignon and Cabernet Franc because it softens them. It can be blended with wines like Baco Noir or Maréchal Foch to help bolster their tannin structure and increase complexity. If fermented on the skins for extended periods, Chambourcin extracts big *vinifera*-type flavors and not the sometimes bitter flavor elements extracted from other French-American hybrids.

The color of Chambourcin wines is a deep crimson red, but it is not generally inky like varieties such as De Chaunac. It ages well, and is color-fast, retaining deep crimson red hues for seven to ten years before changing. In fact, it can sometimes take too long to reach its peak. I have had twenty-year-old Chambourcins that still needed time to reach their peak and they still had those young, bright crimson colors. However, Chambourcin's high acid can clash with immature tannins if the grapes are under-ripe when picked. To expedite the aging of Chambourcin, extended wood aging is required.

Describing the wine profile of Chambourcin can be difficult. Professional wine evaluators and critics have described Chambourcin as similar to a full-bodied Bordeaux, Burgundy, or a Rhône. I believe that Chambourcins, even with their high-acid front and middle, have more of the elements of a soft, light Rhône, Cabernet Franc, or northern Italian red because of its soft tannin structure, flavor profile, subtle nuances, and prominent black pepper aroma and flavor overlay to the grape's basic berry front, most noticeably blackberries and chocolate.

In my experience, Chambourcins have a wide range of fruit flavors, including cherry, black raspberry, black currants, blueberries, cooked mulberries, and prunes, but they can be grapey. The earthy and resinous body has elements of black olives, anise, cloves, black pepper, cinnamon, burnt toast, soy sauce, and flint.

The wines can be aromatic with a bouquet of eucalyptus, spice, smoke, cigar box, tobacco, leather, mahogany, teakwood, and chocolate. They can also be herbaceous with flavors of dill and green peppers. For all of their flavor and body, Chambourcins tend to have muted or closed noses, unless aged for seven to fifteen years. So, in making Chambourcins, adding big-nosed varieties such as Baco Noir, Chelois, or Maréchal Foch into blends is recommended. As you can see, Chambourcins are complex wines that offer much to the winemaker and the consumer alike.

It should be noted that Chambourcins can also make very credible rosés, not unlike an Anjou rosé (which is partially made from Chambourcin). These rosés are not the clawing sweet type of eastern Catawba rosé, which I do like, but steely rosés that have a presence and are easily consumed during the summer months. In the Loire Valley, much Chambourcin is made into locally consumed rosés. These high-acid raspberry-red to peach-colored wines are bright, with elements of cranberries, lemons, watermelons, and Hawaiian Punch with a slate finish. Also, Chambourcin has been made into quality Nouveau wines and soft red wines with the aid of carbonic maceration.

As a big red, Chambourcin has great intensity for a French-American hybrid, with spicy and earthy characteristics. While it is a grape that can handle the retention of all its stems during fermentation, it can also benefit from malolactic fermentation to minimize its pronounced acid profile. Chambourcin stems tend to stay green longer than many other varieties, even when the grapes are ripe which can add bitter flavors. It is recommended that about one-half of the stems should be removed, particularly if they are unripe green stems.

Overall, Chambourcins reminds me of a light Cabernet Franc, a thin Rhône, or an Anjou rosé. The wines are similar to Cabernet Franc, but not as rich. However, they are quite good and better suited to those who like lighter bodied *vinifera* wines. Chambourcins lend themselves to blending with Cabernet Sauvignon, Cabernet Franc, Baco Noir, and Chelois. While generally having a bigger body than other French-American hybrids, it still does not have the body or tannin structure of most *vinifera* varieties. However, adding even 20 percent Chambourcin to other red hybrid wines can really brighten them up and give them substance.

Chambourcin is one of the parents of Regent, a new disease-resistant variety that is increasing in popularity with German grape growers. Chambourcin should have a prominent place in the future of the Hudson Valley wine industry and comes highly recommended both for its attributes in the field and versatility in the cellar.

François Baco

(1865–1947)

François Baco was a teacher from the commune of Belus, department of Landes, in what was the Armagnac province, France. This area is along the Atlantic Ocean just south of Bordeaux and north of Spain. In the past, there has been some confusion as to whether Baco's first name was François or Maurice. Pierre Galet and Lucie T. Morton and Cornell Professor Einset refer to him as François,[82] while the nurseryman Philip Wagner referred to him as Maurice.[83] The New York State Agricultural Experiment Station at Geneva, in *Vineyard and Cellar Notes*,[84] referred to him both as François and Maurice. In the end, the hybridizer's name is François, who is sometimes confused with his son Maurice, who died at the age of seventeen and for whom Baco 22A (Baco Blanc), also known as Maurice Baco, is named.[85] I point this out to suggest that perhaps some of the journal entries of these early French hybridizers may not be as accurate as at first they might appear. Further, if experts in this field can confuse Baco's first name, then there may be other inaccuracies in the reports they provided on these grapes and their genealogy.

What is more certain is that after 1898, François Baco began his work to hybridize grapes against black rot and to improve on the grape variety Noah in Armagnac. Out of the approximately 7,000 hybrids that he created, only Baco Noir (Baco 1) and Baco Blanc (also known as Maurice Baco or Baco 22A) remain commercially important in Europe, the United States, and New Zealand.

Many of Baco's categorized hybrid selections used Noah, Couderc 201, Couderc 4401, and other Couderc hybrids as pollen parents, and *vinifera* varieties such as Folle Blanc, Tannat, and Chasselas as the genetic pool for his hybridization program. His other named hybrids include, in order from the earliest developed to the latest: Totmur (Baco 2-16 (Baco 45A × Baco Noir)); Petit Boue (Baco 7-3 (Baroque Blanc × Couderc 4401)); Baco Chasselas (Baco 7A (Chasselas × Noah)); Rescape (Baco 9-11 (Couderc 4401 × Baco 22A)); Celine (Baco 12-12 (Baco × Couderc 4401)); Baco Blanc (Baco 22A (Folle Blanche × Noah)); Estellat (Baco 30-12 (Baco Noir × Couderc 4401)); Olivar (Baco 30-15 (Couderc 4401 × Baco Noir)); Douriou (Baco 37-16 (Baco 57-1 × Baco Noir)); Caperan (Baco 43-23 (Cabernet Sauvignon × Couderc 4401)); and his last-named variety Cazalet (Baco 58-15 (Baco 45-8 × Noah)).[86]

It is presumed that the first number in each hybrid's name designated the sequential year of Baco's breeding program, which stretched for almost half a century. Yet two of his most commercially successful hybrids were developed early in his attempts: Baco 1, developed in 1902, and Baco Blanc (Baco 22A), perhaps about twenty years later.

Some commentators suggest, however, that Baco Blanc was created as early as 1898. Looking at the sequence of François Baco's work, it was either released in the mid-1920s, or was mislabeled and is really Baco 2-2A, which would indicate it had been developed around 1898. If the latter is true, then Baco's two most successful hybrids were developed in the first four years of his long career. However, looking at the sequential numbering of his hybrids, his early work had used selections numbered much later. This seemingly does not make sense unless the numbering system is not based on the date of hybridization, as believed, but on some other system of categorization. Hence, Totmur (B2-16), for example, (Baco 45A × Baco Noir)—is not that genetic combination, but (Baco 4-5 × Baco Noir).[87] An entire book could be devoted to describing and cataloging the work of these early French grape-breeding pioneers. The point of this discussion is to raise the issue that these early hybridizers were very busy people, who, while hybridizing grapes were also often raising families and working to earn a living at their "day" jobs, and not necessarily keeping proper notations.

None of Baco's work after the mid-1920s was commercially successful. Also, unlike other more sophisticated breeding programs, he relied almost exclusively on the same early, not very genetically complex varieties. He did little cross-breeding of his later hybrids and did not bring in much new genetic material. Also, the seed parents he used were primarily *vinifera* grapes, of which he relied on only a few varieties, such as—Folle Blanche, Tannat, and Baroque.

Before reviewing the attributes of Baco Noir, looking at François Baco's other hybrids may give a glimpse into his grape-breeding philosophy and goals. Baco Blanc is still very extensively grown in Baco's home region of Armagnac. Baco Blanc makes a low-alcohol, high-acid wine that is perfect for distillation and the production of Armagnac brandy. Baco Blanc is a cross of Folle Blanche and Noah. Baco Blanc is less sensitive to black rot than Folle Blanche, so it became popular in Armagnac after it was released early in the 1900s. It seems that much of Baco's work was to produce grapes used for distillation to brandy that were less sensitive to fungal diseases. Further, he extensively used *riparia* grapes in his breeding program to bolster disease resistance.

BACO NOIR (BACO 23-24, BACO 1)[88]

Baco Noir is a thin-skinned grape that is hardy to very winter hardy in the Hudson Valley. It was bred by François Baco in 1902, and commercially named and released in 1910.[89] It is a cross of Folle Blanche (a traditional grape variety used to make brandies in Armagnac) by a mix of pollen of a *riparia* grape called Grand Glabre and *V. riparia ordinaire*.[90]

PARENTAGE
riparia, vinifera

HARVEST DATE
Mid-season

A A– A A A+

Like many other *riparia* hybrids, the vine buds out early, so it is subject to late spring frost damage, but it does produce a much reduced secondary crop. Similar to other *riparias*, the vine is ultra-vigorous and big, with lush vegetative growth that produces long substantial canes. Its growth habit is horizontal and then trailing. Some experts suggest that it needs to be long cane pruned to obtain satisfactory yields because the clusters are relatively light in weight. However, I find that top wire spur pruning works well in producing an ample crop of quality fruit.

Baco Noir is susceptible to black rot and moderately susceptible to powdery mildew, but is resistant to downy mildew. It is susceptible to botrytis, especially if it rains during harvest time, in which case the berries readily crack and botrytis and bunch rot set in rapidly. In wet vineyards, crown gall can be a problem. The plant is not sensitive to sulfur, so it can be used in a fungal treatment program.

Baco can be grown on moderately heavy clay soils. Since Baco is a *riparia* variety, it tolerates excessive soil moisture. Further, locating vineyards on relatively heavy soils for grapes can help to suppress Baco Noir's wild growth, encourage the growth of quality grapes, and allow for better hardening off of canes to avoid winter injury.

Baco Noir is a productive and slightly above-average producing variety. It can have up to three clusters on each shoot, but does not need cluster thinning to produce a quality crop because its vigor can help it sustain a large crop. It ripens consistently by mid-season around the third week of September with sugars of 20° to 23° Brix. One to two weeks before Baco becomes fully ripe for wine production, the birds will descend to harvest the crop, so bird netting is necessary to protect the crop well before its harvest date.

When Baco Noir is ready to pick, it signals that the grape harvest is at least half over. Ideally, the grapes should hang longer to reduce their excessive acid, but the caution is that these thin-skinned grapes crack easily if they are subject to heavy rains at harvest time. The clusters are of medium size, long, tapering, cylindrical, and can be semi-loose to somewhat compact. The jet-black berries are small to small-medium and the flesh is soft. Baco Noir is suitable for the shorter growing seasons and colder parts of the Hudson Valley if it can be protected from late spring frosts.[91]

Overall, while Baco Noir has some issues in the field, particularly for commercial growers, it is great in the cellar. It ferments easily, finishes rapidly, and is clear within one month of fermenting. To make quality wines from Baco Noir, the grapes need to mature to reduce the naturally high malic acids common with most *riparia* hybrids. If picked before maturity, it produces noticeably thin and acidic wines in a "*riparian*" fashion. This can be a challenge because birds also like Baco Noir, so the recommendation is to allow the birds to feast as the grapes ripen to the proper maturity.

Today, Baco Noir is still grown to a limited extent in Burgundy and more so in the Loire Valley, France, but was extensively cultivated before 1968. Today, Baco Noir production in France still far surpasses its production in the eastern part of the United States. Reported acreage of Baco Noir in France in 2008 was 27 acres and 2.5 acres in Switzerland.[92] In 2001, there were 276 acres of Baco Noir grown in New York State, which has declined to 206 acres as of 2006, but which increased to 223 acres in 2011.[93] Baco Noir is also widely grown in Oregon and Ontario, Canada (565 acres in 2006).[94] Baco's one advantage in the Loire Valley is its ability to produce a small secondary crop after a late spring frost.[95]

Over the years, Baco Noir has been identified as one of the signature red grape varieties of the Hudson Valley. The grape grows well in the Valley and makes a wide range of quality diverse wines similar to Burgundian Pinot Noirs, Bordeaux-like Cabernet Sauvignons, light young fall wines or Nouveaus, and even rosés.

In fact, looking at other literature on Baco Noir, you will find that not only was there some confusion on François Baco's first name, but on descriptors of wines made from Baco Noir, as well. Walter S. Taylor and others have suggested that it reminded them of a good Burgundy,[96] while Morton, Mark Miller, and the Fulkersons have said that it reminded them of Cabernet Sauvignon or Bordeaux.[97] Also, commentators have disagreed about whether or not it makes a better light or heavy colored wine.

Baco Noir wines can have deep color, lots of berry and plum fruit, and high-acid levels that stand up well to barbecued meats or other heavy dishes. If picked ripe, it can develop into a big muscle wine, but due to insufficient tannins, it does not fill out adequately. So to round out the wine, it should be blended with between 15 percent to 25 percent Chelois, Cabernet Sauvignon, or other varieties such as Maréchal Foch or Chambourcin. Baco can have a green, stemmy, pin-prickly nose with a bitter edge if made with unripe grapes, but whatever can be said of Baco Noir, it has great aging potential and brings presence to red wine blends. Further, its large leaves are very suitable for making *dolmas* (stuffed grape leaves).

If Baco Noir is left on its skins for more than seven days as it ferments, it can, with age, have many big Bordeaux-like qualities. It can be a deeply colored red wine that is robust with aromatic flavor elements such as cedar, tobacco, leather, and chocolate. It can have complex fruit flavors of black- and chokecherries, blackberries, and prunes. Further, it can have herbal notes of black pepper, licorice, eucalyptus, and cinnamon. As it ages over five to fifteen years, Baco can become a complex, medium- to full-bodied wine that accompanies red meats very well.

Baco Noir can also be made into a slightly lighter wine style reminiscent of a Burgundian Pinot Noir. When made in this style, it has a rich nose reminiscent of raspberries, black raspberries, plums, blueberries, cherries, and strawberry jam. The herbal notes remain, but include more muted flavors of lavender, black pepper, mint, and licorice. Several twenty- to thirty-year-old Baco Noirs I have tasted have reminded me of some better quality red Burgundy and Bordeaux wines of the same age. As Baco Noirs mature, they can take on a toasty, burnt-sugar flavor profile.

Baco Noir has a varietal flavor that is easily distinguished from other French-American hybrids. The wines have a soft tannin structure that needs to be richened to produce superior wines. Further, it can benefit from malolactic fermentation to reduce its high malic acids, which are firm but manageable. To increase complexity and give the wines body and tannin structure, it is highly recommended to barrel age these wines for six to eighteen months.

A rosé can also be made from Baco Noir, however, its acid levels tend to be very high, so some blending is recommended. It is also capable of being made into rich ports due to its pigment, rich flavors, and high acids, as long as it is finished semisweet. When fermented, it does so quickly and cleanly to become a bright purple wine, so it lends itself well as a Nouveau or fall wine.

CONCORD

DELAWARE

IONA

CROTON

EMPIRE STATE

JEFFERSON

BACO NOIR

CHAMBOURCIN

CHELOIS

HUMBERT #3

MARÉCHAL FOCH

SEYVAL BLANC

VERDELET

BURDIN 6055

BURDIN 4672

WHITE BURDIN

COROT NOIR

MELODY

NOIRET

FRONTENAC GRIS

MARQUETTE

RIESLING

LEMBERGER

BUD BREAK: late April to early May

FLOWERING: late May to early June

YOUNG CLUSTERS: late June to early July

VERAISON: mid-August to late August

If not made properly, Baco wines can be herbaceous and weedy. Baco, especially if picked too early and then made into a varietal wine, can be relatively austere and herbaceous, and can develop muscle without completely filling out into a big wine. This is why it is critical to blend it with at least some other French-American hybrid or *vinifera* wines to minimize its somewhat bitter edge and make it into a complete and interesting wine.

With that said, Baco Noir made in the Hudson Valley can truly be of exceptional quality and is considered by many to be the signature red grape variety in the Hudson Valley. It is highly recommended for future plantings.

François Baco. *Detail of the monument in memory of the great hybridizer erected in Belus (Landes), France, where he spent most of his career.*

Eugène Kuhlmann

(1858–1932)

Another of the earlier French hybridizers was Eugène Kuhlmann of Alsace, France. Kuhlmann was appointed director of the Institut Viticole Oberlin[98] at Colmar, Alsace, in 1915, which was also known as the Weinbau-Institut Oberlin (depending on when France or Germany controlled the Alsace region). The Institute Viticole Oberlin,[99] founded by Chrétien Oberlin (1831–1915 or 1916), was a wine school and viticultural institute devoted to research to combat phylloxera's damage to grapes by pursuing the interspecific breeding of vulnerable native Alsatian and resistant American varieties, and by grafting native grapes to resistant American rootstocks.

In addition, the Institut conducted research on which varieties were best suited to different soil types. When the province of Alsace was returned to France after World War I ended in 1918, Kuhlmann put several of his interspecific grape hybrids on the market. As he was one of the earlier hybridizers, his crosses used first- and second-generation American grape varieties with *vinifera*. Wagner states that Kuhlmann's varieties were bred primarily for Alsatian conditions, which makes them promising for northeastern American conditions. The similarity in climate makes the Kuhlmann hybrids well suited for the Hudson Valley as to winter hardiness, vigor, and disease resistance.[100]

The number of Kuhlmann crosses that were catalogued and pursued was fewer than forty-five. Of the seed parents, the vast majority were done with a seedling of Mgt. 101-14 (Millardet 101-14) (a *riparia-rupestris* cross and a completely American hybrid). Also, some seed parents included Oberlin 702A, Saint Laurent, and a *V. amurensis*. The number of pollen parents was also limited primarily to the *vinifera* hybrids Goldriesling, Knipperlé, and Madeleine Royale.[101] Goldriesling or Riesling Doré is supposedly a cross between Riesling and Courtiller Musque or Muscat Précoce de Saumur (a hybrid Muscat).[102] This is a very early ripening grape that is generally blended with other wines to make the new season's flavorful but ester-ridden wines in local Alsatian cafés.[103] It was widely grown in Alsace in the early 1900s, but has since been fast disappearing from that region altogether.

Kuhlmann named most of his hybrid selections, probably to make them more marketable to the general public. The most popular, in order of importance, are Maréchal Foch, Léon Millot, Lucie Kuhlmann, Maréchal Joffre, and Pinard. All are the result of the same *riparia-rupestris* (Mgt. 101-14 (Millardet 101-14)), of Goldriesling parentage, and ripen very early. Other named hybrids include Saint Sauveur, Fraise Sucrée, Millardet, Vosgien, Bon Noir, and Neron.

MARÉCHAL FOCH (KUHLMANN 188-2)

This is a very versatile variety both in the field and in the cellar. It was named after the World War I French general Maréchal Ferdinand Foch (1851–1929).[104] Like Léon Millot below, it is a cross of a seedling of Mgt. 101-14 (*riparia-rupestris*) by pollen of Goldriesling. However, some maintain that it is a cross of Oberlin 595 (Oberlin Noir) and Pinot Noir, Oberlin 595 × Knipperlé, or perhaps O. 595 × Goldriesling.[105] The condition of the vine compared to Maréchal Joffe and Léon Millot disputes this theory.

PARENTAGE
riparia, rupestris, vinifera

HARVEST DATE
Early

A+ A B A– A

Maréchal Foch was developed around 1911 and released as a commercial variety around 1921. In 2001, there were 102 acres of Maréchal Foch grown in New York, which increased to 144 acres in 2006, but decreased to 104 acres in 2011.[106] It is also commonly grown in Ontario, British Columbia, and in Iowa, Illinois, and Oregon.[107] The grape was never very popular in France, covering only 32 acres in 2008 and 30 acres in the eastern part of Switzerland (2009), but was once commonly planted in the Loire Valley.[108]

Maréchal Foch is an early ripening, moderately vigorous to very vigorous variety depending on the soil it is grown on. With that said, Maréchal Foch is not particular to soils and is adaptable. It is a relatively productive variety. However, due to its small to small-medium cluster size, it has only average production. It buds out very early to early on a vine that is smaller than average. It has a small to medium-sized compact cylindrical cluster that is winged. Its berries are small to medium-sized black fruit. The grape is very hardy against winter damage. To attain sufficient yields, it is generally cane pruned. Should a spring freeze occur, it will have a small secondary crop.[109]

The vine is resistant to fungal diseases, particularly downy mildew and botrytis, and slightly less so to black rot and powdery mildew. It is somewhat sensitive to sulfur treatments so they should be avoided. Further, birds are notoriously attracted to its fruit at harvest time.

According to Morton, Maréchal Joffre, another Kuhlmann hybrid with the same genealogy as Maréchal Foch, and a grape called Phantom Foch (or Early Mix Foch) have sometimes been unwittingly mixed into Foch vineyards. The wines from both grapes are similar, but Maréchal Joffre and Phantom Foch can be identified in the field because they ripen a bit earlier. Maréchal Joffre ripens even earlier than Maréchal Foch, so it is better suited to the cold climates north of the Hudson Valley.[110]

Maréchal Foch is very versatile in the cellar. It can be made into standard red table wines, soft and easy to drink Beaujolais-style reds, rosé, Nouveau wines fermented by carbonic maceration, and used in blends to add fruit or soften other wines. Commentators from the 1950s through 1980s described the wines as soft Burgundian-like wines in flavor and balance. The deep to medium-deep colored red-violet wine from Maréchal Foch is light to medium in body. The wine is high in malic acid and relatively low in tannins. It has some limited aging potential and can benefit from oak aging.

Maréchal Foch can be made into many different wine styles, so the fruit flavors it can generate are numerous. Among its flavors are fresh blackberries, blueberries, heavily cooked strawberries, bramble-berry jam, and red cherries. The wine can have a creamy soft fleshy feeling and soft-acid/tannin profile, so it may need to be blended to enhance its presence. Some Maréchal Fochs have a musty, herbaceous, metallic finish. It can also have a herbaceous smell of thyme, however, it can be made complex and perfumey with flavors of chocolate, cloves, nutmeg, leather, black olives, burnt toast, fresh coffee, mocha, and sweet smoke. Overall, it can be made into a quality wine reminiscent of a light Gamay Beaujolais.

The grape is recommended for the Hudson Valley, even in the short season areas in the Catskill and Berkshire Mountains.

LÉON MILLOT (KUHLMANN 194-2)

Reports of the growing habits and wine attributes of Léon Millot are varied because there are either two clones of Léon Millot, or another variety with the same genetic makeup (most probably Lucie Kuhlmann (Kuhlmann 149-3)) has been misidentified as Léon Millot in some vineyards. In 2001, there were 30 acres of Léon Millot grown in New York.[111] In France, especially on its home turf of Alsace where it covered 210 acres as of 2008, Léon Millot is used to add color to pale red wines grown in these cool climate areas. It is also grown in eastern Switzerland (22 acres), and Denmark (10 acres).[112]

PARENTAGE
riparia, rupestris, vinifera

HARVEST DATE
Early mid-season to late mid-season

A	A-	B	A	A+

Léon Millot is a cross of a seedling of Mgt. 101-14 (*Riparia-Rupestris*) by pollen of Goldriesling. This hybrid, along with Maréchal Foch, was developed around 1911 and released as a commercial variety around 1921. Kuhlmann 194-2 was named after Léon Millot who was a local winemaker, tree nursery owner, and president of the Societé Vosgienne de Viticulture.[113] The two Léon Millot clones are called the Foster (nursery) Millot (or Millot Rouge) and Boordy (nursery) Millot (or Millot Noir). The descriptors below will bear out why they are referred to this way. For purposes here, they will be referred to as Millot Rouge and Millot Noir. Both clones provide high sugars of 24° Brix when fully ripe. Note that like siblings that have the same genetic makeup, Maréchal Foch and Léon Millot are still very different.

Millot Rouge is probably what Kuhlmann called Léon Millot. This grape, initially distributed by Foster's Nursery as Léon Millot, has the same genetic makeup as Maréchal Foch (Foch), and is more similar to Foch in its growth habit and the wine that it produces.

This grape ripens very early, sometimes even before Foch, but generally no more than one week after. (There is a difference of opinion among growers and nurserymen as to whether Millot Rouge ripens before or after Foch.) Its early bud break, small cluster size, tight and compact clusters, and shape are similar to Foch, but Millot Rouge is a bit more vigorous in vegetative growth. Although the berry size of Millot Rouge is smaller than Foch, it has the same black color. The variety is very vigorous in the field, more so than Foch, more productive, and it has higher acid and sugar levels. The variety has an early bud break and a poor secondary crop if hit by a late spring frost.[114]

While very winter hardy, its canes are more spindly than Foch and winter dieback is more pronounced (because of the thin canes), hence it is less winter hardy than Foch. Millot Rouge has good disease resistance, but due to its thick canopy, it must be sprayed carefully to minimize fungal diseases, particularly powdery mildew and botrytis (near harvest time). As it is susceptible to botrytis, it must be picked as soon as this disease manifests itself at harvest time. The grape is productive, hence spur pruning is recommended because its yields are satisfactory.

The wine is medium bodied and fruity, with lots of berry, blackberry, and bright prune notes in a Burgundian style. The berry notes of Millot Rouge are more herbal, woody, and complex than Foch, and the color is also darker than Foch. Walter S. Taylor stated that "experts" seem to believe that it is superior to its cousin Foch.[115]

The other possibly misidentified variety or clone of Léon Millot is Millot Noir, which was distributed by Wagner Nursery (the now defunct grape nursery owned by Philip and Jocelyn Wagner of Riderwood, Maryland). It is also a vigorous grower, more like Baco Noir, and more productive than Millot Rouge. It has larger clusters that ripen later than Millot Rouge.

Millot Noir has the same genetic makeup as Millot Rouge and Foch, and while it has some of the attributes of Millot Rouge, there are some significant differences. While the Millot Rouge clusters and plant look more like Foch in many ways, Millot Noir is a very vigorous vine that also has more spindly but longer growing canes, so it does not overwinter as well as Foch. Also, it ripens by mid-season or later, a full two to three weeks after Foch, and about one to two weeks after Millot Rouge. The clusters are much larger than Millot Rouge or Foch and could be classified as medium to medium-large in size, but its berries are small, sometimes very small. The vine is much more productive than Millot Rouge. It has good disease resistance, but is more susceptible to botrytis than Foch. Neither Millot Rouge nor Millot Noir is widely grown in France.[116]

It is the wine that really differentiates Millot Noir from Millot Rouge. Millot Noir is a very big aromatic, chewy, herbaceous, and earthy wine that is more reminiscent of a Rhône or a big Italian or Spanish red, rather than a Burgundy. Its tannin structure is big and its color inky. It is rich, complex with big structure (for a hybrid) with a flavor profile that includes dark flavors of cooked mulberries and blueberries, chokecherries, tobacco, chocolate, black olives, licorice, and charcoal. It is an integrated but layered wine that has elements of leather, bacon, thyme, aged mahogany, eucalyptus, and lots of earth. It has none of the soft berry flavors of Millot Rouge. The mouth feel of Millot Noir is more viscous like a Chancellor.

Both Millot and Foch are solid wines on their own, with good aging potential, but not as good as Baco Noir. They are also good wines for blending. Foch and Millot Rouge can add a nice berry nose to any blend and soften its acid profile, while Millot Noir has a big tannin structure and dark fruits that can give heft and complexity to a blend. Both clones are recommended for the Hudson Valley.

Millot Noir, due to its longer growing season, should be carefully considered before planting in the Catskill and Berkshire Mountains. For these areas, growers should consider planting the shorter growing season varieties such as Maréchal Joffre (K.187-1), Pinard (K.191-1), Lucie Kuhlmann (K.149-3), and Neron (K.296-1).

HUMBERT

(1858–1932)

There is little or no biographical information about the private grape breeder Humbert or details of his work. He—assuming that he was a man—seems to have been born or worked in the Maynal commune, in the Jura department in Franche-Comté in eastern France. Further, he did at least some of his breeding work before World War I. The only grape variety of his work that seems to have survived is Humbert #3, also known as Humbert-Chapon #3. We do not know if Chapon was his business partner or how this individual was associated with Humbert.

HUMBERT #3

PARENTAGE
labrusca, lincecumii, riparia, rupestris, vinifera

HARVEST DATE
Early late to late

A	A-	B	A+	A

There is some dispute as to the genetic makeup of Humbert #3, also known as Humbert-Chapon #3. The Geneva Experiment Station maintains that Humbert #3 is a cross of S.157 × Gaillard 2 and was developed around 1912.[117] It seems to be the only grape variety of Humbert's work to have survived in the United States. The genetic makeup of this grape, if Geneva is correct, consists of, chronologically, a very early Seibel selection. In surveying the genetic history of Seibel's contribution to this variety, the heritage of S.157 is noticeably missing. Perhaps Seibel just gave S.157 to Humbert for his own use. In the S.150 through S.160 series of hybrids, Seibel commonly used seed parents of undisclosed Munson hybrids and Herbemont Touzan and for pollen parents used Sauvignon Blanc, other unidentified *vinifera*, and Ganzin 1.[118] The pollen parent for Humbert #3 is Gaillard 2, a hybrid of Othello × Rupestris with the pollen parent of Noah, which is a *riparia/labrusca* hybrid.

The *Vitis International Variety Catalogue (VIVC)* database which is maintained by The Institute for Grapevine Breeding at Geilweilerhof, in Germany, indicates that Humbert #3 is a cross of Gaillard 157 (Roi des Blancs) by Gaillard 2. Roi des Blancs was bred in 1891 and is a cross of the seed parent Triumph × Eumelan (the Hudson Valley bred grape variety) × the pollen parent S.1. With the growing characteristics of Humbert #3, I am not surprised that one of the grandparents of this grape is Eumelan. In both accounts, the pollen parent is identified as Gaillard 2. It is uncertain whether the seed parent is S.157 or Gaillard 157.[119]

The cluster is medium-loose to loose and of medium-large size. Its berries are of smallish to medium size and a blue-black color. The variety is productive to very productive and of medium winter hardiness in the Hudson Valley. Cluster thinning is recommended to increase quality and to hasten the harvest date. The vine is vigorous with lots of foliage, but its lateral and then downward growth habit facilitates air circulation and sunlight penetration for the plant. This growth habit, with its loose clusters, helps to make Humbert #3 fairly disease resistant, except for powdery mildew, to which it is susceptible. Humbert #3 ripens early late to late in the season around the first or second week of October.

The wine from Humbert #3 can be from a medium crimson red to a deeply colored red wine depending on how it is made. Overall, it can be quite good and well balanced with ripe fruit, otherwise it can be very herbaceous. The wine is integrated with a burnt raspberry nose, cranberry/cherry flavors, black olives, cedar, and smoke. The wine can develop with age, where the fruit remains and begins to add more fruit flavors of blueberries and prunes. The wine can have soft tannins and a medium-bodied flint middle and finish. It can be reminiscent of a light Cabernet Sauvignon, but may need time for these elements to appear in a varietal wine. Hence, Humbert #3 can be used to give body and structure to other softer, less tannic wines or made as a varietal wine.

Humbert #3 has superior performance in the field and solid performance in the cellar. This variety should be considered in future breeding programs to increase wine quality in the cellar.

I like this variety both in the field and cellar. It should be considered for future commercial plantings in the Hudson Valley.

NOTES

1. Hudson Cattell and H. Lee Stauffer, *The Wines of the East: The Hybrids* (Lancaster, Pa.: L & H Photojournalism, 1978), 4–8; See generally Pierre Galet and Lucie T. Morton, *A Practical Ampelography: Grapevine Identification* (Ithaca, N.Y. and London: Cornell University Press, 1979).

2. Galet and Morton, *Practical Ampelography*, 154–55; and Lucie T. Morton, *Winegrowing in Eastern America: An Illustrated Guide to Viniculture East of the Rockies* (Ithaca, N.Y.: Cornell University Press, 1985), 84–85. The conversions from hectares to acres are my own.

3. Cattell and Stauffer, *Hybrids*, 11, 18–19.

4. Galet and Morton, *Practical Ampelography*, 219.

5. See the great work and charts on grape variety genealogy by Carl Camper at the Chateau Stripmine Experimental Vineyard, www.chateaustripmine.info/Breeders.htm.

6. Morton, *Winegrowing in Eastern America*, 69–70; Phytowelt GmbH, *Final Report: "Study on the Use of the Varieties of Interspecific Vines"* (Cologne: Phytowelt GmbH, July 16, 2003), 16–55; see Eurostat, the Statistical Office of the European Union, 1999, www.statistischedaten.de.data.

7. Galet and Morton, *Practical Ampelography*, 154–55.

8. Ibid.

9. Ronald E. Subden. "The Odyssey of DeChaunac (Seibel 9549)," *American Wine Society Journal* 12, no. 4 (Winter 1980): 81–84.

10. The Nelson A. Rockefeller Institute of Government. *2006 New York State Statistical Yearbook* (31st ed., rev. and exp.) (Albany, N.Y.: The Nelson A. Rockefeller Institute of Government, State University of New York, 2006), 598; *New York Fruit Tree and Vineyard Survey*, 2006, United States Department of Agriculture, National Agricultural Statistics Services, and the New York State Department of Agriculture and Markets, 37; *New York Vineyard Survey*, January 2013, United States Department of Agriculture, National Agricultural Statistics Services, and the New York State Department of Agriculture and Markets, 4.

11. Jancis Robinson, Julia Harding, and Jose Vouillamoz. *Wine Grapes: A Complete Guide to 1,368 Vine Varieties, Including Their Origins and Flavor* (New York: Ecco/HarperCollins, 2012), 67.

12. By 1929, S.5279 was commonly being referred to as L'Aurore in France because of its unusually early ripening date and its light white/grayish-pink color. This name, Aurora or Aurore, which was also commonly used in the United States, was ratified by the Federal Bureau of Alcohol, Tobacco, and Firearms in its list of approved names of grape varieties, published in the Federal Register on January 8, 1996. Cattell and Stauffer, *Hybrids*, 11; Hudson Cattell, "Naming the French Hybrids," *Wines East: News of Grapes and Wine in Eastern North America* 35, no. 2 (July–August 2007): 24–31.

13. Robinson et al., *Wine Grapes*, 67.

14. *The Brooks and Olmo Register of Fruit & Nut Varieties* (3rd ed.) (Alexandria, Va.: American Society for Horticultural Sciences Press, 1997), 249.

15. Robinson et al., *Wine Grapes*, 197.

16. Cattell and Stauffer, *Hybrids*, 11; also see Cattell, "Naming the French Hybrids." S.13053 was an unnamed variety in France, but this grape was becoming popular on the East Coast, so the variety was named by the Finger Lakes Wine Growers Association on August 24, 1970, to designate American wines. The name was chosen after consultation with the American Pomological Society and ultimately ratified in 1972 by the Great Lakes Grape Nomenclature Committee, which had both American and Canadian representatives. Ultimately, the name Cascade was ratified by the Federal Bureau of Alcohol, Tobacco, and Firearms by a rule published in the Federal Register on January 8, 1996.

17. Robinson et al., *Wine Grapes*, 197.

18. *Brooks and Olmo Register*, 255–56.

19. Galet and Morton, *Practical Ampelography*, 175.

20. Cattell and Stauffer, *Hybrids*, 11; also see Cattell, "Naming the French Hybrids." S.7053 was an unnamed variety in France, but this grape was becoming popular on the East Coast, so the variety was named by the Finger Lakes Wine Growers Association on August 24, 1970, to designate American wines. This name was chosen after consultation with the American Pomological Society and ultimately ratified by the Great Lakes Grape Nomenclature Committee in 1972. Ultimately, the name Chancellor was ratified by the Federal Bureau of Alcohol, Tobacco, and Firearms by a rule published in the Federal Register on January 8, 1996.

21. Rockefeller Institute, *2006 New York State Statistical Yearbook*, 598; *New York Fruit Tree and Vineyard Survey*, 37; *New York Vineyard Survey*, 4.

22. *Brooks and Olmo Register*, 256.

23. Chateau Stripmine, www.chateaustripmine.info/Breeders.htm; See also Vitis International Variety Catalogue (VIVC) database, www.vivc.de for additional background information.

24. The grape was officially named Chelois by a French official nomenclature committee during a classification process of grapes permitted to be grown in France, decree dated February 27, 1964. The name Chelois was ratified by the Federal Bureau of Alcohol, Tobacco, and Firearms in its list of approved names of grape varieties by a final rule published in the Federal Register on January 8, 1996. See Cattell and Stauffer, *Hybrids*, 11; Cattell, "Naming the French Hybrids."

25. Galet and Morton, *Practical Ampelography*, 178.

26. *2006 New York State Statistical Yearbook*, 598.

27. *Brooks and Olmo Register*, 257.

28. Philip M. Wagner, *American Wines and Wine-Making* (New York: Alfred A. Knopf, 1969), 252.

29. Walter S. Taylor and Richard P. Vine, *Home Winemaker's Handbook* (New York: Harper & Row, 1968), 53.

30. Wagner, *American Wines*, 252.

31. Robert M. Pool, J. Einset, K. H. Kimball, and J. P. Watson. *Vineyard and Cellar Notes, 1958–1973*. Special Report No. 22-A (Ithaca, N.Y.: New York State Agricultural Experiment Station, April 1976), 74.

32. Galet and Morton, *Practical Ampelography*, 177.

33. De Chaunac was an unnamed variety in France, but was becoming popular on the East Coast, so it was named by the Finger Lakes Wine Growers Association on August 24, 1970, to designate American wines. Initially the name chosen for S.9549 was Cameo, which was selected in consultation with the American Pomological Society. However, due to a disagreement about the name, a group of American and Canadian representatives called the Great Lakes Grape Nomenclature Committee was convened to ratify the new name, De Chaunac, in 1972. This new name was ratified by the Federal Bureau of Alcohol, Tobacco, and Firearms in a final rule published in the Federal Register on January 8, 1996. See Cattell and Stauffer, *Hybrids*, 11; Cattell, "Naming the French Hybrids."

34. Morton, *Winegrowing in Eastern America*, 78; Cattell and Stauffer, *Hybrids*, 11.

35. Galet and Morton, *Practical Ampelography*, 177; *2006 New York State Statistical Yearbook*, 598; *New York Fruit Tree and Vineyard Survey*, 37; *New York Vineyard Survey*, 4.

36. *Brooks and Olmo Register*, 259.

37. *Boordy Nursery Grape Book, The Boordy Nursery Grape Vine Catalogue, Grape Vines for Wine Growers* (Riderwood, Md.: Boordy Nursery, 1993), 3.

38. S.9110 was an unnamed variety in France, but was becoming popular on the East Coast so it was named Verdelet by the Finger Lakes Wine Growers Association on August 24, 1970, to designate an American wine. This name was chosen after consultation with the American Pomological Society and ultimately ratified in 1972 by the Great Lakes Grape Nomenclature Committee, which had American and Canadian representatives. The name Verdelet was ultimately ratified by the Federal Bureau of Alcohol, Tobacco, and Firearms in a final rule published in the Federal Register on January 8, 1996. Galet and Morton, *Practical Ampelography*, 176; Cattell and Stauffer, *Hybrids*, 11; Cattell, "Naming the French Hybrids."

39. *Brooks and Olmo Register*, 296.

40. Galet and Morton, *Practical Ampelography*, 177.

41. My telephone conversations with Philip M. Wagner, Athens, New York, December 21, 1990.

42. Morton, *Winegrowing in Eastern America*, 86.

43. Galet and Morton, *Practical Ampelography*, 219.

44. Chateau Stripmine, www.chateaustripmine.info/Breeders.htm.

45. Morton, *Winegrowing in Eastern America*, 85.

46. Chateau Stripmine, www.chateaustripmine.info/Breeders.htm.

47. Wagner, *American Wines*, 253.

48. Morton, *Winegrowing in Eastern America*, 85.

49. Galet and Morton, *Practical Ampelography*, 180.

50. Pool et al., *Vineyard and Cellar Notes*, 34.

51. The grape was named in France, probably by or after the name of its hybridizer Bertille Seyve Jr., it being a contraction of Seyve and Vallier. It was officially named by a French official nomenclature committee during a classification process of grapes permitted to be grown in France, decree dated February 27, 1964. The name was ratified by the Federal Bureau of Alcohol, Tobacco, and Firearms in its list of approved names of grape varieties published in the Federal Register on January 8, 1996. See Cattell and Stauffer, *Hybrids*, 11; Cattell, "Naming the French Hybrids."

52. Robinson et al., *Wine Grapes*, 990.

53. Great Britain had over 250 acres of Seyval Blanc in 1999 and New York State had 330 acres of Seyval Blanc in 2001 and 315 acres in 2011. See *Final Report*, 10; see Eurostat, www.statistischedaten.de.data.1999; *New York State Statistical Yearbook*, 598; *New York Vineyard Survey*, 4.

54. *Brooks and Olmo Register*, 288.

55. Robinson et al., *Wine Grapes*, 990.

56. *2006 New York State Statistical Yearbook*, 598; *New York Fruit Testing and Vineyard Survey*, 37.

57. This widely grown variety in France was commonly called Villard Blanc, after its French hybridizer. Villard Blanc was officially named by the French official nomenclature committee during a classification process of grapes permitted to be grown in France, the decree dated February 27, 1964. This name was ratified by the Federal Bureau of Alcohol, Tobacco, and Firearms in its list of approved names of grape varieties, in a rule published in the Federal Register on January 8, 1996. See Cattell and Stauffer, *Hybrids*, 11; Cattell, "Naming the French Hybrids."

58. Morton, *Winegrowing in Eastern America*, 87.

59. *Brooks and Olmo Register*, 296.

60. Morton, *Winegrowing in Eastern America*, 87; Jancis Robinson, *Vines, Grapes, and Wines* (New York: Alfred A. Knopf, 1986), 238; *Final Report*, 10; Eurostat, www.statistischedaten.de.data; Robinson et al., *Wine Grapes*, 1140.

61. *Final Report*, 13; Eurostat, www.statistischedaten.de.data.

62. Morton, *Winegrowing in Eastern America*, 87.

63. Taylor and Vine, *Home Winemaker's Handbook*, 53.

64. My conversations with Philip Wagner on December 21, 1990.

65. Pool et al., *Vineyard and Cellar Notes*, 84.

66. Chateau Stripmine, www.chateaustripmine.info/Breeders.htm; See also VIVC database, www.vivc.de for additional background information.

67. Galet and Morton, *Practical Ampelography*, 206.

68. Pool et al., *Vineyard and Cellar Notes*, 13; Eurostat, www.statistischedaten.de.data.

69. New York State Fruit Testing Cooperative Association. *A Catalog of New and Noteworthy Fruits, 1991–1992* (Geneva, N.Y.: New York State Fruit Testing Cooperative Association, Inc., 1991), 25.

70. This variety was informally named after its hybridizer and officially named by a French official nomenclature committee during a classification process of grapes permitted to be grown in France by a decree dated February 27, 1964. This name was ratified in the United States by the Federal Bureau of Alcohol, Tobacco, and Firearms in its list of approved names for grape varieties, as published as a rule in the Federal Register on January 8, 1996. See Cattell and Stauffer, *Hybrids*, 11; Cattell, "Naming the French Hybrids."

71. Robinson, *Vines, Grapes, and Wines*, 206; Galet and Morton, *Practical Ampelography*, 184; *Final Report*, 10; Eurostat, www.statistischedaten.de.data.

72. Morton, *Winegrowing in Eastern America*, 87.

73. *Boordy Nursery Grape Book*, 4; Taylor and Vine, *Home Winemaker's Handbook*, 54.

74. Chateau Stripmine, www.chateaustripmine.info/Breeders.htm.

75. Wagner, *American Wines*, 251.

76. Robinson, *Vines, Grapes, and Wines*, 200.

77. Morton, *Winegrowing in Eastern America*, 77; Pool et al., *Vineyard and Cellar Notes*, 60. Sources indicating Seyve Villard 12-417 x Chancellor parentage include *Catalog of New and Noteworthy Fruits*, 26; United States Department of Agriculture, ARS, National Genetic Resources Program, Germplasm Resources Information Network (GRIN), www.ars-grin.gov; Robinson et al., *Wine Grapes*, 218.

78. For current ancestry analysis see the recent DNA study by F. Röckel, O. Trapp, E. Zyprian, L. Hausmann, D. Migliaro, S. Vezzulli, R. Töpper and E. Maul, "A 'Regent' pedigree update: ancestors, offspring and their confirmed resistance loci," *Vitis, Journal of Grapevine Research*, Vol. 60, No 4 (Siebeldingen, Germany: Julius Kühn-Institut, October 2021), 189–93. https://ojs.openagrar.de/index.php/VITIS/article/view/16088.

79. Chambourcin was informally named by its hybridizer in France, but named Chambourcin by an official French nomenclature committee during a classification process of grapes permitted to be grown in France by a decree dated February 27, 1964. The name was recognized in the United States by a Federal Bureau of Alcohol, Tobacco, and Firearms rule that listed approved names of grape varieties published in the Federal Register on January 8, 1996. See Cattell and Stauffer, *Hybrids*, 11; Cattell, "Naming the French Hybrids." Jancis Robinson goes further and states that it was named after the lieu-dit Chambourcin in the village of Bouge-Chambalud in Isère, where Seyve owned an experimental vineyard. See Robinson et al., *Wine Grapes*, 218.

80. Robinson et al., *Wine Grapes*, 219.

81. *2006 New York State Statistical Yearbook*, 598; *New York Fruit Tree and Vineyard Survey*, 37; *New York Vineyard Survey*, 4.

82. Galet and Morton, *Practical Ampelography*, 162.

83. Wagner, *Wine-Grower's Guide*, 208.

84. Pool et al., *Vineyard and Cellar Notes*, 1, 16.

85. Robinson, *Vines, Grapes, and Wines*, 238; Robinson et al., *Wine Grapes*, 79.

86. Chateau Stripmine, www.chateaustripmine.info/Breeders.htm.

87. Ibid.

88. As with many grape varieties developed in France, the variety was identified by the name of the hybridizer; see Cattell and Stauffer, *Hybrids*, 11.

89. Baco Noir was named after its hybridizer François Baco and commonly called by that name in France. It was then officially named Baco Noir by an official French nomenclature committee during a classification process of grapes permitted to be grown in France by a decree dated February 27, 1964. The name was ratified for use in the United States by a Federal Bureau of Alcohol, Tobacco, and Firearms rule that was published in the Federal Register on January 8, 1996. See Cattell and Stauffer, *Hybrids*, 11; Cattell, "Naming the French Hybrids." It was renamed Baco 1 by Prosper Gervais after his visit to M. Baco in 1910. See Robinson et al., *Wine Grapes*, 79.

90. USDA, ARS, National Genetic Resources Program; Robinson et al., *Wine Grapes*, 79.

91. *Brooks and Olmo Register*, 249–50.

92. Robinson et al., *Wine Grapes*, 80.

93. *2006 New York State Statistical Yearbook*, 598; *New York Fruit Tree and Vineyard Survey*, 37; *New York Vineyard Survey*, 4.

94. Robinson et al., *Wine Grapes*, 80.

95. Robinson, *Vines, Grapes, and Wines*, 26, 28, 201; Wagner, *A Wine-Grower's Guide*, 42.

96. Taylor and Vine, *Home Winemaker's Handbook*, 53; Joe Chilberg and Bob Baber, *New York Wine Country: A Tour Guide* (Utica, N.Y.: North Country Books, 1985), 31.

97. Morton, *Winegrowing in Eastern America*, 75; Benmarl Vineyards, *Benmarl Vineyards Catalog, Supplies for Home Wine-makers* (Marlborough, N.Y.: Benmarl Vineyards, n.d., circa 1977), 17; Sayre and Nancy Fulkerson, *Fulkerson's Grape Juice Catalog* (Dundee, N.Y.: Fulkerson Winery and Juice Plant, 2000), 16.

98. Not much is readily available on the history of the Institut Viticole Oberlin at Colmar, probably because it was founded in the Province of Alsace, which was disputed territory between France and Germany after the Franco-Prussian War of 1871 and the end of World War I. However, the Institut was linked to the drive and accomplishments of its founder, Chrétien Oberlin (1831–1915 or 1916).

99. Chrétien Philippe Oberlin was born in Beblenheim in Alsace on July 7, 1831. He was the son of a wealthy vinegrower. He attended school in Beblenheim, then the private college of Ribeauvillé, and then the College of Haguenau. As an engineer, Oberlin first worked at the Department of Highways, and later ended up as an overseer of viticulture at the French Ministry of Agriculture. In 1852, he was commissioned to build the railway line of Selestat to Sainte Marie aux Mines. After this project, Oberlin settled back in Beblenheim to occupy himself with experimental viticulture and his work as an ampelographist. In 1875, as the first signs of the phylloxera infestation became evident in Alsatian vineyards, Oberlin began searching for vines that were resistant to phylloxera. By 1881, he was researching and writing extensively about solutions to the phylloxera problem. Oberlin was especially successful in identifying American grape species that were not only resistant to phylloxera, but also supported the grafting of Alsatian grape varieties in a manner that did not alter the latter's characteristics. Once he identified these American species, Oberlin energetically developed white direct producers for Alsace. By 1893, he bred Goldriesling, a hybrid of Riesling and alternatively the Muscat varieties Courtiller Musque or Muscat Précoce de Saumur, which was released for cultivation in 1921. Chrétien Oberlin was also the mayor of Beblenheim f from 1870 to 1902.

Oberlin established the first experimental vineyards in Colmar in 1895, which is probably shortly after the Institut Viticole Oberlin, also known as the Weinbau-Institut Oberlin, was established. The vineyard still exists and is situated on a flat alluvial plain that extends outward from the city of Colmar. The output of the Oberlin vineyards was primarily devoted to the city of Colmar for municipal banquets and receptions. The Institut used the experimental vineyards to help determine which varieties were best suited to Alsatian soils. Chrétien Oberlin was clearly the driving force behind the work at the Institut, and he continued his experimental work with plant breeding and incorporating phylloxera resistance into his plants at the Institut until he died in either 1916 or 1915.

Upon Oberlin's death, Eugène Kuhlmann, until that time Oberlin's top assistant or protégé, became the director of the Institut Viticole Oberlin at Colmar in 1915. Kuhlmann continued Oberlin's work and concentrated on developing new interspecific hybrids, such as Oberlin 55-56, O.232, O.595 (Oberlin Noir), O.604, O.702A (Blanc Petit), and many others. The Institut seems to have disappeared from the history books after Kuhlmann's death in 1932 and the chaos of World War II. However, in 1980, the Institut Viticole de Colmar and its vineyards were incorporated into or renamed the Domaine de La Ville de Colmar. As of late, the city of Colmar, due to economic issues, seems to have relinquished its control of the Institut's vineyard and shifted it to the Alsatian winery Arthur Metz.

The Institut Viticole Oberlin at Colmar does not seem to be related to the French government–sponsored agricultural experiment station at Colmar called the Institut National de la Recherche Agronomique, which was established in 1868, but which was not made fully operational by 1871 when control of the Alsace territory was shifted to

Germany until the end of World War I. From unpublished article, *Chrétien Oberlin and the Institut Viticole Oberlin at Colmar, Alsace*, by Robert Bedford, November 7, 2012.

100. Wagner, *American Wines*, 251.

101. Chateau Stripmine, www.chateaustripmine.info/Breeders.htm.

102. VIVC database, www.vivc.de; Robinson, *Vines, Grape, and Wines*, 233.

103. Robinson, *Vines, Grapes, and Wines*, 233.

104. Kuhlmann 188-2 was named Maréchal Foch by its hybridizer Eugène Kuhlmann, and it was commonly known by this name in France. It was officially named Maréchal Foch by an official French nomenclature committee during a classification process of grapes permitted to be grown in France by decree dated February 27, 1964. It was recognized in the United States by this name by a rule published in the Federal Register by the Federal Bureau of Alcohol, Tobacco, and Firearms dated January 8, 1996. See Cattell and Stauffer, *Hybrids*, 11; Cattell, "Naming the French Hybrids."

105. *USDA, ARS, National Germplasm Repository*; Robinson et al., *Wine Grapes*, 598.

106. *2006 New York State Statistical Yearbook*, 598; *New York Fruit Tree and Vineyard Survey*, 37; *New York Vineyard Survey*, 4.

107. Robinson et al., *Wine Grapes*, 598.

108. Ibid.

109. *Brooks and Olmo Register*, 273.

110. Lucie T. Morton, "Will the Real Maréchal Foch Please Stand Up?," *American Wine Society Journal* 16, no. 1 (Spring 1984): 14–18; Morton, *Winegrowing in Eastern America*, 79–80.

111. *2006 New York State Statistical Yearbook*, 598.

112. Robinson et al., *Wine Grapes*, 545.

113. Eugène Kuhlmann is probably responsible for naming the grape Léon Millot, commonly known in France by that name. It was officially named Léon Millot by a French official nomenclature committee during a classification process of grapes permitted to be grown in France, decree dated February 27, 1964. The name was recognized in the United States by the Federal Bureau of Alcohol, Tobacco, and Firearms by a final rule published in the Federal Register on January 8, 1996. See Cattell and Stauffer, *Hybrids*, 11; Cattell, "Naming the French Hybrids." 24–31; Robinson et al., *Wine Grapes*, 545.

114. Morton, "Will the Real Maréchal Foch Please Stand Up?"; Morton, *Winegrowing in Eastern America*, 79–80.

115. Taylor and Vine, *Home Winemaker's Handbook*, 53.

116. Morton, *Winegrowing in Eastern America*, 81–82.

117. Pool et al., *Vineyard and Cellar Notes*, 58–59; Chateau Stripmine, www.chateaustripmine.info/Breeders.htm.

118. Pool et al., *Vineyard and Cellar Notes*, 58–59. One variety, S.156, also known as Duc Petit, may have been the seed parent because its genetic makeup is *rupestris*, *lincecumii*, and *vinifera*, hauntingly similar to the genetic makeup assigned to Humbert #3 by Geneva in its *Vineyard Notes* (*labrusca*, *lincecumii*, *riparia*, *rupestris*, and *vinifera*).

119. VIVC database, www.vivc.de; Pierre Galet, *Cépages et vignobles de France* (Montpellier: Charles Déhans (impr.), 1988), 405, 407, 410; Pool et al., *Vineyard and Cellar Notes*, 58–59.

CHAPTER TEN

Later French Hybridizers 1925–1955

UNLIKE THE EARLY FRENCH hybridizers who lived and worked in divergent wine-growing regions across France, such as the lower and upper Rhône Valley, north and south of the Bordeaux region and Alsace, this group of later French hybridizers worked almost exclusively in and around the greater Burgundy area or just to the east and south of Burgundy near Switzerland and the upper Rhône.

Where these hybridizers lived influenced their goals and the kinds of grape varieties that they hoped to breed, what grapes were used in their breeding programs, the climate and region where such grapes were to be cultivated, and their aspirations to develop varieties that possessed either average or high-quality winemaking capabilities. The climate of Burgundy and the surrounding areas is more continental with cool winters and humid summers, while the climate in Bordeaux and the Rhône Valley is a warmer and sunnier Mediterranean climate. Further, the climatic conditions of Alsace are generally much cooler with less intense direct sunlight than Burgundy, Bordeaux, or the Rhône Valley.

The self-perceived high quality of grapes grown in Burgundy for wine production, as opposed to the bulk wine–producing areas where most early French hybridizers lived, influenced the underlying goal of the later French hybridizers to breed in higher wine quality production characteristics in a uniquely Burgundian fashion. Hence, in the tug of war between producing more grapes economically and growing better quality wine grapes, the later French hybridizers were much more concerned about developing grape varieties that could compete with *viniferas* such as Pinot Noir, Gamay Noir, and Chardonnay on a quality level, with the understanding that such new varieties could be grown at a lower per unit cost.

These new hybridizers were generally born around 1900, after the French wine industry had settled into a course of action to fight the invasion of American insect pests and fungal diseases. The later French hybridizers covered in this chapter include Joanny Burdin and his son Remy, Pierre Landot (1900–42), Jean François Ravat (died in 1940),[1] and Jean-Louis Vidal (1880–1976). Their grape varieties are the most commonly grown in the Hudson Valley and the eastern United States.

The early French hybridizers, who were born around 1865, lived at a time when the Franco-Prussian War of 1871 had ended and when phylloxera and new American fungal diseases were rapidly spreading across Europe. The new hybridizers lived during World War I and World War II, when the Viniferists and Hybridists were competing for domination in France. While many of the later French hybridizers' new grape varieties were superior in quality to the older French-American hybrids, the French government was beginning to clamp down on the sale of such newer varieties and directing growers to rip out older French-American hybrids and replace them with local *vinifera* varieties. This is why, for growers in the Hudson Valley and across the eastern United States, we need to evaluate for the first time many of these varieties or reevaluate those that have already been given only a cursory analysis.

Hence, through the work of my colleagues across the United States and in the Hudson Valley, and by the publication of this book, we should continue to evaluate these newer grape varieties developed by the late French hybridizers. We should look to these later French-American hybrids as avenues of growth for the cultivation of economically and environmentally sustainable grape varieties that can produce quality wine in the Hudson Valley and in other cool climate regions. Further, these varieties should be considered for use as building blocks for newer hybrids that have yet to be developed to produce high-quality wine.

Joanny and Remy Burdin

Joanny Burdin and his son Remy Burdin were wine grape growers who began to breed wine grapes beginning around 1925 in the Iguerande commune, Saône-et-Loire department, which is in the middle of the Burgundy region of France, and continued their work after World War II. Burdin hybrids first appeared in 1929, but most came after 1940. Many of their hybrids were released after World War II.[2]

The most prevalent seed parents of Burdin's crosses were widely planted Seibel hybrids grown in France. These high-yielding commonly grown grapes produced *vin de pays* table wines, including, in order of importance to their breeding program, Plantet (S.5455), Subereux (S.6905), and other Seibel hybrids such as Chelois and Chancellor. The pollen parent tended to be from quality *vinifera* varieties that produced superior wines such as Gamay Noir, Gamay Freau, Grolleau Noir, and perhaps Pinot Noir.[3] Generally, these Burdin hybrids are considered to be moderate producers of superior wine. Burdin was clearly a second-generation hybridizer who created, by using later generation Seibel hybrids, genetically very diverse hybrids, which had American ancestor species such as *V. rupestris*, *riparia*, *labrusca*, and *lincecumii*.

By breeding in Gamay Noir and other quality *vinifera* grapes and late Seibel hybrids such as Seinoir (S.8745), this hybridizer was clearly breeding for superior wine quality and not for mass production.[4] Philip Wagner was keen on Burdin's work and over time offered through his nursery business Burdin selections B.8753 and B.11402 (red) and B.4672 (white).[5] In addition to these grapes, other Burdin selections that Mr. Wagner was interested in and wrote about included B.5201 (white), B.7061, B.7705 (Florental), B.8649, and B.10010 (red).[6] Over Joanny Burdin's career, he and his son hybridized over eighty selections, of which only Florental (B.7705) was named.

I believe that since Burdin introduced most of his new grape varieties in the late 1940s to mid-1960s when interest was waning in France for hybrid grape varieties, his work has been unduly ignored. Wagner, Galet, and Morton noted that Burdin selections have retained *vinifera* qualities in growing habit, cold tenderness, and lack of fungal disease resistance.[7] I have not observed those traits in my vineyard with the white varieties B.4672 and B.5201 nor the red varieties B.6055, B.8753, and to some extent B.11402. It seems that many of the Burdin hybrids being grown experimentally in the Hudson Valley are culturally satisfactory in the field, possess superior wine quality, and have crop yields that are about the same or better than Gamay.

The following descriptions of the Burdin grapes for B.4659, B.5957, B.7061, and B.7360 come primarily from the information provided by Galet and Morton, Wagner, and the New York State Geneva Experimental Station, while the remaining grape descriptions are based on my own observations in the vineyard as to their performance in the field and in the cellar. This entire class of hybrids should be explored more systematically for cultivation and wine production in the Hudson Valley. While not high-volume producers (except for B.6055) and somewhat cold tender, these red hybrids are very good in the cellar.

Note: Some of the minor white and red Burdin varieties listed are being reported as either promising or ones that should be further tested in the Hudson Valley. Many of these varieties are being field tested at my farm Cedar Cliff, but it is too early to fully recommend them for commercial cultivation.

BURDIN 6055 (B.6055)

This is a promising red variety that is approximately one quarter *vinifera* in heritage. It is a cross of Plantet (S.5455) × Seinoir (S.8745). B.6055 has a vigorous and upright growth habit and thick three- to four-foot-long canes that grow in an open canopy. It has good fungal disease resistance, even in wet years. Its open canopy, which facilitates spray applications, is resistant to most fungal diseases, and is an easy grape to grow.

PARENTAGE
rupestris, *riparia*, *labrusca*, *lincecumii*, *vinifera*

HARVEST DATE
Late mid-season to late

A A B A A

The vine is winter hardy with strong vigor and heavy to very heavy yields, but it can overproduce, so the variety should be cluster thinned to avoid over-cropping. The medium-sized compact clusters are medium-large in size and cylindrical with one shoulder, and the blue berries are large with thick skins. B.6055 likes a deep, well-drained soil, but will grow in damp, but not wet, clay soils. The grape ripens by late mid-season to late and hangs well after the grapes ripen.

This grape makes a superior wine with aging potential. The wine, at best, can have the color of a medium-dark Pinot Noir or lighter hues of beet juice. It has a pronounced berry nose and a soft, but full, tannin structure. The wine is soft, with a long clean finish. The fruit profile of the wine is of strawberry jam, pomegranates, lots of berries, with a pleasant gamey/grassy undertone. This grape should be grown more widely in the Hudson Valley.

BURDIN 11402 (B.11402)

This grape variety is, chronologically, the last grape variety bred by Burdin. It is only moderately winter hardy to somewhat tender, with some resistance to fungal diseases. The plant has modest vigor with a horizontal to trailing growth habit, but is a shy bearer of fruit. It may have better vigor and production if put on a vigorous rootstock. The grapes ripen very late along with other late-ripening grapes such as Catawba. The clusters resemble slightly longer and narrower bunches like Pinot Noir. Since the variety ripens late, it tends not to get botrytis, but I have still seen this occur.

PARENTAGE
Unknown

HARVEST DATE
Very late

C+	B	C	B	A+

The vine likes a well-drained deep soil, but will grow in damp soils. B.11402 needs cluster thinning to maintain good crop quality and vine health because it overproduces grapes despite only medium vigor. This variety continues to grow until the first fall frost, so its canes may not harden off at the ends, hence its susceptibility to winter injury. The canes are thin, so if unripe, they do not make it through the winter, but base buds on its cordons remain viable in most winters. Spur pruning is recommended to limit production and to remove dead first-year wood that does not ripen during the winter. While not the greatest grape in the field with regard to cultivation or production, it is a reliable producer that makes exceptional wines.

These velvety wines are quite remarkable and can resemble a soft Pinot Noir, with a bright cherry color and floral/berry nose. The flavor profile includes red cherries, black raspberries, licorice, and strawberry jam, with some fruit flavors of a soft Cabernet Franc. B.11402 wine is complex with a structure that is framed by soft tannin, flint like a Chianti, and earth. The wine ages well and reaches its peak within five to ten years. It is a fragile wine that can benefit from some blending, but must be the primary base wine, otherwise its characteristics will be lost. This grape variety is recommended for cultivation in the Hudson Valley.

MINOR WHITE GRAPE VARIETIES OF BURDIN

B.4672

cinerea, labrusca, lincecumii, riparia, rupestris, vinifera

This is a cross of the red grape Plantet (S.5455) with an unknown white *vinifera*. The vine is of medium vigor or better. The cluster is of medium size, compact with a pinkish-green berry that is translucent. The grape, which ripens by mid-season with sugars over 20° Brix, is a reliable producer. Wagner described the wine as delicately fragrant with a touch of *goût d'Alsace* when picked fully ripe. Further, he said that this grape was a "real comer."[8] However, Galet and Morton reported the wine as flat and without character.[9] The grape is not very susceptible to fungal diseases. Based on Wagner's comments and my own observations, B.4672 should be further evaluated in the Hudson Valley for commercial cultivation.

B.5201

aestivalis, cinerea, labrusca, lincecumii, riparia, rupestris, vinifera

This is a cross of Plantet (S.5455) × S.7157. Galet and Morton report this to be vigorous and productive.[10] The cluster is of medium size, compact, and cylindrical. The berries are medium to large in size and pinkish in color. The variety is susceptible to downy mildew. It is a semi-upright grower with a later than average bud break and ripens by mid-season. Based on my own experience, this variety should be further evaluated for future cultivation in the Hudson Valley.

MINOR RED GRAPE VARIETIES OF BURDIN

B.4650

aestivalis, cinerea, labrusca, lincecumii, riparia, rupestris, vinifera

This red hybrid is a cross of Plantet (S.5455) and B.1646. The very small to small vine is tender to winter injury. Its clusters are medium to large with blue berries, and it is reported to have frequent powdery mildew and some downy mildew. Like many other Burdin varieties, it over-crops. It ripens late to mid-late in season. The quality of the wine is superior with deep color and a full body, as well as good tannin structure and balance.[11]

B.5957

aestivalis, cinerea, labrusca, lincecumii, riparia, rupestris, vinifera

This red grape of unknown parentage is reported to be a very small vine, tender to winter injury, has medium-sized compact clusters of blue berries, is not resistant to powdery mildew, and has frequent cracking and poor fruit condition with some sulfur sensitivity. The wine quality was not reviewed favorably.[12]

B.7061

aestivalis, cinerea, labrusca, riparia, rupestris, vinifera

This cross of Chelois and Gamay has about 75 percent *vinifera* heritage. The vine size is very small and it is moderately cold hardy. Its clusters are small to medium in size and compact with black berries.[13] It is not resistant to fungal diseases and its fruit cracks often. The variety ripens late mid-season.[14] Wagner thought the variety was culturally satisfactory and the wines virtually identical to a Gamay Beaujolais.[15]

B.7360

aestivalis, cinerea, labrusca, riparia, rupestris, vinifera

This cross of S.8724 by Gamay is an averaged-sized vine that is tender to winter injury. The cluster is small and compact with black berries, and is susceptible to powdery mildew. This variety ripens late in the season. The wines have good tannin structure and a spicy aroma.[16]

B.7575

Parentage unknown

It is a small vine, but moderately hardy to winter injury and has a medium-sized compact cluster with blue berries. The variety ripens late and is moderately productive.[17] I grow this grape at Cedar Cliff and find it to be a sturdy, moderately vigorous variety that produces a consistent crop every year.

B.7705 (FLORENTAL)

lincecumii, rupestris, vinifera

This cross of S.8365 by Gamay Noir or perhaps Gamay *à jus blanc*[18] has just under 70 percent *vinifera* heritage. Florental[19] was developed around 1925, but has been sold commercially only since 1963. The vine size is very small and is tender to winter injury. The medium-sized cluster is compact with black berries. It is susceptible to fungal diseases and the fruit is subject to cracking and poor condition. The variety is moderately productive, but may not be a reliable producer. Florental ripens by mid-season to late mid-season. The wine quality is reported to be excellent, making a light fruity wine comparable to a light Gamay in a Beaujolais style. The bud break is early and it needs to be grafted to increase vigor.[20]

I have heard very good comments from some growers as to this grape's wine quality. Florental has been grown commercially, to some limited extent, in the Finger Lakes in New York, and Pennsylvania.[21] However, I have found the variety to be very difficult to propagate and grow at my farm at Cedar Cliff. In fact, I have never been able to get this vine to fruit under my care at all. I keep this grape on my list for further evaluation since it has been widely reported on, but my personal experience makes me question it.

B.8753

Parentage unknown

Philip Wagner thought this was a promising variety that produced a wine comparable to a Beaujolais.[22] It is one of the last hybrids to be bred by Burdin chronologically. The vine has moderate vigor, is reasonably hardy to winter damage, but is somewhat susceptible to fungal diseases. The bunches are large with small to medium-sized black berries. The vine needs close pruning to avoid over-cropping. The wine it produces is of a deep color. More field testing is needed before the variety can be fully recommended for large-scale commercial production.

Pierre Landot

(1900–42)

Pierre Landot of the Conzieu commune, Ain department in the Rhône-Alpes region of France north of Lyon, began hybridizing grapes after World War I, in 1924.[23] This second-generation hybridizer made crosses between later developed Seibel and Sevye-Villard hybrids to create more complex grape genealogies.[24] Wagner stated that Landot sought to create varieties that resembled Gamay and Pinot Noir to produce Beaujolais-style wines. Further, while many of these hybrids produced superior wine, many proved to be culturally defective because they included not only the superior wine quality of their *vinifera* parents, but their problematic growing characteristics as well.[25]

The most common seed parents of Landot hybrids included Plantet (S.5455), Rubilande (S.11.803), Seyval Blanc, Pinot Noir, Villard Noir (S.V.18-315), and his own earlier Landot hybrids such as L.244. The pollen parents were mostly Seibel and Seyve-Villard hybrids, including Plantet, Rubilande, Chanaan (S.5163), Villard Noir, Chancellor, and Landot 244 and 506 and very few *vinifera*, including Muscat Hamburg, Gamay, and Pinot Noir. In reviewing Landot's extensive work of at least ninety chronicled varieties, he tried many different combinations of seed and pollen parents and different types of grape varieties. Unlike many other hybridizers, who tended to keep the same seed parent or the same kinds of seed parents (one with high disease resistance) and added quality by the pollen parent, he alternated the same parent varieties as both seed parent and pollen parents to achieve his breeds.[26]

The resulting body of work consists of about 80 percent red and 20 percent white grape varieties. Since Landot produced many of his hybrids in France during the troubled economic and political times between World War I and World War II and died in 1942, many of his hybrids have not been as thoroughly evaluated in the field as they should have been. His work may still hold promise and should be reevaluated in the Hudson Valley, particularly in light of today's advances in viticultural practices.

LANDAL (L.244)

Landal is a cross of seed parent Plantet (S.5455) and pollen parent S.8216. Its ancestry is about 50 percent *vinifera*, which consists of varieties such as Dattier, Aramon, Alicante Bouschet, and Grenache. Landal gets its disease resistance from Ganzin, Noah, Dutchess, Jaeger 70, *rupestris*, *labrusca*, and *lincecumii*. The cluster is medium in size, cylindrical, and loose, with small oval blue-black berries. It is a moderately vigorous, consistently producing variety with average bud break, which matures in mid-season to early late season in the Hudson Valley. This variety is moderately winter hardy.

PARENTAGE
aestivalis, *berlandieri*, *cinerea*, *labrusca*, *lincecumii*, *riparia*, *rupestris*, *vinifera*

HARVEST DATE
Mid-season to early late season

A A C A A

Landal[27] is fairly resistant to fungal diseases, however it is rarely grown on the East Coast of the United States. However, it covers 121 acres of vineyards in France as of 2008, mostly in the Loire Valley and Vendée, but it is losing ground in France.[28] The wine quality is quite good, full of ripe strawberries and red raspberries, with lots of acid. It is very jammy without being tutti-frutti. The wines can be bright, with floral hints of violets, blueberries, and beach plum jam. It is not herbaceous in nose or taste, and is most like a light Beaujolais or heavy Cascade. It has soft tannins, which tends to limit its aging ability. Wagner reports that it yields a deep-colored and "remarkably" Beaujolais-like wine, "One of the best for quality." Further, it is a moderate producer that requires short cane pruning.[29]

This variety needs more evaluation before being unreservedly recommended to commercial growers for cultivation in the Hudson Valley. In my experience, it should be on a short list of grapes to be evaluated.

LANDOT NOIR (L.4511)

Chronologically, Landot Noir[30] was one of the last grape varieties bred by Landot before his early death at the age of forty-two. It is a cross of seed parent Landal (L.244) with pollen parent Villard Blanc (S.V.12-375). It is really a Seibel hybrid that combines L.244 (S.5455 (Plantet) × S.8216) by Villard Blanc, S.V.12.-375 (S.6468 × S.6905) (Subereux). This is a 40–50 percent *vinifera* combination of one of Landot's better known quality red wine hybrids (Landal) with Villard Blanc, which was at one time one of the most popular, productive, and widely grown white hybrid grapes in France. Landot Noir is the seed parent of Frontenac.

PARENTAGE
aestivalis, *berlandieri*, *cinerea*, *labrusca*, *lincecumii*, *riparia*, *rupestris*, *vinifera*

HARVEST DATE
Mid-season

A A B+ A A

Landot Noir is a vigorous to very vigorous grape that ripens by mid-season. This variety may get its vigorous growth habit from Villard Blanc. Bud break is late and it ripens early mid-season to mid-season. It is winter hardy in the Hudson Valley with good disease resistance and is particularly resistant to downy mildew. The growth habit is upright to lateral. Landot Noir is a consistent and heavy producer of fruit that is effortless to grow. The long and relatively narrow cluster is medium-large to large. Further, it has a medium loose cluster, consisting of medium to medium-large-sized blue berries. Wagner recommends spur or short cane pruning to balance growth and to limit its production.[31] The vine has not been widely planted, but preliminary observations are encouraging.

Wagner stated that it was bred to combine the high red wine quality of Landal Noir with the high productivity of Villard Blanc, but its wine quality does not equal Landal.[32] The wine can be made as a fruity Beaujolais-like wine and can make a more traditional Burgundy-type wine in a blend, with full body and good tannin structure. This grape has potential and should be further explored by Hudson Valley growers and winemakers alike.

JEAN FRANÇOIS RAVAT

(?–1940)

Jean François Ravat[33] was a civil engineer from the Marcigny commune, Saône-et-Loire department in Burgundy, France. He bred grapes from 1929 to 1935, and died relatively young in 1940. Ravat's work was continued by his son-in-law, Jean Tissier, after his death.[34] In reviewing his breeding program, Ravat's goal was to breed grapes to increase wine quality, but also presumably to be more disease resistant than traditional *vinifera* grape varieties.[35] Of the twenty-five recorded hybrids that Ravat and Tissier bred, the seed parent was always a disease-resistant Seibel hybrid that Seibel had developed in the middle phase of his career (S.5279 (Aurora)) being the earliest seed parent used by Ravat and S.8724 being one of the latest). The pollen parent was generally a quality *vinifera* grape such as Chardonnay, Pinot Noir (both red and white), and Pinot de Corton. It seems Ravat was trying to impart superior wine qualities to more disease-resistant Seibel hybrids.

While Ravat and Tissier devoted equal resources to the development of both red and white grapes, their success was noted mostly with white wine grapes. To give an example of how Ravat consistently attempted to breed in quality to existing disease-resistant Seibel grape varieties, consider a named variety, such as Ravat Blanc. Ravat Blanc, also known as R.6, is a hybrid of seed parent S.5474 (or some say S.8724) with pollen of Chardonnay, introduced in 1945.[36] Other examples of Ravat's breeding work to increase wine quality by breeding in quality *vinifera* grape varieties include Vignoles (S.6905 (Plantet) × Pinot Noir (white)), Ravat Noir (S.8365 × Pinot Noir), and Tissier Ravat (Aurora × Chardonnay).

VIGNOLES (R.51)

Vignoles is the most widely planted Ravat hybrid grape in the United States. Initially, some historic literature maintained that it was a cross of seed parent S.6905 (Subereux) x some type of Pinot Noir (reports vary from a "white" Pinot Noir, a Pinot de Corton, Chardonnay, or a Pinot Blanc). It has more recently been determined that Vignoles is really a hybrid of S.5455 (Plantet) x S.880.[37] Vignoles was bred around 1930.[38] While this white wine grape variety was never popular in France, it has become one of the mainstays of the eastern North American wine industry.[39] It was named Vignoles in 1970.[40]

PARENTAGE
lincecumii, rupestris, vinifera

HARVEST DATE
Late mid-season to very late

A	A	C	B	A

This versatile grape is similar to Riesling in many ways both viticulturally and in the wines that it produces. It can be made into barrel-aged dry wines, simple semidry country wines, and very complex sweet late-harvest wines. Currently Vignoles late-harvest wines are in vogue. The grapes have a high acid content and can have sugar levels that can reach 30° Brix.

Viticulturally, the grape can be a bit demanding. Vignoles is a winter-hardy grape similar to Seyval Blanc or better. The plant is not vigorous in its growth habit. It is resistant to black rot and moderately resistant to downy mildew. However, it is susceptible to botrytis and powdery mildew because of its very compact cluster. Its susceptibility to botrytis can impart peach, apricot, and honey notes to a wine. Vignoles is not sulfur-sensitive.

The vine has a late bud break, which reduces the risk of late spring freeze injuries. The vine tends to have low to medium vigor. The clusters are relatively small, conical, with a small shoulder and very compact, which makes Vignoles a moderate producer. Further, the light green, pink-tinged berries are small for a white variety. Cluster thinning is not needed for this grape. Vignoles's berries are small with thick skins. In addition, Vignoles's tight clusters make the grape even more susceptible to cracking, bunch rot, and botrytis than would normally be the case. This is particularly true if it is raining during harvest season.[41]

To help minimize botrytis damage, it is best to spray the clusters heavily before the clusters close in on themselves by midsummer so the spray can get into the interior. In addition, the vines should be pruned to open the canopy and increase sunlight penetration to reduce mildew. Cane pruning is recommended to increase production. However, increasing production can reduce the quality of wine produced.

The grape ripens late mid-season for dry wines, but can be picked much later for late-harvest whites wines. Often Vignoles is picked for a late-harvest wine because its sugar levels are high and are well balanced with its high acid content. Vignoles makes a very high-acid wine that can be counterbalanced either by higher sugar levels, lower-acid wines, or blending practices.

Depending on the style of Vignoles made, it can have a variety of tastes and flavor/acid profiles. The color can range from pale straw to golden. If made as a dry wine, it can be clean and crisp with a touch of tartness of green apples, grapefruit,

and bananas. The fruit is citrus in character, with underlying tones of peach and tropical fruit, which works well with oak aging and makes the wine more complex.

If made as a semisweet wine or as a late-harvest wine, Vignoles can have subtle and complex floral tastes of apricots, pineapples, honey, peaches, orange rinds, melons, guava, and orange blossoms with a tart finish. The finish consists of honey, almonds, and Grand Marnier. The residual sugar/acid balance adds to the wines' complexity and mouth feel, while providing a clean crisp edge to the finish. Most compare these wines to German dessert wines, but they can be similar to Sauternes. The body is thick, viscous, syrupy, and pungent like Chinese sweet-and-sour sauce.

Vignoles is also a good grape for blending. It adds complexity, color, and weight to white wine blends. No more than 20 percent of Vignoles should be added to a white blend unless the winemaker wishes to cover up another defective white wine in the blend or convert the wine into a light Vignoles-like wine.

Vignoles is particularly known for its capacity to make very nice late-harvest dessert wines since these grapes have high acid and can attain sugar levels that reach 30° Brix, which is needed to produce quality dessert wines.

This is an important grape whose cultivation should be encouraged in the Hudson Valley.

RAVAT 34

While Ravat 34 has not been commonly grown in the Hudson Valley, it may be a grape suitable for cultivation in the region. The wine quality has been described as good to excellent, clean, light, fruity, and superior in quality to Aurora. Philip Wagner stated that the "wine recalls a Burgundian Aligote."[42] It should be noted that Ravat 34 is not like the more commonly grown Ravat 51 (Vignoles) at all.

PARENTAGE
vinifera × *unknown American cultivar*

HARVEST DATE
Early

A A C A A

This white grape is winter hardy to very winter hardy, more so than grapes such as Cayuga White and Vidal. Ravat 34 ripens early, but after very early-ripening grapes such as Aurora and before Seyval Blanc. The fruit has a pinkish-white cast to amber. Its medium-sized clusters are well filled and not too compact, with medium-sized berries.

The vine is moderately vigorous and productive and needs to be cluster thinned to balance its crop. The Geneva Experimental Station's *Vineyard and Cellar Notes* states that it sustains occasional downy mildew, over-crops, and has moderate sulfur sensitivity.[43] However, the *Double A Vineyards Catalog* maintains that it is resistant to downy mildew and botrytis, moderately resistant to powdery mildew, and not sensitive to sulfur.[44] Ravat 34 is an attractive vine that behaves well and makes abundant crops of lovely grapes that ripen early. This may be a grape that needs more investigation by growers in the Hudson Valley, particularly those located near the colder and shorter-seasoned areas of the Catskill or Berkshire Mountains that border the Hudson Valley.

Jean-Louis Vidal

(1880–1976)

The French hybridizer Jean-Louis Vidal was director of the Institut de Recherches Viticoles Fougerat at Bois-Charentes, France. This research station is located in the department of Charentes, which is located just northeast of Bordeaux and encompasses the commune of Cognac. He studied at the École Nationale Supérieure Agronomique de Montpellier. Vidal bred primarily white grape varieties, hoping to produce new ones suitable for the production of Cognac, a wine that is distilled into brandy and then aged in wood in the Cognac commune of the Charentes region. Vidal had no idea that his work would lead to the development of Vidal 256, which is versatile for the production of many different styles of table wines and of great importance to the eastern United States' wine industry. Vidal 256 (Vidal Blanc) is winter hardy so it tends to be grown in Michigan, Pennsylvania, New York, New England, and Canada.

Vidal worked for many years to breed grapes that could produce a Cognac equal to that produced with the Folle Blanche grape which is susceptible to the mildew diseases of Charente. Vidal was breeding grapes that were more resistant to these fungal diseases. The most common parents of his hybrid crosses were S.4762 (S.405 × S.867), St. Emilion (also known as Ugni Blanc or Trebbiano), and Vidal 49 (a cross of St. Emilion and Folle Blanche) and Rayon d'Or. Of his over fifty recorded grape varieties, all but four are white grapes. Only a few of his selections were ultimately named, including Blanchard 1A, Sainton, Sélect, Rami Vidal Blanc, and Balza Noir.

Vidal's breeding program entailed having one parent being the *vinifera* variety or a hybrid of two different *viniferas* and the other parent being a Seibel hybrid that offered disease resistance. During the course of his work, Vidal tried having the *vinifera* as either the seed parent or pollen parent, but seemed to prefer having the *vinifera* parent being the seed parent, and disease resistance coming from Seibel pollen.[45]

VIDAL BLANC (VIDAL 256)

Genetically, Vidal Blanc is a hybrid of St. Emilion × S.4986 (Rayon d'Or). Vidal Blanc is the only grape from Vidal's work that is grown extensively in the United States, and along with Vignoles is one of the white French-American hybrid mainstays of the East Coast wine industry.[46] Vidal Blanc was among the first one-third of the grapes that Vidal bred fairly early in his work, developed in the 1930s.[47]

PARENTAGE
lincecumii, rupestris, vinifera

HARVEST DATE
Late to very late

B+ B B A+ A

The genetic makeup of Vidal Blanc is about 75 percent *vinifera* (one-half St. Emilion and one-quarter Aramon). Vidal Blanc[48] shares a common parent with Seyval Blanc (Rayon d'Or). This high-acid, versatile white grape can be made into a bone-dry, steely white wine for fish, a barrel-aged wine reminiscent of a Fumé Blanc or Burgundian white, or an ice wine that can rival dessert Rhine wines.

Vidal Blanc is a cross of the *vinifera* grape alternatively called St. Emilion in Cognac, France, Ugni Blanc in the south of France, and Trebbiano Toscano in Chianti, Italy with Rayon d'Or (S.4986). St. Emilion is used in Europe for the production of wine that is later distilled to make Cognac and Armagnac in France, and distilled spirits in Italy. Vidal Blanc is a thick-skinned grape that is moderately hardy to hardy to winter damage in the Hudson Valley, but less so than Baco Noir and about the same as Cayuga White. It is resistant to black rot, botrytis, and fruit bunch rots. It is slightly more susceptible to downy mildew and susceptible to powdery mildew. It is not sensitive to sulfur. Overcropping of Vidal can lead to heavier winter injury.

The plant is vigorous, a consistent producer, and very productive. Vidal Blanc produces greenish-white, medium-small berries on large, compact clusters that are long, narrow, tapering, and cylindrical. The grape can grow in many different soil types and conditions. It buds out late to very late in the spring, but if damaged by a spring frost, it has a secondary crop. Vidal matures late to very late, producing high sugar levels even in cool climates. Close winter pruning and cluster thinning should be considered to ensure that the crop ripens properly. The style of pruning recommended is cordon training with spurs or short cane pruning depending on vine vigor. Vidal lends itself to upright shoot positioning.[49]

As noted earlier, this grape can make a wide range of wine styles, including austere wines for fish, complex Burgundian-style whites, wines like Vouvray, big fat Fumé Blanc–style wines if aged in oak, and Rhine wine–like dessert wines. The Hudson Valley's climate is capable of producing each of these four distinct styles. As a crisp wine without wood aging, Vidal is very clean, metallic, and flinty with floral and resinous notes that include fruit flavors of pineapple, grapefruit, melon, hazelnuts, pears, orange blossoms, spice, dried flowers, and lead pencils. Some have compared it to Riesling in taste and body. It has high acid levels, so sometimes residual sugar is left to balance with the acid so that the wines do not taste too tart. Other times, it is made bone-dry. Further, the wines can have the feel of German May wines in their fruit composition, mineral flinty body, and viscosity.

As a Fumé Blanc or Burgundian-type of wine, the grapes are left on the vine for a longer period of time (one to two weeks longer until the second or third week of October) to increase their sugars and, more importantly, to reduce the grape's acid profile. After it is made, the wine is aged in wood for at least six months to soften its body, brighten its nose, give it more complex smoky notes and rich spice and butter flavors, and to elongate its finish. These wines are much more approachable than the steely Vidals described earlier. This style of Vidal is more nutty, buttery, complex, and "greasy" (like gin). These wines have the flavor of ripe pears, orange rinds, vanilla, and almonds.

As an ice wine, Vidal Blanc has the classic Rhine wine qualities of rich honey, citrus flavors reminiscent of Grand Marnier, and hazelnuts, with an underlying metallic finish like mineral water. The reason these Vidal ice wines, like other ice wines, are so rare and expensive is that the grapes stay on the vines until late November or December. While hanging, the grapes shrivel like raisins, so when they are finally pressed, the grapes yield little juice, though this juice is highly concentrated in flavor and sugar.

Vidal Blanc is also used as a base wine that is blended with other grapes such as Seyval Blanc, Chardonnay, and Vignoles. Vidal on its own has a more muted and neutral nose and taste when compared to other white wine varietals, but it has a solid body and foundation. It can be blended with other whites to brighten its taste and to add complexity. Also, Vidal has proven to be a successful blending component with high-terpene varieties such as Riesling. Further, Vidal can be successfully used in sparkling wine blends.

Overall, Vidal wines, if made properly, have more *vinifera* qualities than most other French hybrids. Vidal in the cellar is easy because it ferments well, is protein stable, and has few hybrid flavors so that it is clean when it finishes fermentation.

Vidal Blanc's consistent yields are due to its winter hardiness, capability of producing a secondary crop after a spring frost, ability to adapt to different soil types, and large average yearly crops. For growers who are considering planting Vidal Blanc, it is a "workhorse" grape in the field and in the cellar. It comes highly recommended.

Vidal Blanc. *Vidal's white hybrid grape is most extensively grown in the U.S. and Canada, and is well-suited for the production of late harvest and ice wines.*

NOTES

1. I did extensive research to find the vital statistics on Joanny and Remy Burdin and the birth date of Jean François Ravat with no success and obtained the assistance of the New York State Legislative Library, the New York State Library, and the LuEsther T. Mertz Library at the New York Botanical Garden, Bronx, New York, but could find no such information. The grape-breeding activities of Burdin and Ravat started after 1929.

2. Pierre Galet and Lucie T. Morton, *A Practical Ampelography: Grapevine Identification* (Ithaca, N.Y.: Cornell University Press (1979), 219.

3. See charts of grape variety genealogy by Carl Camper, Chateau Stripmine Experimental Vineyard, www.chateaustripmine.info/breeders.

4. Philip M. Wagner, *Grapes Into Wine* (New York: Alfred A. Knopf, 1976), 282.

5. *Boordy Nursery Grape Book, the Boordy Nursery Grape Catalog, Grape Vines for Wine Growers* (Riderwood, Md.: Boordy Vineyards,1993), 3, 4.

6. Philip M. Wagner, *American Wines and Wine-Making* (New York: Alfred A. Knopf, 1969), 251; Wagner, *Grapes Into Wine*, 282, 283; Philip M. Wagner, *A Wine-Grower's Guide* (New York: Alfred A. Knopf, 1985), 214, 218.

7. Wagner, *American Wines*, 251; Galet and Morton, *Practical Ampelography*, 163.

8. *Boordy Nursery Grape Book*, 4.

9. Galet and Morton, *Practical Ampelography*, 163.

10. Ibid., 163.

11. Robert M. Pool, J. Einset, K. H. Kimball, and J. P. Watson, *Vineyard and Cellar Notes, 1958–1973*, Special Report no. 22-A (Ithaca, N.Y.: New York State Agricultural Experiment Station, April 1976), 53; my conversations with Philip M. Wagner on December 21, 1990.

12. Pool et al., *Vineyard and Cellar Notes*, 53.

13. USDA, ARS, National Genetic Resources Program, Germplasm Resources Information Network (GRIN), Beltsville, MD.

14. Pool et al., *Vineyard and Cellar Notes*, 54–55.

15. Wagner, *American Wines*, 251.

16. Pool et al., *Vineyard and Cellar Notes*, 55.

17. Ibid.

18. Wagner, *Wine-Grower's Guide*, 218.

19. This variety, B.7705, was commonly referred to as Florental in France before 1964. Very few of the Burdin hybrids were named varieties; they were known mostly by their hybrid names and numbers. A French official nomenclature committee recognized this variety as Florental during the classification process of grapes permitted to be grown in France by decree dated February 27, 1964. The name Florental was ratified for use in the United States by the Federal Bureau of Alcohol, Tobacco, and Firearms by its final rule published in the Federal Register on January 8, 1996. See Hudson Cattell and H. Lee Stauffer, *The Wines of the East: The Hybrids* (Lancaster, Penn.: L & H Photojournalism, 1978), 11; Hudson Cattell, "Naming the French Hybrids," *Wines East: News of Grapes and Wine in Eastern North America* 35, no. 2 (July–August 2007), 24–31.

20. Pool et. al., *Vineyard and Cellar Notes*, 54–55; Galet and Morton, *Practical Ampelography*, 164; Wagner, *Wine-Grower's Guide*, 218; Jancis Robinson, Julia Harding, and Jose Vouillamoz, *Wine Grapes: A Complete Guide to 1,368 Vine Varieties, Including Their Origins and Flavor.* (New York: Ecco/ HarperCollins, 1912), 353.

21. Sixty-nine acres of Florental were grown in France in 2008, mostly in Burgundy and the Loire Valley. Robinson et al., *Wine Grapes*, 353.

22. *Boordy Nursery Grape Book*, 3.

23. Galet and Morton, *Practical Ampelography*, 219.

24. Chateau Stripmine, www.chateaustripmine.info/Breeders.htm.

25. Wagner, *American Wines*, 252.

26. Chateau Stripmine, www.chateaustripmine.info/Breeders.htm.

27. This variety, L.244, was commonly referred to as Landal in France before 1964. It was probably named by or in honor of its hybridizer Pierre Landot. A French official nomenclature committee recognized it as Landal during the classification process of grapes permitted to be grown in France by decree dated February 27, 1964. The name Landal was ratified for use in the United States by the Federal Bureau of Alcohol, Tobacco, and Firearms by its final rule published in the Federal Register on January 8, 1996. See Cattell and Stauffer, *Hybrids*, 11; Cattell, "Naming the French Hybrids."

28. Chateau Stripmine, www.chateaustripmine.info/Breeders.htm; Pool et al., *Vineyard and Cellar Notes*, 64, 65; Robinson et al., *Wine Grapes*, 539.

29. Wagner, *Wine-Grower's Guide*, 219; Walter S. Taylor and Richard P. Vine, *Home Winemaker's Handbook* (New York: Harper & Row, 1968), 53.

30. This variety, L.4511, was commonly referred to as Landot Noir in France before 1964. It was probably named by or in honor of its hybridizer Pierre Landot. It was, interestingly enough, a name that was not recognized as Landot Noir by the French official nomenclature committee during the classification process of grapes permitted to be grown in France by decree dated February 27, 1964. The name Landot Noir was ratified for use in the United Stated by the Federal Bureau of Alcohol, Tobacco, and Firearms by its final rule published in the Federal Register on January 8, 1996. See Cattell and Stauffer, *Hybrids*, 11; Cattell, "Naming the French Hybrids."

31. Wagner, *Wine-Grower's Guide*, 219.

32. Ibid.

33. I did extensive research to find more information on Jean François Ravat, including his birth year, and solicited the assistance of the New York State Legislative Library, the New York State Library in Albany, New York, and the LuEsther T. Mertz Library at the New York Botanical Gardens, Bronx, New York, but with no success.

34. Galet and Morton, *Practical Ampelography*, 219.

35. Wagner, *American Wines*, 252.

36. Galet and Morton, *Practical Ampelography*, 169.

37. The author initially reported the former known heritage in the first edition of this book. This updated information came to the author from Dr. Bruce Reisch of Cornell University, and confirmed by wine journalist Hudson Cattell. See Hudson Cattell, *Wines of Eastern North America: From Prohibition to the Present. A History and Desk Reference*. (Ithaca, N.Y.: Cornell University Press, 2014), 241.

38. Galet and Morton, *Practical Ampelography*, 170; Pool et al., *Vineyard and Cellar Notes*, 22; Wagner, *Grapes Into Wine*, 282; Lucie T. Morton, *Winegrowing in Eastern America* (Ithaca, N.Y.: Cornell University Press, 1985), 87; USDA, ARS, National Genetic Resources Program. *The Brooks and Olmo Register of Fruit & Nut Varieties* (3rd ed.) (Alexandria, Va.: American Society for Horticultural Sciences, 1997), 296.

39. Vignoles covers 148 acres in New York (2006), 85 acres in Michigan (2006), 208 acres in Missouri (2009), and 72 acres in Illinois (2007). See Robinson et al., *Wine Grapes*, 1138.

40. Ravat 51 was an unnamed variety in France, but was becoming popular on the East Coast of the United States, so it was named Vignoles by the Finger Lakes Wine Growers Association on August 24, 1970, so that the name could be used to designate it in American wines. The name Vignoles was chosen after consultation with the American Pomological Society, was ratified by the Great Lakes Grape Nomenclature Committee in 1972, and ratified by the Federal Bureau of Alcohol, Tobacco, and Firearms by a final rule published in the Federal Register on January 8, 1996. Cattell and Stauffer, *Hybrids*, 11; Cattell, "Naming the French Hybrids."

41. *Brooks and Olmo Register*, 296.

42. *Boordy Nursery Grape Book*, 4.

43. Pool et al., *Vineyard and Cellar Notes*, 22–23.

44. *Double A Vineyards, Inc., Catalog* 2000/2001 (Fredonia, NY: Double A Vineyards, 2000/2001).

45. Chateau Stripmine, www.chateaustripmine.info/Breeders.htm.

45. There are 222 acres in New York (2011), 145 acres in Michigan (2006), 118 acres in Missouri (2009), 150 acres in Virginia (2010), and 1,920 acres in Canada (2008). See *New York Vineyard Survey*, January 2013, United States Department of Agriculture, 4; Robinson et al., *Wine Grapes*, 1136.

47. Galet and Morton, *Practical Ampelography*, 186; Pool et al., *Vineyard and Cellar Notes*, 41; Chateau Stripmine, www.chateaustripmine.info/Breeders.htm; Robinson, *Wine Grapes*, 1136.

48. V.256 was an unnamed variety in France and probably not widely grown. Interestingly enough, it was not officially named Vidal Blanc by either the French official nomenclature committee by a decree dated February 27, 1964, the Finger Lakes Wine Growers Association on August 24, 1970, nor the Great Lakes Grape Nomenclature Committee in 1972. During this period, Vidal Blanc was a sleeper of a grape that was probably named in honor of its hybridizer Jean-Louis Vidal as it became more commonly grown on the East Coast. The variety was officially named Vidal Blanc for use in the United States by the Federal Bureau of Alcohol, Tobacco, and Firearms by a final rule published in the Federal Register on January 8, 1996. Also note that while V.256 was not named as a variety by the French nomenclature committee by its decree on February 27, 1964, the variety V.9 was named as Sainton. See Cattell and Stauffer, *Hybrids*, 11; Cattell, "Naming the French Hybrids."

49. *Brooks and Olmo Register*, 296.

Copyright 1905 by the Rotograph Co.
5149 N. Y. Agricultural Experiment Station, Geneva, N. Y.

CHAPTER ELEVEN

Geneva Hybrids

THE NEW YORK STATE Agricultural Experiment Station at Geneva, New York, a division of the New York State College of Agriculture and Life Sciences, a statutory college of the State University of New York at Cornell University, has one of the largest and oldest grape-breeding programs in the United States. The Geneva Experiment Station, commonly referred to as Geneva, was established by the New York State Legislature in 1880, and the Denton family farm, where the Geneva Experiment Station is located, was purchased by the State of New York in 1882.[1] Geneva is the sixth oldest agricultural experiment station to organize its work in the United States.

The efforts to establish an agricultural college in New York State took the collective work of the state's agricultural community and civic and political leaders over sixty years to achieve. Governor DeWitt Clinton, in his governor's message to the New York State Legislature in 1818, urged for the establishment of a State Board of Agriculture and expressed the hope that a chair of agriculture might be included. At its first meeting in 1832, the New York State Agricultural Society took up the matter of establishing an agricultural school.

Throughout the 1830s to 1870s, various schemes were developed and submitted to the New York State Legislature to establish and finance some form of state agricultural college, either on a state, regional, or county basis. Such schemes entailed establishing a research or experimental farm and facilities to advance the various agricultural pursuits in the state, developing methods to quickly disseminate information on new cultivation techniques, and developing or identifying new seed types and plant material to increase the productivity of vegetable and fruit crops.

The New York Agricultural Experiment Station at Geneva, New York. *Top: Circa 1910. Bottom: Aerial view, circa 2012.*

State policy makers focused particular attention on increasing soil fertility and designing soil-improvement projects, such as better drainage systems to increase the amount of productive arable land available to growers. This priority was identified because of the general state of soils in New York. The state's soils became exhausted because such lands had been improperly cultivated for more than 150 years in some areas of the Hudson Valley to scores of years in the much later settled areas of western New York. As a consequence, many local farmers moved further west to cultivate newer and more fertile lands that had not been under cultivation for long periods of time. This trend of moving west hurt New York's agricultural economy because experienced farmers were leaving the state in large numbers. Interestingly, quite a few of the state's leaders who advocated for a New York agricultural school came from transplanted New England farming families who had recently moved west to New York to escape their own even more meager, depleted, and thin soils that had been cultivated for an even longer period of time.[2]

After many years of fits and starts, the Geneva Experiment Station was authorized by the New York State Legislature in 1880. However, due to technical language problems with the 1880 state authorization bill, which revolved around how the board would be constituted to run Geneva, the Station did not begin organized work until 1882, after corrective language was adopted by the New York State Legislature in 1881. The organizations that supported the establishment of the Station included the New York State Agricultural Society, the New York State Grange, the Central New York Farmer's Club, the Elmira Farmer's Club, the Western New York Horticultural Society, and the many local and county farmers' and horticultural societies across New York, such as the Newburgh Bay Horticultural Society. In addition, Ezra Cornell (1807–74), founder of Cornell University, was an active sponsor.[3]

Establishing the Experiment Station

The Geneva Experiment Station, as stated in its legislative enactment clause, was established for the "purpose of promoting agriculture in its various branches of scientific investigation and experiment." Further, the first director, E. Lewis Sturtevant, helped to set out its mission, which was to promote scientific discovery and rapid communication of results to benefit the farmers and consumers of New York. In 1887, the Geneva Experiment Station's charge was expanded to include work on beef cattle, swine, and fruit varieties. In 1923, the originally independent state institution became part of Cornell University and expanded its programs to include crops for canned goods, nursery plants, and raspberries.[4]

The Station currently occupies 870 acres just outside Geneva, New York, with additional lands and buildings at the Hudson Valley Laboratory in Highland, New York, which is used to conduct research on fresh fruits, including apples, grapes, small fruits, and vegetables, and the Vineyard Research Laboratory in Fredonia, New York, which has conducted grape research there for over 100 years.[5]

At Geneva the horticulturalist U. P. Hedrick and his colleagues developed new grape varieties for wine and juice production and table consumption. Hedrick, born in 1870 in Independence, Iowa, grew up in northern Michigan near Harbor Springs. He received his bachelor of science (1893) and master of science (1895) from Michigan State Agricultural College, and his doctor of science from Hobart College in 1913 while working at Geneva. Hedrick came to Geneva in 1905 after teaching at agricultural colleges in Michigan, Utah, and Oregon.

Hedrick became a nationally renowned horticulturalist through his extensive applied research programs and, more importantly, because he was a very prolific writer and chronicler of American grape varieties, detailing techniques on cultivating grapes, apples, and other fruits, and writing books about home winemaking and viticulture.[6] His most groundbreaking book was *The Grapes of New York*, published in 1908. Unlike what its title suggests, it encompassed most of the wild and domestic grapes then existing or being cultivated in all of the United States east of the Mississippi River, and including Texas, Arkansas, and the Great Plains.[7]

The Geneva Experiment Station's fruit program, including grapes, was established in 1887. Geneva has employed many talented individuals and academicians who have helped to advance the scientific identification, categorization, and breeding of grapes, and enhanced cultivation and fungal disease–prevention practices for American and European grape varieties to be used for table consumption, grape juice, and wine production. It has worked to advance the profitable cultivation of grapes in New York and across the

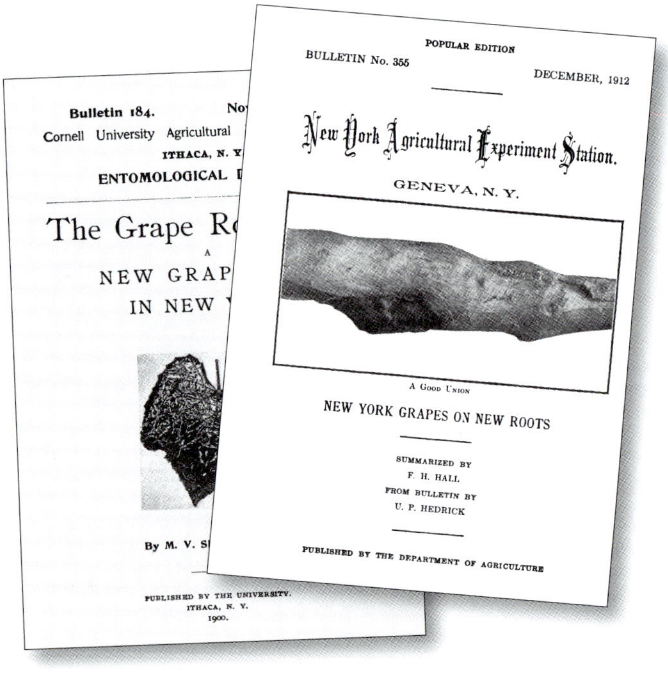

Early bulletins on fruit and grape growing. *Examples of reports published by the New York Agricultural Experiment Station at Geneva in the early twentieth century.*

nation. In addition to Hedrick, who worked at the Station from 1905 to 1937 and ultimately became its director, some of the other faculty members who worked on grape research included Emmett Stull Goff (1882–91); Spencer Beach (1891–1905); Arlow Burdette Stout (1919–48), who worked out of the New York Botanical Gardens, Bronx, New York; Fred Gladwin (1913–40), who worked out of the Fredonia Vineyard Laboratory; Richard Wellington (1920–53); Olav Einset (1924–39); George Oberle (1937–48); John Einset (1942–73), the son of Olav Einset; Nelson J. Shaulis (1941–78); Robert (Bob) Pool (1974–2006); and Bruce I. Reisch (1980 to present). Those who collaborated with wine grape breeding included Willard Robinson (1944–82), Thomas Cottrell (1982–85), Thomas Henick-Kling (1987–2006), and Anna Katharine Mansfield (2008–present).[8]

It is difficult to list all of the professional and technical staff who worked at Geneva over the years without missing a few individuals. To those who have not been mentioned, I apologize for that omission. One individual who seems to have not been highlighted, but who worked for a brief time at Geneva during the early to middle 1950s, was Konstantin Frank (1899–1985). Konstantin Frank vigorously advocated for the exclusive cultivation of *vinifera* grape varieties for wine production in New York, but the staff at the Geneva Experiment Station was skeptical. In 1962, Frank established his own winery, *Vinifera* Wine Cellars, which made wines exclusively from *vinifera* grape varieties.

A Slow Turn Towards Wine Grapes

Due to the national and state prohibitionist forces that strongly advocated against the consumption of all alcoholic beverages, for much of its history, Geneva worked on the development of new table grape varieties, both seeded and seedless, or grapes that could be used in the New York grape juice industry or for home use. The first named Geneva hybrid grape was Goff, named after Professor Emmett Goff, and introduced in 1906. There were forty-two named table, seedless, and juice grape varieties developed and introduced by Geneva from 1906 to 1972, after which the first grape variety devoted exclusively to wine production, Cayuga White, was introduced. Some of the named Geneva hybrids bred after 1906 and up until Cayuga White was introduced are still commercially grown and sold by nurseries such as Miller's/Stark Brothers of Canandaigua and Double A Vineyards Nursery of Fredonia, New York. These grapes were dual-purpose table and wine grapes, and included Alden, Fredonia, Steuben, Seneca, Sheridan, Golden Muscat, and Van Buren.

Geneva's work from the beginning of Prohibition until shortly after World War II did not openly focus on developing grapes to be used for wine production. However, before Prohibition, viticulture and winemaking were one of Hedrick's chief interests, based on his work in the first quarter of the twentieth century. It can be gleaned from the breeding work done at Geneva that, while still complying with state and federal government policies to the letter of the law to restrict research on wine grapes and to develop only more disease-resistant and productive table and grape juice varieties, some Geneva researchers were in fact engaged in a program to develop varieties that could also be used in New York's wine industry.

In 1911, before Prohibition, Geneva conducted an elaborate program of growing *vinifera* grapes to advance quality wine production in New York and to develop at least some hybrids with *vinifera* heritage that could be used to improve the quality of wine grapes being grown in New York. Approximately 100 varieties were planted and by 1917, the results were encouraging enough for Hedrick to declare that such *vinifera* grapes could be grown in New York and survive winter cold injury and the four main fungal diseases that afflicted *vinifera* grapes. Unfortunately, as Hedrick moved forward with his efforts to grow *vinifera* for wine production, the U.S. Constitutional Amendment to institute Prohibition was ratified shortly thereafter in 1919. It was not until the late 1950s and early 1960s that the cultivation of *vinifera* grapes for wine production was taken up again seriously at Geneva.[9]

When Prohibition reared its ugly head and put a complete stop to Hedrick's winemaking and *vinifera* grape cultivation research, he saved the Station's best grapes for himself to be made into wine. Hedrick wrote two books that provided advice to home winemakers and viticulturalists.[10] During Prohibition, he continued to cultivate *vinifera* grapes and spent research time identifying suitable rootstocks for *vinifera* wine grapes and developing new ways to inhibit insect and fungal diseases that attacked *vinifera* grapes.

After the repeal of Prohibition in 1933, Prohibitionist influence was still strong in New York State's capital Albany, and in Washington, D.C. This was evidenced when, in 1933, Geneva's request for a state appropriation of $50,000 for wine research was declined. True, the rejected appropriation may have had something to do with the lack of government funds available due to the Great Depression.

Although wine research was forbidden at Geneva, professors Richard Wellington and George D. Oberle organized a grape-testing and wine-quality improvement program with Canadian and New York wineries. The cellars of these commercial wineries became labs where experimental wines were made and evaluated. When Dr. Oberle moved to Virginia in 1948 and Hedrick died in 1951, Professor John Einset (1915–81) and others at Geneva continued to carry the torch to advance wine grape research. During the next ten years, the Prohibitionists' political forces declined in power and there was increased cultivation of wine grape varieties for New York's expanding wine industry. Finally, in 1962, Geneva's request for state funds to support wine research was granted. After the death of the "dry" Congressman Clarence Cannon (D-Missouri) (1879–1964), who was the chairman of the House Appropriations Committee, the federal government also began to appropriate money for wine and wine grape research. When there was no longer a taboo against conducting wine grape and wine research, Cornell published a bulletin on home winemaking, written by Professor Willard B. Robinson, the first publication of its kind issued by an American university.[11]

In reviewing the work of U. P. Hedrick, the Geneva grape-breeding program, and the descriptors used to describe new Geneva-developed grape varieties from the establishment of the Station to today, the term used to evaluate the quality and use of a new grape hybrid varieties included the word "vinous." Geneva often used this term as a code word to identify grape varieties that could be used to produce wine in addition to being used in the juice industry or for home consumption. "Vinous" means "like wine" and is used to describe wine made from *vinifera* grapes.

The term *vinifera* is from the new Latin-based word *vinifer*, which means "wine-producing," which itself is from the Latin word *vinum*, meaning "wine." It seems to have come into use as part of a scientific classification around 1888 to describe grape varieties that lend themselves to wine production or to describe grape varieties that in fact taste like what we all know to be wine. If wine grape research was forbidden during Prohibition and discouraged until the 1960s, why would the term "vinous" be used so frequently to describe new Geneva hybrids? Further, why did hybridization work continue to breed grapes that included *vinifera* wine grape parents such as Chardonnay, Gamay Noir, and Pinot Noir? The answer is obvious: During Prohibition the researchers at Geneva in fact were conducting work to advance wine grape cultivation and to develop new breeds for the production of wine.

Based on the actual work that has been achieved at Geneva (and not merely what was stated in their articulated goal) since just after World War II and accelerating after the mid-1950s, its grape-breeding program attempted to develop highly productive varieties that could be used by New York's big wineries that existed at that time, such as Taylor, Great Western, Gold Seal, Widmer's, and the Canandaigua Wine Company.

In the interests of developing red wine grapes for the benefit of these large New York commercial wineries, Geneva's goal was to develop varieties that had deep concentrations of color, and which were neutral in taste and body without any *labrusca* flavors. Geneva advanced these types of grape varieties that yielded five to seven tons per acre, had adequate sugar and reduced cultivation inputs such as pesticide and fungicide applications, minimized winter injury, and were able, to some extent, to be mechanically pruned and harvested.

Since 1972, with the introduction of Cayuga White, there have been sixteen newly named grape varieties introduced by Geneva. Of those sixteen, eleven of them have been designated as wine grapes and include, in order of introduction date, Horizon (1982), Melody (1985), Chardonel (1990), Traminette (1996), Geneva Red (2003), Valvin Muscat, Corot Noir and Noiret (2006), and Aromella and Arandell (2013).

The Geneva program, under the current leadership of Dr. Bruce I. Reisch, has tried to breed grape varieties that are generally adapted to the growing conditions of the northeastern United States by breeding in cold tolerance and high resistance to fungal diseases. To this end, Geneva has used the work of French hybridizers such as Albert Seibel and Sevye-Villard and cross-bred their hybrids with their own *labrusca*-based hybrids developed before 1950, as well as using some quality *vinifera* grapes.

In 2010, the Geneva Experiment Station merged with their sister departments at the main Cornell campus in Ithaca, New York, and is currently home to faculty from four departments of Cornell's College of Agriculture and Life Sciences, including Horticulture, Entomology, Plant Pathology and Plant-Microbe Biology, and Food Science. Geneva is also the site of the United States Department of Agriculture's grape germplasm repository, which is dedicated to preserving genetic stock that may be useful for future breeding programs. There are branch stations at the Hudson Valley Laboratory in Highland, New York, and the Cornell Lake Erie Research and Extension Laboratory at Fredonia, New York. In addition, the Station operates the Northeast Center for Food Entrepreneurship, the New York Wine Analysis Lab, the Grape Breeding Program, and the Vinification and Brewing Lab. The reference library of the American Society of Enology and Viticulture, Eastern Section, is also located at Geneva. Cornell University, in association with Geneva, offers graduate and undergraduate courses and programs related to viticulture, plant physiology, entomology, plant pathology, chemistry, food fermentations, microbiology, and chemistry, in addition to supervising cooperative extension activities and wine research programs.

Named White Geneva Hybrid Varieties

Following are five named Geneva Experiment Station white hybrids that are well suited for cultivation and for the production of wine in the Hudson Valley and other cool climate areas. All of these white varieties have been tested, and Geneva is confident that they possess the positive growing characteristics needed to produce commercially acceptable levels of grapes for the production of wine. Hence, these varieties have been named by Geneva and released or introduced to the general public.

CAYUGA WHITE (GW-3, NY 33403)

Cayuga White, along with Traminette, is one of the most popular white hybrids developed by the Geneva wine grape-breeding program.[12] It is a hybrid cross of Seyval Blanc (S.V.5-276) and Schuyler, which is a red table grape developed at Geneva in 1947. The genealogy of Seyval Blanc is detailed in this book.[13] The pollen parent, Schuyler, is a cross of Zinfandel, the red *vinifera* grape used primarily in California, and the white grape Ontario, an early American Geneva grape cross that itself is a hybrid of Winchell and Moore's Diamond,[14] developed in 1908. One of Moore's Diamond's parents is the grape Iona, which is from the Hudson Valley.[15]

PARENTAGE
aestivalis, labrusca, lincecumii, rupestris, vinifera

HARVEST DATE
Mid-season

A A B A A

Cayuga was first bred in 1945, and its fruit first described in 1952. A few years later, Cayuga became part of an initial screening planting and was further selected in 1964 for a more extensive twenty-five-variety trial of Geneva developed hybrids and other French-American hybrids. The grape was finally named Cayuga White in 1972.[16] It comes highly recommended by those at the Geneva Experiment Station and by many growers in the field across New York State.

The vine is winter hardy, along the lines of Seyval Blanc and other Geneva white hybrids such as Chardonel and Melody. However, Cayuga White is less winter hardy than Delaware, Léon Millot, or Maréchal Foch. For Hudson Valley growers, the grape is safe for planting in the entire Valley.

The grape is a reliable producer

and productive to very productive. It is generally disease resistant to black rot, powdery mildew, and botrytis, but is more susceptible to downy mildew. Further, it is not sensitive to sulfur treatments. The vine is vigorous, bearing fruit on a standard to large-sized plant with a somewhat upright growth habit whose shoots can be somewhat subject to wind breakage. It is best summed up as a grape variety that behaves well in the vineyard.

Cayuga White buds out rather late, so tends to avoid late spring frosts. The cluster is large, long, slightly tapering, and well filled, but not too compact which helps with the variety's mildew resistance. The berry is large and roundish to ellipsoidal in shape; it is also resistant to cracking and its flesh is meaty. Grapes ripen by early mid-season to mid-season, depending on the type of wine to be made. Due to its constitution, the fruit hangs well on the vine so later harvests are possible, giving vineyard managers more flexibility in harvesting the crop.[17]

Wines made from Cayuga White can be good to superior. Further, its versatility in the cellar can help a winery's bottom line by being able to be used in dry and semi-sweet wines, sparkling wines, or as a component in blends to brighten up other more stodgy whites.

There are many different flavors associated with Cayuga White, depending on how the wine is made. If picked early, it can be made into attractive sparkling wines with good acidity, structure, and pleasant flavors. When picked ripe, it can be made into a nice clean floral and fruity dry or semidry still wine with medium body and balance. If made semisweet, the sugar gives the wine the body it needs to stand up to its big forward fruit flavors.

If picked late, Cayuga wines can have strong, assertive, fruit-bowl flavors of oranges, ripe pineapples, grapefruit, and honey, which can recall *labrusca* and Muscat flavors, similar to Delaware or Niagara. Also, as these wines age, whether they are made as dry wines or semidry wines, the same *labrusca* and Muscat flavors manifest themselves slightly. The tail end of must obtained after pressing Cayuga tightly can have these same big, luscious, fruit-bowl *labrusca* and Muscat flavors.

When made as a dry wine, it is bright, clean, crisp, and steely. It has only medium body, some spice, and can have fruit flavors of apples, peaches, soft pineapples, spiced pears, grapefruit, and lemons. When made as a semidry or semisweet wine, its more luscious fruit-bowl flavors of oranges, pineapples, honey, citrus, apricots, and some Delaware flavors of guava and Muscat come through. Other Cayugas—like Chenin Blanc—can be more steely. However, it is generally a much bolder and more straightforward wine with grapey flavors, which Chenin Blanc lacks.

The wines made from Cayuga are neither nuanced nor sophisticated. They are big and forward with lots of competing fruit flavors. The grape is recommended for the Hudson Valley because it is good in the field and in the cellar, as long as its limitations are recognized and it is not made into wine styles for which it is not suited. Its sister variety is Horizon.

CHARDONEL (GW-9, NY 45010)

In 1953, Chardonel was cross-bred at the Geneva Experiment Station from Seyval Blanc and Chardonnay. The fruit was first observed in 1958 and propagated in 1960. Chardonel was introduced as a named variety in 1990. The goal of breeding Chardonel was to retain the fruit and winemaking characteristics of Chardonnay with the winter hardiness and some of the fungal disease resistance of Seyval Blanc. Since both varieties can be made into many of the same wine styles, it made sense to marry them.[18] The variety was released based on the urgings of growers and researchers in Arkansas and Michigan because it was viewed then as a better variety to be grown in climates that are milder than are generally found in New York.[19] However, due to the Hudson Valley's generally warmer summer temperatures, Chardonel can do well in most of the warmer parts of the Valley.

PARENTAGE
lincecumii, rupestris, vinifera

HARVEST DATE
Late

A– A– C B+ A

Chardonel has moderate vigor and is moderately productive to productive. My observation at Cedar Cliff is that Chardonel takes a while to establish itself after first being set out. When compared to Chardonnay or Seyval Blanc, it can be as productive. Chardonel may be more productive than Seyval in certain situations because Chardonel is a more vigorous grower, so it has the capacity to yield more grapes per acre. Chardonel is moderately winter hardy and has about the same hardiness as one of its parents, Seyval Blanc. It is hardier than Chardonnay, but less so than Delaware.

In the Hudson Valley, Chardonel should be planted only in the long-season areas of the Valley on deep, preferably sandy, warm, well-drained soils to ensure that the crop can mature properly, so as to provide all the beneficial flavor profiles offered by this grape. If planted on heavy water-retaining soils, Chardonel is susceptible to crown gall and will not ripen properly. With Chardonel, it is important that it is picked only when mature if making a still wine. If not, the variety can be made into sparkling wine.

Chardonel's bud break is average to late, but if the buds are damaged by a late spring frost, it does not seem to have a good secondary crop. It ripens early late to late in the season along with grapes such as Chambourcin and about one week after Chardonnay.

The advantages of Chardonel over Seyval is that it is not as susceptible to bunch rot, does not over-crop, and is more vigorous. The clusters are relatively loose compared to Seyval Blanc, shouldered, and medium-large in size, but smaller than Seyval. Its relatively loose cluster has helped growers to combat bunch rot and other fungal diseases. The medium-sized, round berries are amber with some light green-yellow hues at harvest time.

In general, the disease resistance of Chardonel is better than Chardonnay. It is susceptible to powdery mildew, less susceptible to downy mildew and botrytis, and is not sensitive to sulfur treatments. With that said, Chardonel still needs a vigorous spray program to ensure a healthy crop. Further, the variety is resistant to phylloxera, so it is a direct producer. The growth habit of Chardonel is generally lateral and then down, so high cordon pruning with spurs is recommended to get air and light to the grape clusters to minimize fungal diseases.

The wine quality of Chardonel is good to superior to Seyval Blanc. It is a softer wine that does not have the rough edges that Seyval Blanc can have. It can be made as a bone-dry, clean, and austere white wine similar to a Chablis-style Chardonnay or as a more buttery, fatter California style with more vanilla flavors aged in wood. Also, not fully ripened Chardonel grapes can be used as a base for sparkling wine. Like Chardonnay, Chardonel is a wine that "makes itself." It is an easy grape in the cellar and can be manipulated to winemaker's needs.

The wines resemble both of its parents Chardonnay and Seyval Blanc, but Chardonel tends to be more austere, with more flint in its structure than Chardonnay. The wines can be "hotter" than Chardonnay, but they do not have the off-putting backbite that Seyvals sometimes have in their finish. Its color in the glass is light like Verdelet, a pale light color with light hues of yellow and lime green. Over time, these wines do darken.

The fruit profile can be of green apples, lemons, grapefruit, melons, and some pears, the same fruit descriptors used to describe a Chardonnay or Seyval Blanc, depending on the style they are made in. Chardonels can be made in various styles, but in general it has a medium body and the same smoothness of Chardonnay as opposed to the sometimes disjointed structure of Seyval. These wines are pleasant, simple, and can have herbal/grassy elements. The wines can be crisp or have an oily middle and a long, clean finish.

In sum, the grape is easy to grow, relatively disease free, does not require cluster thinning, and can be made into several different quality wine styles. Chardonel is a variety that growers should consider if they do not have the right vineyard site, time, or patience to grow Chardonnay. It should, however, be emphasized again that Chardonel, to ripen properly, should be grown only in the longer season, warmer portions of the Hudson Valley.

HORIZON (GW-7, NY 33472)

PARENTAGE
aestivalis, labrusca, lincecumii, rupestris, vinifera

HARVEST DATE
Mid-season

A A B A+ B+

Horizon is a sister of Cayuga White (Seyval × Schuyler), which is reported on earlier in this chapter. The variety was bred in 1945, described in 1951, named as Horizon in 1972, and formally introduced by the Geneva Experiment Station in 1982. Horizon is suitable for the production of bulk white wine, which Geneva has ranked above that of Aurora, but below that of Cayuga White in quality. Its low-acid profile makes it useful in blending with other more highly acidic wines, which are commonly found in the Hudson Valley.[20]

The wine is fruity, neutral, and clean with good balance that generally lacks those flavors associated with native American and French-American hybrid grape varieties, but it can show those flavors with some cotton-candy elements. The strength of Horizon is in the field, where it is vigorous, highly productive, and very winter hardy.

While it is more susceptible to powdery mildew and botrytis than Cayuga White, Horizon's winter hardiness and very high yields offset these deficiencies. The vine is almost as winter hardy as Niagara or Concord and is not sensitive to sulfur treatments. It is rated as susceptible to powdery mildew, somewhat resistant to botrytis, and resistant to downy mildew.[21]

The variety ripens a few days after Cayuga White, but is still considered a mid-season ripening variety. The clusters are medium in size, moderately compact, and cylindrical in shape. Its medium-sized and round berries are a light green color. If spring frost damage is sustained, Horizon rarely produces a crop on its secondary or basal buds.[22]

Horizon is a safe alternative white grape variety that should be considered for those very cold sites in the Hudson Valley. Further, the variety produces neutral low-acid wines that are suitable for blending with other more flavorful and high-acid wine produced locally or in the northeastern United States and Canada.

MELODY (NY 65.444.4)

PARENTAGE
aestivalis, labrusca, lincecumii, rupestris, vinifera

HARVEST DATE
Late mid-season to late

A A− C B A+

Melody is a cross of Seyval Blanc (S.V.5-276) × GW 5 (Pinot Blanc × Ontario), which was bred in 1965 at the Geneva Experiment Station. Its fruit was first described in 1969. Later it was designated as NY 65.444.4 and named Melody when it was officially released in 1985.[23] This is a sleeper of a grape that should be explored further in the Hudson Valley.

Melody is a late mid-season to late white wine grape. It is moderately vigorous to vigorous and rated as productive, but not so productive that it needs cluster thinning to regulate its crop. Melody tends to have a late bud break. If adversely affected by late spring frosts, it does not seem to have a large secondary crop. The variety is moderately winter hardy, a bit hardier than Seyval or Cayuga White.[24] However, at my farm Cedar Cliff in Athens, Melody has never sustained winter injury at all. I have not found Melody to be as vigorous a plant or to be as highly productive in yields as suggested by the Geneva Experiment Station bulletins at Cedar Cliff. My observations of Melody are that it has medium vigor and is of medium productivity, but the fruit is of high quality, so that it is worth growing to make a high-quality wine.

The shouldered and compact cluster is of medium size for a white grape. The berries are medium in size, round, and range in color from light green to amber as they ripen. The skin is resistant to cracking and bunch rot. The plant and grape are rated as being resistant to powdery mildew and botrytis, moderately resistant to downy mildew, and susceptible to black rot.[25] However, test plantings at my farm in Cedar Cliff have shown little susceptibility to black rot. Melody is not sulfur sensitive, so this treatment is available for fungus control. The variety is not susceptible to phylloxera, so it can be grown on its own roots, but often the vines are grafted to increase its vigor.

The growth habit is lateral and slightly downward. Due to its growth habit, high cordon pruning with spurs is recommended. Overall, the vine is very easy to grow and manage.

The wines made from Melody are delicate and very pleasant with a light pale green or silvery pale green color in the glass. They are bright, slightly floral, very clean, have good body and balance if made off-dry, and are a nice simple *vinifera*-type wine. The fruit has subtle hints of white peaches, honeydew melons, and light lemons. The wine is of light body and taste, so it is not suitable for wood aging.

Melody may have been overshadowed by Cayuga White within the industry[26] because of Cayuga White's big forward flavors as opposed to Melody's more delicate and subtle flavors. Melody is more reminiscent of a Pinot Blanc, a soft Chenin Blanc, and sometimes a soft, dry Riesling. Because of its soft nature, it is a good addition to a white blend to soften other white French hybrids' forward flavors and high acid. Due to its light body, it tends to be better suited as an off-dry fruity wine because the sugar adds body to this delicate wine.

Overall, this grape, while not as productive in yields as described in the literature, has been overlooked when compared to other Geneva white hybrids and the newer white hybrids coming from Germany or Hungary. Growers and vintners in the Hudson Valley should reexamine the virtues of Melody, both in the field and in the cellar for its ability to soften other high-acid wines produced in the Valley. This grape comes highly recommended.

TRAMINETTE (NY 65.533.13)

This white wine grape is a cross of J.S.23-416 × Gewürztraminer. This cross was made by Herb C. Barrett in 1965 while he was at the University of Illinois in an attempt to produce a large clustered table grape with Gewürztraminer flavor. Seeds from this hybrid were sent to the Geneva grape-breeding program in 1968 for planting, and after a few false starts due to concerns about a possible genetic stem-pitting virus, the grape, which was not introduced in 1982 with Horizon or in 1985 with Melody, was finally introduced over a decade later in 1996.

PARENTAGE
cinerea, labrusca, lincecumii, riparia, rupestris, vinifera

HARVEST DATE
Late to very late

A-	A	B	B	A

The genetics is interesting in that J.S.23-416, a white variety, is a cross of B.S.4825 (Le Colonel, BS 2667, a red variety) × Subereux (S.6905, another red variety) × Chancellor (S.7053). This red cross, which made a white grape, was then hybridized with Gewürztraminer. It is curious and should be pointed out how J.S.23.416 (a white variety), a cross of Le Colonel, Subereux, and Chancellor (all red varieties), became a white variety, but I note that Chancellor and Subereux were often used in the more successful French-American hybrid grape varieties.

The vine is vigorous with medium productivity. The variety buds out late. Its medium to medium-large-sized clusters are shouldered and moderately loose. Its round medium to medium-large-sized berries with tough skin are amber-pink in color when ripe.

Traminette is moderately winter hardy in the Hudson Valley, similar to Seyval Blanc, Chardonel, or Vignoles, but is hardier than Gewürztraminer. The variety tends to have more winter injury and crown gall on heavier soils, where trunk injury occasionally occurs. Its growth habit is semi-upright.[27]

It is more resistant to fungal diseases than its parent Gewürztraminer and moderately more resistant than Seyval Blanc. Traminette is unique in that it is equally moderately resistant to black rot, downy mildew, powdery mildew, and botrytis, unlike most other grapes, which tend to have more or less better resistance to one or more of these fungal diseases. Overall, it is moderately resistant to all fungal diseases similar to Ravat 34 and Maréchal Foch, but less resistant than Cayuga White. Further, it is a direct producer that even with over 50 percent *vinifera* heritage is still resistant to phylloxera and bunch rot.[28]

The grape ripens late to very late. It does not need cluster thinning to bring a crop to harvest with adequate sugars in the fall. This variety generally has good sugar levels and tends to have a good sugar/acid/pH balance for winemaking. Overall, it is an easy and reliable grape to grow on a site that does not have heavy soils. The pruning style for this variety should be cordon trained with spur pruning or vertical and upright shoot positioning.

Fertile, well-drained sites are preferable to heavy, wet soils. Further, if the soil is not fertile, grafting is recommended. It seems that there are two schools of thought on Traminette. One favors direct production without rootstocks, while the other is experimenting with different rootstocks to obtain the fruit they believe is needed to make a quality wine. Further, grafting vines does increase the grape's yield.

The textbook description for Traminette wines is that they are similar to Gewürztraminer as this is a Gewürztraminer hybrid. Further, they produce spicy, fragrant wines of excellent quality similar to its *vinifera* parent and not like its other hybrid parent J.S.23-416 (which I also grow and am evaluating.) Further, Traminette can be superior to either parent because it has a better sugar/acid/pH balance that had not only spice, but nice honey and apricot flavors that linger over time and seem to develop as they age.[29]

For many years I was not fond of Traminette wines because they tended to have clawing, under-ripe Muscat flavors with an underlying greasy/kerosene body and mouth feel with few of the positive Gewürztraminer qualities which the grape was supposed to possess. However, over time I found that there are different ways of making Traminette that significantly alter its taste and, if made properly, it can be made into a high-quality wine.

There are several ways to make wine from Traminette, depending on the wine style one is attempting to achieve. One method is for it to have between twelve to forty-eight hours of skin contact with the must at temperatures that are under 60°F to get nice Gewürztraminer tastes of strong spice and floral aromas, honey, apricot, and melons and a full structure and a long, smooth finish similar to a Gewürztraminer. The wines can be made either dry or semidry. Despite the extended skin contact, there is no bitterness in the wine, nor are there any hybrid flavors extracted. However, with over forty-eight hours of skin contact, sometimes Muscat or orange Muscat flavors come out. As Traminette wines age, they tend to drop the spiciness, add even more honey, apricot and orange flavors, and become similar to an aged Sauterne.

For grapes grown in the warmer parts of the Hudson Valley and the South, increased bitterness and high pH may become a problem. Therefore, shorter skin-contact time should be utilized when making these wines. If not, the floral/spicy flavors can shift to Muscat-like and guava flavors and the body becomes oily/greasy and more one-dimensional.

If a more austere wine with apple and citrus flavors is desired, then a short or no skin contact is recommended. Some prefer this wine style to lush, semisweet Traminettes. No matter the style of wine made, the grape flavors are pretty dominant, so it should be used sparingly in blends, otherwise it will overpower the blend.

Overall, this grape is easy to grow. Further, it is a grape that, with care, can be made into a quality wine. It should be considered for more extensive cultivation in the Hudson Valley.[30]

Named Red Geneva Hybrid Varieties

On the following pages are seven Geneva breeding program red grape varieties that I have grown in the Hudson Valley and describe their attributes in the field and, most importantly, in the cellar. Two of the seven, Corot Noir and Noiret, have been field tested by the Geneva Experiment Station and participating test growers, and have been deemed to have superior growing characteristics and the ability to produce commercial quality wine. Hence Corot Noir and Noiret have been named by Geneva as endorsed grape varieties that have been introduced to the general public as suitable for cultivation and wine production.

The remaining five varieties, listed in this chapter as experimental red Geneva hybrids, were developed by Geneva and distributed or offered for sale by an organization called the New York State Fruit Testing Cooperative Association, Inc. This excellent organization was established in 1918 as a nonprofit cooperative nursery for the purpose of introducing for testing new fruits recommended as worthy for trial by the Experiment Station and distributed to growers to obtain more information on such provisionally recommended varieties as to their characteristics in the field. Unfortunately, due to economic considerations, this organization ceased to operate sometime after 1995. As a replacement for the Cooperative Association, the Double A Vineyards Nursery in Fredonia, New York, has stepped in to offer to the general public hybrid selections that Geneva has deemed ready for limited field testing to determine if they are worthy of being formally named and introduced to the public.

As a participating grower interested in field testing such new grape hybrids offered by the New York State Fruit Testing Cooperative Association, I have been growing the following five experimental varieties for approximately twenty-five years.[31] While some of the varieties have some deficiencies in the field, they all produce wines that have superior winemaking attributes and can either produce very high-quality red wines, or are able to fill a void in the production of lighter, fruitier-style wines. I believe that these experimental varieties, under the right cultivation conditions, can satisfy certain wine-production needs of Hudson Valley wineries and the wineries of other cool climate regions. Hence, they all merit additional evaluation with an eye toward bringing them into commercial production for exceptional or superior-quality wines.

I would like to disclose upfront that the Geneva Experiment Station may not be as enamored with these varieties anymore, and is working on a new generation of grape hybrids for red wine production, such as NY 95.0301.01 (Arandell). While the thrust of this book is to identify grapes suitable for wine production in the Hudson Valley and regions with similar climates, the strong undercurrent has been that there are many undiscovered or underutilized hybrids that were developed by the early and late French hybridizers, and by nineteenth- and early twentieth-century Hudson Valley hybridizers, that should be reexamined and reintroduced to the public as varieties worthy of future consideration. Even with the somewhat deficient growing characteristics of some of these varieties, the listed Geneva experimental red hybrids covered on the following pages fit into this category of grapes, and it would be remiss of me to omit these varieties.

While some of them may not be easily available or not available at all to the general public, it is important to inform the public about them so that both academic and private grape hybridizers may know of the winemaking outcomes and growing characteristics of these varieties. Hopefully, this information can be used to help university and private grape breeders with their future work or their continuing work with genetically similar varieties.

COROT NOIR (NY 70.809.10)

Corot Noir is a cross of S.V.18-307 × Steuben, which was developed by the Geneva Grape Breeding Program in 1970 and introduced in 2006 as Corot Noir.[32] Steuben, the pollen parent of Corot Noir, is a dessert table grape bred by Geneva in 1925 that was a cross of two primarily *labrusca* varieties: Wayne (Mills × Ontario) and Sheridan (Herbert × Worden), which were also bred at Geneva. Steuben was selected in 1931 and introduced as Steuben in 1947. One of Corot Noir's great-great-grandparents is the Hudson Valley–bred grape Iona.

PARENTAGE
aestivalis?, berlandieri, bourquiniana, cinerea, labrusca, lincecumii, riparia, rupestris, vinifera

HARVEST DATE
Late to very late

B+ B B A B

The vine is vigorous and very productive. It has well-formed large clusters that are long and tapering. Its medium-sized black berries have a spicy twang taste. The clusters have wings and are somewhat loose. Also, its berries adhere tightly to the bunch stem.

In selecting red wine grapes, the Geneva Experiment Station tends to like big-clustered grapes such as Corot Noir and Noiret, which have lots of color but are more neutral in taste with little or no *labrusca* flavors. The genetic background for S.V.18-307 is that it is a Seibel-based hybrid of Chancellor and Villard Blanc.

Corot Noir is productive to very productive. The vine buds out relatively late, so it is not hurt by late spring frosts. It does produce a suitable secondary crop if late spring frosts hit, but this crop will not replace the first crop lost.

The vine is moderately winter hardy like Chambourcin, but not as hardy as Baco Noir or Chelois. The variety is resistant to botrytis, slightly susceptible to powdery mildew, and susceptible to downy mildew and black rot. The variety can be treated with sulfur to control fungal diseases. Overall, the variety has about the same or less fungal disease resistance as Chambourcin. The leaves can get galls. The only reason disease pressure may be more so for Corot Noir is due to its highly vegetative lateral cane growth, which can shade fruit. Its vegetative growth can be very high in deep fertile soils.[33] I have found that Corot Noir's high vegetative growth hinders grape quality because it encourages fungal disease growth and decreases sun exposure. Further, it takes additional labor in the vineyard to manage and control its dense leaf canopy.

The variety can be cluster thinned to reduce its production so that its grapes can ripen sufficiently in the Hudson Valley before the first fall frost. Corot Noir is a late to very late ripening variety in the Hudson Valley, so reducing the crop can be an important task to ensure that adequate sugar levels are attained at harvest time. This grape was released in 2006, but its potential range of cultivation on the East Coast may be limited by its attributes in the field. There were 59 acres of Corot Noir in New York in 2006, which declined to 42 acres in 2011, with small amounts being grown in Pennsylvania and Ohio.[34]

The flavor profile of Corot Noir is a combination of its heritage of Steuben and Chancellor. Corot Noir wines retain Steuben's flavor profile of strawberries, bubblegum, and raspberries, which comes from its *labrusca* heritage. It also has its seed parent's flavors of blackberry and prune fruit, which are neutral with an even structure and an acid profile that comes from S.V.18-307 and Chancellor. The breeders of Corot Noir maintain that it is a pleasant, deeply colored red-purple wine that is a light-bodied *vinifera*-type wine with low acidity. Its fruit flavor profile includes attractive cherry and berry fruit aromas, and bell pepper flavors. Further, its tannin structure is complete from the front of the mouth to the back with big soft tannins.[35]

While this overall description of Corot Noir by its breeders is relatively accurate, I do not share as much enthusiasm for Corot Noir as a red table wine. While it can have very dark black cherry and bell pepper flavors and nose, it can also have the dark flavors of chocolate and cooked blackberries, with cherry/strawberry tutti-frutti–like flavors. Overall, this wine is of better quality than Noiret, and can be made into either a varietal wine or used in a blend. For bulk wine production, the grape is fine, but needs to be blended with other grapes to give it more finesse.

NOIRET (NY 73.136.17)

Noiret is a cross of NY 65.0467.08 (NY 33277 × Chancellor) × Steuben, which was made in 1973 and introduced as a named variety in 2006. This is a highly vigorous variety that produces a lot of vegetative growth and is a heavy producer. The volume of heavy vegetative growth hinders the ability to manage this variety in the field and can have adverse consequences for the quality of fruit that is harvested.

PARENTAGE
aestivalis?, bourquiniana?, cinerea, labrusca, lincecumii, riparia, rupestris, vinifera

HARVEST DATE
Late to very late

B B+ A A B

Noiret's large cluster is cylindrical and larger than Concord. Further, its large blue-black berries are larger than Corot Noir, but not as large as Concord. The cluster is semi-compact. Noiret's genetic makeup consists of more *labrusca* varieties (about 65 percent) when compared to Corot Noir (about 50 percent). Surprisingly, even with its higher percentage of *labrusca* heritage, it is still more cold sensitive than Corot Noir and Chambourcin.

Because of the relatively cold tenderness of the variety, its thick canes and downward growth habit, spur pruning to the top wire is recommended. Also, the vine should be given two trunks so that at least one trunk survives the winter. Over time, the trunks tend to get very thick, which can lead to splitting, so trunk replacement is suggested. After particularly cold winters in the Hudson Valley, Noiret's grape production has been sizably reduced due to winter damage.

The bud break usually occurs mid-season to late and at a time that is later than Concord. Its harvest time is late to very late and it rarely achieves sugars higher than 21° Brix in the Hudson Valley. While this variety does not overproduce and may not need cluster thinning, thinning is still suggested not only to improve wine quality, but to ensure that the crop ripens before the first fall frost. Once ripe, the clusters can break off when picked due to brittleness of the rachis. However, the variety does adhere to the cluster or that part of the cluster that does not snap off.[36]

Due to its dense canopy, large leaves, and high vegetative growth, this grape is somewhat vulnerable to fungal diseases, with downy mildew and black rot of particular concern. Further, due to its compact cluster, it can be susceptible to bunch rot and some splitting of grapes in wet years. The variety is not sulfur sensitive, but it has been noted to be sensitive to the fungicide Pristine. However, I have not noticed any harm after such applications. Overall, the variety is no better or worse in fungal disease resistance when compared to other red French hybrid grapes as long as an open canopy can be maintained to increase sunlight and air circulation.

The grape breeders at Geneva state that the wines are richly colored and have notes of green and black pepper along with raspberry, blackberry, and some mint aromas. Further, it has a fine tannin structure that is complete from the front to the back. The tannin structure and the absence of any hybrid aromas strongly distinguish Noiret from other red French-American hybrid grapes. Further, the wine does not have high acidity and malic acid unlike some French hybrid wines. Its tannin structure lends itself to the production of quality wines with good aging potential.[37]

I personally would not be as lavish in praising the wine qualities of Noiret. Noiret's color is a dark ruby red with purple hues, which makes it suitable for port production. Noiret's lineage from Steuben is confirmed by the medicinal tutti-frutti cherry and strawberry flavor, which comes from its high *labrusca* heritage. If made well, Noiret can have a good medium-bodied tannin structure that supports its fruit flavors of chokecherries, blueberries, black raspberries, cooked plums, and grape juice. While the nose can be muted, its structure can be described as flinty, metallic, and like cough syrup, with the spice of black pepper, wood resin, and lots of stewed prunes. While the flavor profile can be good, it is not interesting on its own. Further, its balance is soft, perhaps too soft, one-dimensional, and flat without peaks or valleys.

The general public may not accept Noiret as a table wine that goes well with food, but it can be a suitable addition to a red French-American hybrid or *vinifera* blend, used for bulk or sweet country wines, or used as a solid base for a port wine.[38] Sometimes the average Noiret could be compared to a sort of neutral Zinfandel with lots of color and cooked dark fruit flavors.

Experimental Red Geneva Hybrids

Below are five red Geneva hybrids that merit further evaluation. As noted earlier, I have grown these varieties for over twenty-five years and have found them to have either good or acceptable growing characteristics in the field and, most importantly, valuable attributes for winemaking. Due to their genetic heritage, some may be of value in future breeding programs. These grape varieties were included so that information on the field testing of these grapes may be used by public and private grape breeders in their future work. Further, information on the growing and cellar characteristics of these varieties may be of use to them also.

There is the age-old conflict between the grower who wants to maximize the volume of the crop and the vintner who requires quality fruit to produce exceptional quality wine. New York, due to its growing conditions, is never going to compete against California, Washington State, South America, or Australia based on the quantity of grapes grown per acre; rather, it needs to compete based on the quality of wines produced. The grapes below, after further evaluation, may become some of those varieties that can help New York–produced wines to compete successfully in the international wine market.

All of these varieties have not been officially released or named by Geneva, but during the late 1980s and early 1990s, they were all disseminated to growers via the now-defunct New York State Fruit Testing Cooperative Association which worked in conjunction with Geneva to obtain comments on these then determined-to-be-promising red hybrid wine grape varieties.[39] There are additional red and white varieties that the Geneva Experiment Station is working on, but I have not planted them, and I know of few growers who have planted them, hence, it is much too early for evaluations of them to be published.

FIVE EXPERIMENTAL RED GENEVA HYBRIDS

NY 63.970.7

(Gamay Beaujolais × Chancellor)
cinerea, labrusca, lincecumii, riparia, rupestris, vinifera

This intriguing hybrid was bred in 1963 and has the ability to make superior-quality wines from a grape that is not bad in the field. This variety was informally circulated by Geneva in the late 1980s and early 1990s as a promising New York red wine grape selection. The vine is large with vigorous growth and is moderately productive. Cluster thinning is not needed for NY 63.970.7. It produces medium- to above average-sized blue grapes on small to, at best, medium-sized clusters whose stems are really tough and adherent. The bud break is mid-season to late, so there is little danger of late spring frost damage. This variety does not have a major secondary crop if frost damage is incurred.

Like many *vinifera* varieties, its growth habit is straight up. Its thick canes allow for an open canopy, which increases light and air circulation to its grapes.

The leaves and fruit are resistant to powdery and downy mildew and to some extent to black rot. NY 63.970.7 grapes mature by late mid-season to late, and can hold on much longer provided the weather remains warm and dry. Even in wet weather, the grapes tend not to crack. Its open canopy helps to suppress fungal diseases as the fruit is exposed to air and light.

The vine is, at best, only moderately winter hardy to somewhat tender in the Hudson Valley. Even in very bitter winters, its trunks and cordons remain alive, and while the thick, one-year-old canes may sometimes die back to the last few buds near the intersection with the cordon, that is all of the buds that are needed when spur pruning this variety off cordons, to put on a crop for the coming season.

The wines made from NY 63.970.7 are truly nice. The color of the wines is not deep, more like a dark Pinot Noir or light Beaujolais Nouveau when using ripe fruit. The wines have a nice berry nose of blackberry/raspberry and earth. They are integrated, bright, aromatic wines that have an interesting Burgundian element and soft tannin structure. The fruit flavors are soft, subtle, and not herbaceous with elements of blueberries, blackberries, and raspberries. The wines are approachable and similar in some ways to the Gamay Beaujolais. In all of my tasting notes, the great nose is cited and this nose carries forward to the middle and finish. It is very important to use only ripe fruit, otherwise the wines have strong and very herbaceous/grassy flavors from the nose to the finish. Overall, my tasting notes maintain that NY 63.970 is a combination between a Burgundian Pinot and a Gamay Beaujolais. The Cornell tasting notes suggest that the grape produces a wine that is consistently good. This variety was offered for sale by the New York State Fruit Testing Cooperative Association in the early 1990s and was circulated as a promising New York red wine selection. While producing only modest yields, this variety should be kept around and evaluated further because of the high-quality wines it produces and because of its Gamay Beaujolais heritage.

NY 64.533.2

(S.V.18-307 × Steuben)
aestivalis?, berlandieri, bourquiniana, cinerea, labrusca, lincecumii, riparia, rupestris, vinifera

This variety was bred in 1964 and has the same parents as Corot Noir. This grape was offered for sale by the New York State Fruit Testing Cooperative Association in 1989.[40] The vine is vigorous, but not as lush as Corot Noir, nor is it as productive. However, it is still a productive variety and tends to have better fungal disease resistance than Corot Noir either because of its more open canopy, which lets in more light and air, or because its genetics imparted relatively thick skins on this variety to protect it from fungal diseases. It has a somewhat upright cane growth habit. The grape does not get botrytis, but is susceptible to black rot. Overall, its fungal disease resistance is similar to Chelois, but is not as good as Baco Noir.

The variety is about as winter hardy as Corot Noir and Chelois. It is more winter hardy than Noiret with little loss in wood or buds, but is less winter hardy than Baco Noir.

The somewhat compact, medium-sized clusters have medium to large sized blue berries. NY 64.533.2 ripens late mid-season to late season, but about one week or more before Corot Noir or Noiret. At harvest time, the berries do shell and the stems are brittle. Further, its leaves are green when Corot Noir and Noiret leaves are yellowing. The significance of this is that NY 64.533.2 has a greater capacity to ripen than the latter two varieties, but the berries do fall from the cluster as it ripens.

In the early 1990s, the Geneva Grape Breeding Program had tentatively described the wine as having good tannin, black cherry and bell pepper with some vegetative flavors and good body. The vine is moderately hardy, has a good crop, and is vigorous.[41] Over the years, I have found that its wine is clean and is a light red to medium dark-hued purple-red wine. While pleasant, it is truly the daughter of Steuben in its acid profile and taste. It is a softer version of Steuben of darker color. It has metallic and medicinal qualities in the nose and taste with some green bell pepper. But the wine is somewhat complex with very floral and forward fruit flavors of strawberries predominantly, some cherry bubblegum, lots of blueberries, roses, licorice, and some light raspberries. As a young wine, it can have the feel somewhat of a soft Gamay Beaujolais, but as it ages, it goes back to its Steuben roots. It is not a traditional dinner wine, but is enjoyable to drink with its big forward fruity and clean nose and taste. NY 64.533.2 should have a place in blends to brighten them up or soften a harsher wine's acid profile. In spite of its tendency to shell when ripe, this variety should be reconsidered and merits further evaluation for the production of soft, very fruity red wines.

NY 66.709.01

(Pinot Noir × Colobel (S.8357))
cinerea, labrusca, lincecumii, riparia, rupestris, vinifera

This grape variety, based on its parentage, should remain in play for future evaluation, particularly in the Hudson Valley. The vine is moderately winter hardy in a manner similar to Riesling. While it does not really have good fungal disease resistance to either downy or powdery mildew, it has some resistance to black rot, which is important in the Hudson Valley. The vine has only medium vigor, with a lateral growth pattern of long canes. Most plants are bushy in appearance with space between vines on the trellis. The wet clay site on which my NY 66.709.01 has been planted is not the best, which may be part of the reason for its lack of vigor. The cluster size is large and loosely filled, and consists of small- to medium-sized blue berries. The grape ripens late to very late with sugars of no more than 20° Brix.

With all that said, this Pinot hybrid has real potential in the cellar. The purple-red wines, when made with ripe grapes, are big and full, yet soft in a manner similar to Pinot Noir. The color is very dark and much deeper than Pinot Noir. The integrated fruit flavors of this wine are of blackberries and some light non-herbaceous black currants. The mouth feel is pleasant and soft with a firm underlying tannin base. It is very important to use ripe fruit, otherwise the wines are very herbaceous and grassy. Even with its mediocre abilities in the field and, at best, its only moderate crop yields, it is a solid grape in the cellar, so it is recommended for further study and experimental plantings.

NY 66.717.6

(S.128 (Pate Noir) × Chancellor)
cinerea, labrusca, lincecumii, riparia, rupestris, vinifera

Developed in 1966, this is a plant of average to above-average vigor and vine size that has an upright growth habit. It is moderately productive with small, bright, light-green leaves and an open canopy, which helps to minimize fungal diseases. The leaves can get galls, but that does not seem to hurt its ability to produce grapes. NY 66.717.6 is an early late-season-ripening grape variety that can hold well on the vine without dropping its berries. The variety has medium- to large-sized clusters on big, tough, bright-green stems with small to very small blue berries. Even when ripe, the stems remain bright green. The variety does not need cluster thinning.

The variety is winter hardy in the Hudson Valley. The fungal disease resistance of this variety is above average for black rot, but it is susceptible to powdery mildew.

The Geneva Grape Breeding Program evaluators stated that NY 66.717.6 has some good tannin, hints of sour cherry and fruity flavors, with some vegetative and French hybrid-like qualities. Further, it was rated first among forty red wines evaluated in the 1989 vintage.[42] This variety was offered for sale by the New York State Fruit Cooperative Association in 1995.

My tasting notes suggest that this deeply dark and thick purple-red wine has lots of berry and chocolate flavors, is complex but soft, and is similar to a dark Malbec with a few approachable soft vegetative/herbal flavors. The wine is big and complex, and yet pleasant and interesting with no bubblegum flavors that many Geneva red hybrids can have. Further, the wine has great aging potential. This may be because there is little *labrusca* in this variety's genetic history. This is one of my favorite red varieties to come from the Geneva Grape Breeding Program.

NY 70.816.5

(Steuben × S.V.18-307)
aestivalis?, berlandieri, bourquiniana, cinerea, labrusca, lincecumii, riparia, rupestris, vinifera

This Geneva red hybrid has the same genetic makeup as Corot Noir, which was bred in 1970, only the order of the pollen and seed parent of Steuben and S.V.18-307 were reversed. However, like other cousins or siblings, it is very different from Corot Noir, its immediate family member.

The variety is productive, at least as productive as Corot Noir. It is relatively easy to manage in the field because it does not have the high vegetative growth of Corot Noir. Its canopy is open and its cane growth is lateral. The canes are relatively short and thick, and the wood ripens even with green leaves on it. The variety is as winter hardy as Baco Noir.

The straggly clusters are slightly smaller than Corot Noir, but the berries are much bigger. Further, the blue berries are more slip-skinned and berries can fall off after the variety ripens, so at harvest time the loose clusters can have missing berries either due to disease (mostly black rot) or droppage. However, with that said, the variety can still be as productive as Corot Noir.

On the issue of fungal disease susceptibility, NY 70.816.5 is as susceptible to black rot as Corot Noir, perhaps even more so, but after being infected with black rot, some berries remain on the cluster to harvest. The variety does not seem susceptible to other fungal diseases or botrytis because of its thick skin.

The wine produced from NY 70.816.5 is very different from Corot Noir. It is light, clean, and pleasant with berry fruit and very soft tannins. In my tasting notes of it over the years, I have found flavors of lightly cooked blueberries, blackberries, strawberry jam, prunes, and soft cherry/red raspberry that are not cloying like tutti-frutti bubblegum. While it is a soft wine, it does have darker elements of light chocolate, plums, and the slight spice of black pepper. Further, its acid/fruit profile is balanced with an underpinning of charcoal and light herbal tobacco, but it has little tannin to support its structure. In that way, it is like a Steuben and other *labrusca* wines.

This variety tends to have a more *vinifera* taste similar to a light Beaujolais Nouveau. In making wines from NY 70.816.5, the must can be left on the skins for a longer period than with either Corot Noir or Noiret without imparting the adverse strawberry, bubblegum, and *labrusca* flavors of the latter. This may lead some to believe that there are good flavors in the skins that do not need to be suppressed as with Noiret and, to some extent, with Corot Noir. This variety was also informally disseminated for comment to growers in the late 1980s to early 1990s.

In sum, it is an integrated, fun, and a relatively uncomplicated wine that is truly a table wine made for consumption with food. It is ideally a light summer red to have with lunch on the patio. This grape may be good on its own or in a blend because it would not dominate such a blend, but could reduce the acid profile of other red hybrids or soften a red *vinifera* wine.

This chapter was devoted to interspecific grape hybrids that can be grown in the Hudson Valley and most other cool climate areas of the United States and Canada. As the book has progressed, we started with an explanation of wild American grape species that were used in French and American breeding programs to incorporate enhanced fungal disease resistance, winter hardiness, and drought resistance to the newer varieties. Subsequent chapters detailed: chance *labrusca* varieties that were widely cultivated, such as Concord and Delaware; the work of Hudson Valley hybridizers to incorporate *vinifera* qualities to improve the taste of table grapes and wine quality; the work of French hybridizers to incorporate American fungal and insect resistance into their vulnerable European *vinifera* grape varieties; and the work of Geneva to develop more productive and better-tasting table grapes and grapes that can be used in the wine industry.

The next chapter will detail the work of that grape-breeding pioneer, Elmer Swenson of Osceola, Wisconsin, and the work of the grape-breeding program at the University of Minnesota. The goal of these programs was to develop wine grapes that could thrive in the truly cold climates of the interior of the United States in areas such as Wisconsin and Minnesota, but which are also well-adapted to the colder portions of the Hudson Valley.

Cornell's Hudson Valley Laboratory. *A branch research station was established in the Hudson Valley in 1923, and moved to the Highland, NY, location in 1963. Professor Emeritus of Plant Pathology Dave Rosenberger (left) and New York State Senator William J. Larkin, Jr. (right) in the Highland vineyard, circa 2005.*

NOTES

1. See Chapter 592 of the Laws of 1880 and Chapter 702 of the Laws of 1881 adopted by the New York State Legislature, and Assembly Bill No. 98, dated March 6, 1883, which accepts a Report of the Board of Control of the York [sic] Experiment Station, Robert J. Swan, president of the Control Board.

2. For example, Dr. E. Lewis Sturtevant of South Framingham, Massachusetts, the first director of the Station (1882–87), was a Bostonian who prepared for college at Blue Hill, Maine, and attended Bowdoin College; and Dr. Whitman Howard Jordan, the fourth director of the Station (after whom Jordan Hall was named), was a native of Raymond, Maine, who began work at the first established agricultural experiment station in Connecticut, then at the Pennsylvania and Maine Stations until he came to work at Geneva from 1896 until 1921.

3. U. P. Hedrick, *A History of Agriculture in the State of New York. The New York State Agricultural Society.* (Albany, N.Y.: J. B. Lyon Company, 1933), 412–24.

4. Chapter 592 of the Laws of 1880 by the State of New York and the Cornell University, College of Agriculture and Life Sciences, New York State Agricultural Experiment Station, www.nysaes.cornell.edu.

5. Edward Neiles, *The New York Red Book* (Albany, N.Y.: New York Legal Publishing Corp., 2011), 700.

6. Some of Hedrick's popular books include: U. P. Hedrick, *The Grapes of New York. Report of the New York Agricultural Experiment Station for the Year 1907. Department of Agriculture, State of New York. Fifteenth Annual Report, Vol. 3 – Part II.* (Albany, N.Y.: J. B. Lyon Company, 1908); U. P. Hedrick, *The Plums of New York* (Albany, N.Y.: J. B. Lyon Company, printers, 1911); U. P. Hedrick, *The Cherries of New York* (Albany, N.Y.: J. B. Lyon Company, printers, 1915); U. P. Hedrick, *The Peaches of New York* (Albany, N.Y.: J. B. Lyon Company, printers, 1917); U. P. Hedrick, *Manual of American Grape Growing* (New York: The Macmillan Company, 1919); U. P. Hedrick, ed., *Sturtevant's Notes on Edible Plants, Report of the New York Agricultural Experiment Station for the Year 1919, II* (Albany, N.Y.: J. B. Lyon Company, 1919); U. P. Hedrick, *Cyclopedia of Hardy Fruits* (New York: Macmillan, 1921); U. P. Hedrick, *The Pears of New York* (Albany, N.Y.: J. B. Lyon Company, printers, 1921); U. P. Hedrick, *Systematic Pomology* (New York: Macmillan, 1925); U. P. Hedrick, *The Small Fruits of New York* (Albany, N.Y.: Department of Farms and Markets, State of New York, 1925); U. P. Hedrick, *The Vegetables of New York*, vols. 1, parts I–IV (Albany, N.Y.: Education Department, State of New York, 1928, 1931, 1934, 1937); U. P. Hedrick, *A History of Agriculture in the State of New York. The New York State Agricultural Society.* (Albany, N.Y.: J. B. Lyon Company, 1933); U. P. Hedrick, *Fruits for the Home Garden* (New York: Oxford University Press, 1944); U. P. Hedrick, *Grapes and Wines from Home Vineyards* (London: Oxford University Press, 1945); U. P. Hedrick, *A History of Horticulture in America to 1860* (New York: Oxford University Press, 1950), and countless bulletins issued by the Geneva Experiment Station.

7. Leon D. Adams, *The Wines of America* (Boston: Houghton Mifflin Company, 1973), 105–7.

8. As always, a core group of research specialists and assistants at Geneva helped advance work there; they include Keith Kimball (1950–57 and 1966–82), John Watson (1950–88), G. Remaily (1960s–1970s), K. Maloney (1988–90), Mary Howell Maloney (1983–93), P. Wallace (1996–present), and R. S. Luce (1994–present).

9. Thomas Pinney, *A History of Wine in America: From the Beginnings to Prohibition, vol. 1* (Berkeley: University of California Press, 2007), 369–70, 377; Adams, *Wines of America*, 106.

10. See Hedrick, *Manual of American Grape-Growing (New York: The Macmillan Company, 1919)*; Hedrick, *Grapes and Wines from Home Vineyards (London: Oxford University Press, 1945)*.

11. Adams, *Wines of America*, 105–7

12. Cayuga White covered 404 acres in New York in 2006 and 385 acres in 2011. It is also grown in Pennsylvania, Missouri, and Michigan. See Jancis Robinson, Julia Harding, and Jose Vouillamoz, *Wine Grapes: A Complete Guide to 1,368 Vine Varieties, Including Their Origins and Flavor.* (New York: Ecco/HarperCollins, 2012), 212; *New York Vineyard Survey*, January 2013, United States Department of Agriculture, National Agricultural Statistical Service, 4.

13. See page 141 on the genealogy of Seyval Blanc.

14. See page 77 for more information on the grape Diamond (also called Moore's Diamond).

15. See page 98 for more information on the grape Iona.

16. J. Einset and W. B. Robinson, *Cayuga White, the First of a Finger Lakes Series of Wine Grapes for New York, New York's Food and Life Sciences Bulletin no. 22* (August 1972), New York State Agricultural Experiment Station, Geneva, N.Y., a Division of the New York State College of Agriculture and Life Sciences; Robinson et al., *Wine Grapes*, 212.

17. T. D. Jordan, R. M. Pool, T. J. Zabadal, and J. P. Tomkins, *Cultural Practices for Commercial Vineyards*. New York State Agricultural Experiment Station, Cornell University, Misc. Bulletin no. 111 (Geneva, N.Y.: 1981), 64;

Bruce I. Reisch, Robert M. Pool, David V. Peterson, Mary-Howell Martens, and Thomas Henick-Kling, *Wine and Juice Grape Varieties for Cool Climates*, Information Bulletin no. 233, a Cornell Cooperative Extension Publication (1993), 12; *The Brooks and Olmo Register of Fruit & Nut Varieties* (3rd ed.), (Alexandria, Va.: *American Society for Horticultural Sciences*, 1997), 256.

18. Chardonel covered 190 acres in Missouri (2009), 24 acres in Michigan (2006), and is grown to some extent in Pennsylvania, Arkansas, and Ohio. See Robinson et al., *Wine Grapes*, 220.

19. Bruce I. Reisch and Robert M. Pool, *Chardonel Grape*, New York's Food and Life Sciences Bulletin no. 132 (1990), New York State Agricultural Experiment Station, Geneva, NY, a Division of the New York State College of Agriculture and Life Sciences; Reisch et al., *Wine and Juice Grape Varieties*, 6; New York State Fruit Testing Cooperative Association, Inc., *A Catalog of New & Noteworthy Fruits*, 1991–1992, 25; *Brooks and Olmo Register*, 256–57; Robinson et al., *Wine Grapes*, 220.

20. Bruce I. Reisch,W. B. Robinson, K. Kimball, Robert M. Pool and J. Watson, *Horizon Grape*, New York's Food and Life Sciences Bulletin no. 96 (1982), New York State Agricultural Experiment Station, Geneva, N.Y., a Division of the New York State College of Agricultural and Life Sciences.

21. Reisch et al., *Wine and Juice Grape Varieties*, 6, 12.

22. Reisch et al., *Horizon Grape*; *Brooks and Olmo Register*, 268

23. Bruce I. Reisch and Robert M. Pool, *Melody Grape*, New York's Food and Life Sciences Bulletin no. 112 (1985), New York State Agricultural Experiment Station, Geneva, N.Y., a Division of the New York State College of Agriculture and Life Sciences.

24. Ibid.; *Brooks and Olmo Register*, 274; Robinson et al., *Wine Grapes*, 623.

25. Double A Vineyards, Inc., *Double A Vineyard 2003/2004 Catalog*, Fredonia, N.Y., 16–17.

26. Melody's acreage in New York was only 14 acres in 2009, while Cayuga White covered 404 acres in 2006 and 385 acres in 2011. See Robinson et al., *Wine Grapes*, 623; *New York Vineyard Survey*, 4.

27. Bruce I. Reisch and Robert M. Pool, *Traminette Grape*, New York's Food and Life Sciences Bulletin no. 149 (1996), New York State Agricultural Experiment Station, Geneva, N.Y., a Division of the New York State College of Agriculture and Life Sciences; "Two New Grape Varieties Named and Released," *Grape Research News* 7, no. 2 (Summer 1996), published by the New York State Agricultural Experiment Station and the New York Wine and Grape Foundation; Carl Camper, Chateau Stripemine Experimental Vineyard, www.chateaustripmine.info/Breeders.htm; Robinson et al., *Wine Grapes*, 1073–74.

28. *Double A Vineyard 2003/2004 Catalog*, 16–17.

29. *Grape Research News* 7, no. 2 (Summer 1996), published by the New York State Agricultural Experiment Station, Geneva, N.Y., and the New York Wine and Grape Foundation.

30. Traminette is becoming popular on the East Coast, with 65 acres grown in New York (2006), increasing to 139 acres by 2011, 34 acres in Pennsylvania (2008), 105 acres in Missouri (2009), 80 acres in Virginia (2008), 30 acres in Michigan (2006), 20 acres in Ohio (2008), 63 acres in Illinois (2009), and it even grows in Nebraska and Iowa. See Robinson et al., *Wine Grapes*, 1074; *New York Vineyard Survey*, 4.

31. These new varieties were released on a limited basis so that they could be evaluated in the field. Before being distributed, participating growers sign a non-distribution/testing agreement so that Geneva can obtain royalties should these varieties ultimately be deemed ready for sale to the general public.

32. Bruce I. Reisch and R. S. Luce, *Corot Noir Grape*, New York's Food and Life Sciences Bulletin no. 159 (2006), New York State Agricultural Experiment Station, Geneva, N.Y., a Division of the New York State College of Agriculture and Life Sciences.

33. Ibid.; *Double A Vineyards 2003/2004 Catalog*, 16–17.

34. Robinson, *Wine Grapes*, 268; *New York Vineyard Survey*, 4.

35. Reisch et al., *Corot Noir Grape*; "Cornell Releases, Names Three New Hybrids," *Vineyard and Winery Management News* (November 2006), 57; *Double A Vineyard 2000/2001 Catalog*, 10; Robinson et al., *Wine Grapes*, 268.

36. Bruce I. Reisch and R. S. Luce, *Noiret Grape*, New York's Food and Life Sciences Bulletin no. 160 (2006), New York State Agricultural Experiment Station, Geneva, N.Y., a Division of the New York State College of Agriculture and Life Sciences; Robinson et al., *Wine Grapes*, 734–35.

37. Reisch et al., *Noiret Grape*.

38. There were 39 acres of Noiret in New York (2006), which increased to 57 acres in 2011, and there were 12 acres in Pennsylvania (2006). See Robinson et al., *Wine Grapes*, 735; *New York Vineyard Survey*, 4.

39. These new varieties were released on a limited basis so that they could be evaluated in the field. Before being distributed, participating growers sign a non-distribution/testing agreement so that Geneva can obtain royalties on these varieties should they be formally named and released for sale to the general public.

40. New York State Fruit Testing Cooperative Association, Inc., *A Catalog of New & Noteworthy Fruits*, 1989–1990, 26.

41. *Catalog of New & Noteworthy Fruits*, 1989–1990, 26; and as described in *Promising New York Red Wine Selections* in 1991.

42. New York State Fruit Testing Cooperative Association, Inc., *A Catalog of New & Noteworthy Fruits*, 1995, 23.

Minnesota Hybrids

THIS CHAPTER COVERS the cool climate Minnesota hybrids that have been developed by Elmer Swenson (1913–2004) of Osceola, Wisconsin, and by Peter Hemstad, James Luby, and Patrick Pierquet of the University of Minnesota. Some of these varieties—such as Frontenac, St. Croix, St. Pepin, La Crosse, La Crescent, Marquette, and Swenson White—are being grown in the Hudson Valley communities of Ashland, Castleton, Marlboro, Middle Hope, and New Paltz. In addition, these grapes are being made into quality wines by several wineries in the Hudson Valley.

All of the Swenson and University of Minnesota hybrids can be grown in the coldest parts of the Hudson Valley and in the nearby Catskill and Berkshire Mountains, which border the Hudson Valley. Many of these Minnesota hybrids were bred by grape-breeding pioneer Elmer Swenson. Swenson worked on his 120 acre farm near Osceola, Wisconsin, which he inherited from his maternal grandfather. Swenson acquired his interest in grape and fruit culture from his Grandpa Larson, who died when Elmer was only five years old. On the farm, there were two acres of grapes, apples, plums, and cherries. Among Larson's books was T. V. Munson's *The Foundations of American Grape Culture*, which interested Swenson.

Beginning of the Hybridization Program

Swenson began breeding grapes on his own in earnest around 1943, after his brothers left the farm to serve in the armed forces during World War II. During most of his grape-breeding career—until he was hired by the University of Minnesota in 1969 as a gardener, at the age of fifty-six—Swenson was a private grape breeder. He accomplished his breeding work in addition to performing his routine dairy farm chores, however dairy work did not seem to interest him intellectually. He also married, raised four children, and was an elder in the Lutheran Church.

Elmer Swenson. *One of the primary developers of Minnesota hybrids in the twentieth century, Elmer Swenson is considered by many to be the father of the cool climate grape hybridization movement in the United States.*

When he began his grape-breeding program, Swenson obtained University of Minnesota–developed hybrids, such as Minnesota 78 (MN 78), which became one of the foundations of his breeding work. MN 78 was used extensively because it reliably contributed winter hardiness, vigor, and early maturity of fruit and wood. MN 78 is a cross of Beta (Concord × *riparia*) × Jessica (*labrusca* × *vinifera*). One of the other University of Minnesota hardy, wild grapes that Swenson used later on in his breeding program included MN 89 Riparia, which was not a purposely developed hybrid but a grape that was identified in the wild.[1]

Early on in Swenson's breeding program, he cross-bred French-American hybrid grapes and *labrusca*-based grape varieties developed at the Geneva Agricultural Experiment Station prior to World War II with the hardy, local

MN 78. His goal was to produce high-quality varieties for table consumption and later for wine production, varieties that could survive the harsh cold climate and short growing season of the upper midwestern part of the United States. At the time, few varieties could consistently be relied upon to survive Wisconsin and Minnesota's harsh cold winters and short growing season. He was inspired by the great American horticulturalist and grape hybridizer T. V. Munson of Denison, Texas.

Swenson's Methodology

In Swenson's grape-breeding program, he relied on quality French-American grape varieties such as Villard Blanc, Humbert #3, Landot 4511, Couderc 299-35, J.S.23-416, S.V.23-657, Seyval Blanc, Baco Noir, and Seibel 11803; *vinifera* such as Muscat Hamburg (also relied upon by Hudson Valley breeders) and Chasselas Doro; older *labrusca*-based Geneva hybrids such as Dunkirk, Ontario, Buffalo, and Golden Muscat; and Carman, that was developed by T. V. Munson, to impart superior table or wine grape qualities to his hybrids. For cold and disease resistance and sufficient early ripening capability for Swenson's short growing season, he relied primarily on local wild *riparia* grapes and *riparia* hybrids, which he developed or obtained from the University of Minnesota, such as MN 78, MN 89 Riparia, and Riparia WI #2.

Swenson tended to use locally obtained cold-hardy and disease-resistant hybrids, which he either developed or received from the University of Minnesota, as the seed parent and "quality" French-American hybrids or *vinifera* as the pollen parent to impart quality for table and wine grapes. Chronologically, later on in his breeding program, he used as his "quality" pollen parent those varieties that he bred, which genetically possessed within them a lot of "quality" characteristics from varieties such as Seyval Blanc, Landot 4511, and Couderc 299-35, but which also had in their heritage MN 78 or other *riparia* hybrids that imparted hardiness. He then bred these quality hybrids that had such cold-hardiness heritage back into his hybrids as the pollen parent to impart quality and resistance.

His grape-breeding program, much like Albert Seibel's, was very extensive and he kept breeding and cross-breeding his own varieties with each other to maximize the genetic makeup of the plants and to reconfigure the plants to obtain the desirable traits that he was looking for—that is, use as a table grape, cold hardiness, fungal disease resistance, early wood hardening-off times, and the ability to grow in short seasons. Further, his breeding program, which started in 1943, lasted for over half a century and consisted of more than 700 catalogued grape hybrids. Few other grape hybridizers, except for Albert Seibel and T. V. Munson, can claim such a lengthy time in the vineyards or the breadth of the hybrids that they created. His efforts will long be remembered and used because his work is used by many currently practicing hybridizers who have taken up his torch.

Throughout the 1950s and 1960s, Swenson provided some of his most promising hybrids to nurseries for sale. However, even though the plants were sold, Swenson received little credit for his work. In 1967, in order to more widely distribute his selections, he went to the University of Minnesota fruit field day in an effort to have his new plants recognized by a university program. Initially, the program officials at the university were only mildly enthusiastic.

University of Minnesota

In 1969, Swenson took a job caring for fruit crops at the University of Minnesota and planted some of his hybrids there, in an effort to increase the interest of university faculty in the merits of his work. The university, like many other state agricultural experiment stations, had a very small grape-breeding program as early as 1908. Its limited breeding program during the 1920s to 1940s used Beta as one of its breeding parents, which created MN 78, later used extensively by Swenson. Other varieties developed early on by the University of Minnesota included the *labrusca*-based hybrids Blue Jay (MN 69), Moonbeam (MN 66), Red Amber (MN 45), and Bluebell (MN 158).[2]

Swenson's work paid off when the university slowly began to revitalize its grape-breeding program in the latter half of the century. This interest, which was a nationwide trend, was based in part in the enactment of farm winery license laws in many states. The enactment of these laws increased the interest in breeding grapes among a growing number of small but locally based wineries throughout the upper Midwest. The stagnant or declining commodity prices for locally produced meat, grain, and milk may also have increased interest in grape culture because farmers were looking for new crops to grow.

The university's first two re-releases, jointly released in 1978, were Edelweiss (MN 78 × Ontario) and Swenson Red (MN 78 × S.11803), which were some of the very first hybrids that Swenson had developed.[3] Ontario is the same variety that is in the genetic heritage of Cayuga White. In 1979, Swenson retired from the university, but he was provided with a grant to continue his breeding work. The following grape varieties, in chronological order of their breeding, were developed by Swenson: Edelweiss (E.S.40, early 1950s), St. Pepin (E.S.282, in the 1950s), La Crosse (E.S.294, in the 1950s), Swenson Red (E.S.439, around 1960), Kay Gray (E.S.1-63, mid-1970s), Sabrevois (E.S.2-1-9, 1978), St. Croix (E.S.2-3-21, 1978), Prairie Star (E.S.3-24-7, 1980), Louise Swenson (E.S.4-8-33, 1980), Swenson White (E.S.6-1-43, between 1981 and 1983), and Brianna (E.S.7-4-76, 1983). It is interesting to note that some of these hybrids were released very shortly after they were bred, while others were not released for ten years or more so that they could be evaluated further.

Swenson's numbered hybrid selections went up to E.S.14-4-100.[4] This means that many of the Swenson selections developed after Brianna (E.S.7-4-76), bred in 1983, may have not been fully evaluated for their attributes in the field or in the cellar. Since Swenson's retirement in 1979, more and more of the breeding work has been conducted by the University of Minnesota under the direction of Peter Hemstad, who joined the university in 1985.

Swenson had a very generous policy of sharing breeding material and grape variety selections, sending them to anyone who requested them. Hence, some of his more significant hybrids were utilized by other growers and, after obtaining permission from Swenson, were able to name, use, and promote these hybrids. By sharing his material widely with others, he benefited from the experience of other growers across the country and was then able to accelerate his own breeding program by relying on the fieldwork of others. This should be a lesson to many of our current university-based grape-breeding programs, which seem to want to control the products developed, but in doing so they limit the scope of the field research that can be done by not widely disseminating their plant material for comment.

Elmer Swenson continued his breeding work until around 2000. After 2000, it became more difficult for him to maintain his five-acre experimental vineyard. While his breeding work was coming to an end, there were still many promising cultivars that had not been evaluated. With the cooperation of volunteers from the Minnesota Grape Growers Association, and with support from the Quebec Winegrowers Association and the Seaway Wine and Viticultural Association of New York, this group and its volunteers helped to maintain his experimental plot and conduct the Swenson Preservation Project, which continued for three years so that the newer selections from his vineyard could be evaluated and cuttings taken to preserve Swenson's later developed hybrids. In addition, almost fifty of Swenson's hybrids have been included to be preserved in the grape collection at the National Clonal Germplasm Repository in Geneva, New York.

Elmer Swenson died on Christmas Eve in 2004. His main experimental vineyard was ultimately removed because the Swenson family had no desire to maintain it, however, the original house vineyard established by Grandpa Larson is still being maintained. Swenson's work continues to be carried on by a new generation of dedicated private and university-based grape hybridizers.[5]

Swenson's Legacy

Swenson helped to revitalize the dormant grape-breeding program at the University of Minnesota during the 1970s. The university formally initiated a grape-breeding program for wine grapes by the middle to late 1970s. Under the leadership of Peter Hemstad and James Luby,[6] this program evaluates the viticultural and enological potential of newer hybrids that are being developed by the university. The program has approximately 10 acres of research vineyards with approximately 10,000 experimental vines and new seedlings produced and evaluated each year. Since the program was formalized, the university has officially named and released Frontenac (1996), La Crescent (2002), Frontenac Gris (2003), and Marquette (2006).[7]

In previous chapters, I have described the winter hardiness of each grape variety. I have not done so in as much detail in this chapter because all of the varieties cited are either super-winter hardy or of sufficient hardiness. They can unquestionably be grown and will thrive in the coldest places in the Hudson Valley watershed area, which includes the top of the Catskill Mountains, such as Windham or Round Top, or in the foothills of the Berkshire Mountains.

Minnesota White Varieties

Below are eight white wine grape varieties developed by Swenson and the University of Minnesota that should be considered for cultivation by growers located in the colder parts of the Hudson Valley and in the nearby foothills of the Catskill and Berkshire Mountains, which border the Hudson Valley.

By extension, these varieties should be evaluated by New York growers in the cooler parts of this state, especially in the Mohawk River Valley, those areas north of the Mohawk River, and in Michigan, New England, and Canada.

LA CRESCENT (MN 1166)

La Crescent is a white hybrid bred in 1988, selected in 1992, and introduced and named in 2002 by the University of Minnesota. Genetically, its heritage includes only Swenson hybrids that have 45 percent *vinifera* heritage and 28 percent *riparia* heritage. Some of its ancestors include Seyval Blanc, Rosette (S.1000), and Muscat Hamburg. This grape is a cross-breed of St. Pepin × E.S.6-8-25 (*Riparia* 89 × Muscat Hamburg).

PARENTAGE
aestivalis, labrusca, riparia, rupestris, vinifera

HARVEST DATE
Late early to mid-season

A+ A– B B+ A

This variety, increasingly popular in the upper Midwest of the United States, has a moderately loose to compact and long conical cluster with one wing of small berries that are yellow-gold with some amber colors. The cluster looks similar to Baco Noir in form and shape with the same vigorous, sprawling growth habit. The vine is moderately vigorous to vigorous, depending on the fertility of the soil it is grown in. La Crescent is easy to grow and somewhat resistant to downy and powdery mildew, black rot and bunch rot, but its spray program should not be neglected. Its bud break is relatively early, but it has only a small- to medium-sized secondary crop if hit by a late spring frost. The grape ripens by late early to mid-season and has high sugars of around 24° Brix. The yields are moderate on lighter soils and heavier on richer soils.

The wine is quite perfumey and fruity with a nose and flavor of apricots, but it can also have flavors that range from apples to white peaches, tropical fruits, pineapples, mangoes, and honey, which can lend interest and character when blended with more neutral wines. The wine has good body and is balanced with elegant fruit flavors that are not aggressive. Its flavor profile has been compared to Vignoles or Riesling. It is good for making semi-sweet wines with solid, but not too high acid levels, which balance high sugars that average between 24° and 25° Brix. Its body is good, but its acid can range from firm to high, so vinification techniques may be needed to offset its acidity.[8]

LA CROSSE (E.S.294)

La Crosse is a sibling of St. Pepin, listed below. It was developed very early in Swenson's breeding program in the 1950s. In 1983, it was named after the Wisconsin city on the Mississippi River. It has Seyval Blanc and Rosette (S.1000) in its genetic background (E.S.114 (MN 78 × Rosette (S.1000)) × Seyval Blanc).

PARENTAGE
labrusca, lincecumii, riparia, rupestris, vinifera

HARVEST DATE
Early mid-season

A+ A C+ A A

The medium-sized cluster is slightly loose to well filled with medium-sized, yellow-green to light golden berries. It looks similar to its parent Seyval Blanc with a somewhat more open and semi-upright growth habit. La Crosse is moderately vigorous to vigorous and a productive variety. It has overall solid fungal disease resistance, but is susceptible to black rot and bunch rot, but less so for downy and powdery mildew. It has a mid-season bud break, but has a secondary crop if frost damage occurs. The variety ripens by early mid-season a bit before Seyval Blanc and about two weeks after St. Pepin with sugars between 12° and 19° Brix. It is reported to have higher acid and is less aromatic than St. Pepin.

The grape is balanced for wine in both sugar and acid level in a manner similar to Seyval Blanc. These Seyval Blanc–like fruity wines have been known for being barrel fermented with a malolactic fermentation to reduce its acid further. Nose and flavor descriptors include floral notes with rich fruit flavors of pears, apricots, melons, grapefruit and citrus, with some floral, Muscat, and spice elements in the nose. This grape can stand on its own or be used in a blend to add body to thinner wines. For a Minnesota hybrid, it has a wide range of wine styles that it can be made into, ranging from a dry barrel-aged white to a semi-dry white wine.[9]

ST. PEPIN (E.S.282)

St. Pepin is also one of Swenson's first hybrids that he bred during the 1950s, but which was introduced much later in 1983, with a patent assigned in 1986 to Swenson Smith Vines. It is a sister seedling of La Crosse and has Seyval Blanc and Rosette in its genetic background. St. Pepin is a cross-breed of E.S.114 (MN 78 × Rosette (S.1000)) × Seyval Blanc. It makes some of the best white wines of the Minnesota hybrids.

PARENTAGE
labrusca, lincecumii, riparia, rupestris, vinifera

HARVEST DATE
Early mid-season

A+ A B A A

St. Pepin's moderately large conical cluster is moderately compact to loose and made up of medium-sized slip-skin berries that look similar to one of its parents, Seyval Blanc, in color. The juice that it yields is a very light pink.

The vine grows well and is moderately vigorous to very vigorous and moderately productive to productive, depending on the fertility of the soils that it grows in. It has an upward growth habit with a somewhat open canopy. St. Pepin is somewhat resistant to most fungal diseases, except for powdery mildew, to which it is susceptible. It grows easily with little maintenance and prefers to be pruned to the top wire.

It ripens by early mid-season, anywhere from one to two weeks before its sibling La Crosse, and is fruitier in flavor, similar to Riesling, with a mild *labrusca* flavor. At harvest, the sugars range from 20° to 24° Brix and its acids are relatively low, but they are balanced for wine production. It can be used either as a varietal wine or in a blend.

The plant is pistillate (it has only female flowers), so it needs to be planted near other grape varieties, such as La Crosse, Frontenac, or Frontenac Gris so that it can cross-pollinate for fruit production. (St. Pepin was used to breed La Crescent.) The grape can also be used as a table grape.[10]

MINOR WHITE GRAPE VARIETIES OF SWENSON

BRIANNA (E.S.7-4-76)

(Kay Gray × E.S.2-12-13)
aestivalis, berlandieri, cinerea, labrusca, lincecumii, riparia, rupestris, vinifera

This cross, made in 1983 by Swenson, was chronologically bred during the latter part of his grape-breeding program. It was selected as a table grape in the fall of 1989. Genetically, it has MN 78, Golden Muscat, Villard Blanc, and Swenson Red in its heritage. It was named Brianna by Ed Swanson of Cuthills Vineyard, Pierce, Nebraska, in 2001 to be used as a wine grape. It has medium to high vigor, depending on the fertility of its soils. It is moderately resistant to all major fungal diseases, a bit less resistant to black rot and botrytis, except for crown gall, to which it is susceptible. It is not sensitive to sulfur applications. Brianna can be grown in a wide variety of soils and is easy to manage in the field. Further, it can take both hot and cold temperature extremes. It has a secondary crop if hit by a late spring frost. It is a medium-sized, semi-compact cluster with medium to large berries that are thick-skinned and greenish-gold to gold when fully ripe. It ripens by early mid-season.

When the grapes are picked fully ripe and made into a semisweet wine, it has a nice pineapple aroma and flavor. When made into a dryer wine, it can have additional flavors of grapefruit, bananas, and mangoes. Brianna is a good blender without overpowering the other grapes in the blend. With its refreshing acid levels and flavor profile, it could be useful in dessert wines, but is generally made as a semisweet wine. This variety can benefit from cool fermentation to help retain its fruit flavors.[11]

KAY GRAY (E.S.1-63)

(E.S.217 (MN 78 × Golden Muscat) × Open Pollinated or perhaps Onaka)
labrusca, riparia, vinifera

This is one of the earliest hybrids bred by Swenson and was named by him. It is one of his more reliable producers of white wine that can also serve as a table grape. The possible pollen parent, Onaka, is an old South Dakota hybrid that is a cross of Beta × Salem. One of Kay Gray's grandparents is Golden Muscat. The grape is named after the wife of Dick Gray, the director of the Minnesota Freshwater Biological Institute. It was released in 1981 with a plant patent assigned to Swenson Smith Vines in 1982.

The variety is resistant to all fungal diseases, and is very vigorous and productive. The relatively compact cluster is small to medium in size, cylindrical, and uniform. Its yellow-gold slip-skin berries are medium to large. The grapes ripen early to early mid-season and attain sugars of 20°–22° Brix with rather low acidity.

The wines are neutral and can be used for blending, but it can be made into a varietal wine that has a flowery and mild *labrusca* nose and taste when made in good vintage years. The wine oxidizes very easily, so it should remain topped up and not racked too frequently. Also, if overripe fruit is used, heavy undesirable Muscat flavors can result, so one may wish to harvest the fruit earlier to avoid such flavors.[12]

LOUISE SWENSON (E.S.4-8-33)

(E.S.2-3-17 × Kay Gray)
labrusca, lincecumii, riparia, rupestris, vinifera

This was developed by Swenson during the middle-late period of his breeding program in 1980, selected in 1984, and introduced in 2001. It has Seyval Blanc, Seneca, Golden Muscat, and Rosette (S.1000) in its genetic heritage. It is named for Elmer Swenson's wife Louise. It is a moderately loose to well-filled, small- to medium-sized conical cluster with small- to medium-sized light-green and translucent berries. It has good disease resistance on a vine of moderate vigor, but may be susceptible to anthracnose. Its production is moderate, but it does not like droughty conditions or light sandy soils. There are few problems in the field with Louise Swenson, but its sugars tend to remain at or below 20° Brix with only moderate acidity. It buds out late, so is not affected by late spring frosts. Further, it has an orderly trailing to semi-upright growth habit, so it is easy to maintain in the field. It ripens by early mid-season.

The variety is one of the better Swenson white wine grapes in the cellar with soft, fruity aromas and flavors of flowers, pears, and

Minor white grape varieties, cont'd.

honey, which have no hybrid characteristics. The wine is consistently good and balanced from year to year, but because of its light body and presence, it needs to be enhanced by other wines to augment it. It can be blended with higher-acid wines to reduce their acid profile. It blends well with Prairie Star because while Louise Swenson has delicate aromatics and lighter body, Prairie Star has the requisite body and finish to complement it.[13]

PRAIRIE STAR (E.S.3-24-7)

(E.S.2-7-13 (E.S.5-14 × Swenson Red) × E.S.2-8-1 (E.S.5-14 × Swenson Red))
aestivalis, berlandieri, cinerea, labrusca, lincecumii, riparia, rupestris, vinifera

This variety was developed by Swenson during the middle-late period of his breeding program in 1980, selected in 1984, and named by Tom Plocher and Bob Parke in 2000.[14] The grape has Villard Blanc in its genetic background.

The cluster is long, thin, and moderately loose to well filled, with small-medium-sized berries that are a yellow-gold color. The cluster looks similar to Vidal Blanc. This variety is generally resistant to fungal diseases, except for black rot and anthracnose, to which it is only moderately resistant. The plant is vigorous with a semi-upright growth habit. The bud break is by mid-season and it produces a modest secondary crop after a late spring frost, but it can be susceptible to poor fruit set in some seasons. Early in the season, its shoots have a tendency to break off in high winds, so low cordon pruning may be needed with catch wires to protect its shoots in those areas that are prone to such winds. Prairie Star ripens by mid-season, around the time of Baco Noir or a bit beforehand. It is noticeably more productive on heavier soils or if grafted.

The grape has good sugars at 20°–22° Brix and acid that is balanced for winemaking. Its wines are well balanced and somewhat floral, but generally neutral with no *labrusca* flavors. It does have good body, mouth feel, and a finish that can be used in blends to beef up thinner white wines. It blends well with Louise Swenson because Louise Swenson has delicate aromatics and Prairie Star has the body and finish needed to complement it. Prairie Star and Louise Swenson were each named around the same time in 2000 and 2001 because the grapes complement each other in blended wines.[15]

SWENSON WHITE (E.S.6-1-43)

(Edelweiss (MN 78 × Ontario) × E.S.442 (MN 78 × S.11803))
aestivalis, berlandieri, cinerea, labrusca, lincecumii, riparia, rupestris, vinifera

This grape has a medium to large loose to well-filled conical cluster of large berries of a light green color with some yellows. It was bred late in Swenson's breeding program, around 1981 to 1983. It was selected in 1988 and introduced in 1994. One of its grandparents is the Geneva-developed white hybrid grape Ontario, which is also a grandparent of the Geneva-developed Cayuga White.

The unmethodical way in which some of the Swenson hybrids were named is evidenced by the naming of Swenson White. Lon Rombough (1949–2012), the horticulturalist and nurseryman from Oregon, sent cuttings of this variety to a friend in Colorado for testing. After his friend took a job at a nursery in Boulder, these cuttings were propagated and reviewed by a garden writer who was visiting the nursery. The tag on the grape was "A white grape from Swenson," which was shortened to "Swenson White." Once the name of the grape was published in a national magazine as Swenson White, that became its "official release."[16]

The skins are thick, which allows the variety to hang on the vine so that it can be made into late-harvest or ice wines. Further, the thick skins seem to help minimize bird and insect damage. It has good fungal disease resistance, except for downy and powdery mildew, to which it is somewhat susceptible. It is, however, a healthy grower. The bud break is by mid-season, but it has little secondary crop if hit by a late spring frost. Due to its large cluster size, sometimes cluster thinning is needed. It is good in the field and in the cellar, ripening by late mid-season to late and achieves sugars of no more than 22° Brix.

The must is well balanced in sugar with moderate acidity so that it can make quality wine. The wines have a nice pronounced flowery nose that is similar to St. Pepin. The wines have lots of fruit in the body and finish, with some elements of nice *labrusca* flavors.[17]

Minnesota Red Varieties

Below are five cold-hardy red wine grape varieties that should be considered by growers who live in the cooler parts of the Hudson Valley and adjacent foothills in the Catskill and Berkshire Mountains. By extension, these varieties, like Swenson's white varieties, should be evaluated by other cool climate areas of New York State, including the Mohawk River Valley and those areas that are north of the Mohawk River and in Michigan, New England, and Canada.

FRONTENAC (MN 1047)

Frontenac is a cross-breed of MN 89 Riparia × Landot 4511. This variety was bred in 1978, selected in 1983, and named and introduced by the University of Minnesota in 1996.[18] The grape is named for the village of Frontenac, Minnesota, which is located on Lake Pepin, just southeast of St. Paul. Incidentally, the city of Frontenac, Minnesota, had been named for Louis de Buade, Comte de Frontenac (1622–98), who was the governor general of the French-Canadian colony of New France from 1672–82 and 1689–98.

PARENTAGE
aestivalis, berlandieri, cinerea, labrusca, lincecumii, riparia, rupestris, vinifera

HARVEST DATE
Late mid-season

A+ A B+ A+ A–

Frontenac is the most widely planted of the Minnesota hybrids in the Hudson Valley and is popular in the Midwest and Plains states.[19] Some of its ancestors include the highly productive French-American hybrid Villard Blanc and the equally productive red French-American hybrid grape Plantet.

This variety is vigorous to very vigorous, consistently very productive, and does well in most soils. Its conical clusters are loose to moderately loose, long, and medium to large in size. Its small- to medium-sized berries are blue-black. It has a slightly upward growth habit and then droops. It buds out by mid-season. The variety has good fungal disease resistance, except for having only moderate resistance to black rot. Its moderately loose clusters negate bunch rot or berry-splitting problems. Due to its high productivity, cluster thinning is recommended to maintain crop quality. Frontenac ripens late mid-season to late, about one week after Baco Noir with sugars of 24°–27° Brix or more. To curtail its very high acid levels, it is best to harvest the grapes only after it reaches 24° Brix or higher.

The wines can be very dark and inky or a bright purple, depending on how the wine is made. Even with all of its color, the wines are only of medium body. Due to its *riparia* heritage, the wines can be very high in total acidity, low in tannins, and are herbaceous if harvested too early. To help reduce its acids, malolactic fermentation should be considered.

Even with ripe fruit, Frontenac can be a bit one-dimensional with flavors of very pronounced dark cherries, plums, cooked elderberries, and wood. At times, these cherry, plum, and elderberry flavors can become too fruity and grapey with cotton candy, bubblegum, and cough syrup–like overtones like wines made from Steuben or Noiret grapes. It can also have a distinctive mint/wintergreen flavor that sometimes integrates with the fruit and at other times does not.

The Frontenac grape is very versatile in the cellar and is capable of producing a dry table wine, blush, rosé, or port. Barrel aging is recommended to take the rough edges off this wine if it is made as a dry table wine. If made as a rosé, the grape's acids are more in keeping with a white wine. The grape also lends itself to dessert port production because of its big forward fruit flavors and high acidity, which can handle fortification with brandy. Growers and winemakers in the Hudson Valley seem to like this variety.[20]

FRONTENAC GRIS (MN 1187)

Frontenac Gris was identified at the University of Minnesota in 1992, introduced and named in 2003 by the university, and originates from a single mutant gray cluster of Frontenac (which is amber-colored). In a manner similar to Pinot Gris, which is a mutant of Pinot Noir, Frontenac Gris is a mutant of Frontenac. It has similar viticultural characteristics as Frontenac, but seems to have slightly better fungal disease resistance. Genetically, Frontenac Gris is a cross-breed of MN 89 Riparia × Landot 4511.

PARENTAGE
aestivalis, berlandieri, cinerea, labrusca, lincecumii, riparia, rupestris, vinifera

HARVEST DATE
Late mid-season

A+ A B+ A+ A–

Like Frontenac, Frontenac Gris ripens late mid-season, about one week after Baco Noir, reaches 24°–25° Brix, and grows well in most soils. It is used to make very fruity white to copper-colored rosé wines. Frontenac Gris wines tend to have aromas and flavors of peaches and apricots, with hints of pineapple, honey, grapefruit, and some tropical fruits, such as passion fruit, with a good balance between its fruit and acidity. It is a clean wine with little or no herbaceous or *labrusca* aromas or flavors, which is not similar to Frontenac at all. Frontenac Gris, due to its high acid, should be finished semidry, but can also be made into dessert wines.[21]

MARQUETTE (MN 1211)

Marquette was bred in 1989 by the University of Minnesota, selected in 1994 for further propagation and study, and introduced and named by the university in 2006. It is a genetically complex hybrid with Pinot Noir as its grandparent and a shared recent ancestor to Frontenac, Landot 4511 (MN 1094 (MN 1019 × MN 1016) × Ravat 262 (S.8365 × Pinot Noir)). Other grapes in Marquette's genetic gene pool include Cabernet Sauvignon, Merlot, Landot 4511, and the highly productive French-American hybrids Villard Blanc and Plantet. The grape is named after Father Marquette, a Jesuit missionary and explorer in the second half of the seventeenth century.

PARENTAGE
aestivalis, berlandieri, cinerea, labrusca, lincecumii, riparia, rupestris, vinifera

HARVEST DATE
Mid-season

A+ A C B+ A

The variety has very good resistance to downy mildew, powdery mildew, bunch rot, and black rot, and needs only a minimal spray program. Resistance to infestation by foliar phylloxera is only moderate. Marquette produces a moderate to good crop. Its bud break is somewhat early, so it is at risk of late spring frost damage, but it does produce a smaller secondary crop. Marquette has moderate vigor and an open and orderly, somewhat upright growth habit that facilitates fungicide applications and exposes the fruit and leaves to air and sunshine to minimize fungal disease. Its shoots tend to have only two small- to medium-sized clusters with small-medium-sized black berries, so it does not overproduce or need cluster thinning to enhance crop quality. It ripens by mid-season, a few days before Frontenac and about the same time as Baco Noir. When harvested, the sugar levels can be higher than 25° Brix. It has high acids, but they are lower than Frontenac.

The grape has been touted to be of excellent quality, with attractive deep ruby-red color, nice flavors of cherry, black currants, black pepper, spice, and berry in the nose and palate, with only a moderate amount of body and tannin structure that is better than Frontenac. The wine's flavors and aromas are integrated with a clean finish. I have found the wines to have elements of southern Rhônes with white pepper and raspberries like Grenache, smooth tannins and plum flavors like Syrah, without the thin body that French-American hybrids sometimes have. The wine benefits from barrel aging and malolactic fermentation to reduce its high acids.

This variety was released in 2006, so it would be prudent to wait for more evenhanded evaluations of this wine over time. Growers in cold climate areas such as New York, Michigan, New England, Minnesota, and Quebec are showing interest in this grape, and it is being planted widely in the Hudson Valley.[22]

MINOR RED GRAPE VARIETIES OF SWENSON

SABREVOIS (E.S.2-1-9)

(E.S.283 (E.S.114 × Seyval Blanc) × E.S.193 (MN 78 × Seneca))
labrusca, lincecumii, riparia, rupestris, vinifera

This sister of St. Croix was made toward the middle of Swenson's breeding program in 1978. It is hardier and a more reliable and productive variety than its sibling. It was named by Gilles Benoit, of Vignoble de Pins, a grape grower and winemaker from Sabrevois, Quebec. The grape is named after Sabrevois, Quebec, which is near the Richelieu River, south of Montreal, Canada, where this is the most widely cultivated red grape variety in the area. It does not seem to be as popular in the upper Midwest of the United States.[23]

This Swenson hybrid has a small- to medium-sized, semi-loose cluster. The black berries are of small to medium size. It is a vigorous plant, with an upright and then drooping growth habit, that can be a good producer if pruned properly to set a good crop, otherwise it is only a moderate producer. The variety has high resistance to fungal diseases and anthracnose, but a preventative spray program is prudent to maintain clean fruit. Phylloxera may be a concern. The variety ripens by early mid-season about one week after St. Croix, and is a consistent and reliable producer.

The acid is higher than St. Croix, but it is still workable in the cellar and its sugar content rarely goes above 20° Brix. Malolactic fermentation should be considered to offset the wine's lack of sugar to balance its high acids. However, it cannot hang too long on the vine because its pH will increase to unacceptably high levels, so often it is picked when the sugar is around 19° Brix. The color of the wines can be from electric blue-purple to very dark with a nice berry-like fruitiness in nose and taste, but it tends to lack tannin and body. In Quebec it has also been made into aromatic rosés and fairly full-bodied reds not unlike Baco Noir, but just not as overtly fruity and with a muted nose. It has been described as complex, powerful, and having upfront, plummy Zinfandel and black pepper flavors that do well when aged for a few years. While good on its own, it is best used in blends with other high-sugar varieties such as Baco Noir, Frontenac, Maréchal Foch, and St. Croix. The variety ages well, especially when blended with Frontenac.[24]

ST. CROIX (E.S.2-3-21)

(E.S.283 (E.S.114 x Seyval Blanc) × E.S.193 (MN 78 × Seneca))
labrusca, lincecumii, riparia, rupestris, vinifera

St. Croix is a sibling of Sabrevois, which was, as noted earlier, also bred in 1978. It was introduced in 1981 with a plant patent assigned to Swenson Smith Vines in 1982. It is commonly grown and vinified by wineries in Minnesota and Quebec.

The clusters are of medium size and compact, with thin-skinned, medium-large blue berries. Like Sabrevois, it works hard to achieve 20° Brix, but its acid levels are lower. The St. Croix vine is vigorous with very heavy vegetative growth, but is only moderately productive, so it must be pruned to maximize its crop and increase sunlight to enhance fruit quality. It has been reported to be very precocious in bearing, but it is moderately disease resistant, with good resistance to black rot. The fruit ripens early to early mid-season, about one week before Sabrevois, so it may be vulnerable to bird damage.

St. Croix produces attractive, light- to medium-bodied red wines that have soft tannins and flavors of leather, tobacco, cooked or jammy berries, black currants, and other dark fruit flavors. The wines can be dark, somewhat lacking in tannins, and have a vegetative nose and flavor of tobacco, but is otherwise rather neutral, so can be used in blends with other red wines. Vinification techniques, such as carbonic maceration, can be utilized to enhance the wine's fruit flavors. The grape can be vinified as a varietal or used in a blend.[25]

NOTES

1. According to Tom Plocher, in the mid-1970s, University of Minnesota Prof. Cecil Stushnoff and graduate student Patrick Pierquet conducted several "*riparia*-hunting" trips, which set the stage for much of the subsequent University of Minnesota grape breeding work. They discovered MN Riparia 89 (a parent of the Frontenacs) on an island in the Minnesota River while canoeing and looking for elite *riparia*. They also hiked Riding Mountain National Park, in Manitoba, Canada (hundreds of miles northwest of Winnipeg) and discovered the very northern range of native *riparia*, and MN 37 Riparia and 64 Riparia were collected. These two *riparias* are sensitive to photoperiod and start going dormant in August, when the length of day shortens to a certain number of hours. Riparia 64 ended up in the parentage of both Marquette and Petite Pearl. Conversations with Tom Plocher, early 2015.

2. *The Brooks and Olmo Register of Fruit & Nut Varieties* (3rd ed.) (Alexandria, Va.: American Society for Horticultural Science, 1997), 251–52, 275, 281.

3. "Elmer [Swenson] did the breeding and testing of Edelweiss and Swenson Red on his Wisconsin farm. Because he agreed to jointly release those two varieties with the U. of Minnesota, my advisor (Cecil Stushnoff) agreed to plant a few rows of each at the HRC, for further testing and for propagation material. At that time, Elmer also gave us a few vines of other non-named selections for testing (those selections are now known as St. Croix, LaCrosse and Esprit)." Correspondence with Patrick Pierquet, early 2015.

4. When Elmer Swenson planted a University of Minnesota selection at his Osceola farm, he always assigned it a Swenson vineyard location number. That way he could always identify it by its position in the vineyard – for example, Brianna (E.S.7-4-76) refers to Block 7, Row 4, Plant 76. Conversations with Tom Plocher, early 2015.

5. Bruce Smith, "History of the Early Swenson Hybrids," a presentation at the Nebraska Grape and Wine Conference, November 11, 2006, http://agronomy.unl.edu/c/document; Tom Plocher, "Swenson Preservation Project," November 2003, www.northernwinework.com; "History of the University of Minnesota grape-breeding program," University of Minnesota (2012), http://enology.umn.edu/history.

6. James Luby joined the university in 1983 and is currently the director of the Horticulture Research Center (HRC), where he oversees all fruit breeding research. Patrick Pierquet, a graduate student at the university from 1975–78 and Research Fellow from 1978–80, left the program in 1980, and is currently an Enology Research Associate at Ohio State University at Wooster.

7. "History of the University of Minnesota grape-breeding program," http://enology.umn.edu/history.

8. Tom Plocher and Bob Parke, *Northern Winework: Growing Grapes and Making Wine in Cold Climates* (Hugo, Minn.: Northern Winework, Inc., 2001), 160; Richard Leahy, "Minnesota Hybrids Make Headlines: Cold-Climate Wine Competition Proves Their Potential," *Vineyard & Winery Management* (January–February 2010): 78–83; Richard Leahy, "East Coast Watch, Meetings Full of Optimism," *Vineyard & Winery Management* (July–August 2008): 25–29; Richard Leahy, "East Coast Watch," *Vineyard & Winery Management* (March–April 2007): 41; B. I. Reisch and Stephen Luce, "Less Risky Cold Climate Grape Varieties," *Wine East*, No. 33 (May–June, 2005): 20–24; Anna Katherine Mansfield, "Winemaking in Minnesota," *Wine East*, No. 33:1 (May–June 2005): 24–29; Lisa Smiley, "La Crescent," Iowa State, Dept. of Horticulture (2008), http://viticulture.hort.iastate.edu/cultivars/cultivars.html.

La Crescent is becoming popular in the upper Midwest with 160 acres in Minnesota (2007), 19 acres in Iowa (2006), and is grown somewhat in the Hudson Valley, Nebraska, Michigan, and Ontario, Canada. See Jancis Robinson, Julia Harding, and Jose Vouillamoz, *Wine Grapes: A Complete Guide to 1,368 Vine Varieties, Including Their Origins and Flavor* (New York: Ecco/HarperCollins, 2012), 523–24.

9. Plocher and Parke, *Northern Winework*, 159–60; Double A Vineyards, *Double A Vineyards 2011/2012 Catalog*, 16; Bruce I. Reisch, Robert M. Pool, David V. Peterson, Mary-Howell Martens, and Thomas Henick-Kling, *Wine and Juice Grape Varieties for Cool Climates*, New York State Agricultural Experiment Station, Cornell University, Information Bulletin no. 233 (Geneva, N.Y.: Cornell Cooperative Extension, 1979); Mansfield, "Winemaking in Minnesota"; Lisa Smiley, "La Crosse," (2008), http://viticulture.hort.iastate.edu/cultivars/cultivars.html; *Brooks and Olmo Register*, 271. This variety is somewhat popular in the upper Midwest, which covers 33 acres in Iowa (2006), 28 acres in Nebraska (2007), 16 acres in Minnesota (2007), 10 acres in Illinois (2007), and is grown to some extent in New York and Pennsylvania. See Robinson et al., *Wine Grapes*, 524–25.

10. Plocher and Parke, *Northern Winework*, 159; *Double A Vineyards 2011/2012 Catalog*, 18; Reisch et al., *Wine and Juice Grape Varieties*, 6; Lisa Smiley, "St. Pepin," Iowa State, Dept. of Horticulture (2008), viticulture.hort.iastate.edu/cultivars; *Brooks and Olmo Register*, 285. The variety covered 17 acres in Minnesota (2007), 15 acres in Iowa (2006), and 10 acres in Illinois (2007), with some limited plantings in the Hudson Valley. See Robinson et al., *Wine Grapes*, 1011–12.

11. "Around the East, Elmer Swenson Hybrids Names in Nebraska, Quebec," *Wine East*, No. 33 (May–June, 2005): 6-7; *Double A Vineyards 2011/2012 Catalog*, 13; Matt Maniscalco, "Brianna Grape Is Midwest's New Tropical Fruit," *Midwest Wine Press* (May 3, 2012); Lisa Smiley, "Brianna," Iowa State, Dept. of Horticulture (2008), http://viticul-

ture.hort.iastate.edu/cultivars/cultivars.html; Robinson et al., *Wine Grapes*, 137.

12. Plocher and Parke, *Northern Winework*, 158–59; *Double A Vineyards 2011/2012 Catalog*, 16; Lisa Smiley, "Kay Gray," Iowa State, Dept. of Horticulture (2008), http://viticulture.hort.iastate.edu/cultivars/cultivars.html; *Brooks and Olmo Register*, 270; Robinson et al., *Wine Grapes*, 500.

13. Plocher and Parke, *Northern Winework*, 161–62; *Double A Vineyards 2011/2012 Catalog*, 16; Lisa Smiley, "Louise Swenson," Iowa State, Dept. of Horticulture (2008), http://viticulture.hort.iastate.edu/cultivars. Louise Swenson is grown on a limited basis in Vermont, Wisconsin, and Minnesota; Robinson et al., *Wine Grapes*, 554.

14. Tom Plocher and Bob Parke, authors of *Northern Winework*, worked closely with Elmer Swenson in his grape-breeding activities over the years.

15. Plocher and Parke, *Northern Winework*, 161; *Double A Vineyards 2011/2012 Catalog*, 17; Mansfield, "Winemaking in Minnesota," 20–29; Lisa Smiley, "Prairie Star," Iowa State, Dept. of Horticulture (2008), http://viticulture.hort.iastate.edu/cultivars/cultivars.html. There are 40 acres of Prairie Star in Minnesota (2007) and it is grown to some extent elsewhere in the upper Midwest in Iowa, Nebraska, and Quebec, Canada; See Robinson et al., *Wine Grapes*, 842–43.

16. Lon Ronbough, *The Grape Grower: A Guide to Organic Viticulture* (White River Junction, Vt.: Chelsea Green Publishing, 2002), 192.

17. Plocher and Parke, *Northern Winework*, 160–61; Lisa Smiley, "Swenson White," Iowa State, Dept. of Horticulture (2008), http://viticulture.hort.iastate.edu/cultivars/cultivars.html. Swenson White is not a widely planted variety commercially; See Robinson et al., *Wine Grapes*, 1021.

18. Patrick Pierquet recalls the origins of Frontenac: "As part of my grad student research work, I collected superior specimens of wild *Vitis riparia* growing along the Minnesota River in east central Minnesota. One of my best selections, Riparia #89, was the seed parent to Frontenac (the pollen parent being, of course, Landot 4511). I made this cross in 1977. The next year, I grew out and planted the seedlings from this cross, and after my departure from the University of Minnesota (1980), Frontenac was selected and named by Prof. Jim Luby and grape breeder Peter Hemstad." Correspondence with Patrick Pierquet, December 2010.

19. Frontenac is a popular variety with 116 acres in Minnesota (2007), 58 acres in Illinois (2007), 47 acres in Iowa (2006), 14 acres in Nebraska (2007), and 52 acres in Quebec, Canada (2009). It is widely planted in New York State north of the Mohawk River and in the Hudson Valley; See Robinson et al., *Wine Grapes*, 369.

20. Plocher and Parke, *Northern Winework*, 157; *Double A Vineyards*, 2011/12 Catalog, 15; Leahy, "Minnesota Hybrids Make Headlines," 78–83; Leahy, "East Coast Watch," 40–41; University of Minnesota, *Cold Hardy Wine Grapes: Expertise Results in High Quality Wine Grapes*, Brochure no. 08216; Reisch and Luce, "Less Risky Cold Climate Grape Varieties," 20–24; Mansfield, "Winemaking in Minnesota," 24–29; Lisa Smiley, "Frontenac," Iowa State, Dept. of Horticulture (2008), http://viticulture.hort.iastate.edu/cultivars/cultivars.html; *Brooks and Olmo Register*, 265; Robinson et al., *Wine Grapes*, 369–70.

21. *Double A Vineyards, 2011/2012 Catalog*, 15; Leahy, "Minnesota Hybrids Make Headlines," 78–83; Leahy, "East Coast Watch," 40–41; University of Minnesota, *Cold Hardy Wine Grapes*; Maniscalco, "Frontenac Gris Grape Moves Beyond Minnesota"; Reisch and Luce, "Less Risky Cold Climate Grape Varieties," 20–24; Mansfield, "Winemaking in Minnesota," 24–29; Lisa Smiley, "Frontenac Gris," Iowa State, Dept. of Horticulture (2008), http://viticulture.hort.iastate.edu/cultivars/cultivars.html; Robinson et al., *Wine Grapes*, 369–70.

22. *Double A Vineyards 2011/2012 Catalog*, 17; Leahy, "Minnesota Hybrids Make Headlines," 78–83; Richard Leahy, "Vineyard, East Coast Watch, Meetings Full of Optimism," *Vineyard & Winery Management* (July–August 2008), 25–29; Leahy, "East Coast Watch," 40–41; University of Minnesota, "Cold Hardy Wine Grapes"; Peter Hemstad and James Luby, "'Marquette', a New Wine Grape, Named in Minnesota," *Wine East* 33, no. 4 (2005): 7; Mansfield, "Winemaking in Minnesota," 24–29; Lisa Smiley, "Marquette," Iowa State, Dept. of Horticulture (2008), viticulture, http://hort.iastate.edu/cultivars. Even with its recent release date, there are 214 acres of Marquette in Minnesota (2007), 10 acres in Indiana (2010), six acres in Iowa (2006), and some acreage in the Hudson Valley. See Robinson et al., *Wine Grapes*, 599–600.

23. There are 26 acres of Sabrevois in Minnesota (2007) and 16 acres in Quebec, Canada (2011). See Robinson et al., *Wine Grapes*, 935–36.

24. *Double A Vineyards, 2011/2012 Catalog*, 18; Plocher and Parke, *Northern Winework*, 157; "Around the East, Elmer Swenson Hybrids Named in Nebraska, Quebec," *Wine East*, No. 33:1 (May–June, 2005), 6–7; Lisa Smiley, "Sabrevois," Iowa State, Dept. of Horticulture (2008), http://viticulture.hort.iastate.edu/cultivars/cultivars.html.

25. Plocher and Parke, *Northern Winework*, 156–57; Reisch and Luce, "Less Risky Cold Climate Grape *Varieties 2005*," *20–24; Reisch et al., Wine and Juice Grape Varieties for Cool Climates*, 6; *Double A Vineyards 2011/2012 Catalog*, 18; Mansfield, "Winemaking in Minnesota," 24–29; Lisa Smiley, "St. Croix," Iowa State, Dept. of Horticulture (2008), http://viticulture.hort.iastate.edu/cultivars/cultivars.html; *Brooks and Olmo Register*, 285. St. Croix is popular in the upper Midwest with 29 acres in Iowa (2006), 20 acres in Illinois (2007), 18 acres in Minnesota (2007), six acres in Pennsylvania (2008), 36 acres in Quebec, Canada (2011), and is becoming popular in upstate New York, particularly above the Mohawk Valley. See Robinson et al., *Wine Grapes*, 1011.

CHAPTER THIRTEEN

Central European *Vinifera* and Hybrid Grapes

WHILE CENTRAL EUROPEAN COUNTRIES such as Germany, Austria, and Hungary grow the traditional "classic" grape varieties such as Chardonnay, Cabernet Sauvignon, Merlot, and Riesling, central Europe specializes in the cultivation of many other lesser known *vinifera* grape varieties and interspecific hybrids. These lesser known varieties and the wines they produce have often been overlooked by wine writers and grape variety survey books. However, some of these varieties are covered in this book because the Hudson Valley's cool climate, like that of central Europe, is suitable for their cultivation.

Central Europe has a long history of cultivating *vinifera* grapes and developing new wine grape varieties at their distinguished horticultural research institutes and associated grape-breeding programs. These institutions, which have developed hundreds of new grape varieties, include the Geilweilerhof Institute for Grape Breeding at Siebeldingen, Rhineland-Palatinate region of western Germany; the Geisenheim Grape Breeding Institute in Geisenheim, Rheingau, Germany; the Bavarian State Institute for Viticulture and Horticulture in Veitshöchheim, Würzburg, Bavaria; the State College and Research Institute for Viticultural and Horticulture at Weinsberg, in the southern German state of Baden-Württemberg; the State Institute of Grapevine Breeding at Alzey in Rhineland-Pfalz; the Federal Institute for Viticulture and Pomology at Klosterneuburg in Niederösterreich, Austria; and the Weinbauinstitut Kecskemét, Kutató Intézete in Katona, Hungary. At these institutions, groundbreaking work has been conducted since the 1860s by viticulturalists and grape breeders such as Dr. Friedrich Zweigelt (1888–1964), Dr. Georg Scheu (1879–1949), the Swiss botanist Hermann Müller (1850–1927), and Peter Morio (1887–1960) to hybridize new grape varieties of pure *vinifera* heritage and interspecific hybrids that incorporated American and Asian grape species to increase fruit quality, cold hardiness, productivity, and fungal disease resistance.

Throughout Germany's turbulent history from the 1870s to today—from the German Empire, Weimer Germany, and Nazi Germany to the Federal Republic of Germany (West Germany) and a now-unified Germany—this country has consistently supported extensive horticultural research programs to develop many scores of new grape varieties, particularly whites, to replicate Riesling, and reds to replicate a darker version of Pinot Noir. The goal of these breeding programs has been to create hybrids that increased yields, possessed better fungal disease resistance, had cold hardiness to expand the range of cultivation sites, and that were easier to cultivate. Some of the grapes developed by these breeding programs include Müller-Thurgau, Morio-Muskat, Faber, Huxelrebe, Heroldrebe, Helfensteiner, Ortega, Elbling, Optima, Reichensteiner, Perle, Nobling, Kanzler, Wurzer, Comtessa, Noblessa, Rieslaner, Freisamer, Schonburger, Albalonga, Findling, Septimer, Domina, Mariensteiner, Rotberger (not to be confused with Zweigelt), Gutenborner, and Forta. The point is that by developing new and better hybrids, those interested in producing better wines or wines that require less inputs, both economically and environmentally, will advance the practice of viticulture and enology.

The cultivation of grapes should not be a static pursuit that allows only for the propagation and growing of a few select "classic" grape varieties such as Chardonnay, Cabernet Sauvignon, Merlot, and Riesling. Many of the cool climate grape varieties listed on the following pages should be suitable for cultivation in the warmer parts of the Hudson Valley.

White European Varieties

BIANCA (EC 40)

Bianca is a chance seedling of Villard Blanc (S.V.12-375) × Bouvier. This variety, also known as Egri Csillagok 40, is genetically of more than one-half *vinifera* heritage and was developed by Jozsef Csizmazia and Lazlo Bereznai in Hungary in 1963. The breeder is the Weinbauinstitut Kecskemét, Kutató Intézete, Katona, in the Eger region of northern Hungary. Bianca is becoming a popular and more widely grown variety in Hungary that covered approximately 2,185 acres in 2001 and increased to 3,163 acres in 2011. Most of Bianca is grown in the Kunsag area in the center of Hungary (2,810 acres). There are 6,750 acres of Bianca in the Krasnodar Krai region of Russia (2009). Bianca is grown in Minnesota and Denmark, so it should be sufficiently winter hardy for the Hudson Valley.

PARENTAGE
berlandieri, labrusca, lincecumii, rupestris, vinifera

HARVEST DATE
Early to early mid-season

B A- B A A

Bianca buds out by early to mid-season and is reported to have good fungal disease resistance. It tolerates most soil types, but prefers moist, well-drained soils. The variety is a vigorous grower with high yields, but is sensitive to droughty conditions. Further, it has a sensitive flower set in the spring. Bianca is medium hardy against winter injury in the Hudson Valley.

It has medium- to large-sized clusters of small- to medium-sized berries. This short-season grape ripens by early to early mid-season with sugars around 19°–20° Brix and suitable acidity for wine production. The variety produces loosely formed and straggly clusters of medium size. The berries turn yellow-green when ripe.

The wines are described as neutral, with body and substance that can give a boost to lighter-bodied whites. The few Biancas that I have had were quality wines of a light honey color and not "hybrid-like" in taste. The body was solid and almost viscous. Its fruit flavors are similar to a Sauvignon Blanc or Vidal Blanc in having some hazelnuts, bananas, mead, and vanilla with hay or straw-like overtones. This variety should be further explored for cultivation in the Hudson Valley.[1]

GRÜNER VELTLINER

Grüner Veltliner is a *vinifera* variety that is native to Austria and is the most widely planted grape there. It is also grown in nearby Slovakia, Hungary, and the Czech Republic.[2] The variety is productive, healthy, and adaptable to many different soil types. While it tolerates over-cropping to produce everyday table wines, if the crop is properly managed, it can make quality, full-bodied wines. It has small greenish-yellow berries. The variety is about as winter hardy as Riesling, with medium vigor and an upward growth habit. Grüner Veltliner is as susceptible to fungal diseases as any other *vinifera* grape variety, except that it is less susceptible to botrytis. It is not sensitive to sulfur treatments, so it can be used to combat fungal diseases. It has a late harvest date.

PARENTAGE
vinifera

HARVEST DATE
Late

C	B-	C	A	A+

Grüner Veltliner is versatile in the cellar. It can produce ordinary pale green dry and clean fruity table wines that are spicy, with almost musky peach aromas. In addition, these aromatic Vidal Blanc–like wines can be flinty with fruit flavors of green apples, lemons, pink grapefruit, and citrus flavors with a tangy acid profile and medium body if they are consumed when young. It can have an aroma that is reminiscent of an Alsatian white. They are also capable of becoming big, full-bodied (but minerally) Chardonnay-like wines that can handle barrel aging and sparkling wines.

The wines have been described as complex, full of exotic tropical fruits, white pepper, and lentils that go well with Asian food. The variety is very food-friendly and is now beginning to be grown on a small scale in New York, New Jersey, Oregon, and Maryland.[3]

PINOT BLANC

Pinot Blanc is a mutation of Pinot Gris, which, in turn, is a mutation of Pinot Noir. It originated in northern France, but is now widely grown in central Europe, where it goes by the names Weissburgunder or Clevner in Germany and Austria, Fehér burgundi in Hungary, Rouci Bile in the Czech Republic, and Pinot Bianco in Italy.[4]

PARENTAGE
vinifera

HARVEST DATE
Mid-season

C	B-	C	B+	A+

The vine is nearly as winter hardy as Chardonnay, with compact small-medium to medium-sized conic clusters that have small- to medium-sized berries. It has often been compared to Chardonnay in the appearance of the vine and cluster, their proximity in harvest times, and the fact that both are tolerant of many different soil types. A further comparison is that Pinot Blanc tends to be more productive than Chardonnay, but has lower sugar levels.

This productive variety has only moderate vigor that buds relatively early and ripens by mid-season. It has better than average resistance to fungal diseases for a *vinifera*, but can get bunch rot due to the compactness of its cluster. To promote vigor in the vine, it does well on relatively deep and damp soils and should be planted on either SO4 or 5C rootstock in deep soils, or otherwise planted on 125AA or 5BB. The pruning style should be cordon training.

The wines made from Pinot Blanc tend to be much lighter with a muted nose than Chardonnay. Also, because Pinot Blanc is capable of having firm acid levels, it is used in sparkling wines. The wines have a light green color. As a young wine, they tend to have soft fruits of pears and melons. It is a soft, approachable, clean wine, with a muted nose. While Pinot Blanc is not generally made to age, unlike Chardonnay, as it gets older, the wines become very French tasting, with lots of melon, green apples, flint, lemons, and sometimes almonds and vanilla. They are not complicated wines, but have a great structure that complements their fruit. While Pinot Blancs are underrated, I like them a great deal.[5]

OTHER VARIETIES

BACCHUS (GF 33-29-133)

((Sylvaner × Riesling) × Müller-Thurgau)
vinifera

This variety, bred in 1933 at the Geiweilerhof Institute, was named in honor of Bacchus, the Roman god of wine. It was authorized for cultivation in 1972. It has a compact to very compact medium-sized conical cluster with medium-sized yellow-green berries. This highly vigorous and productive variety is moderately tender to cold damage in the Hudson Valley, about as hardy as Riesling or Gewürztraminer. The variety buds out relatively late, but can be susceptible to late spring frosts. The fruit ripens by early late to late season, is susceptible to botrytis rot, and is easily sunburned. Bacchus has about the same resistance to other fungal diseases as other *vinifera* varieties.

This is a widely planted grape in Rheinhessen, and somewhat grown in Franken, Nahe, Pfalz, and Baden, Germany, where it is planted on marginal or low lands not suitable for Riesling or Silvaner.[6] This is because it is not particular as to the soils that it grows on and is generally an adaptable grape variety.

A large portion of Bacchus production traditionally goes toward making ordinary Liebfraumilch wines, which are fragrant and spicy semisweet white wines. The grape has high sugars and lots of body, but is low in acid. It is a very big, aromatic, forward fruity and floral wine with almost Muscat-like flavors, so it is used a lot in blends. If picked too early, it does not exhibit its big fruit flavors.[7] Sometimes it can be grassy in a manner similar to Sauvignon Blanc, but it is much fruitier with hints of elderflowers.

EHRENFELSER (GM 9-93)

(Riesling x Knipperlé)[8]
vinifera

This variety was developed at the Geisenheim Grape Breeding Institute at Rheingau, Germany, in 1929. It is about as cold tender as Gewürztraminer or slightly more so. Its name comes from the ruined Schloss Ehrenfels on the Rhine, at Rudesheim in the Rheingau region.

The well-filled to compact clusters are small and cylindrical with small-medium white-green berries. Due to its compact cluster, it is susceptible to botrytis rot, otherwise it has similar susceptibility to fungal diseases like other *vinifera* varieties. The moderately vigorous vine has fruit that ripens in late season, at about the same time as Riesling or slightly earlier. Viticulturally, it produces slightly more than Riesling, is less finicky about where it grows, and ripens slightly earlier and more fully under less favorable conditions. Due to its adaptability and quality of wine, plantings of Ehrenfelser are common, particularly in the northernmost reaches of Germany's wine-producing areas.[9]

In New York, it makes pleasantly fruity wines that have full body and the fruit of its parents Riesling and Sylvaner, with a long, clean finish. In Germany, it is reported to make a slightly less intense Riesling-like wine that has less acidity, but is still quite good. This variety merits further evaluation.[10]

KERNER (WE S.25-30)

Trollinger (also known as Black Hamburg) × Riesling
vinifera

This variety is a cross-breed of the commonly grown and productive central European red variety of Trollinger with the classic white variety Riesling and one of the most widely grown hybrids developed in Germany. Black Hamburg was used extensively in the breeding programs of some Hudson Valley breeders, especially James H. Ricketts. Kerner was bred by August Herold in 1929 at Weinsberg, Württemberg's State College and Research Institute for Viticulture and Horticulture. The variety is named after Justinius Kerner, who was a local nineteenth-century poet and writer of drinking songs and who was reputedly a friend of Herold's. He was also a physician who recommended that his patients drink a glass of wine as the best natural medicine. Kerner received protection as a separate breed and was released for general cultivation in 1969. It is widely planted in Palatinate, Rheinhessen, the Mosel, and Württemberg, Germany, Austria, Switzerland, England, and northern Italy.

The grape is popular for many reasons. For growers, it is not

Other varieties, cont'd.

finicky on the soils it grows on, is very productive, buds out late to avoid late spring frosts, and ripens reliably with high sugars. The vine is as winter hardy as Gewürztraminer or Pinot Noir. It is a vigorous growing variety with an upward growth habit that ripens late in the season. It has medium-sized berries on a medium-sized, compact to well-filled cluster whose color is green-white. The fungal disease resistance is about the same or a bit better than other white *vinifera* varieties.[11]

For winemakers, Kerner has many benefits in the cellar as it can make either dry, semidry, or sweet, quality Riesling-like wines. Its high sugars and total acidity facilitate the ease with which it can make traditional Rhine wines that are light to golden yellow in color and are rich, but have a fresh acidity and fruity character. Its flavors include apple, pear, and citrus profiles that can be layered with light peach and Muscat flavors. The wines have lower acidity and are slightly coarser than Riesling. It can stand on its own or be used in a blend. Kerner can also age well when making sweet dessert wines.[12]

REFORM (RF 48)

(Perle Csaba × Aurora (S. 5279))
lincecumii, rupestris, vinifera

This white interspecific hybrid variety, bred at the Research Institute for Viticulture and Enology, in Kecskemét, Hungary, is about as winter hardy as Seyval Blanc, with medium-sized, semi-loose, conical clusters and medium-sized berries. It is reported to be a reliable producer with disease resistance at least as good as Aurora, one of its parents. The wine is reported to be good with elements of ripe oranges when fully ripe and does not have *labrusca* flavors. Reform is useful when utilized as a neutral base wine when building a blended white. The variety ripens early by late August to Labor Day in the Hudson Valley, with sugars of 20°–22° Brix.[13]

SCHEUREBE (ALZEY S.88)

(Riesling × unknown variety (open-pollinated))[14]
vinifera

This grape was bred by Dr. Georg Scheu[15] in 1916 at the State Institute of Grapevine Breeding at Alzey, Rheinland-Pfalz, in southwestern Germany. At one point, it was thought that the seed parent was Sylvaner, but DNA testing has now refuted that assumption. It was named in honor of Dr. Scheu in 1945, and released under that name sometime between 1956 and 1958. The term *rebe* simply meaning "vine" in German.

Scheurebe is a widely grown white grape in Rheinhessen and Rheinpfalz, Germany, Austria, and Switzerland,[16] however it was bred for the sandy soils of Rheinhessen. This small-medium-clustered variety is compact and conical. The medium-sized berries are very light green to white when ripe.

Scheurebe is one of the more cold tender of the varieties listed in this chapter, less than Gewürztraminer in cold resistance, but hardier than Pinot Noir. The variety is moderately vigorous and grows like Riesling. It buds out late, so avoids late spring frosts and ripens by early late to late season. The vine is more productive than Riesling and has higher sugars, but lower acids. Some have cautioned that it may need to be harvested earlier than is optimum to avoid botrytis rot in some years. However, if picked too early, it does not exhibit its nice Riesling-like qualities.

This superior wine is similar to Riesling, especially if made as an Auslese or late harvest-style wine. Scheurebe is versatile in the cellar since it is capable of producing quality dry to very sweet dessert wines. It is an aromatic and floral wine with refreshing acidity that has herbal hints and intense flavors of pink grapefruit and honeysuckle. Some suggest that it has black currant and herbal flavors. It is capable of being aged as other quality sweet German dessert wines.[17]

SIEGERREBE (ALZEY 7957)

(Madeleine Angevine × Savagnin Rose (a version of Gewürztraminer))
vinifera

Siegerrebe was developed by Dr. Georg Scheu's grape-breeding program in 1929 at the State Institute for Grapevine Breeding in Alzey, but not officially released until 1958. It is relatively cold tender for a *vinifera* variety, about the same or a bit less tender than Gewürztraminer. The medium-large loose clusters are conical and generally have one moderately sized wing. The medium-large berries have a pink-green to amber color, similar to fully ripe Iona or

Other varieties, cont'd.

Jefferson grapes or partially ripe Delaware grapes. Siegerrebe has been reported to bud out early to late.

The vines are productive, but vigor is low to moderate, so pruning must be done in such a way that the vine does not overbear. Cluster thinning may be needed to balance the crop to the vine's capacity to produce a quality crop. Siegerrebe is susceptible to coulture, so its yields are light. The grape ripens early in the season, around Labor Day or slightly afterwards, so netting is required to minimize bird and bee damage.

The fruit and wine are reported to be reminiscent of one of its reputed parents, Gewürztraminer, in that the wines have spicy and Muscat-like fruit, with hues of yellow-green to golden yellow. The flavors of Siegerrebe are very rich and concentrated, with sugars that are very high (about 30° Brix), and its acid is relatively low. In the Rhine, it is used sparingly as a flavor enhancer as it adds flavor, richness, and sugars to Riesling blends, but is used sparingly due to its concentration of flavors and sugar. The variety may be considered for white blends because of its high sugar levels and the body that it can give them.[18]

RED EUROPEAN VARIETIES

DORNFELDER (WE S.341)

PARENTAGE
vinifera

HARVEST DATE
Mid-season

C C C+ A A

Dornfelder is a relative of Pinot Noir. Genetically, it is a cross-breed of Helfensteiner (Fruhburgunder × Trollinger (also known as Black Hamburg)) × Heroldrebe (Blauer Portugieser × Lemberger). This red variety was bred by August Herold (1902–73) in 1955 or 1956 at the State College and Research Institute for Viticulture and Horticulture at Weinsberg, in the southern German state of Baden-Württemberg (Weinsberg S.341), and is cultivated very extensively in Rheinpfalz and Rheinhessen.[19] The grape is named after Immanuel August Ludwig Dornfeld (1796–1869), one of the founders the Württemberg Viticultural School in the 1860s.

In Germany, the grape is popular because it is good in the field and has a deeper red color than the lightly colored Pinot Noir clones grown locally. It is vigorous, does not rot easily, is easy to grow, its canopy is robust, it ripens fairly early, and it is productive to very productive. In Germany, it is the variety that is replacing Blauer Portugieser and various Pinot Noir clones to make more robust and deeply colored red wines.

In New York, it is a bit less winter hardy than either Riesling or Cabernet Franc. The variety has medium vigor and an upright growth habit. The bud break is late-early to mid-season, so it may have some late spring frost damage issues, but it ripens early to by mid-season. As with most *viniferas*, it is susceptible to most fungal diseases, except for botrytis and bunch rot, to which it is only somewhat susceptible. It is not sensitive to sulfur treatments, so sulfur can be used in a fungus treatment program. The medium-large, thick-skinned berries are blue-black in color on clusters that are large and loose to well filled. It grows best on well-drained, fertile soils and not dry or shallow soils since it is a heavy feeder. Dornfelder should be cordon pruned with spurs across the wire.

The wines made from Dornfelder are interesting, with a floral and fruity nose, and fruit flavors of cherries, red berries, plums, and darker elderberry elements. The wine has a medium deep-red color that can have either a somewhat full or soft body, soft tannins, with some complexity. It has good sugars and acid balance with the ability to age in wood. Due to its solid acid backbone and floral/berry nose, it can be made successfully into a rosé. To produce quality wines, this productive variety's yields must be controlled.

Dornfelder should be more closely evaluated for cultivation in the Hudson Valley as a sturdier substitute for Pinot Noir.[20]

LEMBERGER

PARENTAGE
vinifera

HARVEST DATE
Late

C	B-	B	A	A

Lemberger is one of the signature red wine grapes of central Europe due to the scope of its production and range throughout the region. It is called Lemberger or Limberger in Germany, Blaufränkish in Austria, and Kekfrankos in Hungary, Crna Moravka in southeastern Europe, and Gamé in Bulgaria. It is grown, to some extent, in New York and Washington State, where it is called Lemberger, so we call it Lemberger for our purposes.[21] It is a *vinifera* variety that probably originated in Austria or in the lower portions of the Danube River, which may be a cross-breed of Gouais Blanc × unknown central European red variety. Gouais Blanc is one of the parents of Chardonnay.

Lemberger can be grown only on the warmest sites in the Hudson Valley, since it buds out early and ripens late-mid-season to late. Sheltered sites that protect it from late spring frosts and that can help it to ripen sufficiently are best for producing a quality crop. The variety is a vigorous and productive grower, with an upright growth habit that tolerates many different soil types, but is noted to do especially well in deep, fertile, loam soils. In some years, cluster thinning may be needed to increase crop quality and to ensure that its wood hardens off for the winter. Overall, it is not a difficult grape to grow and consistently has sugar levels of between 21° and 23° Brix.

The large and long clusters have medium-sized blue-black berries. It is one of the more winter hardy red *vinifera* varieties, along with Zweigelt or Riesling. It has better fungal disease resistance than most other red *vinifera* grapes, and very good resistance to botrytis and bunch rots. It is not sensitive to sulfur treatments.

Lemberger is attractive in the cellar because of its versatility. Because of its relatively high total acidity and bright purple-blue colors, it is suitable to make a light and fruity rosé. In addition, it can make a relatively medium-bodied but complicated and deeply colored red, or used in blends in a manner where it does not dominate the blend.

As a medium-bodied red, its aroma and flavor profile is of red cherries, blackberries, cooked mulberries, chocolate, smoke, and spice, with an underlying minerally taste and components of black pepper, herbs, earth, gaminess, and soft to rustic tannins. These wines are a light red color with berry fruit, red cherries, anise, strawberry jam, violets and black raspberry. The wines can be very soft and aromatic in a manner similar to Pinot Noir or Cascade, although its high total acidity can overshadow its soft tannins. It is similar in many ways to the color, nose, taste, and finish of a hearty Chancellor, but does not have Chancellor's sweet finish. Barrel aging helps these wines along quite a bit by integrating its varietal tendency toward high total acidity.[22]

ZWEIGELT (KLOSTERNEUBURG 71)

PARENTAGE
vinifera

HARVEST DATE
Late mid-season

C+	B-	B	A	A+

Zweigelt was bred by Dr. Friedrich Zweigelt in 1922 at the Federal Institute for Viticulture and Pomology at Klosterneuburg, Austria.[23] The original cross was known as Klosterneuberg 71, and the variety was initially named Rotburger. However, due to the confusion that this caused with the similarly named grape Rotberger, in 1975, the influential Austrian winemaker Lenz Moser renamed the grape Zweigelt in honor of its originator. Today, Zweigelt is one of the most widely grown red grape varieties in Austria and is also grown in Canada, New York, and Washington State.[24] Zweigelt is a cross of St. Laurent × Lemberger. St. Laurent is a highly aromatic dark-skinned variety that comes from the same family as Pinot Noir and Lemberger, a medium-bodied red with cherry and blackberry flavors described above. Zweigelt ripens earlier than Lemberger and buds out later than St. Laurent, making it a superior grape for growers.

Overall, the grape is healthy in the field, ripens by late mid-season, and has good fungal disease resistance for a red *vinifera* variety with no sensitivity to sulfur. Zweigelt is probably one of the most winter hardy of the red *vinifera* grapes, has good vigor, and brings in yields that are so high that the crop needs to be controlled to increase wine quality. Because of its high vegetative vigor and yields, a solid canopy management and cluster-thinning program may be needed to produce high-quality wine. The vine grows in a wide range of soils and buds out relatively late, so it can avoid late spring frosts.

The well-filled to compact cylindrical cluster is medium in size with prominent wings. Its small- to medium-sized berries are a dark blue-black color.

The style of wines made from Zweigelt seems to have settled on medium-light to medium-bodied reds with soft tannins that have some aging potential. The overarching theme is that they are medium violet-red to ruby-red wines that sometimes have deeper colors with lots of spice, herbal flavors, and black pepper. Further, the fruit profile, which competes with the spice, includes flavors of red cherry and raspberry at its core, with ancillary flavors of plummy fruits and earth. The wines benefit greatly by oak aging. Zweigelt produces wines that tend to be more like St. Laurent rather than Lemberger. They can be a solid component to anchor a red blend.[25]

OTHER RED VARIETIES

REGENT (GF 67-198-3)

Diana[26] *(Silvaner Gruen × Müller Thurgau) × Chambourcin (J.S.26-205).*
berlandieri, labrusca, lincecumii, riparia, rupestris, vinifera

Two of Regent's grandparents are the diverse grape varieties Chancellor (red) and Müller Thurgau (white). In addition, one of its ancestors is Gouais Blanc, one of the parents of Chardonnay. The variety was hybridized in 1967 by Professor Gerhardt Alleweldt at the Geilweilerhof Institute. Since its introduction in 1996, its acreage has been rapidly expanding in Germany.[27] Regent was named after the famed diamond, *Le Régent*, which was found in India, but brought back to Europe in the 1770s to become part of the French crown jewels. The diamond is now housed in the French Royal Treasury at the Louvre Museum in Paris.

The variety is blue-black in color, has good resistance to fungal diseases (as do most French-American hybrids), and is not sensitive to sulfur treatments. The cluster is a bit smaller than Chambourcin, but is more productive with a berry size that is large and thick-skinned like Chambourcin. It buds out rather late, similar to Cayuga White and Chambourcin. Regent is cold hardy for most of the Hudson River Valley, has an upright growth habit, is moderately vigorous, and ripens by mid-season with sugars around 22° Brix.

The wines are deeply colored. It is full bodied with an earthy nose with spicy hints and dark berry aromas and flavors of black currants, red currants, and cherries. The tannin levels are good, not unlike a soft red Bordeaux or Rhône. The wines are integrated from nose to finish. It has moderate total acidity, but firm tannins, so barrel aging helps to round off the wines.[28]

RONDO (GM 6494-5)

(Saperavi Severnyi (Früher Malingre × V. amurensis) × St. Laurent)
amurensis, vinifera

One of Rondo's parents, St. Laurent, is a deeply colored red wine grape widely cultivated in Austria and Germany. Rondo was bred in 1964 by Professor V. Kraus

Other red varieties, cont'd.

in what was then Czechoslovakia. He offered this variety to Dr. Helmut Becker (1927–90) at the Geisenheim Institute, and it was authorized for cultivation in Germany in 1999.

The compact cluster is medium-large with medium- to above- average-sized black-blue berries. This is a vigorous and productive variety that can over-crop. It has a somewhat trailing growth habit. Rondo buds out early at about the same time as Léon Millot (Millot Rouge) and Maréchal Foch, and ripens early like Foch with sugars of between 21° and 23° Brix. The vine is grown in Minnesota and Denmark so, like Foch and Léon Millot, it is winter hardy in the Hudson Valley.[29] The only caution is that Rondo can lose its dormancy during midwinter thaws. The fungal disease resistance is better than that of most *vinifera* grapes and about the same as French-American hybrids, except for powdery mildew, to which it is susceptible. Rondo is partially susceptible to downy mildew and bunch rot.

The wine is a deep ruby color without hybrid blue hues. The wine is reported to be a good *vinifera*-like wine with intense fruit, raspberry flavors, and wild berries, which is earthy with nice tannins. However, it can be a bit one-dimensional, so it may need to be blended and aged in wood to give the wine some interest. It has been compared to Zweigelt in flavor profile and tannin structure.[30]

The next and final chapter covers the traditional classic European *vinifera* varieties: Chardonnay, Gewürztraminer, Riesling, Cabernet Franc, Gamay Noir, and Pinot Noir. All of these varieties originated in France, except for Gewürztraminer and Riesling, which originated in the northern border areas of France that is just inside Germany, in the Rhine Valley. While these varieties may have come from France and Germany, they are now commonly grown throughout much of Europe, North America, South America, and Australia, with outposts in South Africa.

NOTES

1. Tom Plocher and Bob Parke, *Northern Wineworks: Growing Grapes and Making Wine in Cold Climates* (Hugo, Minn.: Northern Winework, 2001), 19, 145, 162. See charts on grape variety genealogy by Carl Camper, Chateau Stripmine Experimental Vineyard, http://www.chateaus-tripmine.info/Parentage.htm; www.winegrowers.info/varieties/Vine_varieties/Bianca; Joe Masabni and Dwight Wolfe, *Eastern European Grape Cultivars*, Cooperative Extension Service, University of Kentucky, College of Agriculture, Horticultural Department, HortFact—3007 (2007); *Final Report, Study on the Use of the Varieties of Interspecific Vines*, 7/16/03 Contract #AGR 30881 of December 30, 2002, Phytowelt GmbH, Germany, 10, 14; Eurostat, the Statistical Office of the European Union, www.statistischedaten. de.data. 1999; Jancis Robinson, Julia Harding, and Jose Vouillamoz. *Wine Grapes: A Complete Guide to 1,368 Vine Varieties, Including Their Origins and Flavor.* (New York: Ecco/HarperCollins, 2012), 101.

2. There are 42,092 acres of Grüner Veltliner grown in Austria, almost one-third of its total vineyard area. It is also grown in the Czech Republic (4,230 acres), Slovakia (9,400 acres), and Hungary (3,556 acres). See Robinson et al., *Wine Grapes*, 449–50.

3. Jancis Robinson, *Vines, Grapes, and Wines* (New York: Knopf, 1986), 255–56; Double A Vineyards, *Double A Vineyards 2012/2013 Catalog*, 22, 44–45; www.jancisrobinson.com/articles/20070301_3; www.winepros.org/wine101/grape_profiles/gruner; Robinson et al., *Wine Grapes*, 449–50.

4. Pinot Blanc is very widely grown in Europe, with 3,190 acres in France, mostly Alsace (2009), 12,667 acres in Italy (2000), 9,220 acres in Germany (2008), 4,930 acres in Austria (2010), 494 acres in Hungary, 1,334 acres in Slovenia (2009), 2,000 acres in the Czech Republic (2009), and 369 acres in British Columbia. See Robinson et al., *Wine Grapes*, 821–23.

5. Robinson, *Vines, Grapes, and Wines*, 160–63; Robert M. Pool et al., *Growing Vitis Vinifera Grapes in New York State, I: Performance of New and Interesting Varieties* (Geneva, N.Y.: New York State Agricultural Experiment Station,

Cornell University, 1990), 25; Bruce I. Reisch, Robert M. Pool, David V. Peterson, Mary-Howell Martens, and Thomas Henick-Kling, *Wine and Juice Grape Varieties for Cool Climates*. New York State Agricultural Experiment Station, Cornell University, Information Bulletin no. 233. (Geneva, N.Y.: Cornell Cooperative Extension, 1979), 10,13; www.ars-grin.gov/cgi-bin; Robinson et al., *Wine Grapes*, 821–23.

6. This is a widely grown grape in Germany and England. Currently, there are 4,980 acres of Bacchus grown in Germany, 321 acres grown in England (2009), and 62 acres in Japan (2009). However, worldwide acreage has been substantially reduced from a high of 8,650 acres in 1990. See Robinson et al., *Wine Grapes*, 77.

7. Robinson, *Vines, Grapes, and Wines*, 249–50; Pool et al., *Growing Vitis Vinifera Grapes in New York State*, 1; Robinson et al., *Wine Grapes*, 77.

8. The 1990 Cornell circular booklet, *Growing Vitis Vinifera Grapes* in New York State by Pool et al. (p. 8) maintains that Ehrenfelser is a hybrid of Riesling x Sylvaner. However, the Vitis International Variety Catalogue (VIVC) database maintains that Ehrenfelser is a hybrid of Riesling Weiss x Knipperlé. Recent DNA profiling in Montpelier, France, has shown that Silvaner is not one of its parents, as previously believed, and in fact the second parent is Knipperlé, a sister grape to Chardonnay. See VIVC database, www.vivc.de; Nancy Sweet, "Ehrenfelser," *FPS Grape Program Newsletter*, October 2010 (Davis, Calif: Foundation Plant Service, UC Davis, 2010), 4.

9. There are 225 acres of Ehrenfelser grown in Germany, mostly in Rheinhessen and Pfalz regions, but this acreage is gradually declining due to the increasing popularity of Kerner. There are 74 acres grown in British Columbia, with some cultivation in California and New York.

10. Robinson, *Vines, Grapes, and Wines*, 252; Pool et al., *Growing Vitis Vinifera Grapes in New York State*, 8; Robinson et al., *Wine Grapes*, 323–24.

11. Currently, there are 9,170 acres of Kerner grown in Germany, down substantially from 12,350 acres grown in 2003. There are also 882 acres of Kerner in the northern region of Hokkaido Island, Japan, and 33 acres in British Columbia. See Robinson et al., *Wine Grapes*, 502–3.

12. Robinson, *Vines, Grapes, and Wines*, 249; *Double A Vineyards 2012/2013 Catalog*, 22, 44–45; Robinson et al., *Wine Grapes*, 502–3.

13. Plocher and Parke, *Northern Wineworks*, 19, 21, 164; Tom Plocher, "Tasting Grapes and Wines at Geilweilerhof, Notes From the North," Minnesota Grape Growers Association, *Quarterly Newsletter Winter* (1999).

14. Scheurebe was originally believed to be a crossing of Silvaner and Riesling, but DNA profiling in the late 1990s revealed that Scheurebe is probably a crossing of an unknown wild grape and Riesling. The VIVC database, however, indicates that markers confirm the pedigree as Riesling x Bukettrebe. See VIVC database, www.vivc.de; Wines of Germany: www.germanwineusa.com/press-trade/scheurebe.html.

15. Dr. Georg Scheu (1879–1949) was a pioneer in German viticulture. Among the grape varieties that he bred, in addition to Scheurebe, were the varieties Huxelrebe, Kanzler, Septimer, and Siegerrebe. Dr. Scheu was the founding director of the Alzey Research Institute, in the Rhineland-Pfalz region of southwestern Germany from 1909 to 1947. The Alzey Research Institute closed in the late 1990s and incorporated into the Oppenheim State College and Research Institute in Oppenheim, though this later closed in 2003. The site was sold in 2006, and is currently occupied by a winery, Weingut Trinkbare Landschaften.

16. Scheurebe covers 4,132 acres in Germany (2011), mostly in Rheinhessen and Pfalz, but this acreage is in decline, falling from 5,436 acres as recently as 2003. A similar decline has occurred in Austria, where it is grown in Neusiedler See district near the Hungarian border. There were 1,263 acres there in 2007. There are 14 acres in British Columbia. See Robinson et al., *Wine Grapes*, 969–70.

17. Robinson, *Vines, Grapes, and Vines*, 250–51; Pool et al., *Growing Vitis Vinifera Grapes*, 36; Robinson et al., *Wine Grapes*, 969–70.

18. Robinson, *Vines, Grapes, and Wines*, 253; Pool et al., *Growing Vitis Vinifera Grapes*, 37. There are 255 acres of Siegerrebe in Germany, mostly in Rheinhessen and Pfalz, 38 acres in British Columbia (2011), and 25 acres in England. See Robinson et al., *Wine Grapes*, 997–98.

19. Officially authorized for cultivation in Germany in 1980, this variety rapidly increased in acreage to its now more stable acreage of 20,020 acres, mostly in Rheinhessen and Pfalz. There are 40 acres in England (2011), and it is grown to some extent in New York, Pennsylvania, and Japan. See Robinson et al., *Wine Grapes*, 307–8.

20. Robinson, *Vines, Grapes, and Wines*, 219–20; *Double A Vineyards 2012/2013 Catalog*, 22, 42–43; Wines of Germany, www.germanwineusa.com/german-wine-101/dornfelder; DVIT 2705 via USDA Agricultural Research Service on: www.ars-grin.gov/cgi-bin/npgs/acc/display.pl?1564721; Robinson et al., *Wine Grapes*, 307–8.

21. The VIVC database lists this variety with the primary name of Blaufränkisch, as Austria is the acknowledged country of origin. Lemberger is so widely grown in central Europe that the VIVC lists 113 synonyms for this variety. See VIVC database, www.vivc.de. There are 4,272 acres in Germany (mostly Württemberg (2011), 8,250 acres in Austria (2008), 19,770 acres in Hungary (2011), 4,305 acres in

Slovakia (2011), and 2,175 acres in Croatia (2009). In the United States, 73 acres of Lemberger are grown in Washington State (2011), and it is grown somewhat in New York; See Robinson et al., *Wine Grapes*, 116–18.

22. Robinson, *Vines, Grapes, and Wines*, 220–21; *Double A Vineyards 2012/2013 Catalog*, 23, 44–45; Lisa Smiley, "Lemberger," Iowa State University (2008), 1-3; Wines of Germany, www.germanwineusa.com/german-wine-101/lemberger.html; DVIT 797 via USDA Agricultural Research Service on: www.ars-grin.gov/cgi-bin/npgs/acc/display.pl?1008926; Blue Danube Wine Company, www.bluedanubewine.com/resources/grape_varietals/blaufrankisch; Robinson et al., *Wine Grapes*, 116–18.

23. Dr. Friedrich Zweigelt was born in Hitzendorf, near the city of Graz, in Styria, Austria. He completed his undergraduate and graduate studies at the University of Graz, where he obtained a doctorate in natural sciences in 1911. During this time, he was an assistant at the Plant Physiology Institute at Graz, after which he obtained his advanced degree in Agricultural Plant Health at the Vienna University of Agricultural Science. In 1912, Zweigelt joined the staff at the world's first and oldest viticultural and oenology school in Europe, the Federal Institute of Viticulture and Pomology at Klosterneuburg, Austria. In 1921, along with Paul Steingruber and Franz Voboril, he help to found the Federal Grape Breeding Station at the Klosterneuburg Institute. Between 1921 and 1938, he held various positions at the Klosterneuburg Institute, and ultimately became its director in 1938, after the National Socialists Party took over Austria, and remained its director until 1945, when he was forced to retire at the end of World War II. After leaving the Institute, Zweigelt remained active as a consultant on viticultural and entomology issues and continued to write articles on these subjects. He died in Graz in 1964. See Robert Bedford, "*What is Zweigelt?*" and "*Dr. Fritz Zweigelt,*" at The Zweigelt Project, http://zweigeltproject.com.

24. There are 16,089 acres of Zweigelt grown in Austria, 2,125 acres in the Czech Republic (2011), 247 acres in Germany, and it is grown to some extent in New York, Washington State, and Japan (571 acres). See Robinson et al., *Wine Grapes*, 1177.

25. *Double A Vineyards 2012/2013 Catalog*, 26, 44–45; The Zweigelt Project: http://zweigeltproject.com; http://wineinstituteofnewengland.com/zweigelt-zwei-not; Robinson, *Vines, Grapes, and Wines*, 221; Austrian Wine, www.austrianwine.com/our-wine/grape-varieties/red-wine/zweigelt-blauer-zweigelt-rotburger/; Blue Danube Wine Company, www.bluedanubewine.com/resources/grape_varietals/zweigelt; Palate Press, palatepress.com/2012/06/wine/try-zweigelt-this-summer/; Robinson et al., *Wine Grapes*, 1177.

26. Not to be confused with the Hudson Valley–developed *labrusca* variety called by the same name, Diana.

27. Regent covers 5,340 acres in Germany (2011), mostly in Rheinhessen, Pfalz, and Baden, 99 acres in northeast Switzerland (2011), and 37 acres in England. It is being considered by some for cultivation in New York. See Robinson et al., *Wine Grapes*, 881–82.

28. See Chateau Stripmine, www.chateaustripmine.info/Breeders.htm; *Research Report for 2009 to the Viticultural Consortium-East*, Lake Erie Regional Grape Program; New York Wine and Grape Foundation, *Breeding and Evaluation of New Wine Grape Varieties with Improved Cold Tolerance and Disease Resistance* (January 22, 2010); Double A Vineyards, www.doubleavineyards.com/p-772-regent; Backyard Vineyard & Winery, www.backyard vineyardand winery.com/2009/01/regent-wine-grape.html. See also *Final Report, Study on the Use of the Varieties of Interspecific Vines*, 15; Robinson et al., *Wine Grapes*, 881–82.

29. There are 96 acres of Rondo in England (2011), and it is being grown in cool climate countries such as Belgium, Poland, Netherlands, Denmark, and Sweden. See Robinson et al., *Wine Grapes*, 907.

30. Plocher and Parke, *Northern Winework*, 19, 21, 166; Melissa Hansen, "Grapes for Puget Sound," Good Fruit Grower (June 2011), www.goodfruit.com; Robinson et al., *Wine Grapes*, 907.

CHAPTER FOURTEEN

Classic *Vinifera* Varieties

Block Two West Chardonnay at Millbrook Vineyards & Winery. *More than half of Millbrook's 35-acre estate in Dutchess County is planted with Chardonnay. The rest of the vineyard is planted with Riesling, Pinot Noir, Tocai Friulano, and Cabernet Franc.*

THIS CHAPTER COVERS the classic European *vinifera* wine grape varieties: Chardonnay, Riesling, and Gewürztraminer for the whites, and Cabernet Franc, Gamay Noir, and Pinot Noir for the reds. While the growing characteristics and winemaking capabilities of these grapes have been widely written about both in the United States and Europe, this chapter attempts to give at least a few practical considerations concerning their growth characteristics and capabilities of producing quality wines in the Hudson Valley.

There are other *vinifera* grape varieties that are not covered by this book, but which also might show promise. I have tried to limit the book's coverage to grape varieties that, as of right now, have demonstrated their ability to be grown in the Hudson Valley and similar cool climates in the eastern United States. However, this should not negate the efforts of future pioneers to push the envelope and expand the number of new *vinifera* grapes that can be grown locally. Cold tender and sensitive grape varieties that should be considered by these future pioneers include Melon for the whites, and Pinot Gris and Gamay Beaujolais for the reds.

Grapes that noticeably did not make the list include Cabernet Sauvignon and Merlot. These grapes are simply too risky to grow in the Hudson Valley due to their cold tenderness and inability to ripen reliably in the fall. Further, due to the region's cooler climate and humid summers, the quality of the grapes grown may not consistently be suitable for the production of quality wines.

White Vinifera Varieties

While there are many white *vinifera* grape varieties that could have been covered by this book, the number of grapes included was limited to those varieties that are either growing in the Hudson Valley to some extent or which—due to their suitable growing characteristics and success in other nearby wine-growing regions such as the Finger Lakes, Niagara Peninsula, or Pennsylvania—have a higher probability of success in the Hudson Valley. Based on these criteria, winter-tender varieties that have significant ripening and fungus-control issues, such as Furmint and Sauvignon Blanc, were not included. Other varieties, such as Melon (as it is known in Burgundy, or Muscadet, as it is known in the Loire Valley) could have been included because they may be suitable for cultivation in the Hudson Valley on a few select sites, but such sites wold be the exception and not the rule. However, there is little experience with these varieties in nearby wine-growing regions, so there was very little reliable information to report on.

CHARDONNAY

PARENTAGE
vinifera

HARVEST DATE
Mid-season

C+ B− C+ B+ A+

Chardonnay is one of those "trifecta" grapes—good in the field, good in the cellar, and good in the market. So it's surprising that there is not more of it grown in the Hudson Valley. Chardonnay is a hybrid of a Pinot Noir clone and the bulk white wine/table grape Gouais, which originated anywhere from Lebanon to France or somewhere in between, perhaps in Croatia. It is agreed that Chardonnay is an ancient variety that has been identified, isolated, and initially cultivated in Burgundy, France, since the Middle Ages.[1] Further, it is one of the most widely planted white *vinifera* grape varieties in the United States, Europe, and throughout the world.[2]

Of all the *viniferas*, Chardonnay is probably the easiest to grow. It is tolerant of most soil types, and its mid-season ripening time makes it easy to harvest with little difficulty. In the cellar, Chardonnays practically make themselves. For example, while it is very difficult to make a good Pink Catawba or "Pink Cat" (because the fruit's sugars and acids are not well balanced for wine), almost anyone can make a good Chardonnay. Having personally made Chardonnays on and off again for the past thirty years, I would say that the caliber of a good winemaker should be gauged on how he or she makes a good Pink Cat, and not a Chardonnay. The bottom line is that Chardonnays have good balance, sufficient sugars, and nice flavors that are not too weak or overbearing.

Chardonnay is not only easy to make and forgiving in the cellar, it is also very versatile and can be made into many different styles of superior wines. Like Seyval Blanc and Vidal, it can be made into sparkling wines; big, fat, buttery wines full of vanilla; austere, steely, flinty wines of great intellect; or easy-drinking table wines. I once had a nice Hudson Valley Chardonnay that Eric Miller made during his second stint at Benmarl Vineyards, just before the winery was sold by the Millers to the Spaccarelli family in 2006. It was fermented cool and made in the style of a Germanic Riesling, which was a fruit-forward wine with lots of citrus, pineapples, and apricots similar to a Riesling. The point is that Chardonnay is a very malleable grape in the cellar.

Because of the wide range of wines that can be made from Chardonnay, it is a wine that is liked by the general public. Chardonnay, like strawberries, corn-on-the-cob, or macaroni and cheese, is a crowd pleaser. The Hudson Valley has the proper climate and soils to make leaner, more French-style Chardonnays.

Along with Riesling, Chardonnay is one of the most winter hardy of the white *viniferas* for the Hudson Valley. It is moderately winter tender, about the same or less winter hardy than Riesling or Cabernet Franc, but is much less winter hardy than Seyval Blanc, Vidal, or Vignoles. Unlike Riesling, which ripens late to very late, Chardonnay ripens by mid-season along with Baco Noir and Concord, around the third to fourth week of September. Its relatively early ripening time helps it to avoid the inevitable autumn rains and add the specter of fungus problems to the growers' already sometimes stressful "harvest experience." Also, if the season is delayed, there is always time for Chardonnay to ripen properly.

It is moderately productive to productive, but consistent in yields and sugar levels. The variety has medium-sized clusters that are cylindrical and moderately compact to compact. The cluster is winged with round, smallish berries that become yellow and amber when ripe. It is important to emphasize that, unlike Pinot Noir, Chardonnay is a reliable producer of quality fruit in good years and bad. As with many *vinifera* grape varieties, there are different clones of Chardonnay that should be examined and utilized based on the vineyard site and types of wines that the winemaker hopes to produce. There are clones named after Dijon, Geisenheim, New York vineyards or sources, and California vineyards or sources.

Chardonnay is a moderately vigorous to vigorous variety whose early budding time increases the danger of crop damage from a late spring frost. The variety, for a *vinifera*, is somewhat resistant to black rot, a problem in the Hudson Valley, but it is susceptible to powdery and downy mildew, and very susceptible to botrytis. It is not sensitive to sulfur treatments, so sulfur may be used in the spray program. To minimize fungal diseases, vertical training, combined with summer pruning and leaf removal, is recommended to increase sunlight and air circulation.

To avoid losing a season's crop due to winter injury, Chardonnay should have two to three trunks, two lower replacement spurs, and be long-cane pruned for optimum production. Further, it is recommended to wait until late March or April to prune Chardonnay to get an accurate read on winter damage. Also, since Chardonnay buds out relatively early, during a long and warm March spring thaw, Chardonnay can be fooled that spring has arrived and begin to push its buds too early, with resultant bud damage. By waiting until late March or early April to prune, there will be more buds to push, hence, on average, they will remain closed and more resistant to cold damage. The rootstock selected for Chardonnay should match the fertility and the water-holding capacity of the land.[3]

As stated earlier, Chardonnay, like Seyval Blanc from the Hudson Valley, can be made into a wide variety of wine styles. They can be like French Chablis–style-like wines that are steely, crisp, lean, flinty, and acidic with fruits of green apples, lemons, and grapefruit. Or, they can be like a more subdued version of those big California round, toasty, and oaky wines with rich textures and intense aromas and flavors of butter, vanilla, melons, pears, tropical fruits, old bananas, slight orange rinds, and nuts like almonds and hazelnuts. Or, they can be made into barrel-fermented *sur lie* wines that are layered, yeasty, warm, creamy, and brightly flavored, with elements of ripe apples, melons, pears, soft lemons, and vanilla.

In addition, Chardonnays, both in France and in the United States, are an important component used to produce quality champagnes and sparkling wines. The grape combination is traditionally made up of Pinot Noir, Chardonnay, and Pinot Meunier.

As the types of wine made from Chardonnay grapes are wide and varied, so are the colors of its wines. They range from pale, light yellow, and pale green-yellows to straw and light gold colors.

It is my hope that more Chardonnay will be grown in the Hudson Valley. With more grapes available, the Valley's winemakers can begin to create their own unique contribution to the development of new Chardonnay styles or to augment already existing ones.

GEWÜRZTRAMINER

Gewürztraminer, sometimes called Gewürz, or Traminer, is the other classic white grape grown in the Rhineland in Germany, Alsace in France, and to some extent in the United States.[4] Because of its distinctively perfumey and highly floral nose and taste, it is not commonly produced or consumed in the United States, however, it may have a place in the Hudson Valley, after other white *vinifera* varieties such as Chardonnay and Riesling.

PARENTAGE
vinifera

HARVEST DATE
Late

C– C B+ B- A+

Pierre Galet maintains that the grape originated in the Pfalz region of Germany and was introduced into the Alsace region after that province, a borderland area between France and Germany, was transferred to the newly established German Empire after the Franco-Prussian War in 1871. After World War I, Alsace was transferred back to France, and the popularity of Gewürztraminer in Alsace remained.[5] However, Jancis Robinson had maintained that the grape came from the Italian Tryol region and was first identified around the year AD 1000.[6] Since these views were documented on the derivation of Gewürztraminer, Galet and Robinson now believe that this grape is a mutation of Savagnin Rose, and that this mutation occurred near Rheingau, Germany.[7]

Gewürztraminer is vigorous to very vigorous in growth habit and produces an abundant amount of vegetative growth. This variety has an early bud break, so it is vulnerable to late spring frosts. Further, it has poor fruit set if there is cool weather during flowering. Its cluster is small, conical, and varies from compact to somewhat loose. Its thick-skinned berries are small, somewhat oval, and pink to pink-red in color, but it makes a good-quality white wine. Its grape yield is at best modest, but the quality of wines is good to excellent.

The variety can be less winter hardy—some say significantly less winter hardy—than either Chardonnay or Riesling, but it can still be grown on favorable sites in the Hudson Valley. Its vigorous vegetative growth and long growth cycle can cause uneven wood maturity, which can lead to the production of unripe wood, which is more vulnerable to winter damage. The variety's vegetative growth needs to be curtailed early in the fall so that its wood can harden off to survive winter cold. This variety should be classified as tender to winter damage in the Hudson Valley, so it should not be grown on very fertile lands, and cultivation practices for it should be altered to minimize unwanted vegetative growth so that it does not sustain winter damage.

Gewürztraminer is susceptible to black rot and downy mildew, more susceptible to powdery mildew, and slightly less susceptible to botrytis damage, however, with a careful spray program, it can do well in the Hudson Valley. It is not sensitive to sulfur fungicide applications. Training systems that increase sunlight and air circulation to minimize fungal diseases are recommended, as well as summer hedging and some leaf pulling. This is particularly important due to the high vegetative growth of Gewürztraminer. The grape ripens late mid-season in late September or very early October in the Hudson Valley, similar to varieties such as Cayuga White, Chancellor, and Chelois.

Due to its more than above-average susceptibility to winter injury, the vine should have two or three trunks, depending on the vineyard site, so that there are replacement trunks available to compensate for possible winter injury. Some growers cane prune Gewürztraminer to an upright trellis. However, to reduce its vigor, it is recommended that the vines have cordons placed about three feet off the ground; fruiting buds are created by spurs of two to three buds.[8]

The wines are distinctively very aromatic, perfumey, and flowery with varietal flavors that are unique to Gewürztraminer. It is a delicate wine with medium body and mouth feel, which is enhanced by finishing the wines off semisweet. Unlike Riesling, it does not have extraordinary high levels of acid and tends to be more perfumey in its nose and taste. Further, unlike Riesling, it can have high alcohol levels and, coupled with low acidity, is structurally very different than Riesling.

The wine's color can range from pale straw to yellow-gold with a tinge of pink to rose-colored amber. The wine color is similar to another pink wine grape, Delaware. While there is some soft spice, it has very floral aromas and tastes of tropical fruits such as guavas, lemon/lime, cloves, nutmeg, ginger, baked spiced apples, honeysuckle, roses, cinnamon, citrus and citrus blossoms, citrus zest, pine forests, and Muscat. Like Muscats, Gewürztraminers can be both grapey and intense.

While this grape may be a bit risky for cultivation in the Hudson Valley, with a shaky ability to be sold in the market, it may be a worthwhile grape to be pursued by those who want to produce quality *vinifera* wines unique to the region.

RIESLING

The Hudson Valley has often been called "America's Rhineland," as its beautiful river, shorelines, mountains, and buildings perched high on nearby cliffs are reminiscent of the Rhine in Germany. Areas of the Hudson Valley have geological formations of shale, slate, and schist rocks topped by well-drained clay soils, which are similar to the Rhineland where fine Riesling wines are produced. This grape is extensively cultivated in the Finger Lakes and produces some of that region's most notable wines. It only makes sense that the Valley should have more Riesling planted here.

PARENTAGE
vinifera

HARVEST DATE
Late to very late

C+ C+ C B+ A+

The grape Riesling, also known as Johannisberg Riesling or White Riesling, originated, or was at least identified, in the Rhineland as early as the 900s and definitely by the 1400s. It was probably a wild *vinifera* vine that was discovered by local growers, propagated, and then distributed to other growers.[9] Genetically, Riesling has a parent-offspring relationship with Gouais Blanc, just as Gouais Blanc has the same relationship with Chardonnay, Gamay Noir, and Furmint.[10] Like Chardonnay, it adapts to many different types of soils, but unlike Chardonnay, it tends to retain its varietal wine flavors and structure better when grown in cool climate areas.[11]

Riesling is a moderately vigorous variety whose yield is more dependent on growing conditions than most other grapes. On suitable sites, it has good yields. Lately, the trend has been to plant Riesling with a closer vine spacing of four to five feet between plants in rows of eight feet or even much closer instead of the commonly planted six feet by nine feet configuration.[12]

Its bud break is late, so spring frost damage is rare. Riesling's cylindrical-conical and compact clusters are small to small-medium with small round berries. When ripe, they can be clear green to golden yellow color.

Of the white *viniferas*, it is probably the most winter hardy, but is still considered somewhat cold tender and not as hardy as some of the more cold tender French-American hybrids such as Chambourcin and Villard Blanc. It is susceptible, like all *vinifera*, to black rot, downy mildew, and powdery mildew. Further, due to its late to very late harvest date, it is susceptible to botrytis. However, if controlled, an infection of botrytis cinerea mold that does not have secondary pathogens associated with bunch rot can be beneficial to the winemaker. While reducing the volume of wine produced, the botrytis mold concentrates Riesling's flavors and allows for the production of big sweet to very sweet, intensely flavored dessert wines that will be described later in more detail. The variety is not sulfur sensitive, so these treatments can be made. In sum, with a vigorous spray program, Riesling can be a large producer of excellent quality fruit.

To minimize winter damage and ensure an adequate crop the following year, two or three trunks should be retained. Like Pinot Noir, generally, it is best to leave two low spurs to replace trunks or cordons. The variety is generally pruned to leave four fruiting canes of eight to twelve buds per cane for a total number of no more than twenty-five to thirty buds. The number of buds should be adjusted to the level of soil fertility, water-holding capacity, and the underlying geological substrata of each vineyard. Sometimes due to winter injury, the vine does not have four sound canes for fruiting, so spur pruning of an existing cordon can be used to obtain the appropriate number of buds. The variety has an upward to lateral growth habit and can be vertically trained to increase sunlight and air penetration to minimize fungal diseases. Also, summer hedging and leaf pulling is recommended to increase sunlight and air.

There are several scores of Riesling clones. Some are used in the Finger Lakes and Lake Ontario wine regions, but many of these clones have not been systematically tested in the Hudson Valley. Those clones are 90, 198, 239, and 356, but this is not an exhaustive list of Riesling clones that should be considered. The rootstocks that have been used include SO4, 1613, 5BB, and C3309. With Riesling, as with most *viniferas* grown in the Hudson Valley, choosing the right clone and rootstock for the vineyard and style of wine to be made is very important, more so than with other grape varieties.

Rieslings can be made into dry, semidry, sweet, and very sweet dessert wines. Unlike many other wine grapes, Riesling develops its fruity aromas at relatively low sugar levels of 17° Brix. If a more lean Riesling is intended, the variety can be picked as early as late mid-season with lower sugars to produce dry, austere Rieslings. Or, more importantly, if the season is late due to cold or rain, Riesling can be picked at 17° to 19° Brix and still be made into a quality wine with varietal flavors. Also, due to its high acidity and low sugar content, Riesling can be successfully made into sparkling wines.

No matter what type of Riesling is made, it tends to be a distinctively varietal wine that is aromatic, forward, fruity, spicy, and very flowery, so it lends itself to stainless steel or glass aging, and not wood aging. However, I have heard of winemakers who have aged the wine in lighter woods such as beech with positive results. Overall, it is not as versatile a grape as Chardonnay or Seyval in the cellar. However, Riesling does have a cost advantage over Chardonnay in its production as it does not require wood aging or malolactic fermentation. Unlike Chardonnay, a Riesling made in the fall can be ready for sale by the following May.

In dry and semidry Rieslings, you will find that its crisp high acid is balanced by its big, forward floral fragrant smells and flavors of white peaches, citrus, pineapples, green apples, lime/lime peel, lemon, dried flowers, melons, hints of honey, spice, and herbs that finish clean with a minerally aftertaste. The substantial body of these wines is minerally like flint with a herbal taste of pine resin or juniper berries similar to that found in gin, which has a greasy taste. With that said, it has a solid palate weight to balance all of its fruit and acid.

For the sweeter Rieslings and dessert or ice wines, the very big aromatic flavors of ripe apricots, peaches, pineapples, honey, oranges, and orange rinds, similar to that found in the liquor Grand Marnier are more evident. This big opulent fruit is balanced by the heavy viscosity of the wine and its high acid, which is not noticed due to the high sugar content. As Rieslings get sweeter and more viscous, the flinty/minerally body and taste tend to lessen and are replaced by the acids and tastes found in raw honey. Not to diminish the appeal of these dessert Rieslings, but similar flavors can be found in dessert wines made from Vignoles and sometimes Delaware and Vidal.

Riesling is a quality grape that should be grown more commonly in the Hudson Valley.

Red Vinifera Varieties

This section details three red *vinifera* grape varieties grown in the Hudson Valley that produce superior table wines. As previously mentioned, grape varieties that were not included are Cabernet Sauvignon, Malbec, Merlot, and Shiraz. While some of these varieties are being successfully grown in the nearby Long Island, Finger Lakes, and Niagara Peninsula wine regions, there is insufficient information on their growing characteristics or success in the Hudson Valley to include them in this book. With both Cabernet Sauvignon and Merlot, the growing conditions in the Hudson Valley are not conducive to producing quality wines because for the former, the growing season is too short to consistently and properly ripen fruit, and for the latter, it is too cold tender and buds out too early for successful cultivation. Sensitive *vinifera* varieties that some more adventurous growers in the Hudson Valley may wish to consider in the future, but which the book does not cover, include Gamay Beaujolais (a Pinot Noir clone) and Pinot Gris.

CABERNET FRANC

Cabernet Franc can be delightful, serious, light, or complex. It is known as Breton in the Loire Valley and Bouchet in St. Emilion and Pomerol, France. In d'Anjou, it is often the base of dry or semidry rosés.[13] The curse of Cabernet Franc is that it was once primarily grown in Bordeaux and used as a blender to soften and brighten the large, tannic, omnipresent Cabernet Sauvignon.

PARENTAGE
vinifera

HARVEST DATE
Late

C+ B– B B A+

Cabernet Franc, commonly called Cab Franc, was once compared to Cabernet Sauvignon, with Cabernet Sauvignon being the prosperous, wise uncle and Cabernet Franc the poor clueless nephew. Highlighting the differences between the two can illuminate Cabernet Franc's many virtues for those who are more familiar with the cultivation and cellaring characteristics of Cabernet Sauvignon. It is now known that Cabernet Franc is one of the parents of both Cabernet Sauvignon and Merlot, the other parent of Cabernet Sauvignon being the white Bordeaux grape Sauvignon Blanc.[14] There are several theories on where Cabernet Franc came from, ranging from the Basque area of Spain, further north in Bordeaux, or even further north in the Loire Valley or Brittany.[15]

In the Hudson Valley, there is a real need to cultivate a high-quality red *vinifera* grape. Cabernet Franc fits that bill, in that the grape offers quality fruit that can be reliably grown in the Valley and that it gives winemakers versatility in the cellar. It is often argued that the cultivation of Cabernet Sauvignon in the Hudson Valley is problematic because the region's climate cannot reliably ripen fruit to consistently make world-class wines, yet, as in the Loire Valley, the Hudson Valley can consistently produce quality Cabernet Franc.

In the field, Cabernet Franc is a vigorous plant that produces an abundant amount of vegetation and lateral shoots. The bud break is mid-season, so it can sometimes be susceptible to spring frosts. This moderately productive grape has loose cylindrical-conical clusters that are small to medium in size, and has a yield that is heavier than Cabernet Sauvignon. Berries in the cluster are small and the color is blue-black.

This variety, like all *vinifera*, is susceptible to black rot and downy and powdery mildews, so a vigorous spray program is recommended, but it is not susceptible to botrytis or bunch rot. Its loose cluster facilitates good spray penetration and maximizes air circulation to reduce fungal disease pressure. Further, it is not sensitive to sulfur treatments. Some commentators maintain that leaf-roll virus, which reduces production, is prevalent in some Cabernet Franc nursery stocks, so be sure to purchase certified disease-free nursery stock.

To increase production, Cabernet Franc can be long-cane pruned. Other growers suggest spur pruning to suppress production and avoid any need to cluster thin. Its shoots grow strongly upward, so upright shoot–positioned training and trellis systems should be used. If horizontal or downward pruning techniques are used, this variety will push lateral shoots. Also, its tendency to excessive vegetative growth should be considered when pruning this variety to avoid fruit shading. Leaf pulling is an important practice to increase sun to the clusters because of the variety's already strong bent toward producing off-putting grassy or weedy wines.

Cabernet Franc, like Merlot, grows better on heavier clay soils than Cabernet Sauvignon because it metabolizes nutrients better in clay soils. In the Loire Valley, Cabernet Franc is grown in its lower elevation clay soils, which are similar to those found in the Hudson Valley. However, due to the Valley's colder climate, which has both higher and lower temperature extremes, Cabernet Franc should be grown only on the better well-drained sites that can avoid spring frosts and can attain full fruit maturity.

The grape matures late, around the first week of October to as late as mid-October. This is still about one week to ten days before Cabernet Sauvignon, when it is grown in the Hudson Valley, so it is a grape variety that is more suitable than Cabernet Sauvignon for cultivation in the Hudson Valley.[16]

There are two important decisions that need to be made when selecting Cabernet Franc vines. First, the rootstock should suppress the plant's pronounced vegetative growth. The past practice of using clone CA2 grafted on vigorous rootstocks, such as 5BB and SO4, is being discarded. This combination produced Albert Einstein hair-like canopies of wild growth with high yields of unripe fruit that led to the production of weedy wines. To reduce over-vigorous vegetative growth, rootstocks such as 101-14 and Riparia Gloire should be considered on fertile sites. Also, suppressing vegetative growth can help the vines harden off and enhance winter hardiness.

Second, growers who are considering this variety should review the various clones available to ascertain their suitability for each vineyard site for both yield and the style of wines that the winemaker wants to produce. Inconclusive research suggests that the following clones should be considered: Clone 214 is noted for having raspberry and violet aromas and flavors; Clone 327 is a low-vigor variety that produces wines of solid structure; and Clone 623 makes a deeply colored, good wine.[17]

The big advantage of Cabernet Franc is that it is probably one of the most winter-hardy red *vinifera* varieties. Depending on the vineyard site, it is almost as hardy as Chambourcin or Riesling, and can be more hardy than Chardonnay, but is not as hardy as Baco Noir or Chancellor.

Wines produced from Cabernet Franc tend to be more herbaceous, lower in tannin, and lighter in color than Cabernet Sauvignon, and have colors of purple and light ruby, as opposed to Cabernet Sauvignon's darker hues. Cabernet Franc is still heavier in body and presence than most French-American hybrids, with the exception of Chambourcin. Its flavor profile, unlike the serious, big, and dank flavors of Cabernet Sauvignon, is far more approachable and aromatic. Cabernet Franc is full of perfumey aromas and tastes of red and black raspberries, rose petals, violets, ripe plums, cooked and fresh blueberries, fresh strawberries, and, yes, cranberries. Importantly, there are big underlying herbal notes of rosemary and licorice, and a spiciness that Cabernet Sauvignon often lacks. Many Cabernet Francs have been described as being spicy with elements of nutmeg, cloves, and black pepper with an herbal or vegetative quality. Superior Cabernet Francs have complex layers of soft herbal textures, while lower quality ones possess a grassiness that may not be not pleasing to some palates.

Unlike Cabernet Sauvignon, the relatively lean tannin structure and medium body of Cabernet Francs have a very welcoming presence of soft earth, cigar box, and light chocolate. While the tannin structure and body are light for a *vinifera* red, it is bigger than most red French-American hybrids, except for Chambourcin, S.V.18-307, and NY 66.717. To maximize color and flavor extraction, many winemakers let the skins macerate for three weeks to a month before pressing. With this long maceration time, the cap can fall, so carbon dioxide is added to minimize oxidation.

Cabernet Franc wines can have many layers of berries, cranberries, some light red cherries, and a round body that makes them interesting to drink over the course of an entire dinner. The aging potential of Cabernet Franc is good, with lighter ones peaking in three to five years, while more "serious" ones peak in seven to ten years.

Since the 1970s, the cultivation of Cabernet Franc has extended further north in France away from warmer Bordeaux districts to the cooler Loire Valley. The cooler growing conditions bring out the finer qualities of Cabernet Franc. This discovery has encouraged growers in other cool growing regions, such as the Hudson Valley, to begin planting Cabernet Franc in earnest and to make more wines from this supporting character of Bordeaux.[18]

Hudson Valley Cabernet Francs tend to be medium-bodied wines with floral aromas of red and black raspberries, cranberries, violets, and muted herbal flavors. In good growing years, these wines can have a hefty body with more depth and undertones of chocolate and earth. Cabernet Franc is also a very good blender that gives more presence and complexity to already complex red grapes such as Cabernet Sauvignon, Merlot, Chambourcin, Chelois, or Chancellor. In fact, in the Loire Valley, Chambourcin is a companion grape that is grown and blended with Cabernet Franc to make either light summer rosé wines or medium-bodied reds.

The Hudson Valley is uniquely suited to produce both big, complex Cabernet Francs and the kinds of lighter-bodied reds that are currently being made in the Loire Valley. While Cabernet Francs can be made into these lighter-style reds, it is recommended that heavier-styled wines be made in the Valley because other commonly grown grapes such as Baco Noir, Cascade, and Maréchal Foch can be used instead to make lighter-styled reds. Cabernet Franc should fill the Hudson Valley's need for a grape that can produce big, forward wines with a big tannin structure and body. This is a grape that comes highly recommended if the right vineyard site, rootstock, and clone are identified.

GAMAY NOIR

Gamay Noir, the grape known primarily for making Beaujolais-Village Nouveau, has the potential to be so much more. This variety, also known as Gamay Noir à Jus Blanc, is a hybrid that was identified by growers in Burgundy probably by the early 1300s. It is a hybrid of one of the Pinot Noir clones growing locally in the area and a heavy-producing white *vinifera* grape called Gouais, which was one of the parents of that classic white grape of Burgundy, Chardonnay.[19] Despite being banned in Burgundy in 1395 by Philip II, Duke of Burgundy (also known as Philippe *le Hardi* or Philip the Bold), Gamay has remained in the area to provide everyday red table wines.[20]

PARENTAGE
vinifera

HARVEST DATE
Mid-season

C-	C-	C+	B	A+

There is little track record for growing Gamay Noir in the Hudson Valley or for making wines from these grapes. However, due to its reported superior growing characteristics in climates similar to the Hudson Valley, such as the Niagara Peninsula, in Ontario, Canada, where it is grown extensively, Gamay Noir should at least be mentioned to have a place in the production of quality red wines in the Hudson Valley.[21]

Gamay Noir, should not be confused with Gamay Beaujolais, a Pinot Noir clone grown in California.[22] Gamay Noir is a medium vigorous to vigorous variety that ripens by mid-season. The vine is more productive to much more productive than Pinot Noir. While its early bud break exposes it to late spring frosts, it can produce a small secondary crop to even out its overall production level over the years. Its cluster is medium in size, compact, and cylindrical, and can be slightly winged. Its berries are medium in size, black, and somewhat oval.[23]

Compared to Pinot Noir, it is about as susceptible to black rot, downy mildew, and powdery mildew, but it is slightly less susceptible to botrytis and bunch rot. Further, it is not sulfur sensitive.[24] To obtain true Gamay Noir vines, it is safer to obtain them from nurseries in Ontario, Canada, where they were imported directly from Beaujolais, France, and not, as stated earlier, from California where there are several varieties that although they are called Gamay, can be Gamay Beaujolais, a Pinot Noir clone, or Napa Gamay, known in France as Valdiguie.[25]

Gamay Noir, as is the case with Pinot Noir, has different clones, but not nearly as many identifiable clones. Clones such as ENTAV[26] clones 222, 358, 509, and 565 have been noted for their winter hardiness and quality wine production. It can be as winter hardy as Chardonnay, but not as hardy as Riesling if grown correctly on the right sites, but it can still take a significant hit during a bad winter.

While Gamay Noir may have somewhat more fungal disease resistance than Pinot Noir, especially for botrytis, it still must have a rigorous spray program and be pruned to create an open canopy to increase sunlight and air circulation to mitigate fungal diseases and to ripen its fruit. It is generally spur pruned, but can be cane pruned on fertile soils. The proper mix of clones and rootstocks needed to maximize the production of quality fruit has not yet clearly been identified on the East Coast.

The textbook Gamay Noir grape made as a Beaujolais Nouveau wine is an uncomplicated fruit-forward purple wine, with high acid, low tannins, of light to medium body, with a nose and taste of fresh blackberry or black raspberry juice. It is hard to overemphasize the big berry flavors both in the nose and taste of these wines.[27] These wines are made to be consumed while very young, generally within two months to one year of bottling. It is similar in some ways to Baco Noir or Maréchal Foch. The technique of fermentation by carbonic maceration (whole-berry fermentation) is what creates this style of wine.

Even though Gamays are known for the Nouveau wines they produce, the variety is capable of producing big, dark, complex wines with some tannin and great aging potential, as are also being produced in Beaujolais. These dark wines have more body, black cherry, chocolate, and black pepper flavors.

For the Hudson Valley, winemakers should strive for this style of big, complex Gamays. In France, Gamay and Pinots are commonly blended together. In the Hudson Valley, it may be advantageous to blend Gamay with Baco Noir or Maréchal Foch and age in wood for at least six months to a year to produce uniquely Hudson Valley red wines.

This is a variety that should be closely examined and its acreage expanded in the Hudson Valley. Even considering Gamay Noir's more difficult growing characteristics, susceptibility to fungal diseases, and winter tenderness, it should be further evaluated as a possible red *vinifera* grape variety that can produce reliable crops of top-quality red wines.

PINOT NOIR

"Pinot Noir is a minx of a vine. … So alluring is the goal of making even the faintest shadow of great red burgundy in the newer wine regions that … [the] phrase Holy Grail crops up often in discussions about cultivating Pinot Noir outside of Europe." This very insightful observation was made by Jancis Robinson.[28] Based on personal experience, and with all of the research I have conducted on Pinot Noir, this description, I believe, best sums up this elusive grape variety, which comes from Burgundy, France.

PARENTAGE
vinifera

HARVEST DATE
Mid-season

C– D C C+ A+

While superior Pinot Noir wines are being made in Napa and Sonoma counties in California and now in Oregon, the Hudson Valley has the potential to produce superior Pinots that more closely resemble those firm but velvety aromatic wines produced in Burgundy. The Hudson Valley's climate, soil composition, and underlying geological substrata are conducive to producing French-style Pinot Noirs. For instance, in the late 1980s, the French Champagne makers Charbaut et Fils, under the direction of Tim Biacalana, made excellent sparkling and red still Pinot Noir wines. And, like the French, several Hudson Valley wineries have produced very fine Pinot Noirs over the years, such as Brotherhood, Millbrook, El Paso, Oak Summit and Riverview.

To produce quality Pinot Noir wines, there must be the right combination of terroir, climate, mix of Pinot clones, rootstocks, growing conditions, and cultivation techniques. From literature available in Europe, Pinot Noir is not classified as a truly finicky grape—difficult and frustrating, yes, with lots of growing constraints—but not finicky. However, the literature in the United States emphasizes over and over again that it is a finicky grape both in the field and in the cellar. This may be because so much of production in the United States is mechanized.[29]

Part of the inability to isolate the winemaking qualities of Pinot is that it is made from scores of very different clones. Pinot Noir is probably one of the first identifiable grape varieties that may have been cultivated in Burgundy before the Roman Empire. It has the ability to genetically mutate quickly, hence the high number of clones available to produce wines. Pinot Noir has also mutated into grape varieties such as Pinot Blanc and Pinot Gris. It is grown throughout Europe by its own local names such as Spätburgunder, Blauburgunder, Nagyburgundi, Noirien, Pineau, and Savagnin Noir, among others. It is also a parent of the other famous grapes of Burgundy, Chardonnay and Gamay, and varieties such as Aligote and Melon. DNA testing has been unable to identify the parents of Pinot Noir, so it is probably a wild variety that was later cultivated.[30]

Pinot Noir is winter tender but can still be grown in the Hudson Valley on its better sites for the production of both sparkling and still red wines. It is not as winter hardy as Riesling, Cabernet Franc, or Chambourcin, but it is hardier than Cabernet Sauvignon or Merlot. It is slightly more winter tender than the white Burgundian grape, Chardonnay. Of the classic European *vinifera* reds, Pinot Noir, Cabernet Franc, and Gamay Noir are the only ones that can be consistently grown in the Valley on its better sites. Since it is solidly an early mid-season to mid-season ripening variety, along with Baco

Noir and Léon Millot, once it gets through winters it can easily ripen to maturity in the Valley's climate, but getting it to harvest can be perilous. Pinot Noir's bud break is early, so it is susceptible to late spring frosts. If damaged by spring frosts, it has little or no secondary crop. Further, cold weather during its bloom will result in poor fruit set.[31]

While it is difficult to generalize about Pinot Noir due to its many clones, the variety is generally delicate and only moderately vigorous. Further, it is only of low to moderate productivity, rarely yielding more than three tons an acre. If more grapes are produced, vine health declines, and the wines produced are mediocre. Also, for the best-quality wine, it must be grown on marginal sites where the vine has to struggle against cool weather. Its special wine characteristics are lost in hot, dry climates. Pinot Noir's small to small-medium cylindrical clusters, with their thin skins, are very compact and hence very susceptible to botrytis and bunch rot. The blue-black berries are also small.

As with most other *vinifera*, it is also susceptible to black rot, downy mildew, and powdery mildew. For fungal treatments, it is not sensitive to sulfur. Due to the vulnerability of Pinots to fungal diseases, the training system used is important to maximize sunlight and air circulation to minimize fungal disease pressure. Some have suggested that vines should be trained in a vertical low-head fashion, but most cane prune it. There is further disagreement on whether cane or cordon pruning is best.[32]

Winston Churchill once said that "Democracy is the worst form of government, except for all those other forms that have been tried from time to time."[33] The same can be said of Pinot Noir. While it is a more perilous grape to grow than most others, it is best suited to cool climates due to its early budding, harvest date, and propensity to succumb to fungal diseases and does not thrive at all in more Mediterranean climates.

Pinot Noir produces its best wines in cool climates. The grape is grown successfully in Burgundy, France, and to a lesser extent in cooler parts of Europe, such as Austria, Germany, Hungary, and northern Italy.[34] It can be crushed slightly at pressing to yield a flinty white wine that is the base of sparkling wines made in Champagne. However, different clones of Pinot Noir are used to make the famous red Burgundies of France. While red wines produced in Bordeaux are blends of the grapes Cabernet Sauvignon, Merlot, Cabernet Franc, Malbec, and Petit Verdot, red Burgundies use just one grape, Pinot Noir, however, there are scores of major Pinot Noir clones that are cross-blended with each other so that the finished wine has good color, tannin structure, and flavor profile. Hence, there is no real one description of Pinot Noir, but more on this later.

As previously mentioned, this difficult-to-grow grape is highly subject to botrytis and bunch rot, which can turn the grapes to mush within a week, so growers are inclined to pick the grape too early to ensure that they obtain a crop. This is why there are more than a few lightly colored, green underripe Pinots on the market. With all of its risks, growers should wait and pick Pinot when it is ripe.

In the cellar, it is also very finicky, but if made well, it is a superlative wine. Like true Burgundies, Hudson Valley Pinots can have a soft but firm underlying body and structure, like a steel girder wrapped in velvet. They are soft to the touch, but underlying that softness is an inner strength supported by an ever-present tannin structure.

The layered flavors in Pinot Noir wines can include strawberry jam, red cherries, raspberries, plums, raisins, black pepper/spice, vanilla, chocolate, and "barnyard flavors" (which sound bad, but are actually quite good), licorice, and earth. They can also exhibit plum and eucalyptus flavors, which layer nicely with cedar and sweet smoke. Great Pinots, while integrated from the nose to the finish, display themselves in layers that make the wine warm and interesting. All great Pinots can be described as possessing the three S's—Soft, Sophisticated, and Subtle.

Depending on the clone used, the wines can have either a rather light red color or medium-red to garnet color, which is a good segue to a discussion on Pinot Noir clones, which was touched upon earlier. For the Hudson Valley to become a producer of superior Pinot Noirs, it will need to undertake the laborious task of matching soil types to rootstocks to clones with the right cultivation practices. There are significant differences among the clones with regard to winter hardiness, wine quality and flavors, susceptibility to bunch rot, color, and tannin structure. In making Pinot Noir wines, it is important to use several clones to arrive at the right balance of color, nose, flavor, and tannin structure. Dr. Robert Pool at Geneva had suggested[35] that it seems the better wine quality came from bunch rot–susceptible varieties.

The list of some of the Pinot Noir clones below should be considered for cultivation in the Hudson Valley and perhaps in the rest of the East Coast and Canada. It is a compilation of some of the industry articles that evaluated clones for the East.[36] It will be a long, slow process to match the right soils with the appropriate Pinot Noir clones and rootstocks needed to produce superior wines in the East. It is important to note that it took several decades of dedicated work to isolate those clones and rootstocks to produce superior wines in Napa and Sonoma counties and other parts of California, Oregon, and Washington State.[37]

Even with the numerous difficulties outlined for the production of quality Pinot Noir, both in the field and cellar, it is important for growers and winemakers to make at least one signature classic red *vinifera* wine. That wine should be Pinot Noir. It should also be considered in producing quality blended wines with grapes such as Baco Noir and Chelois.

PINOT NOIR CLONES

CLONE 7

A clone of unknown origin possessed by Dr. Konstantin Frank, this is cold hardy with moderate bunch rot resistance. It is used in Dr. Frank's Vinifera Cellars' Pinot Noir wines, which have been rated as good to superior.

CLONE 13 (MARTINI)

The "jack of all trades, master of none" clone. Dr. Bob Pool referred to this clone as "a utility infielder—never leads the league, but won't make an error."[38] It has pleasant cherry-berry fruit, medium tannin structure, moderate yield and color, good tannin, and cooked-fruit aromas. I think Bob liked this clone, so it is a keeper.

CLONE 27

This clone from Geisenheim, Germany, has good winter hardiness, but is susceptible to botrytis.

CLONE 29

Also known as one of the Jackson clones, Clone 29 was imported from France to California in the nineteenth century. There is a difference of opinion on its bunch rot resistance, but it is cold hardy. Wines are considered to be good with strong berry aromas, and good color and body. This clone has low to moderate yields and performs well for a sparkling wine, but deserves more scrutiny for red wine production.

CLONE 113

From a selection program in Dijon, France, this produces wines with a floral nose, flavors of cherries and spice, and a good structure. It is regarded in Europe as disease resistant. When compared to Clone 115, it seems to have more intense fruit flavors and good strong tannin structure.

CLONE 115

This has better-than-average winter hardiness, good color, ripe plum and berry flavors, a medium structure, and low tannin.

CLONE V

Obtained from France and deemed to be relatively disease resistant with good tannin structure, its wines have been described as having violets and floral notes, red berry flavor, and wet-earth aromas that are well structured.

GAMAY BEAUJOLAIS CLONE

The clone has moderate wine quality, but its resistance to bunch rot allows the clone to be harvested when it is ripe and able to make a solid wine. The grape has good winter hardiness and high yields, but is reported to have light color, thin body, and some berry and cherry aromas.

GENEVA CLONE

This has good winter hardiness, but is still susceptible to bunch rot. The wine quality is reported to be mediocre with low to moderate color, moderate berry fruity aromas, and low tannins. Its yields are high.

MARIAFELD CLONE (CLONE 17)

This clone originates from Switzerland. For a Pinot, it has relatively good resistance to bunch rot (probably due to its relatively large berry size and loose cluster), relatively high yields, and ripens early, but it is below average in winter hardiness. Because of its ability to somewhat resist bunch rot, it has the ability to ripen in the field better so that clean ripe fruit can be picked. While the wine quality has been reported as being lower than some, it can have some strawberry and berry aromas. Its dark color and firm tannin structure make it a solid addition to blends.

MEUNIER

This has good winter hardiness and production and, some maintain, is good for red wine production in good years, but is used primarily for sparkling wine production due to its light color.

PERNAND

This grape has good winter hardiness and wine quality, but only slight bunch rot resistance. The yield is moderate. Its color is moderately good and it has good fruit aromas, but sometimes a coarse tannin structure.

NOTES

1. Pierre Galet and Lucie T. Morton, *A Practical Ampelography: Grapevine Identification* (Ithaca, N.Y. and London: Cornell University Press, 1979), 63–64; Jancis Robinson, *Vines, Grapes, and Wines* (New York: Alfred A. Knopf, 1986), 108; Baudouin Neirynck, PhD, *The Grapes of Wine: The Fine Art of Growing Grapes and Making Wine* (Garden City, N.Y.: SquareOne Publishers, 2009), 48-49; Jancis Robinson, *Julia Harding, and Jose Vouillamoz. Wine Grapes: A Complete Guide to 1,368 Vine Varieties, Including Their Origins and Flavor* (New York: Ecco/HarperCollins, 2012), 221–27.

2. Chardonnay is one of the most widely grown white grape varieties in the world. It grows in cool and temperate climates. In France there are 110,190 acres of Chardonnay (2009), with greater Burgundy having 36,455 acres and Champagne 24,169 acres (2009). It is also commonly grown in Italy, Hungary, Croatia, Chile, Argentina, Australia, New Zealand, and South Africa. In the cool climate states of the United States, there were 7,650 acres in Washington State (2011), 1,008 acres in Oregon (2008), 981 acres in New York (2006), 474 acres in Virginia (2010), and 2,180 acres in Canada (2009). See Robinson et al., *Wine Grapes*, 221–27.

3. The descriptions of Chardonnay are based on my experience, but here are some citations in the literature on Chardonnay: Galet and Morton, *Practical Ampelography*, 63–64; Lucie T. Morton, *Winegrowing in Eastern America* (Ithaca, N.Y.: Cornell University Press, 1985), 77; Robinson, *Vines, Grapes, and Wines*, 106–13; R. M. Pool et al., *Growing Vitis Vinifera Grapes in New York State: I Performance of New and Interesting Varieties* (Geneva, N.Y.: New York State Agricultural Experiment Station, Cornell University, 1990), 5; *Double A Vineyards, 2003/2004 Catalog*, 11, 17; Bruce I. Reisch, Robert M. Pool, David V. Peterson, Mary-Howell Martens, and Thomas Henick-Kling. *Wine and Juice Grape Varieties for Cool Climates*. New York State Agricultural Experiment Station, Information Bulletin no. 233 (Geneva, N.Y.: Cornell Cooperative Extension, 1993), 9, 12; Robinson et al., *Wine Grapes*, 221–27.

4. Gewürztraminer is not as widely grown as Riesling and tends to be grown in more specialized areas. In France, there are 7,618 acres of Gewürztraminer (2009), mostly in Alsace, in Germany 2,063 acres (2008), Austria 200 acres (2011), Hungary 1,779 acres (2008), and in the Czech Republic 1,483 acres (2009). In North America, there were 775 acres grown in Washington State (2011), 217 acres in Oregon (2008), 143 acres in New York, mostly in the Finger Lakes (2006), 45 acres in Michigan (2006), 15 acres in Virginia (2008), and 644 acres in British Columbia, Canada. See Robinson et al., *Wine Grapes*, 966–68.

5. Galet and Morton, *Practical Ampelography*, 72.

6. Robinson, *Vines, Grapes, and Wines*, 166.

7. Robinson et al., *Wine Grapes*, 966.

8. Galet and Morton, *Practical Ampelography*, 72–73; Morton, *Winegrowing in Eastern America*, 80; Robinson, *Vines, Grapes, and Wines*, 166–70; Pool et al., *Growing Vitis Vinifera Grapes in New York State*, 12; *Double A Vineyards, 2003/2004 Nursery Catalog*, 11; Reisch et al., *Wine and*

Juice Grape Varieties, 10, 12; Robinson et al., *Wine Grapes*, 966–68.

9. Robinson, *Vines, Grapes, and Wines*, 97; Oz Clarke, *Oz Clarke's Encyclopedia of Grapes: A Comprehensive Guide to Varieties and Flavors* (New York: Harcourt, 2010, 191–93; Robinson et al., *Wine Grapes*, 888–93.

10. Robinson et al., *Wine Grapes*, 888.

11. Rieslings tend to do better in cool climate areas and that is reflected in where it is commonly grown. In Germany, there are 55,436 acres (2011), in Alsace, France, there are 8,357 acres (2009), Austria 4,631 acres (2011), and in Hungary 3,170 acres (2011).

In North America, there are 6,320 acres in Washington State (2011), 683 acres in New York (2008), mostly in the Finger Lakes, and 367 acres in British Columbia, Canada (2011); Robinson et al., *Wine Grapes*, 888–93.

12. Galet and Morton, *Practical Ampelography*, 84–85; Morton, *Winegrowing in Eastern America*, 83–84; Robinson, *Vines, Grapes, and Wines*, 97–105; Pool et al., *Growing Vitis Vinifera Grapes in New York State*, 40–41; *Double A Vineyards 2003/2004 Nursery Catalog*, 17; Reisch et al., *Wine and Juice Grape Varieties*, 9; Neirynck, *Grapes of Wine*, 62–63; Clarke, *Encyclopedia of Grapes*, 191–93; Robinson et al., *Wine Grapes*, 888–92.

13. Robinson, *Vines, Grapes, and Wines*, 134.

14. See original work of John E. Bowers and Carole P. Meredith, "The Parentage of a Classic Wine Grape, Cabernet Sauvignon," *Nature Genetics* 16 (April 30, 1997): 84–87; Robinson et al., *Wine Grapes*, "Pinot Pedigree Diagram," 806-807

15. Robinson et al., *Wine Grapes*, 153–55.

16. Galet and Morton, *Practical Ampelography*, 60; Morton, *Winegrowing in Eastern America*, 75; Robinson, *Vines, Grapes, and Wines*, 132–34; Pool et al., *Growing Vitis Vinifera Grapes in New York State*, 4; *Double A Vineyards 2003/2004 Catalog*, 17; Wine and Juice Grape Varieties for Cool Climates, 11–12; Clarke, *Encyclopedia of Grapes*, 44; Neirynck, *The Grapes of Wine*, 117–18; Robinson, *Wine Grapes*, 153–57.

17. David Falchek, "Saving Eastern Cabernet Franc: Better Clones and Rootstocks Key to Improvements, Say Researchers," *Vineyard & Winery Management News* (January/February 2008): 35–38.

18. Robinson, *Vines, Grapes, and Wines*, 132–34; Clarke, *Encyclopedia of Grapes*, 44; Neirynck, *Grapes of Wine*, 117–18. In France, there were 91,300 acres of Cabernet Franc grown (2009). The core of Cabernet Franc cultivation is in the Loire Valley, where 14,710 acres were grown (2011). In the mid-Val de Loire, in the Pays de Loire region, there were an additional 22,675 acres grown (2011). In the rest of France, it is used as a blending grape to lighten and soften other wines. In the cool climate areas of North America, there were 498 acres in New York (2008), 86 acres in Michigan (2006), 300 acres in Virginia (2011), and 1,565 acres in Ontario, Canada (2011). See Robinson et al., *Wine Grapes*, 155–56.

19. Dan Yi Yang, Yukio Kakuda, and Ronald E. Subden, "Higher Alcohols, Diacetyl, Acetoin, and 2,3-Butanediol Biosynthesis in Grapes Undergoing Carbonic Maceration," *Food Research International* 39 (2006): 112–16.

20. Galet and Morton, *Practical Ampelography*, 70; Robinson et al., *Wine Grapes*, 384–85.

21. In France, there were 75,225 acres of Gamay Noir (2009), with most of it grown in the Rhône Valley, and very high concentrations in Beaujolais. In Ontario, Canada, there were 185 acres of it (2006), and it is grown to some extent in Oregon, Michigan, and New York. See Robinson et al., *Wine Grapes*, 385–86.

22. Galet and Morton, *Practical Ampelography*, 70; Morton, *Winegrowing in Eastern America*, 80.

23. Galet and Morton, *Practical Ampelography*, 70; Robinson, *Vines, Grapes, and Wines*, 135–37.

24. Galet and Morton, *Practical Ampelography*, 70; Reisch et al, *Wine and Juice Grape Varieties*, 11–12; Robinson et al., *Wine Grapes*, 383–86.

25. Morton, *Winegrowing in Eastern America*, 80.

26. ENTAV (Etablissement National Technique pour l'Amélioration de la Viticulture) and ITV France (Institut Technique de la Vigne et du Vin) merged in 2007 to create the Institut Français de la Vigne et du Vin (IFV). ITV is now the chief network for research and development, with an 11 million Euro annual budget. See http://www.vignevin.com/.

27. Robinson, *Vines, Grapes, and Wines*, 135; Clarke, *Encyclopedia of Grapes*, 91; Neirynck, *Grapes of Wine*, 119.

28. Robinson, *Vines, Grapes, and Wines*, 81; Clarke, *Encyclopedia of Grapes*, 175.

29. Robinson, *Vines, Grapes, and Wines*, 81–86.

30. Ibid., 81–86; Robinson et al., *Wine Grapes*, 805–7.

31. Galet and Morton, *Practical Ampelography*, 83.

32. Ibid., 82–83; Robinson, *Vines, Grapes, and Wines*, 81–86; Reisch et al., *Wine and Juice Grape Varieties*, 10, 11, 13; Morton, *Winegrowing in Eastern America*, 83; Pool et al., *Growing Vitis Vinifera Grapes in New York State*, 29; *Double A Vineyards 2003/2004 Catalog*, 17; Clarke, *Encyclopedia of Grapes*, 175–79; Neirynck, *Grapes of Wine*, 85–86; Robinson et al., *Wine Grapes*, 808–15.

33. Quote from a speech made by Winston Churchill before the House of Commons on November 11, 1947.

34. Unlike its reputation, Pinot Noir is actually grown extensively and in a wide range of areas in Europe and the United States. Pinot Noir's first home is France, where 72,600 acres were grown (2009), with 31,875 acres in Champagne (2010) and 26,418 acres in greater Burgundy (2011). In Italy, there were 8,190 acres (2000), in Germany 29,158 acres (2008), in Austria 1,604 acres (2009), and in Hungary 2,380 acres (2008). In the cool climates of North America, Oregon leads Pinot Noir acreage with 11,200 acres (2008), New York 335 acres (2009), Michigan 135 acres (2006), Ontario 1,025 acres (2006), and British Columbia, Canada, 793 acres (2008). See Robinson, *Wine Grapes*, 811–15.

35. *Vineyard & Winery Management News* (March/April 1998): 32.

36. Hudson Cattell, "Clonal and Varietal Evaluation at Geneva," *Wine East* (May/June 1999): 10–15; David A. Falchek, "Pinot Noir Clones: Hot Choices for the Cool East," *Vineyard & Winery Management News* (March/April 1998): 30–33; Linda Jones McKee, "Commentary on Eastern Wines: Pinot Noir Tasting Notes," *Wine East* (2000), 24–27, 53; Robinson et al., *Wine Grapes*, 806.

37. Ted Rieger, "Pinot Noir Variety Focus at UC Davis," *Vineyard & Winery Management* (September/October 2005): 48–58; Mike Marcley, "A Leap of Faith," *Practical Winery and Vineyard Journal* (May/June 1999): 15–20; Paul Franson, "Pinot Noir Pioneer: Carneros Creek Winery," *Practical Winery and Vineyard Journal* (May/June 1998): 62–67; Eleanor and Ray Heald, "Varietal Review: Making Quality Decisions for Oregon Pinot Noir," *Practical Winery and Vineyard Journal* (September/October 1996): 30–33; Jim Wolpert, "An Overview of Pinot Noir Clones Tested at UC-Davis," *Vineyard & Winery Management News* (July/August 1995): 18–21; Eleanor and Ray Heald, "Varietal Review, Pinot Noir," *Practical Winery and Vineyard Journal* (March/April 1995): 50–53; Melissa Moravec, "Matching Pinot Noir Clones to Site: Clonal Evolution at Carneros Creek," *Vineyard & Winery Management News* (May/June 1995): 36–39; Eleanor and Ray Heald, "Pinot Noir—Shedding its Formidable Reputation," *Practical Winery and Vineyard Journal* (May/June 1993): 52–61; Stan Hock, "Clonal Selections: Obstacles and Opportunities with Pinot Noir," *Practical Winery and Vineyard Journal* (May/June 1989): 37–42; Julie Morrison, "Carneros Comes of Age," *Practical Winery and Vineyard Journal* (November/December 1988): 14–19.

38. David A. Falchek, "Pinot Noir Clones: Hot Choices for the Cool East," *Vineyard & Winery Management News* (March/April 1998): 32, quoting Dr. Robert Pool of the Geneva Agricultural Experiment Station.

MASS. HORTICULTURAL SEED STORE.

NEW ENGLAND GRAPE BREEDERS

First Horticultural Hall of the Massachusetts Horticultural Society on School Street, Boston. *While this society was named after the Commonwealth of Massachusetts, it was, in fact, not a regional society, but the foremost recognized horticultural society in the U.S. in the nineteeth century. Circa 1845.*

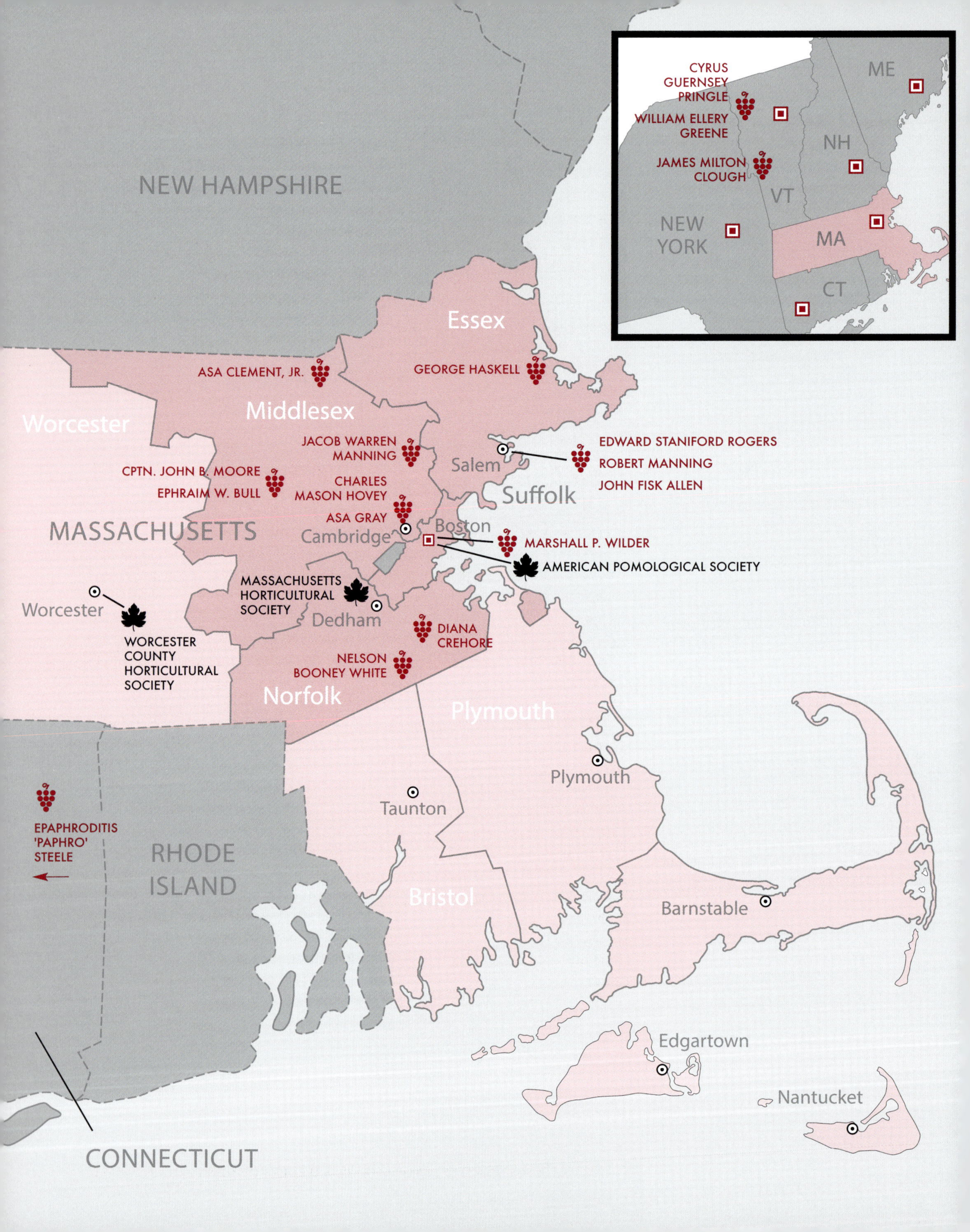

CYRUS GUERNSEY PRINGLE
WILLIAM ELLERY GREENE
JAMES MILTON CLOUGH
ME
NH
VT
NEW YORK
MA
CT
NEW HAMPSHIRE
Essex
ASA CLEMENT, JR.
GEORGE HASKELL
Middlesex
Worcester
JACOB WARREN MANNING
Salem
EDWARD STANIFORD ROGERS
ROBERT MANNING
JOHN FISK ALLEN
CPTN. JOHN B. MOORE
EPHRAIM W. BULL
CHARLES MASON HOVEY
Suffolk
ASA GRAY
MASSACHUSETTS
Cambridge
Boston
MARSHALL P. WILDER
AMERICAN POMOLOGICAL SOCIETY
MASSACHUSETTS HORTICULTURAL SOCIETY
Worcester
Dedham
WORCESTER COUNTY HORTICULTURAL SOCIETY
DIANA CREHORE
NELSON BOONEY WHITE
Norfolk
Plymouth
Plymouth
Taunton
EPAPHRODITIS 'PAPHRO' STEELE
RHODE ISLAND
Bristol
Barnstable
Edgartown
Nantucket
CONNECTICUT

CHAPTER FIFTEEN

Nineteenth Century Grape Breeders of New England

WHILE NEW ENGLAND MAY NOT have a reputation for producing noted grape breeders, history tells us a different story. During the nineteenth century, there were more than a dozen active New England grape breeders who communicated regularly with each other on their fruit breeding activities. In addition to grape breeders such as Ephraim Bull, who developed Concord and Cottage; Edward S. Rogers, who developed the Rogers series of grape hybrids; and Captain John Moore, who developed Moore Early, there were many other New England grape breeders who either created, or identified, new and exciting grape varieties. These nineteenth century grape varieties have added significantly to America's horticultural legacy, to its inventory of fruits, and to the genetic material that continues to be used as breeding material to create even newer grape hybrids.

New England's grape and other fruit breeders, horticulturalists, growers, writers, nursery owners, and botanical chroniclers resided not only in eastern Massachusetts, but throughout New England. These individuals participated in New England's prestigious horticultural and agricultural societies such as the nationally renowned Boston-based Massachusetts Horticultural Society and the American Pomological Society;[1] regional societies founded in Worcester, Middlesex, Essex, Franklin, Hampshire, and Norfolk counties in Massachusetts; and those established throughout the rest of New England such as the Hartford County Horticultural Society and the Lake Champlain Horticultural Society. It was in the context of these horticultural and agricultural societies and nurseries that the bulk of communication occurred. Their members collaborated and shared information on the latest growing techniques and grape varieties.

Grape breeding in New England. *During the nineteenth century, grape breeding programs were developed throughout New England, including Massachusetts, Vermont, and Connecticut.*

Early Grape Breeding

Between 1840 and 1890, the Mid-Hudson River Valley, eastern Massachusetts, and the St. Louis, Missouri, area were the three centers of American grape breeding. In Massachusetts, breeders like Edward Staniford Rogers, John Fisk Allen, and George Haskell were wealthy New England Brahmins whose families made their fortunes either as merchants or in mercantile shipping. These gentleman farmers engaged in horticulture for intellectual stimulation and social comradery.

From the late eighteenth century to the end of the nineteenth century, several forces compelled these Massachusetts horticulturalists to develop new hybrid fruit varieties. Many desired to create new plant material for their suburban and rural homes, emulating the British landed gentry and securing greater social prestige within their community. Others desired to develop fruits that were more productive, disease resistant, and with earlier ripening times, for profit. An underlying theme was the uniquely New England quasi-religious, social-ethical belief among business, social, and religious leaders that one's devotion to and worship of God could be manifested by service to community. Work had a moral component. The highest divine calling was to be productive, not by trading goods, but by engaging in agriculture and manufacturing activities.[2]

By 1800, the region's thin agricultural soils had become even more depleted due to more than one-hundred and fifty years of extensive, but unwise, cultivation practices. A movement arose to study agriculture, hybridization, and plant sciences so that local farmers could revitalize their increasingly poor and over-cropped soils.

The business community supported these agricultural research initiatives so that farmers would remain in Massachusetts, and continue to be loyal customers, rather than see them being forced to move further west in search of more fertile soils. The business community's support was evidenced by the founding of the Massachusetts Society for Promoting Agriculture earlier in 1792. Its membership was clearly mercantile in composition, including most of eastern Massachusetts's prominent families, along with attorneys and a few physicians and clergy, most of whom were Harvard College alumni.[3]

As early as 1800, nursery owner Charles Mason Hovey facilitated communication between nearby breeders and regularly corresponded with the national fruit breeding community either in person or through the many plant catalogs, pamphlets, and horticultural books that he wrote from 1835 until 1885.[4] In addition, Hovey helped to direct, along with Marshall Pinckney Wilder, the nationally recognized Massachusetts Horticultural Society and American Pomological Society. Local agricultural and horticultural societies were also actively evaluating new grape varieties in most of eastern Massachusetts.

As a result of all this agricultural activity, New England produced a number of notable grape breeders, though many have been forgotten today. The work and lives of New England's most noted grape breeders and the resultant grape varieties that they created are detailed on the following pages.

Rogers's Prominent Role

The most prominent and prolific of the New England grape breeders was Edward Staniford Rogers. Rogers was one of the first to successfully breed *labrusca-vinifera* hybrids in the United States and inspired many after him to do the same.

These varieties, because of their winter hardiness, vigor, diseases resistance, and interesting flavor profiles, should be considered by those who are interested in pursuing more sustainable grape culture. Growers should pay closer attention to the Rogers hybrids, especially in our rapidly changing climate, where we are experiencing higher incidences of violent weather patterns with wider swings in temperature and precipitation throughout the year.

Since the 1930s, rarely have Rogers's hybrids, as a class, been closely studied or written about. It is my hope that by concentrating this chapter on the work of E. S. Rogers and his hybrids that more light can be shed on the horticultural work and advances that Rogers and his contemporary colleagues made in grape breeding and cultivating more resilient grape varieties that can be reliably grown.

Edward Staniford Rogers

(1826–1899)

Edward Staniford Rogers, Esq. was born on June 28, 1826 into a prosperous mercantile shipping family in Salem, Massachusetts. His father Nathaniel Leverett Rogers (1785–1858) and his uncles, John Whittingham Rogers (1787–1872) and Richard Saltonstall Rogers (1790–1873), began their careers as shipping factors (agents) and advanced to become supercargoes and sea captains, who then founded the firm N. L. Rogers & Bros. The firm opened American trade with Zanzibar and Australia, and sailed their ships often to Sumatra, India, Thailand, Tahiti, and Europe.

The Rogers family was closely related to the Dodge, Saltonstall, and Leverett families. For readers outside of New England, these families were the elite Brahmin families of Boston, and many of their descendants became the Commonwealth's first governors, U. S. Senators, members of Congress, prominent business leaders, and clergy. An indication of the Rogers family's high economic, political, and social status, Nathaniel L. Rogers was a member and onetime president of the East India Marine Society, the predecessor to the now prestigious Peabody Essex Museum in Salem. In sum, the Rogers family were socially, politically, and economically one of the most prominent families of New England.

Nathaniel Leverett Rogers married Harriet Wait (1790–1882) and had four children, Augustus Dodge Rogers (1823–1896), Henry Whittingham Rogers (1824–1855), Edward Staniford Rogers (1826–1899), and Harriet Wait Rogers (1828–1859). The family lived in a mansion at 376 Essex Street in Salem, Massachusetts. Their home on Essex Street, between Flint and Monroe Streets, was near the current Salem Public Library. Essex Street, one of Salem's main thoroughfares, is located near the city's center near the wharves of Salem Harbor, then one of the busiest seaports on the Atlantic seaboard. This area of Salem contained most of the palatial residences where the local elite resided.

Rogers was educated in Master Ira Cheever's School of Salem. After completing his education, he made several voyages on his father's ships employed as a clerk and supercargo. A supercargo is a representative of the merchant on a ship who is responsible for overseeing the merchant's cargo, its sale, and all commercial matters related to such cargo. After his sea voyages, Rogers was employed in the firm's counting room in Salem.

Rogers ceased to work in the counting room after his father's death in 1858, and was free to pursue his long-held horticultural passions. Rogers continued to live in the old family mansion with his mother and brother Augustus, where for fifteen years he withdrew almost completely from the public eye so that he could devote his full time to his garden.[5]

Fruit Cultivation and Breeding

In the mid-nineteenth century, it was in the backyard of the Rogers home on Essex Street, that Rogers pursued his passion for horticulture and conducted his extensive grape breeding activities. While the garden behind his house was quite large for a city lot, it was only one-half acre in size.[6]

Since Rogers came from one of the wealthy Massachusetts' Brahmin families, he had the time, money, and social standing to successfully pursue horticulture and botanical studies. Further, to help finance his expensive passion for horticulture, he continued to conduct business to earn an income, even though it was not in the family shipping and merchant business. He earned at least some of his income by purchasing and subdividing land in the Town of Rockport, Massachusetts.[7]

Edward Rogers was, by temperament, quiet and retiring. According to horticulturalist U. P. Hedrick, Rogers possessed a "literary ability" and was an extensive reader, but "could rarely be drawn into conversation excepting among his most intimate friends who were wont to "drop in" at his long, low greenhouse in the garden or at his office, extemporized in the old colonial barn at the rear of the house." In sum, he was an introvert, a recluse. He was very generous about giving his grape cuttings and plants to friends and visitors, and so he made little money from his breeding activities or the sale of his vines. In addition, as his one-half-acre garden was not large enough to establish a commercial nursery business and he did not have sufficient room for proper facilities to propagate more plant material, he tended to freely distribute his plant material widely.[8]

Of this garden and the man, Marshall P. Wilder, President of the Massachusetts Horticultural Society (1841–1848) and the American Pomological Society (1850–1886),[9] said "[h]ow much can be done with little is illustrated by the fact that all of the 45 Rogers' [sic] hybrids were produced by a lame man in a half-acre city lot 150 years in cultivation." Further, he said of the garden, "[i]t is 150 years old; a cold matted soil, filled with old apple and pear trees, currant bushes, flax, and everything mingled in together. It is in a close, hived up place in the City of Salem, and it is a wonder that he ever had a bunch of grapes to show."[10]

Rogers maintained his vineyard, orchard, and garden behind the family home until his mother's death in 1882. When she died, the Rogers home was sold and the two brothers, Edward and Augustus, moved to another house in Salem. No one knows what happened to the garden behind his former home or what plant material, if any, was removed and transplanted to an alternate site. Upon the death of his brother in 1896, Rogers moved to his last home in nearby Peabody, Massachusetts. Here, he cultivated trees, roses, and flowers for pleasure and experiment until his death three years later in 1899. Horticulturalist U. P. Hedrick reports that a personal injury accident, at some undisclosed time towards the end of his life, resulted in a permanent injury to his body, which prevented much physical labor and may have hastened his ultimate death. However, other sources maintain that he was or became "lame" at a much earlier point in his life.[11]

Grape Breeding Program

In 1851, Rogers began his breeding program in Salem. What sparked Rogers's interest in fruit hybridization was an article written by the English botanist, Dr. John Lindley (1799–1865)[12] of the University of London, which was originally printed in the London Horticulturalist and reprinted in A. J. Downing's[13] *Horticulturalist & Journal of Rural Art and Rural Taste* in September 1847. The article was entitled "Remarks on Hybridizing Plants" and was a general discussion of the results of this practice at that time. Ultimately, Rogers named one of his grapes in honor of Lindley. In addition, Rogers must have noticed the grape breeding work of John Fisk Allen[14] since they lived only two blocks from each other. The Allen family resided at the townhouses located at 31–35 Chestnut Street.[15]

Starting in 1851, Rogers made crosses between a local native-American *labrusca (– vinifera?)* variety called Carter, also known as Mammoth Globe, with the European *vinifera* grape varieties Black Hamburg and White Chasselas. Carter, being very similar to Sage, a local large-fruited red variety that is self-sterile, was the seed parent of all of Rogers's first-generation hybrids and the only generation of his hybrids to exist today. The pollen used to fertilize Carter were Black Hamburg and White Chasselas, two of the most commonly available European *vinifera* varieties grown in Massachusetts. The pollen from these *vinifera* grapes was obtained from vines growing in a cold grapery nearby, possibly from the grapery of John Fisk Allen or George Haskell, who resided in nearby Ipswich. A cold grapery is a building where grapes grow under glass, but without a heat source such as in a hothouse or greenhouse. The glass structure does, however, provide enough shelter from the cold winter winds and elements and the glass roof and walls retain heat to moderate its temperature.

To cross pollinate, the Carter blossoms were emasculated and fertilized with the *vinifera* pollen and small cotton bags were placed over the Carter female flowers. In addition to repeated applications of pollen to the stigmas (seed or female part of the flower), Rogers placed clusters of *vinifera* blossoms in the enclosed cotton sacks. From this cross-pollination work, he secured new clusters of grapes that ultimately consisted of about one hundred and fifty seeds. These seeds were then planted in his backyard garden that fall. The following spring, many of these seeds germinated, but cutworms and other misfortunes reduced the number of plants to forty-five fruiting vines. These forty-five vines grew upward on poles for three years, where they were originally planted.

However, due to overcrowding, twenty-five of such plants needed to be transplanted to other parts of the garden to give them enough room to grow. The untransplanted vines started to bear fruit in 1856, and the transplanted varieties fruited a few years later.

Rogers's Seedlings

Rogers's seedlings were numbered one to forty-five and were known by such numbers for over fifteen years, at which time, some of them were given names. Of these vines, one to five, inclusive, were of a Carter x Black Hamburg cross; six to fourteen, inclusive, were Carter x White Chasselas hybrids; and the remaining numbers from fifteen to forty-five were Carter x Black Hamburg crosses. Subsequently, there was one Rogers hybrid that was assigned a new number of 53, but this was done because there was confusion with the original Rogers No. 22, which ultimately became known as "Salem."

During the years 1858 and 1859, Rogers sent cuttings of these numbered varieties to growers and horticulturalists for further testing. He disseminated these varieties, probably due to the small size of his backyard and because the common practice, then, as it is now, is to share plant material with colleagues to get comments on the growing attributes, strengths, and weaknesses of such plants in a wide range climates and soil types.

In 1867, the first Rogers hybrid, No. 22 (or the redesignated No. 53) was given the name "Salem" by Rogers, in honor of his own hometown. Two years later, at the request of the Lake Shore Grape Growers' Association, Rogers gave names to eleven of his other hybrids. The names selected were to recognize the contributions of persons noted for their horticultural or literary achievements, or for local places or persons of historical note. The named varieties, in numerical order by their hybridization number are as follows: Goethe (No. 1); Massasoit (No. 3); Wilder (No. 4); Lindley (No. 9); Gaertner (No. 14); Agawam (No. 15); Merrimac (No. 19); Requa (No. 28); Aminia (No. 39); Essex (No. 41); Barry (No. 43); Herbert (No. 44); Salem (No. 22 or 53).[16] As noted earlier, of these named Rogers hybrids the overwhelming majority of them were of Carter x Black Hamburg progeny; only Lindley and Gaertner were of Carter x White Chasselas progeny.

Nursery Business

After 1860, Rogers had a very large hobby or, alternatively, a small nursery business. He advertised in national and regional horticultural journals of the day. He shipped his hybrid vines to growers and nurseries across the United States and Canada, with most being shipped to the Northeast and Midwest. Some of his vines were shipped to prominent horticultural commentators and nursery owners such as Charles Downing,[17] A. J. Caywood, the Parsons Nursery,[18] George Haskell, and Samuel Miller, in addition to his main supporters, Charles Hovey (below) and Marshall Wilder. From 1866 to 1869, the number and quantity of his vine shipments that he was offering for sale was sizable – he was sending out shipments from as little as one sample each of his thirteen varieties, to quantities of between twenty-five to over 200 vines. After 1870, it seems that his nursery activity was subsiding as other nursery owners now had his new grape creations and were offering them for sale through their own nursery businesses. From Rogers's records, his most popular selling varieties where Agawam, Salem, Wilder, Massasoit, and Lindley.[19]

Charles Hovey

Between 1851 and 1867, when Rogers officially introduced his first generation of hybrids to the public, he continued his grape breeding program. His personal papers indicate that the Rogers hybrids would be "a foundation for by again [sic] crossing with the foreign species varieties which shall have all excellent qualities of the fruit of the latter with the hardihood [sic] and adaptation to our climate which belongs to our indigenous vines. And from these kinds already introduced – chance seedlings may be obtained some of which… will have a tenderness return to the foreign parents with the hardiness of the native… ." It is possible that Rogers waited so long to release his first generation of hybrids in 1867 because he was working on his second and third generation of hybrids which he had hoped to release to the public instead.[20]

From the Rogers papers, there are notes dated February 10, 1868 which detail his hybridization work conducted between 1862 and 1868. It seems that Rogers was satisfied with the disease resistance, hardiness, and early ripening time of his first generation of Rogers hybrids. This is evidenced because all of his next generation hybrids used as its seed parent the Rogers hybrids Agawam, Massasoit, Merrimac, and Rogers No. 2 and 12. The pollen parent, as was the case with his first generation of hybrids, were all *vinifera* varieties, including Black Hamburg (black), Rose Chasselas (pink), Muscat of Alexandria (white), Queen of Nice (white), Malaga, and Syrian. The resultant hybrids were called Rogers No. 46–82 and were laid out in his outdoor garden and greenhouse. From Rogers activities, he was breeding in more *vinifera* to improve the flavors of his grapes and probably to make their flesh more like European varieties. These hybrids, while detailed in the Rogers papers, do not seem to have been promoted to the public at large.[21]

Today, commercially grown grape varieties are all self-fertile; so they do not need another variety to be nearby to cross-pollinate it to produce a crop. Unfortunately, Rogers hybrids tend to be self-sterile as was their seed parent, Carter. Except for Agawam (which is self-fertile), Goethe (which is more than partially self-fertile), and Lindley, Wilder, Essex, and Requa (which are not completely self-sterile), as a class, the Rogers hybrids are all self-sterile. This means that other varieties must be planted nearby for them to cross-pollinate and produce fruit. Further, the Rogers hybrids do not cross-pollinate each other. Upon reflection on the contents and condition of the Rogers garden, he must have had several or many other grape varieties planted there, otherwise, his hybrids would not have fruited in his garden. One of the major reasons why commercial growers did not plant Rogers hybrids, except for Agawam (self-fertile) and Salem (self-sterile), was due to their self-sterility. Further, some varieties were noted more than others for not being adaptive to various kinds of American soils.

Rogers did not write extensively about his breeding work, so it is best to quote him at length in order for the reader can get a feel for Rogers, both as a person and his thought processes as a hybridist:

> "When I commenced experimenting I had no knowledge of any one who had raised grapes by this process, though I had heard of flowers, pears, &c., and I had attempted crosses of pears. Reading articles in the London Horticulturalist, it occurred to me that I could get a new grape by this process; combining the qualities needed for open culture, it would be more valuable than any other fruit. Accordingly, in the spring of 1851, searching for a parent vine, I selected a Labrusca, called the 'Mammoth' for the native mother plants: this a large, brownish red or amber grape, one of the earliest and largest I could find. I chose as the staminate or foreign parent the Black Hamburg and White or Golden Chasselas, two of the hardiest foreign varieties, growing in a cold grapery near by. When the Mammoth was ready, several clusters were chosen. The petals, which were at the top, were cut away, and the stamens. Then the pistil of each blossom was touched with pollen, the vines in the grapery being retarded for the purpose, and the bunches protected by tying over them little bags of fine cotton cloth, enclosing in each bag a flower cluster, the same kind used as a fertilizer. The bags were opened the next day and the pistils retouched, after which the bags were left unopened a number of days. When opened, the berries were nearly all set. As soon as the grapes were ripe, the seeds were taken out and planted at once in a piece of ground protected by boards, in rows six inches apart, a few leaves being put over them as cold weather came on, as a winter protection.
>
> In the spring the covering was removed, and nearly every seed vegetated, coming up regularly as planted. But about fifty of the one hundred and fifty plants were raised, the others having been destroyed by the cut-worm soon after coming up. In the fall most of them were a foot or more high and were again protected for the winter by leaves and boards. The following spring there were many of them transplanted. A few vines from the Golden Chasselas were kept by themselves; the remainder, by Black Hamburg cross, were put by themselves, and each plant staked and numbered, the final numbering being from one to forty-five.
>
> The foliage of the young plants early showing plainly the foreign intermixture, they were

protected each winter, and in 1856 began to show fruit, and every one since had proved fruitful; in the fruit showing that they partook of the foreign elements, in the intermixture of the two species, combining the desired qualities, having large clusters, and much of the delicate flavor of the foreign sorts, with the earliness and hardiness of the native. The vines are even more vigorous than the parents, and more exempt from diseases, and more hardy than most out-door varieties."[22]

Hybridization and Breeding Goals

E. L. Lord, in a Florida State Horticultural Society publication, Isadore Bush, through his *Bushberg Catalogue*, and, to some extent U. P. Hedrick in his writings, maintained that Rogers's hybridization of native *labrusca* grapes with European *vinifera* varieties could obtain fruit for wine making that had the quality, subtle flavors, and high sugar content of European grapes and the disease resistance and winter hardiness of American varieties.[23] As opposed to American *labrusca* varieties, European *vinifera* varieties could bring to its progeny a greater variety of subtle flavors, high sugars for wine production, bunches and berries that are large and suitable for table grapes, and the absence of *labrusca*'s often slip skin berries. *Labrusca* varieties, on the other hand, brought to its progeny greater resistance to American fungus diseases, such as powdery mildew, downy mildew, and black rot, tolerance of cold winter temperatures, earlier ripening times, big berry and cluster size, high acid for juice production, and "grapey-foxy" flavors that are suitable for juice production and table use. Unlike *vinifera* varieties, *labrusca* varieties have the ability to adopt to and thrive in many different American soil types.[24]

Rogers's goal was to create grapes that could grow in Massachusetts and that had the dual purpose of being primarily high-value table grapes which could also produce quality wine if they could not be sold as fresh fruit. He hoped these dual-purpose grapes would have the American *labrusca* big berry and cluster size, early ripening time, and hardy fungus disease and winter resistance, while having, for table consumption, the European *vinifera* qualities of more subtle flavors and attractive fleshy consistency, unlike the American *labrusca* slip-skin varieties.

Rogers Hybrid No. 8

Another goal was to have his hybrids ripen much earlier to accommodate the shorter growing season of New England as opposed to the Middle Atlantic and Midwestern states and the South. This was because, between 1830 to 1850, the most common grape varieties grown in the United States were late-ripening grapes such as Catawba and Isabella. By 1830, these varieties had come to Massachusetts from the Middle Atlantic or Southern states of Maryland, Virginia, or North and South Carolina, and their place of origin had a much longer growing season.[25]

Rogers pursued grape breeding in a location where there was little commercial fruit or grape cultivation to speak of, except by families of means. This is not to say that there was no fruit cultivation in eastern Massachusetts. Yeoman farmers and those with small village lots did cultivate apples, pears, grapes, and bramble fruits in limited quantities for home consumption. Rogers pursued this endeavor with gusto in northeastern Massachusetts, where little grape growing occurred. However, as was pointed out earlier, Rogers lived close to George Haskell of Ipswich, and John Fisk Allen of Salem, who wrote a book on grape cultivation, and both of whom bred grapes; and the pomologist Robert Manning[26] of Salem. In addition, Marshall P. Wilder, of nearby Dorchester, and Charles Hovey[27] of Cambridge, actively publicized and promoted the efforts of Rogers, Allen, and Haskell.

In the end, Rogers bred grapes because it was the vogue thing to do for men of means and culture. In the 1840s and 1850s, there was a "grape mania" that overtook the American horticultural and scientific community. This mania manifested itself in breeding, publicizing results, evaluating new grape hybrids, and passing judgment on hundreds of new grape varieties. Rogers, who was fond of horticulture, and shy in public, conducted his pioneering work alone in the security of his backyard. He was among the first to do so successfully in America from the late 1850s to the 1890s and led the way for other Americans.[28]

Rogers was also one of the first in the United States to systematically, and on a large scale, successfully hybridize new grape varieties from *vinifera* and *labrusca* parents. Rogers chose as his genetic material locally-grown or found *labrusca* varieties and high-quality *vinifera* varieties that were being grown under glass. This shy man was a trend setter in this horticultural movement. Through Rogers's efforts—and that of Charles Hovey and Marshall Wilder, who promoted his work through the activities of the Massachusetts Horticultural Society[29] and the American Pomological Society—many dedicated American horticulturalists, botanists, and amateur gardeners answered the call to develop hundreds of new grape varieties. Rogers's use of Black Hamburg and White Chasselas as his breeding material to impart quality *vinifera* attributes led many other grape hybridizers to follow suit, including some of the most successful American breeders of the time, such as James H. Ricketts of Newburgh, New York; Dr. Richard T. and Stephen Underhill of Croton Point, New York; George W. Campbell[30] of Delaware, Ohio; and Jacob Moore of Attica, New York.[31]

The Horticultural Legacy of Rogers

The legacy of Rogers's work is two-fold. First, by being one of the first to successfully breed *labrusca-vinifera* hybrids in the United States, he inspired many other American grape breeders to do the same and to use the same kinds of genetic material. Along these lines, his pollen parent choices of Black Hamburg and White Chasselas, whether selected because that was the convention or his innovation, was replicated by many successful American breeders. Second, his first series of grape hybrids and their progeny are still being used by grape breeders as building blocks for the development of new grape varieties.

In addition to Rogers, American breeders such as Elmer Swenson of Osceola, Wisconsin[32] (1913–2004); George W. Campbell (1817–1898); James H. Ricketts (1818–1915); Nelson Bonney White[33] (1824–1914) of West Norwood, Massachusetts; the New York State Agricultural Experiment Station;[34] Jacob Moore[35] (1835–1908), and Dr. Richard T. Underhill (1802–1871) and his nephew Stephen W. Underhill[36] (1837–1925), all used Black Hamburg and White Chasselas as the primary pollen parents in their hybridization work. These hybridizers created hybrids such as ES 6-5-33, ES 6-6-50, and 6-8-25 (Swenson); Campbell Early[37] and Triumph (Campbell); Downing and Secretary (Ricketts); Norwood and Amber Queen (White); Diana Hamburg (Moore); and Croton and Irving (Underhill).

However, the nurseryman and fruit specialists Charles Downing (1802–1885) of Newburgh, New York, did not like the Rogers hybrids, because as a rule they were too coarse.

Building for the Future

Rogers hybrids have been used as building blocks for the development of future grape varieties. For example, the New York State Agricultural Experiment Station at Geneva developed the grapes Buffalo (Herbert x Watkins) and Sheridan (Herbert x Worden) which are essentially Rogers hybrids, and which ultimately became the foundation of the Geneva grape breeding program. Some of the more prominent later Geneva hybrids that contained Rogers progeny include, Steuben (Wayne x Sheridan), Corot Noir (S.V. 18-307 x Steuben), Noiret (NY 65.467.8 x Steuben), and GR-7 (Buffalo x Baco Noir).

Thomas Volney Munson[38] (1843–1913) of Denison, Texas, extensively used Rogers hybrids in his breeding program. Rogers hybrids were used for the early development of Brilliant (Lindley x Delaware) and Delago (Delaware x Goethe) which became a foundation of the Munson breeding program. From Brilliant, used as the pollen parent, came Dixie, Headlight, Hidalgo, Manito, Rupert, Valhallah, and Wapanuka, and a score of other hybrids. From Delago, as the seed parent only, came Amethyst, Augustina, Hidalgo, Meladel, and Melassum, and approximately twenty other Munson hybrids. Lindley and Herbert were used extensively as a seed parents, combined, for over ten of his new hybrids, while Goethe was used often as a pollen parent. Among Munson's more commonly used Herbert hybrids today are Captivator, a hybrid of (Herbert x Meladel (Delago x Brilliant)), Hernito, and President.

Many other public and private grape breeders used Rogers hybrids in their breeding programs. Examples of these breeders include Joseph Bachman of Altus, Arkansas, who developed Banner using Lindley as a seed parent (Lindley x Delaware); N. E. Hansen of the South Dakota Agricultural Experiment Station, who extensively used Agawam, Salem, Massasoit, Lindley, Merrimac, and Wilder; Clifford Ambers of Lynchburg, Virginia, who used Gaertner, Herbert, and Lindley as the seed parent, and Goethe, and Sheridan (a Rogers hybrid) as a pollen parent; Paul H. Shepard of the Missouri State Fruit Experiment Station, who used Agawam, Merrimac, and Herbert as seed parents; Elmer Swenson, who often used Buffalo (Herbert x Watkins) as a pollen parent; J. K. Stayman of Leavenworth, Kansas who used Goethe to breed Daisy (Goethe x Open Pollinated), and many others such as W. H. Lightfoot of Springfield, Illinois, and C. Engle of Paw Paw, Michigan who almost exclusively used Salem as a seed parent.[39]

* * *

When Rogers died on March 29, 1899, his obituary was widely printed in scores of major newspapers across the United States and Canada, especially in those areas of North America where grape cultivation was an important industry, and in major cities such as Boston, New York, San Francisco, Philadelphia, Albany, and Hartford. These obituaries uniformly cited Rogers's skill, achievements, and world fame in breeding a whole new class of hybrid grape varieties and pears. Further, they noted that Rogers deserved the credit for first artificially hybridizing grapes, and received the only gold medal ever awarded by the Massachusetts Horticultural Society for his breeding work.

Funeral services for Rogers were held at the North Church in Salem, a Congregational Unitarian house of worship on Essex Street, just one block away from his former home where he conducted his breeding work. Rogers is buried at the Broad Street Cemetery in Salem in the Wait Tomb.[40]

Breeding Material Used by Rogers

Before we discuss the Rogers hybrids, we will look at the breeding material that Rogers used to create these new dual-purpose table/wine grapes.

BLACK HAMBURG

Black Hamburg was used extensively by Rogers, American grape breeders such as Ricketts, Moore, Swenson, the Cornell University breeding program, and other American and German grape breeders in the nineteenth and early twentieth centuries. It is a pure *vinifera* variety that from the 1820s to early 1900s was commonly grown in heated and unheated greenhouses in New England and New York. In Rogers's breeding program, Black Hamburg was used exclusively as a pollen parent. For more information on Black Hamburg, see Chapter Eight, page 102, which outlines its growing habits and horticultural attributes.[41]

PARENTAGE
vinifera

HARVEST DATE
Mid-season

C C B A+ A

CARTER/MAMMOTH GLOBE

This native-American *labrusca* variety was the seed parent for all of Rogers's first generation inter-specific grape hybrids. The origin and genetic makeup of Carter/Mammoth Globe, herein after referred to as Carter, is not completely clear. What is clear is that it is a large-fruited red *labrusca* variety that was either found growing in the forests or fields near Salem or was brought to Salem by a local horticulturalist for cultivation. The exact origin and genetic ancestry of Carter may have been kept close to the vest by Rogers as its disclosure would have made it easier for competing breeders to replicate Rogers's breeding work.

PARENTAGE
labrusca or labrusca-vinifera

HARVEST DATE
Late early to early mid-season

A A B A C

In *The Grapes of New York* and *the Bushberg Catalogue* there are two very different ancestries and descriptions on the origins of Carter. Carter was first mentioned around 1831, the *Bushberg Catalogue* maintaining that the first of these alleged Carter varieties (*labrusca-vinifera*) was a seedling of Isabella. Isabella is a (*labrusca* x *vinifera*?) x (open pollinated) hybrid. The reference and description of Carter in the *Bushberg Catalogue* is very sparse, and not of great value considering that it was assembled by Isadore Bush, who had compiled this very comprehensive grape catalog for all American grape varieties.

What we do know about the possible, but very improbable, origin of Carter as being a seedling of Isabella is what can be pieced together from various sources. In 1816, the grape Isabella was found by the noted nurserymen William Prince[42] (1766–1842) of Flushing Landing, Long Island, New York, in the garden of Mrs. Isabella Gibbs, the wife of a prosperous merchant who lived in Brooklyn, New York. Her husband, Col. George Gibbs, was a very active member and on the Board of Trustees of the New York Horticultural Society. The origin of Isabella has been disputed as coming from alternatively North or South Carolina, Virginia, Delaware, or Europe.[43]

Because Isabella came from the South and was found in Brooklyn in 1816, it most likely did not venture up to Eastern Massachusetts in time to have been bred to create Carter which was first reported to have come from this part of the Commonwealth. So Isabella is most likely not one of the parents of Carter. The description of Isabella does not even fit the description of Carter, as Isabella is often described as having a moderately large cluster size, shouldered, with a medium-large dark-purple to black oval berry, with a heavy bloom (as opposed to Carter's lighter red color).

The second description of Carter (*labrusca*), from Hedrick's *The Grapes of New York*, labels it as a large-fruited, red *labrusca* variety that is local to the Salem area and is also known as Mammoth Globe. It was very similar, but not identical, to Sage. Further, Hedrick states that this was the Carter variety, the vine being of four or five years of age, that Rogers used in breeding his hybrids.[44]

With so much having been written about the Rogers hybrids and their extensive use by breeders, it is surprising that so little has been written about the seed parent of all of his first generation hybrids, which was most likely Mammoth Globe or a seedling of Sage (*labrusca*). If Carter/Mammoth Globe was the variety used as the seed parent, then a description of Sage will provide an idea of the true characteristics of the seed parent of the Rogers hybrids.

Hedrick states that Sage (*labrusca*), also known as Mammoth Sage, Mammoth, and Globe, was of interest because it represented a type of large-fruited, early ripening *labrusca* that had been used by breeders as the native seed parent in *vinifera* crosses. This Carter/Mammoth Globe seed parent variety, so similar to Sage, was used in breeding by Rogers, and by Nelson Bonney White (1824–1914) who was influenced by Rogers's work.

Sage is a chance seedling found by Henry E. Sage around 1811, on the banks of a small stream near Portland, Connecticut. In 1848, the variety was made known to the public by the horticulturalist John Fisk Allen, a neighbor of Rogers in Salem. This is at about the same time that Rogers started his own grape breeding activities. The Sage variety was described as being very foxy in aroma, berries very large, with a light chestnut or mahogany color. The cluster is small with three to six berries in a ball-like cluster. The grape is sweet, has hard pulp, ripens early, and is hardy like any wild grape.[45] This description of Sage indicates to me that Carter/Mammoth Globe is either one and the same as Sage, or was very closely related, being either a seedling of, or direct hybrid of, Sage.

It makes sense that Rogers may have been vague about the exact identity and history of the seed parent of all of his named hybrids or perhaps he did not know. In a manner similar to A. J. Caywood, who used his own undisclosed White Concord hybrid to develop the highly-prized grape variety Dutchess, Rogers either obtained or developed Carter for his series of hybrids.

To help confirm this theory on the seed parent of the Rogers hybrids, in the 1895 *Bushberg Catalogue* there is absolutely no mention of the grape varieties Globe, Mammoth, or Sage. Further, while it mentions Carter, the only entry is to cross-reference its description to the description of Isabella, where such entry, as aforementioned, states that Carter is one of many seedlings of Isabella, by the same name or by other names. However, there is also a very short one-sentence description for Mammoth Sage, which states that it is a *labrusca* variety from Onondaga County, New York, and is the mother of Rogers's hybrids.[46]

There are also other reliable descriptions of the seed parent that indicate that Carter is really Sage, Mammoth Sage, or a seedling that closely resembles Sage. In the *Report of the Commissioner of Agriculture for the Year 1864*, issued by the United States Government it stated that Rogers,

"... took the Mammoth or Sage grape for the pistillate vine. This is said to be a vigorous grower, clambering over rocks and trees in a wild state in eastern New England. It bears bunches, oval-shaped, on a stout stem, with three to six berries in a bunch, but these few berries very large. So that little else except hardiness, earliness, and size exist in it. It has much acidity to condemn it. It is considered a fair representation of the Mammoth or Sage, leaf, stem, berry, and bunch. It is thus described by one familiar with it:

"It is much like most of the wild Fox grapes of this vicinity, but the berries are much larger, light chestnut or mahogany color, and they have a flattened or compressed shape, instead of being round, frequently an inch in diameter. The bunch is small, with three to six berries in a round, ball-like cluster, with sometimes a side stem with one berry at the end of it for a shoulder. The stem of the bunch is not very long. The leaves usually are 'entire' with a short pointed termination at the end of the midrib. and two other points of the other divisions into which all American leaves are divided, making always either plainly, or in the rudimental state, five lobes. Thus the leaves are not much lobed, scarcely toothed, and have a rusty, woolly appearance. The young wood, last season's growth, is hard and wiry [?], and covered with bristles. The grape itself is sweet, but has a hard pulp, that some compare to a piece of India rubber when eating it. It is early, and perfectly hardy, as much so as any wild grape in this vicinity."

Later on in the same above quoted passage, it notes that the "broad, nearly square coarse leaf of the Mammoth, without distinct lobes, has become more acuminate, and with lobed leaves at the base of each leader, and the coarse wild leaf assumes a more delicate attenuation; yet, with the native hardness perfectly retained, large, finely moulded [sic.], and shouldered bunches, with two and only two types of color and several flavors, have been produced in the hybrids."[47]

In November of 1856, there was another mention of a grape called Carter in Hovey's *The Magazine of Horticulture*. "The Carter Grape – This is the name which is given to the new variety cultivated in Milford, Mass., and recently shown at the annual exhibition of the Massachusetts Horticultural Society. It is a very fine grape, earlier than the Isabella, and superior to it in every quality. The berries are round, of large size, and of a deep reddish color, approaching to black at full maturity. What its origin is we do not know, but if it can be traced we shall give a full account of it at another time." From this description and entry of the grape Carter, this too could be a variation or seedling of Mammoth Globe that was used as the seed parent by Rogers.[48]

In Rogers's obituary, published in the *Boston Evening Transcript*, it stated that Rogers purchased a Mammoth Globe vine (*labrusca*) from a man in Lowell in 1846.[49] In the end, it is most likely that Carter/Mammoth Globe, the seed parent of all of Rogers's first-generation hybrids, was Mammoth Globe, Sage, or a seedling of Sage that closely resembled Sage in all aspects. The reason for the lengthy discussion on this topic is that up to now, the literature has not been clear on the origin of the only seed parent that Rogers used in his first generation of hybrids, its growing habits, fungus disease resistance, winter hardiness, quality, or description of its fruit. This is an important point to be made to properly evaluate these hybrids.

Today, Isabella is rarely grown in the United States, but is grown extensively in the Republic of Moldovia in Eastern Europe to produce semi-sweet dark red wines. These wines are quite good and resemble ruby ports in flavor profile, acid balance, and sweetness. Carter/Sage/Mammoth Globe is no longer grown, but it lives on through its many Rogers hybrids and those hybrids developed by Munson and the New York State Agricultural Experiment Station.

WHITE CHASSELAS

White Chasselas is a white grape that most likely originated in Switzerland, around Lake Geneva. It was identified as a separate variety as early as the 1500s. In the nineteenth and early twentieth centuries, it was used extensively in American fruit-breeding programs to develop table and wine grapes. Rogers used White Chasselas as one of the pollen parents for his hybrids. Other American grape breeders that used White Chasselas included Underhill, Ricketts, and Campbell.[50]

PARENTAGE
vinifera

HARVEST DATE
Early mid-season

C+ C B A A

Cultivation of White Chasselas spread from Switzerland to eastern France where approximately 6,000 acres is grown mostly in Alsace, Pouilly-sur-Loire, and Haute-Savoie. Switzerland grows approximately 9,900 acres in the cantons of Vaud and Valais, with 2,800 acres being grown in the Baden region of Germany. It is also widely grown in Austria and Hungary. White Chasselas is known by many other local names, including: Fendant Blanc (Switzerland), Amber Chasselas, Golden Chasselas, and Chasselas Dore (U.S.), Royal Muscadine (England), and Gutedel (Austria and Germany).[51]

White Chasselas is a popular early-ripening grape, and its juicy thin-skinned berries makes it ideal as a table grape, as used in France, but it is also used to make soft wines in Switzerland. It is a vigorous variety, which if grown on weak rootstocks, encourages early maturity. The cluster hangs well on the vine at maturity, so its growing season can be extended. It also ships well, which is a desirable trait for a table grape. White Chasselas is very productive but is susceptible to downy and powdery mildew. The cluster is medium in size, cylindrical, loose, with round berries. The color is a clear green to amber yellow, with a thin, but tough skin.[52]

In the United States, between 1820 and 1900, it was commonly grown in greenhouses by amateur and professional gardeners and horticulturalists. It was used extensively for breeding new grape varieties that would wed subtle European flavors with the superior fungus disease resistance of American grape varieties and their early ripening times, big berry size, and winter hardiness.

Major Grape Varieties of Rogers

Here we outline the growth habit, winter hardiness, fungus disease resistance, flavor profiles, acid balance, and potential uses of Rogers's hybrids. The endnotes for each grape variety cite mostly older evaluations that were written in the latter nineteenth and early twentieth century. The reason for this is twofold.

First, it was during the 1870 and 1930 time period that the Rogers hybrids were actively being considered by growers across the United States and Canada for cultivation and for table use, juice, or wine production. It was during this time that scores of professional and amateur horticulturalists state agricultural experiment stations, and horticultural societies from across the continent were seriously evaluating these grape varieties. Because of this high interest, there were many expert observations being made by plant scientists which led to very accurate evaluations and descriptions of these Rogers hybrids. Those evaluations are still useful today.

Second, by surveying the horticultural and plant science literature during the late nineteenth and early twentieth centuries, one can begin to understand why the Rogers hybrids were so extensively evaluated and considered for use by so many sophisticated state grape breeding programs at state agricultural experiment stations across the United States and Canada, and by pioneer American grape breeders such as T. V. Munson, Elmer Swenson, and Clifford Ambers. The scope and breadth of those individuals, state and local horticultural societies, and state fruit experiment stations that grew and reported on the Rogers hybrids was wide and came from both fruit-growing states and states that were assessing if they could develop a viable grape industry. Such evaluations came from diverse states such as Massachusetts, Vermont, New York, Ohio, Missouri, Michigan, Illinois, Florida, Alabama, Tennessee, Colorado, South Dakota, and Kansas, and the Province of Ontario, Canada.

To confirm the evaluations made by these plant scientists of the late nineteenth to early twentieth centuries, this author has been growing, observing, and making wine from many of these Rogers hybrids for the past fifteen years. I currently grow Agawam, Aminia, Barry, Goethe, Herbert, Lindley, Massasoit, Merrimac, Salem, and Wilder. My observations on the growth habit, disease resistance, cold hardiness, and wine making potential of these hybrids has been incorporated into these variety's evaluations and descriptions. Importantly, they also make lovely and nuanced wines that the general public should and would like to consume.

As a rule, I have found that the Rogers hybrids are much more winter hardy than their *vinifera* pollen parent, and more like the winter hardy *labrusca* seed parent. However, the Rogers hybrids, while more resistant to fungus diseases than the *vinifera* pollen parent, are not as resistant to fungus diseases as Carter. As a class, the Rogers hybrids are somewhat susceptible to powdery mildew in very hot, humid, and dry weather, but are less susceptible to powdery mildew in wet years, and more susceptible to black rot. Overall, while there are some differences in winter hardiness and fungus disease resistance between the Rogers hybrids, as a class, they are remarkably similar. Further, in comparing Rogers hybrids, they are generally high in sugar, in a manner similar to *vinifera* varieties, and not like the less sweet *labrusca* seed parent. As to the taste of each variety, there are differences; some being more European in flavor, while others being more native-American.

Agawam *(Rogers No. 15)*

It is my hope that more of my colleagues in New England, New York, and Northeast Asia will consider growing these promising Rogers hybrids. Importantly, they make lovely and nuanced wines that the general public should and would like to consume.

AGAWAM (ROGERS NO. 15)

This variety, like all Rogers's hybrids, was bred in 1851 and first fruited in 1856. It was one of the most widely grown of the Rogers grape varieties in the latter nineteenth and early twentieth centuries. Agawam is a hybrid of Carter as the seed parent x Black Hamburg as the pollen parent. It was first mentioned as a possible quality variety in 1861 and placed in the *American Pomological Society Fruit Catalogue* in 1867. Rogers named Rogers No. 15 "Agawam" in 1869 at the request of the Lake Shore Grape Growers' Association who preferred grapes to have names and not hybrid numbers so that growers and the public could readily identify and utilize these varieties, and so that nurseries could more easily sell them. Agawam is the English name for a local Native-American tribe that had lived north of Salem, Massachusetts, on Cape Ann and to the west.[53]

PARENTAGE
vinifera

HARVEST DATE
Mid-season to late mid-season

A A- B+ A- A-

Agawam is a red grape, not a blue or black grape. The color is a maroon to a dull purplish-red with a lilac bloom, somewhat like Catawba; colorwise not as handsome as that of other red Rogers hybrids. Agawam is a vigorous to very vigorous growing variety that is hardy with thick blonde canes. It has an upright growth habit to an upright and then lateral growth habit, which indicates that it possesses *vinifera* characteristics. Agawam is winter hardy to very hardy, like Delaware, Iona, or Foch whose wood hardens off early. The variety is productive to very productive, but needs a careful spray program because it is moderately susceptible to fungus diseases, but less so than Salem.

Agawam prefers a somewhat heavy soil, growing better on clay than on sand or gravel. Further, it is chlorosis-resistant or nearly chlorosis-free in high lime soils. Its flowers bloom early mid-season, a few days to a week after Concord and at about the same time as Baco Blanc, Empire State, Chelois, and Winchell, and around the same time as Iona. Agawam is harvested around mid-season to late-mid season, soon after Concord.

Of the Rogers hybrids, Agawam is the only completely self-fertile variety with perfect flowers and upright stamens. It does not need another variety to cross-pollinate it to yield fruit, but its production is enhanced when placed in a mixed variety vineyard with compatible pollinators. Overall, it is a good grower and bearer of fruit, and was liked by growers in the late nineteenth to early twentieth centuries. However, Salem has been deemed to be better overall in quality, even though it is similar to Agawam in cluster and color.

The berry size is large to very large, like Concord, its shape being roundish to slightly oval. The attractive cluster is medium-large to large and somewhat loose to moderately compact. The shape of the cluster is short, rather broad, and tapering with one shoulder.

The grape itself has a coarse solid texture, with a thick tough skin, but it is not a slip skin variety like Concord. The sugar levels on Agawam are approximately 19° to 20° Brix, similar to Catawba, with acid levels like Concord; it has medium-high sugars for a *labrusca* hybrid, with medium acidity in its balance. It is a dual-purpose grape that is good for the table and for wine production.

Agawam makes an approachable quality white wine that is floral, aromatic, and fruity with a soft Muscat flavor and herbal finish that has some *labrusca* undertones. It blends well with other white wines. The flavor profile has fruit of Muscat, bananas, some ripe pineapple and melons, and is round and surprisingly complex with a firm, but nice acid/fruit balance. As a table grape, it is very rich and sweet, not vinous, but not overly foxy either, with a particularly aromatic Muscat quality. The grape improves in flavor after it has been picked, will adhere to its stem, and can shrivel to become raisin-like.[54]

Agawam was commonly grown in Chautauqua County, New York, in Oregon, and on the Niagara Peninsula of Canada, as late as the 1960s. It is currently being grown in Florida and Ohio. I have been growing Agawam at my farm Cedar Cliff in Athens, New York, and find it to be good in the field and in the cellar. This variety should be reexamined by cool climate grape growers because it is a solid heirloom variety.

BARRY (ROGERS NO. 43)

Barry is one of Rogers's best black grapes. This variety is named after Patrick Barry,[55] a prominent Rochester, New York, nurseryman, pomologist, viticulturalist, and horticultural writer. Barry was bred in 1851, first noted separately in 1864, and named in 1869 by E. S. Rogers at the request of the Lake Shore Grape Growers' Association. It was first listed by the *American Pomological Society Fruit Catalogue* that same year.

PARENTAGE
labrusca, vinifera

HARVEST DATE
Late mid-season to early late season

A	B	B	B+	B

Barry is a Carter x Black Hamburg hybrid. The color is black to dark purplish-black, glossy with a heavy blue bloom. Barry is a vigorous variety that is winter hardy to very winter hardy in a manner similar to Delaware, Iona, or Foch. The variety is productive to very productive and somewhat susceptible to fungus diseases, especially powdery mildew. It has a modest upward growth habit and open canopy on long canes. Its flowers bloom by mid-season, around the same time as Concord, and it is sterile with reflexed stamens. Its best pollinators include other mid-season, self-fertile varieties such as Baco Blanc, Campbell Early, Chambourcin, Concord, Corot Noir, Cottage, Delaware, Diana, Empire State, and Noiret. Barry is harvested late mid-season to early late-season, after Concord. Barry will readily cross pollinate in a mixed variety vineyard.

The berry is large to very large, resembling Black Hamburg, its pollen parent, and has an oval to spherical shaped berry with a thin skin that adheres to the pulp. The attractive cluster is compact, medium to large, short, very broad, slightly tapering to cylindrical, that sub-divides a lot with double shoulders. Hedrick liked the variety, saying that it was attractive and "one of our best black grapes." It somewhat resembles Black Hamburg in berry and in flavor, which is delicate and sweet, flesh tender, vinous and pleasant. I have been growing Barry for over ten years and finds that it has a soft and appealing Concord-like taste, with little or no Muscat flavors.[56]

GOETHE (ROGERS NO. 1)

Goethe is Rogers hybrid No. 1 and is named after Johann Wolfgang von Goethe[57] (1749–1832), a leading German literary figure, playwright, novelist, botanist, naturalist, and statesman. Only three of Rogers's grapes were named in honor of Europeans, the others being Lindley, the English botanist and horticulturalist, and Gaertner, the German pioneer in the study of hybridization. Goethe is a Carter x Black Hamburg hybrid, and was bred in 1851, along with the other first generation of Rogers hybrids. This is a flavorful, high-quality grape. Of the Rogers's hybrids, it shows more *vinifera* Muscat flavors and less of the "wild" *labrusca* flavors that meld well together in a unified flavor profile.

PARENTAGE
labrusca, vinifera

HARVEST DATE
Late to very late

A	B+	B-	B	A

Goethe's color is like Catawba, red-brown to brick red to pale red covered with a slight gray-lilac bloom. It was first listed by the American Pomological Society in 1867, two years before it was named. Descriptions of its vigorousness range from being a medium difficult grower to being vigorous to very vigorous in growth. This varied description of its growth habit may be because Goethe is sensitive to the soils that it grows in.

Goethe is reported to grow best in sandy and gravel soils. In very rich clay soils it grows too much foliage and is too vigorous to yield a quality crop. Overall, a sandy loam soil is best. The roots can be weak and degrade in heavy clay soils and it may be susceptible to phylloxera. Further, it is chlorosis-resistant or nearly chlorosis-free in high lime soils.[58] Goethe is productive, but only on the soils in which it grows best. The variety can be too productive in overly rich soils and needs crop thinning to balance its crop. It is not productive enough in lean soils.

The variety is hardy to very winter hardy. It is fairly immune to somewhat susceptible to fungus diseases; however, it is more sensitive to powdery mildew. It was universally liked both in the field and in the quality of the grape by commentators such as Isadore Bush, U. P. Hedrick, George Husmann, and Lucie Morton. The variety blooms by mid-season and is one of the few Rogers hybrids that is partially self-fertile with stamens upright, but it should have another pollinator nearby. Some of the more effective self-fertile varieties that can be used to pollinate it are Campbell Early, Chambourcin, Concord, Cottage, Delaware, Diana, Humbert #3, and Noiret. The harvest date for Goethe is late to very late, along with Catawba and Diana, and in more northern vineyards it may not ripen properly.[59]

The berry size is above medium to very large, with an oval to round berry. The skin is thin and adheres moderately to the pulp in a manner similar to a *vinifera* variety. The cluster is attractive, of medium length, but broad and tapering, with a single shoulder that is somewhat compact.

Overall, Goethe is one of the finer Rogers hybrids in the field being vigorous, productive, of good winter hardiness, and somewhat resistant to fungus diseases. Its only downside being its late harvest date in the Hudson Valley and eastern Massachusetts. It is not likely to ripen adequately at all north of these two regions. Hedrick stated that Goethe, of all the Rogers hybrids, "shows most of the Vinifera characters, resembling in appearance to a marked degree the White Malaga of European fame and not falling far short of the best Old World grape in quality" but needs to be grown well.[60] Bush said that it was a "very valuable variety ... and shows in its fruit more of the character of the European species than any of Rogers [sic] other sorts...."[61] Both Hedrick and Bush liked the wine; Morton liked the grape and wine, believing it to be both hardy and disease resistant with red fruit that has fine *vinifera* character.[62] Goethe does require cluster thinning to prevent over bearing. Morton recommends that this grape should be reintroduced to our home vineyards. However, due to its late maturity date, it is best adapted to the Mid-Atlantic states and along the lower Ohio, Hudson, and Mississippi River Valleys.

Of the Rogers hybrids, Goethe seems to be one of the best quality grapes for the table and for wine production. As a table grape, it keeps very well. Its flesh is tender and has a pleasant soft *labrusca* taste that is aromatic, but with elements of European flavor and less harsh native *labrusca* flavors.[63] For wine, it is successfully grown in southern Brazil in the Urussanga region in the state of Santa Catarina. Here,

Goethe has mutated into a white and pink/red clone. It is used to make still and sparkling wine that reportedly has a very vinous *vinifera*, and sprightly taste.[64] Hedrick reported Goethe reaching sugar levels of 18.8° Brix, which means that in the nineteenth century it was considered as a wine grape. Goethe has sugar levels like that of Captivator at around 18° to 19° Brix, and low acid levels like Delaware, Diamond, or Salem, and so it is a relatively high sugar, low acid *labrusca* hybrid. I planted this variety in 2016 at my farm in the Hudson Valley, and from my experience I have to say the descriptions cited above look to be accurate.[65]

This grape had enough promise that Munson and Clifford Ambers of Lynchburg, Virginia, used Goethe extensively in their grape breeding programs. Munson used Delago (Delaware x Goethe) as a seed parent to create Hidalgo and Meladel, of which its progeny includes Captivator.[66] This grape should be re-examined to be used as an heirloom grape of historic note for the garden in the long season growing regions of the lower Hudson Valley, Long Island, Massachusetts coast, and the warmer parts of the Middle Atlantic States.

HERBERT (ROGERS NO. 44)

Herbert is a quality black heirloom grape that can be used for wine or the table. Its most noted feature is that it was used extensively by the New York State Agricultural Experiment Station grape breeding program and by other grape breeders. From Herbert came direct descendant hybrids Buffalo (Herbert x Watkins) and Sheridan (Herbert x Worden), which led to the development of Steuben (Wayne x Sheridan) and then to Corot Noir (S.V. 18-307 x Steuben), Noiret (NY 65.467 x Steuben), and Geneva Red (GR-7) (Buffalo x Baco Noir).

PARENTAGE
labrusca, vinifera

HARVEST DATE
Mid-season

A- B+ B- B B

Like most other Rogers hybrids, it is a Carter x Black Hamburg cross. Its color is dull blue-black to black with a thick blue bloom. It was first listed in the *American Pomological Society Fruit Catalogue* in 1869, the same year that it was named. Unlike the other hybrids that Rogers named, this author has not been able to determine why, or for who, the grape "Herbert" was named. However, it was named by Rogers in 1869 at the request of the Lake Shore Grape Growers' Association, so the name must have some significance to Rogers.

Herbert is vigorous to very vigorous for a *labrusca-vinifera* hybrid and is productive. It produces a quality black grape, and its growth habit is open and airy which helps to keep fungus diseases in check. Its canes are long and thick with wood that grows laterally and hardens off early. The variety is winter hardy to moderately winter hardy. Its reported fungus disease resistance is good. My own observations in growing this variety is that it has good cold hardiness and fungus disease resistance. Herbert was deemed to have possessed sufficiently favorable genetic characteristics and so it was extensively used by the New York State Agricultural Experiment Station in its breeding program. It is chlorosis resistant or nearly chlorosis-free in high lime soils. The bloom date is early mid-season to mid-season, about one day or more before Concord or Goethe. As with most Rogers hybrids, it has a self-sterile female flower with reflexed stamens. Its best pollinators are other early mid-season to mid-season, self-fertile flowering varieties such as Campbell Early, Chambourcin, Concord, Cottage, Delaware, Diana, and Humbert #3, but not any of the self-infertile Rogers hybrids. It pollenates well in a mixed variety vineyard to produce full clusters.[67]

The berry is round oval, and very large to medium-large, but is irregular in size. The skin adheres to the flesh of the berry somewhat. The cluster, like most first generation Rogers hybrids, is large to medium-large, broad and tapering. The cluster can be rather long, with one shoulder and moderately compact to loose. The harvest date is mid-season to a bit earlier, along the lines of Concord and earlier than Barry.

For a grape that was not really popular with growers, it had a favorable following with nineteenth and early twentieth century grape commentators. Hedrick placed it high as a garden or home vineyard type grape that almost equals Barry in taste, but Herbert is a handsome fruit nearly equal to Black Hamburg.[68] Isadore Bush of *Bushberg's Catalogue* called it "probably the best of the black varieties of Rogers's hybrids,"[69] and Campbell of Ohio liked the grape.[70] As noted earlier, Herbert was liked by grape breeders. It was used by Munson as a seed parent, and used to develop Captivator (Herbert x Meladel (Delago x Brilliant)), Hernito, and President. Clifford Ambers of Virginia and Paul H. Shepard of Missouri State Fruit Experiment Station both used Herbert as seed parents, and Elmer Swenson used Buffalo (Herbert x Watkins) often as a pollen parent.

Herbert was a true dual-purpose grape, with an emphasis on wine with sugar levels measured at between 18.8° to 20.7° Brix. Herbert is a sweet grape, for a *labrusca* hybrid, like Ontario or Ulster, with relatively high acid levels that are above Iona and similar to Ives, making it a high-sugar, high-acid grape similar to Catawba.[71] It does not have strong *labrusca* flavors or nose and its *labrusca* and musky flavors were suppressed. Overall, its taste is of a soft Concord variety with few Muscat flavors.[72] While it is sprightly and refreshing, it is not cloying in taste. Herbert retained the high acids of its *labrusca* seed parent and not of its *vinifera* pollen parent. All of these favorable flavor and acid profiles are possessed by a grapevine that has desirable growing characteristics in the field. Based on my own observations, I feel this grape variety should be reexamined and considered for future plantings as an heirloom of curiosity or garden type grape for home consumption.

LINDLEY (ROGERS NO. 9)

Lindley is a high quality dual-purpose red grape that was named after John Lindley (1799–1865), an English botanist, educator, horticulturalist, author, and botanical illustrator. Rogers named the grape Lindley at the request of the Lake Shore Grape Growers' Association in 1869. Lindley is a hybrid of Carter x White Chasselas, bred in 1851, first mentioned as a variety in 1862 and was included in the *American Pomological Society Fruit Catalogue* in 1867. The variety is acknowledged as one of the best red grapes developed by Rogers, and is unique in that it is one of only two named Rogers hybrids that has as its pollen parent White Chasselas, instead of Black Hamburg.

PARENTAGE
labrusca, vinifera

HARVEST DATE
Mid–season

B B+ B- B A

Lindley is a very attractive red, dark red to dark reddish-brown or brick red-colored grape, with a lilac to light-blue bloom, unlike Catawba. The variety is vigorous to very vigorous and healthy, but has only medium winter hardiness and so can be injured when placed in exposed locations that experience damaging cold winter winds. Lindley has also been reported to be a precarious producer which is on average less productive than other Rogers hybrids, but which can produce a suitable crop if placed near a compatible pollinator. The leaf pattern is similar to the beautifully pointed Delaware leaf. It has been reported as being fairly healthy and only somewhat susceptible to fungus diseases. As the vine gets older, its production can decline. Some viticulturalists maintain that it can be very susceptible to fungus diseases and vulnerable to attacks by leafhoppers and thrip, so a good spray program should not be neglected. Lindley does not adapt well to many different soil types, and is more productive on loam and well-drained clay soils than it is on gravel or sand-based soils; it has a light to medium incidence of chlorosis in high lime soils.

In contrast to the rather negative comments about Lindley by earlier commentators, I personally find the grape to be good in the field, productive, and disease resistant, and produces quality fruit. It is not as precocious a grower as some nineteenth and twentieth century commentators have suggested, and seems to pollinate well to produce an ample crop.

Lindley blooms by mid-season to late–mid-season, a few days after Concord. It has a self-sterile female flower with reflexed stamens that some evaluators believed was hard to find a pollinator to cross it with. Studies had indicated that it does not pollinate well and that its chances to be pollinated by Catawba were measured at 63%, Worden at 70%, and Niagara 77%, and these varieties are some of its best pollinators. I believe that the earlier cited pollinators may not be very suitable for Lindley. In addition, other suitable self-fertile pollinators include Baco Blanc, B. 6055, Chambourcin, Chardonel, Chelois, Concord, Corot Noir, Cottage, Delaware, Empire State, and Noiret.

The berry of Lindley is large to medium-large and is uniformly well formed roundish to slightly oval. Its skin is thick and rather tough, but not objectionable, and does not shatter with the skin adhering to the pulp. The attractive cluster is somewhat loose because it is generally not well-filled due to fertilization issues. The pretty cluster is narrow and long, but inclined to be tapering to cylindrical, often with a single shoulder. Its harvest date can be mid-season, around the same time as Concord, depending on its crop load.[73]

The wine and table quality of Lindley is very good to excellent, along the lines of Delaware. It has rich aromatic flavors and is a true wine grape with Brix measured at 19.5° to 20.5°. It has low acids, well below Concord, Agawam, or Jefferson, and is more like Delaware, making it a relatively high sugar, low acid grape for a *labrusca* hybrid.[74] Lindley and Lindley hybrids were used extensively in T. V. Munson's breeding program to create new table and wine grapes, with Brilliant (Lindley x Delaware) being the most common pollen parent. This was used to create varieties such as Captivator, Headlight, Hidalgo, Manito, Valhallah, and Wapanuka. Lindley was also used by N. E. Hansen of the South Dakota Agricultural Experiment Station, and by East Coast viticulturist Clifford Ambers as a seed parent for his hybrids.

Overall, it is a quality table grape that makes a quality *labrusca* type table wine that is slightly musky. The flavors include guava, mixed tropical fruit juices, mangoes, pineapple, with other wild, but pleasant tropical flavors. U. P. Hedrick maintained that Lindley was the best of the red Rogers hybrids.[75] Marshall Wilder of the Massachusetts Horticultural Society called Lindley and Jefferson the "Muscats of America," and the *Bushberg Catalogue* called it "beautiful and valuable."[76]

I planted Lindley in 2016 and look forward to fully evaluating this variety, though what I have seen so far is encouraging. Lindley, because of past favorable evaluations despite its pollination issues, should be looked at and kept in play for wine, table grapes, or for breeding purposes.

MASSASOIT (ROGERS NO. 3)

Rogers named Massasoit in 1869 in honor of the local historical Native American chief of the Wampanog tribe.[77] It is a Carter x Black Hamburg hybrid. The color is like Catawba – dark-brownish red with lilac bloom to a red color that is lighter than Delaware. It was first listed in the *American Pomological Society Fruit Catalogue* in 1867. Massasoit is strong, hardy, and a vigorous to very vigorous grower for a *labrusca-vinifera* hybrid which grows like Concord.

PARENTAGE
labrusca, vinifera

HARVEST DATE
Mid–season

A C+ A- B- B-

Massasoit is a productive variety with a growth habit that is lateral to slightly downward with thick canes. It is winter hardy because its wood hardens off early. However, it is fairly susceptible to fungus diseases. Its flowers are self-sterile with reflex stamens that bloom by mid-season. It will occasionally set out very partially filled clusters of small to very small seedless grapes. Because of this phenomena, Massasoit has also been referred to as Williams Seedless. This characteristic may be of value to seedless grape breeders. This author has grown and tasted these seedless Williams clusters and they are quite a treat. Its flowers open up about the same time as Concord or Barry and can be pollinated by other self-fertile, mid-season varieties such as Baco Blanc, B. 6055, Campbell Early, Chambourcin, Concord, Cottage, Corot Noir, Delaware, Diana, Empire State, Humbert #3, and Noiret.

The berry is round to oval, and large to medium-large in size. Its thin skin adheres mostly to the pulp in a manner similar to a *vinifera* grape variety, but it also possesses some slip-skin qualities. The texture of the fruit is good, firm, tender, and juicy, with a rich sweet flavor unlike Concord. The cluster size is medium and short, rather broad, that is cylindrical to tapering with a single shoulder. The harvest date is early for a Rogers hybrid, comparable to other early ripening varieties, such as Delaware and Foch. The sugars of Massasoit average 19.5° Brix, like Catawba or Agawam, with acids like Concord, Ontario, or Agawam, making it a relatively sweet grape for a *labrusca* hybrid, with medium acidity. Nineteenth and early twentieth century grape commentators, such as U. P. Hedrick, thought it to be inferior to other Rogers hybrids in the field and on the table.[78] If used, it should be picked before complete maturity as it develops heavy *labrusca* and foxy flavors as it ripens, but it is rich and sweet. In growing Massasoit I find that it can have pollination problems yielding only four to ten berries per cluster, and it has a high susceptibility to black rot.

SALEM (ROGERS NO. 22 AND 53)

Salem was one of Rogers's finest grapes, and of the grapes that he developed it is the one that he liked the most. It was bred in 1851, as all first generation Rogers hybrids were, and is a hybrid of Carter x Black Hamburg. It was the first hybrid named by Rogers in 1867, two years before his other varieties were named. Salem was named after Rogers's own home town of Salem, Massachusetts, and was listed in the *American Pomological Society Fruit Catalogue* in 1869. Rogers believed that Salem satisfied his breeding goal of a European grape on an American vine.

PARENTAGE
vinifera

HARVEST DATE
Mid–season

A B+ B A A

Salem has two hybrid numbers due to some confusion by outside parties who were incorrectly sending out plant material under the designation of Rogers No. 22 (which was a black grape). So to reduce the confusion on the release of the incorrect No. 22, Rogers renamed this grape Rogers No. 53. Since his numbered hybrids only went up to No. 45 at that time, it was clear that No. 53 was the renamed variety.[79]

The color of the Salem grape is an attractive dark red, with a medium blue bloom to dark chestnut, like Catawba or Agawam. It has been described as a dark dull red. Salem is a vigorous variety that is moderately productive. It can be variable in its production; it needs to be cluster thinned not only to increase cluster and berry size, but so that it will not overbear to the detriment of the health of the vine. The variety can be susceptible to fungus diseases without a good spray program. On the plus side, it has an open canopy which increases air flow and sunshine penetration to help combat fungus diseases, and which helps to facilitate effective fungicide spray applications. Salem's canes grow upward and then arch downward towards the ground like Chambourcin. It does not thrive at all in hot and humid conditions. It is one of Rogers most winter hardy varieties, and as hardy as Delaware, or perhaps more so.

Salem is not particular as to the soils that it will grow in. It can grow throughout the eastern United States, and is praised, like Concord, for its ability to grow in any soil type. This is one of the reasons for its popularity with growers. Further, it is chlorosis-resistant or nearly chlorosis-free in high lime soils. Its self-sterile flowers bloom by mid-season around the same time as Concord and Massasoit, and it has reflexed stamens, so it needs a pollinator nearby. Self-fertile pollinators that bloom at the same time as Salem include Baco Blanc, Chambourcin, Concord, Cottage, Delaware, and Noiret.[80] Its harvest date is early for a Rogers hybrid and it ripens mid-season along with Concord or even before.

Overall, Salem is a true commercial grape that was one of the most extensively grown Rogers hybrids in the northeastern United States and the West Coast. The berry size is large to very large. Its cluster and berry is somewhat similar to Black Hamburg, with a cluster that is medium to large, compact, short and broad, tapering to cylindrical, that is shouldered. The skin of the berry is a rather thick, non-slip skin which adheres to the pulp and does not shatter or crack.

This is a true dual-purpose grape for table consumption or wine production. It is a handsome fruit of quality for the table and makes a high-quality white wine for a *labrusca*-type grape. It has a soft *labrusca*/Muscat nose and flavors, not the taste of a heavy *labrusca* variety, but with a delicately rich aromatic, sprightly, and vinous flavor. Salem is relatively low in sugar for a *labrusca* hybrid, similar to Concord, at 17° Brix, with low acidity like Delaware. It was a commonly grown grape used as a varietal wine by Widmer's Wine Cellars, in Naples, New York, up until the 1970s and early on by Brotherhood Winery in the Hudson Valley.[81]

I have grown Salem for over ten years and like it both for its attributes in the field and in the cellar.

This is a grape that should be re-examined as a heritage grape for the Hudson Valley and Massachusetts. Grape breeders such as N. E. Hansen of the South Dakota Agricultural Experiment Station, and C. Engle of Paw Paw, Michigan, used Salem extensively. The Tennessee Agricultural Experiment Station deemed Salem to be a "highly successful" vine and regarded it as "valuable for general cultivation."[82]

This next segment outlines five minor named Rogers hybrids that did not rise to the level of prominence of the major Rogers hybrids that were covered on the previous pages. These varieties, however, do merit coverage and some discussion as they have positive cultivation and cellar attributes that may be useful to future generations of growers and breeders.

MINOR NAMED GRAPE VARIETIES OF ROGERS

AMINIA (ROGERS NO. 39)

labrusca, vinifera

This variety was a favorite of Isadore Bush, and was highly praised in his *Bushberg Catalogue*. Aminia was bred in 1851 and is believed to be a Carter x Black Hamburg hybrid. There was some confusion as to whether this grape is really the original Rogers No. 39. Bush secured three vines of Rogers No. 39 in 1867 from different sources, and when they came into bearing fruit they were all different varieties. As the original Rogers No. 39 was destroyed in Rogers's backyard garden in Salem,[83] it was impossible to determine which one was the correct No. 39. In response to this, Bush selected the best of the three presumed Rogers No. 39 candidates, and with the approval of Rogers, named it Aminia. There seems to be no record of why Bush named this grape Aminia. It is a name of Arabic or African origin that was used as a girl's name and means trustworthy, honest, or faithful. The name Aminia has a Biblical connotation that was popular at that time.

Bush thought this grape to be so promising that he propagated and promoted it. The other two "No. 39" hybrids were destroyed to eliminate any confusion with the newly designated Aminia. According to the description of Aminia contained in the *Bushberg Catalogue* and Hedrick's *The Grapes of New York*, the color is dull black with blue bloom to dark purple to nearly black. The variety is vigorous to moderately vigorous, but of moderate productivity, so much so that it is not really a commercial variety. There is no mention of its fungus disease resistance, but that it is only moderately winter hardy.

Its bloom time is by mid-season and it is self-sterile with reflexed stamens. Aminia can be pollinated by mid- to late-season self-fertile varieties similar to those varieties that can pollinate Barry, Lindley, or Massasoit. It is a very early to early ripening black grape that ripens with Moore Early and Hartford. Hedrick called this variety one of the best early black table grapes. The berry is of variable size from small to large, roundish and it ripens unevenly. The cluster is small to medium and variable in size, not well formed, rather loose to moderately compact, broad and irregular, and sort of conical, generally not shouldered, though sometimes single-shouldered. It has been reported to be a good quality table grape with melting flesh, foxy in taste, and with skin that adheres to the flesh.[84]

ESSEX (ROGERS NO. 41)

labrusca, vinifera

This is a minor Rogers hybrid of Carter x Black Hamburg. It was named Essex in 1869 by Rogers after Essex County, Massachusetts, which is the home county where he resided. Essex was listed in the *American Pomological Society Fruit Catalogue* from 1865–1895, after which it was removed from the

Minor named grape varieties of Rogers, cont'd.

recommended list. The color is reddish-black with a blue bloom to dark purplish-black with a blue bloom.

The variety was described as vigorous, but not so winter hardy and of medium productivity. It blooms mid-season or slightly later to medium late. It is self-sterile or nearly so, being one of the few Rogers hybrids that is not completely self-sterile, but it does not set fruit well at all, even in mixed variety vineyards. Its stamens are reflexed but can self-pollinate to a limited extent with some of the better self-fertile pollinators being the same as those that are suitable for Barry, Massasoit, and Lindley.

Essex ripens mid-season with Concord or later. The berry size is large to above medium, round and somewhat flattened like its seed parent Carter. The usually compact cluster when properly fertilized is medium to large, shouldered, broad with one short shoulder. The flesh of the grape is tender and sweet, highly aromatic, and vinous. Compared to other Rogers hybrids, it is similar to Barry, Wilder, and Herbert, but it is not their equal on many fronts.[85]

GAERTNER (ROGERS NO. 14)

labrusca, vinifera

Gaertner is one of only two other named Rogers hybrids that was fertilized by the pollen of White Chasselas instead of by Black Hamburg, the other variety being Lindley. Rogers No. 14 was first mentioned in 1865, and was given the name Gaertner by Rogers in 1869. This grape was probably named in honor of the German botanist Karl Friedrich von Gaertner (1772–1850), and not his father Joseph Gaertner (1732–1791), both of whom were pioneers in the study of hybrids.[86]

The grape is a dark reddish brown/brick-red with an ash-gray tinge, with a thin blue bloom. Gaertner is described by various sources as light to dark red, rather glossy covered with a lilac bloom. It is a vigorous to medium vigorous grower, and as all Rogers hybrids, produces a fair to good crop, but is not a super big producer. The variety is at best only winter hardy to moderately hardy and more capricious to soils than other Rogers hybrids; and so it requires good care and suitable soils to do well. Further, it has light to medium incidence of chlorosis in high lime soils. From the literature of the era, little was said of its fungus disease resistance. The bloom date is moderately late, mid-season to late mid-season to even late season, with sterile flowers with reflexed stamens. It pollinates with other mid-season and late-season blooming varieties similar to Lindley and self-fertile varieties such as Chardonel, Iona, or Jefferson.

Both nursery owner Patrick Barry and U. P. Hedrick liked Gaertner, and said it was like Massasoit in fruit characteristics, but of much better quality. The berry is of variable size from large to below medium, and roundish to slightly oval. The cluster is pretty, large to medium, attractive, short to medium in length, rather loose, cylindrical to slightly tapering, with a single or double shoulder that can have many shot berries. The variety ripens unevenly around mid-season and has a thin skin, so it does not ship well.

Ultimately, Hedrick thought more of Gaertner than Barry and recommended that it should be more widely planted. The grape while never popular, is agreeably vinous and has a pleasant aromatic flavor. Gaertner is relatively high in sugar for a *labrusca* hybrid with low acids, and reaches 20° Brix, like Diamond, Lindley, or Ulster. It is reported to be of excellent quality and good for the home garden. The East Coast grape-breeder Clifford Ambers thought enough of the variety that he used it as a seed parent in his breeding program. Based on past interest by several state agricultural experiment stations, Gaertner should be re-examined for future cultivation or use in breeding programs.[87]

MERRIMAC (ROGERS NO. 19)

labrusca, vinifera

Merrimac was bred in 1851, Carter x Black Hamburg, and named Merrimac in 1869 after the local North Shore Merrimac River.[88] It was granted a premium by the Essex Agricultural Society in 1859, and was first listed in the *American Pomological Society Fruit Catalogue* in 1867. Merrimac is a good, strong, and very vigorous grower, more so than other Rogers black hybrids. Merrimac has long canes that grow laterally and which produce lots of foliage. Further, it is rated hardy to very winter hardy, more so than the other Rogers black hybrids.

Minor named grape varieties of Rogers, cont'd.

While early commentators maintained that it had higher resistance to fungus diseases when compared to other Rogers hybrids, I have not found that to be the case. It is especially susceptible to black rot. It is rated as being a productive variety as long as it does not succumb to fungus diseases.

Merrimac's bloom time is mid-season, around the same time as Concord and Massasoit, and it is self-sterile with reflexed stamens. Its best pollinators are the same as those used for Massasoit and Lindley, somewhat, which is unique for a Rogers hybrid. This is because, as a rule, Rogers hybrids do not pollinate other Rogers hybrids at all. The harvest date is mid-season, around the time of Concord, which is a bit earlier than that of the other Rogers black hybrids.

The color is blue-black and glossy with a blue bloom. The berry is large to medium, with a cluster that is medium in size, being tapering to cylindrical, which is smaller than other black Rogers varieties. The skin of the berry is tough and adheres slightly to the skin, and it is more like a slip-skin variety. While Merrimac may have some sound viticultural attributes when compared to other Rogers hybrids, such as Barry, Herbert, or Wilder, its quality in taste is not as good. However, I have found that it has some nice soft *labrusca* flavors.[89] The Missouri State Fruit Experiment Station and South Dakota Agricultural Experiment Station sometimes used Merrimac in its grape breeding program as a seed parent. This was probably because of its above average vigor in the field.

Having grown Merrimac for many years I find that its dense canopy encourages fungus diseases, and it must be pruned in a way to increase air flow and sunshine into the canopy to combat such diseases. It has lighter hued green leaves, unlike the dark olive green leaves of Concord or Cottage.

WILDER (ROGERS NO. 4)

labrusca, vinifera

This variety never became a major Rogers hybrid. Rogers No. 4 was bred in 1851 of a Carter seed x Black Hamburg pollen. The first notes on this grape were published in 1858; it was first listed in the *American Pomological Society Fruit Catalogue* in 1867. The name Wilder was conferred on this grape by Rogers in 1869, and it was named in honor of Massachusetts politician and horticulturalist Marshall P. Wilder since he himself selected this variety to be so named. The color is purplish-black to black, not glossy and with a heavy blue bloom to a dark purple to nearly black color.

Wilder is vigorous to very vigorous and winter hardy. It is somewhat susceptible to fungus diseases to being susceptible, but overall, is more resistant than many other Rogers hybrids. In sum, the variety is reliable and fairly productive. Its self-sterile female flower blooms by mid-season, around the time of Concord, with reflexed stamens. Its best pollinators are the same varieties as those used to pollinate Goethe. It ripens by early mid-season to mid-season just before Concord. The size of the berry is large and slightly oval with a thick skin. The skin adheres somewhat to the pulp with bright red pigment that is astringent. The flesh is tender with some *vinifera* flavors. The size of the cluster is variable but not large, which is to say short, broad, medium-to-small and tapering. It often has a single shoulder and is loose.

The quality of Wilder is surpassed by most other Rogers hybrids, but it is a reliable producer in the field, hardy and productive. It is a sweet grape for a *labrusca* hybrid, at over 20° Brix, like Ontario, Lindley, or Herbert, with low acidity like Jefferson, Worden, or Campbell Early. The variety has some *vinifera* sprightliness, it is not foxy or *labrusca* in taste, and keeps and ships well. Hedrick stated that Wilder did not make quality wine, nor was it suitable for juice production, and in 1908 it was considered solely as table grape.

Today, there are many better table grapes in cultivation. Of historical interest is the fact that Wilder is a great, great, grandparent of the cool climate Minnesota hybrid Marquette.[90] I have planted Wilder at my Cedar Cliff farm in the Hudson Valley, and am currently evaluating it.

NOTES

1. The American Pomological Society was organized in 1848 and incorporated in the Commonwealth of Massachusetts in 1887. This Society was principally organized by Marshall P. Wilder, with the support of other New England horticulturalists and nurserymen such as Joseph Sebastian Cabot, Charles Mason Hovey, and Jacob Warren Manning; the Prince and Parsons families of Flushing Landing, New York, the Downings of Newburgh, New York, and Patrick Barry of Rochester, New York. The goals of the society were to: exchange ideas and information on the successful cultivation, propagation, and worthiness of fruits; to cull out of existence worthless sorts of fruits; to systematically classify and name worthy varieties that should be propagated by nurseries and growers; to furnish reliable information on the cultivation of fruits in America's varied climates; encourage the breeding of improved varieties of fruits adapted to the various soils and climates of the United States; and to maintain a spirit of cordial intercourse with kindred agricultural and horticultural state and local societies. In short, to extend and improve the culture of fruit throughout the United States and Canada. The Society especially concentrated on the breeding of new fruits and encouraged breeders to exchange information on the best seed and pollen parents and breeding techniques for such new hybrid varieties. Its proceedings and publications served as examples and methods of work to be adopted by other state and local pomological and horticultural societies. Today, the society remains a strong and vibrant organization that continues to advance the above goals. Its greatest contribution to American fruit culture is the biennial publication of its *Catalogue of Fruits*, which lists superior fruits that should be cultivated, and those inferior sorts that should be abandoned. The society, national in scope, established a system of state reports by panels of expert local horticulturalists and pomologists, who wrote on the cultivation of fruit in each jurisdiction. See articles of incorporation of the American Pomological Society, 1887; Marshall P. Wilder, pres., Address at the Twentieth Session of the American Pomological Society, held in Grand Rapids, MI, September 9–11, 1885 (Boston: American Pomological Society, 1885).

2. Charles Mason Hovey established Hovey & Co., a large national nursery in Cambridge, Massachusetts. Through his work at the nursery, the Massachusetts Horticultural Society, and his writings about fruit culture and hybridization, Hovey supported the work of new fruit breeders by being a conduit to exchange information by such horticulturalists.

3. See generally Tamara Plakins Thorton, *Cultivating Gentlemen: The Meaning of Country Life among the Boston Elite* 1785–1860 (New Haven: Yale University Press, 1989);

4. J. Stephen Casscles, "E. S. Rogers and the Origins of American Grape Breeding," *Arnoldia*, Vol. 77, No. 2 (Boston: Arnold Arboretum of Harvard University, 2019); 30–33; see also generally Thorton, *Cultivating Gentlemen; Victoria Johnson, American Eden: David Hosack, Botany, and Medicine in the Garden of the Early Republic* (New York: W. W. Norton & Co., 2018); Myra B. Young Armstead, *Freedom's Gardener: James F. Brown, Horticulture, and the Hudson Valley in Antebellum America* (New York: New York University Press, 2012); David Schuyler, *Apostle of Taste – Andrew Jackson Downing 1815–1852* (Baltimore: Johns Hopkins University Press, 1996); Emily Pawley, "Cataloging Nature: Standardizing Fruit Varieties in the United States, 1800–1860," *Business History Review*, Vol. 90, No. 3 (Cambridge, Cambridge University Press, Fall 2016), 405–29; B. June Hutchinson, "A Taste for Horticulture," *Arnoldia*, Vol. 40, No. 1 (Boston: The Arnold Arboretum of Harvard University, Jan. 1980), 31–48.

5. Rogers Family Papers, 1792–1901, MSS 87, Box 8, Phillips Library at the Peabody Essex Museum, Salem, Massachusetts; J. Stephen Casscles, "E. S. Rogers and the Origins of American Grape Breeding," *Arnoldia*, Vol. 77, Issue 2 (Boston: The Arnold Arboretum of Harvard University, 2019), 30–34. *U. P. Hedrick, The Grapes of New York. Report of the New York Agricultural Experiment Station for the Year 1907*. Department of Agriculture, State of New York. Fifteenth Annual Report, Vol. 3 – Part II. (Albany, N.Y.: Lyon Company, 1908), 390; *Who Was Who in American History – Science and Technology: A Component of Who's Who in American History* (Chicago: Marquis Who's Who, 1976), 519; "Obituary of Edward Rogers, Originator of the Rogers Grape," *Boston Evening Transcript* (Boston: The Boston Transcript Company, March 30, 1899).

6. Rogers Family Papers, 1792–1901, MSS 87, Box 8; Hedrick, *Grapes of New York*, 390–92.

7. Rogers Family Papers, 1792–1901, MSS 87, Box 8, Phillips Library at the Peabody Essex Museum in Salem, contains several deeds for large tracts of land purchased by Rogers after 1870 which was then subdivided in Rockport, Massachusetts; Casscles, "E. S. Rogers and the Origins of American Grape Breeding," *Arnoldia*, 30–39.

8. Hedrick, *Grapes of New York*, 390; Rogers Family Papers, 1792–1901, MSS 8, Box 8; Casscles, "E. S. Rogers," *Arnoldia*, 30–34.

9. Marshall Pinckney Wilder (1798–1886) was a pre-eminent American pomologist and horticulturalist; one of his most important contributions to horticulture and pomology was his leadership of the Massachusetts Horticultural Society and the American Pomological Society.

Wilder was a successful Massachusetts politician, militia officer, and genealogist, a thrice-married and twice-widowed father of a very large family. In 1832, he purchased a suburban farm in Dorchester, just outside of Boston, called Hawthorn Grove. He was a founding member of the Massachusetts Horticultural Society, a member of that group for 56 years, and President from 1841–1848. Wilder was one of the founders of the American Pomological Society, and was its first president from the organization's founding in 1850 until his death in 1886, except for one term. He was also one of the founders of the Massachusetts Board of Agriculture, Massachusetts Agricultural College (now part of University of Massachusetts Amherst), and the U.S. Agricultural Society. A beautifully constructed red brick building with arches has been dedicated to Wilder at University of Massachusetts Amherst.

See Massachusetts Horticultural Society, *For the Year 1887, Part 1, Memorial of Marshall P. Wilder by Pres. Walcott* (Boston: Printed for the Society, 1887), 1–40; Liberty Hyde Bailey, *The Standard Cyclopedia of Horticulture*, Vol. 3 (New York: Macmillan, 1915), 1603; John Ward Dean, *New England Bibliopolist: Containing the 'Book Notice' in the Historical Genealogical Register, Notices of Books on American History*, Vol. 1, No. 1 (Boston: David Clapp & Son, January 1880), 10–11; Marshall P. Wilder Collection 1848–1886, Call No. R6 002/3 W55, Special Collections & University Archives, University of Massachusetts Amherst Libraries; U. P. Hedrick, *The Pears of New York* (Albany, N.Y.: Department of Agriculture, State of New York, 1921), 128; Thomas Pinney, *A History of Wine in America, From the Beginnings to Prohibition*, Vol. 1 (Berkeley: University of California Press, 2007), 216–17.

10. Hedrick, *Grapes of New York*, 390–91; "The Grapes of New York," *The Country Gentleman*, Vol. 74, No. 2933 (Albany, NY: Luther Tucker & Son, April 15, 1909), 376.

11. Hedrick, *Grapes of New York*, 390–92.

12. John Lindley (1799–1865) was a famed English botanist, educator, horticulturalist, editor, author, and botanical illustrator. Lindley was born near Norwich, England, one of four children of George and Mary Lindley. His father, George, was a nurseryman and pomologist who ran a commercial nursery, and while skilled in his profession, the business was not profitable. Through John Lindley's social acquaintances which he had made in the horticultural and gardening world, he became associated with the botanist William Jackson Hooker and Sir Joseph Banks at an early age. John Lindley began publishing literature on botany and was elected as a fellow of The Linnean Society of London in 1820, at the age of 21. He was appointed assistant secretary to the Royal Horticultural Society in 1822, and in 1829, was appointed to the Chair of Botany at the newly found University of London, where he remained until 1860. Lindley retained both jobs and continued to write books and illustrate them to support his family and to pay off his father's past debts. While he specialized in managing various horticultural libraries, herbariums, and gardens, he simultaneously wrote and illustrated many books on topics such as orchids, roses, and other exotic plants, fruits and vegetables. With respect to his work with grapes, Lindley wrote *An Outline of the First Principles of Horticulture* (1832), *The Theory of Horticulture* (1840), and the largely revised *The Theory and Practice of Horticulture*, 2nd Ed. (1855). Lindley married Sarah Freestone (1797–1869) in 1823, and they had two daughters, who were also artists, and one son, who became a lawyer. Rogers named one of his hybrids after Lindley (Rogers No. 9). See generally, Hugh Chisolm, ed., "Lindley, John" entry, *Encyclopaedia Britannica*, 11th Ed., Vol. XV (New York: Encyclopaedia Britannica, Inc., 1911), 178; and G. S. Boulger, "Lindley, John" entry, *Dictionary of National Biography*, Vol. XXXIII (New York: Macmillan and Co., 1893), 277–79.

13. Andrew Jackson Downing (1815–1852) of Newburgh, New York, was a nationally known horticulturalist, landscape designer, and rural architect. For more on A. J. Downing see Chapter Eight, pages 84–85. See also the full-length biography by David Schuyler, *Apostle of Taste: Andrew Jackson Downing*, 1815–1852 (Baltimore: Johns Hopkins University Press, 1996).

14. John Fisk Allen (1807–1876), a horticulturalist who lived in Salem, Massachusetts, specialized in the cultivation and breeding of new grape varieties. See the following Chapter Sixteen for more on J. F. Allen and his work. See also *Transactions of the Massachusetts Horticultural Society for the Year 1876, Part II, Resolution adopted on November 1, 1876 on the Death of John Fisk Allen* (Boston: Printed for the Society, 1877), 8–10; Pinney, *History of Wine*, Vol. 1, 206–07.

15. Pinney, *History of Wine*, Vol. 1, 206; Hedrick, *Grapes of New York*, 390; "Grapes of New York," *Country Gentleman*, Vol. 74, 376; Philip M. Wagner, *A Wine-Grower's Guide* (New York: Alfred A. Knopf, 1985), 33.

16. Hedrick, *Grapes of New York*, 392; Lucie T. Morton, *Wine Growing in Eastern America: An Illustrated Guide to Viniculture East of the Rockies* (Ithaca, N.Y.: Cornell University Press, 1985), 180.

17. Charles Downing (1802–1885), one of the unsung heroes of nineteenth century American horticulture, was often overshadowed by his younger brother A. J. Downing. For more on the Downing brothers see Chapter Eight, pages 84–85; and Schuyler, *Apostle of Taste: Andrew Jackson Downing*, 15, 56, 243–44.

18. Parsons Nursery, like the Prince, Bloodgood, and Garretson Nurseries, were all located in the nineteenth century horticultural hub of Flushing Landing, Long

Island, New York. The Parsons Nursery was established in 1838 by Samuel Parsons (1771–1841) for his two sons. Parsons was fluent in French due to his association with French émigrés. He retired comfortably from his business and became a Quaker minister in Flushing. The Parsons Nursery was split into two nurseries, one for each son, Robert Bowne Parsons (1821–1898) and Samuel Bowne Parsons (1819–1906). The Parsons specialized in ornamental tree propagation, many of which were planted in the Village of Flushing to enhance the community. They also propagated grape vines, growing over 800,000 vines annually between 1862–65. Like the nearby Prince Nurseries, the Parsons published extensive plant catalogs to list their many plant material offerings. See generally, W. W. Munsell, *History of Queens County, New York, with Illustrations, Portraits, & Sketches of Prominent Families and Individuals* (New York: Press of George MacNamara, 1882), 92–94.

19. *Horticultural Order Book 1866–1873*, Rogers Family Papers, 1792–1901, Box 8, MSS 87, Phillips Library at the Peabody Essex Museum, Salem, Massachusetts.

20. *Rogers Garden Chart*, dated February 10, 1868, lists grape hybrids No. 46 through 82. Rogers Family Papers, 1792–1901, MSS 87, Box 8, Phillips Library at the Peabody Essex Museum, Salem.

21. Ibid. The Rogers Family Papers at the Peabody Essex Museum contains a chart that Rogers created by hand, with his own notes on the grapes that he crossed while making new breeds.

22. United States Department of Agriculture, *Report of the Commissioner of Agriculture for the Year 1864* (Washington, D.C.: Government Printing Office, 1865), 132–36.

23. E. L. Lord, "Grape Varieties," *Proceedings of the Thirty-Fifth Annual Meeting of the Florida State Horticultural Society, Held at Lakeland, Fla., May 2, 3, 4, 5, 1922*, Vol. 35
(Deland, Fla.: The Society Office of Publication, 1922), 127–30; U. P. Hedrick, *A Manual of American Grape-Growing* (New York: The Macmillan Company, 1919), 274–77; Wagner, *A Wine-Grower's Guide*, 32–33, 205–07; Philip M. Wagner, *Grapes into Wine* (New York: Alfred A. Knopf, 1976), 40–41; Philip M. Wagner, *American Wines and Wine-Making* (New York: Alfred A. Knopf, 1965), 36–37.

24. Casscles, "E. S. Rogers and the Origins of American Grape Breeding," *Arnoldia*, 30–36.

25. Ibid., 33–35.

26. Robert Manning (1784–1842) was a nationally known pomologist and horticultural writer from Salem, Massachusetts, who specialized in pears and other fruits. Manning started collecting fruit plants in 1823, when he established his "Pomological Gardens" at Salem so that he could test and introduce new varieties of fruit in the United States. He was an accurate identifier and describer of fruit and plant material, and an expert in fruit nomenclature. He wrote *Book of Fruits* (1838), which describes tree fruits, small fruits, grapes, trees and shrubs. He was one of the founders of the Massachusetts Horticultural Society. See Hedrick, *Pears of New York*, 53, 162.

27. Charles Mason Hovey (1810–1887), along with his older brother Phineas (1803–1885), established Hovey & Co., a large national nursery located in Cambridge, Massachusetts, that sold fruit trees, grapes and small fruits, in addition to hardy perennials, ornamental trees and shrubs, and flower and vegetable seeds. Their fruit catalogues gave descriptions to a wide assortment of pears, apples, peaches, apricots, cherries, grapes, plums and small fruits that they offered. His brother Phineas joined the Massachusetts Horticultural Society when it was founded in 1829, while Charles joined the Society four years later in 1833. Beginning in January 1835, the brothers launched *The American Gardener's Magazine and Register of Useful Discoveries and Improvements in Horticulture and Rural Affairs*. The magazine was published monthly and contained articles of interest on all aspects of gardening, horticulture and growing plants; reporting on new plant and fruit varieties; providing information on new and improved equipment; and the kinds of out-buildings needed to grow such plants or operate a garden or a farm. By the third issue, the name had been changed to *The Magazine of Horticulture* and its title page listed only C. M. Hovey as the editor. It was one of the first commercially successful magazines devoted to horticulture in America. Hovey wrote *The Fruits of America, Notes on Apples, Peaches, Plums, Grapes, and Pears* (1852, 1856), a manuscript for the Massachusetts Horticultural Society, and many notes and remarks for the Society on strawberries, descriptions and engravings of selected cherries, and other articles on gardening. In addition to breeding and hybridizing strawberries and other fruits, Hovey created new breeds of flowers such as camellias, azaleas, lilies, and orchids. See generally, B. June Hutchinson "A Taste for Horticulture," *Arnoldia*, Vol. 40, Issue 1 (Boston: The Arnold Arboretum of Harvard University, 1980), 31–48; U. P. Hedrick, *The Small Fruits of New York* (Albany, N.Y.: Department of Agriculture, State of New York, 1925), 367.

28. Pinney, *History of Wine*, Vol. 1, 216–17; see generally, Tamara Plakins Thorton, *Cultivating Gentlemen: The Meaning of Country Life Among the Boston Elite, 1785–1860* (New Haven, Conn.: Yale University Press, 1989).

29. The Massachusetts Horticultural Society, founded in 1829, is one of the oldest horticultural institutions in the United States and continues to operate today. The Society, then and now, provides information on horticulture and related plant sciences; disseminates botanical information through its library; conducts educational programing; sponsors exhibitions and community outreach initiatives; and operates its botanic gardens at Elm Bank, in Wellesley, Massachusetts. In the nineteenth century, the Society was the prime conduit for communication between the horticultural leaders of Boston and the North Shore with those who were working throughout the rest of Massachusetts, New England, New York, and the United States and Canada. The Society was formed by a group of concerned and academically interested farmers, gardeners, horticulturalists, academicians, and business leaders in the Boston area. Some of the founding members of the Society included: Gen. H. A. S. Dearborn, Zebedee Cook, Enoch Bartlett, Dr. Jacob Bigelow, and Dr. John Lowell of Boston; Robert Manning and John Ives of Salem, and Dr. Thomas Nuttall of Harvard College. Beginning around 1825, the Society's members proposed to establish a combined cemetery, botanical garden, arboretum, and experimental garden that would be open to the public for their enjoyment, and to provide a suitable place of repose for the deceased and their bereaved family members to visit. This dream became a reality with the purchase of 70 acres of land in 1830 that would become the Mount Auburn Cemetery in Cambridge. The initial committees of the Society included nurseries, fruit trees and fruits, ornamental trees, shrubs, and flowers, products of the kitchen garden, and library. The Society sponsored professorships at Harvard College that included Dr. John W. Webster, Prof. of Horticultural Chemistry, Dr. Jacob Bigelow, Prof. of Botany, and Dr. Thaddeus William Harris, Prof. of Entomology. Through its annual meetings, horticultural journals, and plant material exhibitions, the Society was one of the leading centers of horticultural, pomological, and plant science learning in the United States. Further, it was the primary vehicle for horticulturalists to communicate with each other and their associated regional and state horticultural societies in the United States, Canada, and Europe. See *History of the Massachusetts Horticultural Society, 1829–1878* (Boston: Rand, Avery & Co. for the Massachusetts Horticultural Society, 1880), 55–84; see generally, Thorton, *Cultivating Gentlemen: The Meaning of Country Life Among the Boston Elite, 1785–1860*; and https://masshort.org/ the website of the Massachusetts Horticultural Society.

30. George W. Campbell (1817–1898) was a leading writer and speaker in the northern United States on the culture of grapes and grape breeding. Campbell was born in Cortlandville, New York, in 1817. His family moved to Ohio in 1821 when he was four years old. Early in his life George Campbell was a printer and editor, as was his father. In 1849, he moved from Sandusky, Ohio (an area that grew grapes), to Delaware, Ohio (the town whose namesake, the grape Delaware, was named after by Charles Downing). It is here that he began to engage in horticulture as a paying profession, but he had been interested in this subject much earlier. Campbell was a member of the American Pomological Society from its beginning in 1850 until his death. He created thousands of seedling grapes and introduced the famed Campbell Early, Concord Chasselas, Concord Muscat, and White Delaware. See Hedrick, *Grapes of New York*, 198.

31. F. E. Gladwin, "A History of Grape-Growing in the Eastern United States: Part VI," *Rural New Yorker*, Vol. XC, No. 5161, May 23, 1931 (New York: Rural Publishing Co., 1931), 613–15.

32. Elmer Swenson (1913–2004) was one of the most prolific American breeders of superior quality cool and cold climate grapes in the United States. See Chapter Twelve, pp195-197 for more on Swenson and his Minnesonta hybrids; and for more details on his personal life, see J. Stephen Casscles, "Elmer Swenson: Pioneer Cool Climate Grape Breeder (1913–2004)," *AWS Wine Journal*, Vol. 52, No. 4 (Scranton, Pa.: American Wine Society, Winter 2020), 6–9.

33. Nelson Bonney White (1824–1914) was born in Putney, Vermont, in 1824. When he was young, White lived for a time in Ohio and New York before settling in Norwood, Massachusetts. He came under the influence of E. S. Rogers and took up grape breeding as his hobby. See Chapter Sixteen, pages 284–285, for more on White.

34. The New York State Agricultural Experiment Station at Geneva, New York, is one of the largest and oldest grape breeding programs in the United States. In 1908, Geneva horticulturalist Ulysses Prentiss Hedrick wrote the definitive work on cool climate grapes for the United States east of the Mississippi, *The Grapes of New York*, which has on its frontispiece a prominent portrait of E. S. Rogers. See Chapter Eleven, pages 175–179, for more information on this institution.

35. Jacob Moore (1835–1908) was born in Brighton, New York, and did most of his breeding work in Attica, New York. He was a careful and exacting grape breeder. For more biographical information on Jacob Moore, see Chapter Seven, page 77.

36. Dr. Richard T. Underhill (1802–1871) and his nephew Stephen W. Underhill (1837–1925) jointly worked on breeding new grape varieties. They were successful in breeding native American grapes, using Delaware and Concord as the seed parent, with European vinifera varieties as a pollen parent. See Chapter Eight, pages 110–112.

37. Campbell Early is a hybrid created by George Campbell of Ohio (Moore Early x (Belvidere x Muscat Hamburg)). It is among the most widely-grown grapes in South Korea and Japan today. See Chapter Sixteen, pages 281 and 292 endnote 25, for more on Campbell Early.

38. See Chapter Six, pages 65–66, for a summary of the work and life of Thomas Volney Munson.

39. See generally, the "Breeders" works, charts, and genealogy by Carl Camper, at Chateau Stripmine, an experimental Vineyard in Southeastern Montana, http://chateaustripmine.info/Parentage.htm for a good background of hybrid grapes, progeny, and of breeders activity; See also Sherrie S. McLeRoy and Roy E. Renfo, "Appendix III, Grape Varieties Created by T. V. Munson," *Grape Man of Texas: Thomas Volney Munson & the Origins of American Viticulture* (San Francisco: The Wine Appreciation Guild, 2008), 247–69.

40. "Obituary of Edward Rogers, Originator of the Rogers Grape," *Boston Evening Transcript*, March 30, 1899. A similar obituary was featured in scores of newspapers across the United States and Canada. See Rogers Family Papers, 1792–1901, MSS 87, Box 8, Phillips Library at the Peabody Essex Museum, Salem, Massachusetts.

41. See Chapter Eight, page 102 for more on Black Hamburg.

42. William Prince, Jr. (1766–1842) was the third of five generations of the Prince family who operated the Prince Nursery and founded the Linnaean Botanic Garden and Nurseries of Flushing Landing, Long Island, New York. The Prince family nursery was established fifty years before the American Revolutionary War and closed just after the end the American Civil War.

The Prince family came to Flushing Landing, New York, from Salem, Massachusetts, sometime prior to 1720. For over 125 years, the Prince family helped to advance horticultural and pomological studies in the United States and Europe. During the first half of the nineteenth century, the nursery was one of America's primary centers of horticultural and pomological learning and plant breeding. William Prince, Jr., and his son William Robert Prince (1795–1869), worked closely with horticulturalists in eastern Massachusetts and throughout New England and the South. The Princes trained generations of horticulturalists, gardeners, and landscape designers, who then practiced their craft throughout the United States, especially in New England and the Middle Atlantic states. The Prince family nursery annually published over ten separate catalogs that offered plants such as strawberries, grapes, and fruit trees, ornamental plants, evergreen trees and shrubs, roses, iris, phlox, and chrysanthemums, greenhouse plants, bulbous plants, offerings from East Asia, annual flowers and vegetables, and wholesale plants for other nurseries. Not only did these catalogs offer a wide variety of plant material, but included detailed gardening and botanical treatises on how to plant and maintain such plants. The Prince nursery was where most European, Asian, and African trees, shrubs, and plants were first imported into the United States for distribution in America. Further, the Prince Nursery became the central location from which many American plants were collected and exported to Europe. The role that the Princes provided to American horticulture have since been taken over by federal- and state-sponsored agricultural universities and experiment stations. William and his son, William Robert Prince, were especially active writers of horticultural texts on the subjects of tree fruits, grapes, brambles, and ornamental plants. See Chapter One, pages 3–4; and generally, U. P. Hedrick, *The Plums of New York: Report of the New York State Agricultural Experiment Station for the Year 1910, II* (Albany, N.Y.: J. B. Lyon Co., 1911), 389; Rodney K. Waldron, "Prince's Plants," *The Call Number*, Vol. 20, No. 1 (Portland, Ore.: University of Oregon Library, Fall 1958), 7–12; C. Stuart Gager, ed., "The First Botanic Garden on Long Island," *Brooklyn Botanic Garden Record*, Vol. 1, No. 4, October, 1912 (New York: Press of L. Middleditch Co., 1912), 97–99; Abigail Jacobsen and J. Dustin Williams, "Tales from the Archives, Prince Family Nurseries (circa 1737–post 1851)," *Bulletin of the Hunt Institute for Botanical Documentation*, Vol. 21, No. 1 (Pittsburgh: Carnegie Mellon University, Spring 2009), 4–7; Peggy Cornett, "Encounters with America's Premier Nursery and Botanic Garden: The Prince Nursery of New York City," *Twinleaf Journal*, January 2004 (Charlottesville, Va.: Th. Jefferson Monticello, 2004), 1–12; Also see the online version of the article at: https://www.monticello.org/house-gardens/center-for-historic-plants/twinleaf-journal-online/prince-nursery/ .

43. Hedrick, *Grapes of New York*, 307–10, 445.

44. Ibid., 390–91, 395–96, 445; *The Bushberg Catalogue, A Grape Manual* (Bushberg, Jefferson City, MO: Bush and Son & Meissner, 1895), 141.

45. Hedrick, *Grapes of New York*, 395–96.

46. Bushberg Catalogue, 141, 151.

47. U. S. Dept. of Agriculture, *Report of the Commissioner of Agriculture 1864*, 134.

48. C. M. Hovey, ed., *The Magazine of Horticulture, Botany, and All Useful Discoveries and Improvements in Rural Affairs*, Vol. XXII (Vol. II, Third Series) (Boston: Hovey and Co., November 1856), 507, and Vol. XXIII (Vol. III, Third Series), January 1857, 6, and February 1857, 86.

49. Rogers Family Papers, 1792–1901, MSS 87, Box 9, Phillips Library at the Peabody Essex Museum, Salem, Massachusetts; "The Obituary of Edward Rogers, Originator of the Rogers Grape," *Boston Evening Transcript* (Boston: The Boston Transcript Company, March 30, 1899).

50. See generally, the work of Carl Camper, Chateau Stripmine Experimental Vineyard, http://chateaustripmine.info/ Breeders.htm.

51. Jancis Robinson, Julia Harding, and Jose Vouillamoz, *Wine Grapes: A Complete Guide to 1,368 Vine Varieties, Including Their Origins and Flavor* (New York: HarperCollins, 2012), 229–32; Pierre Galet and Lucie T. Morton, *A Practi-*

cal Ampelography: Grapevine Identification (Ithaca, N.Y.: Cornell University Press, 1979), 124–25.

52. Galet and Morton, *Practical Ampelography*, 124–25; Hedrick, *Manual of American Grape-Growing*, 347–48.

53. The coastal town of Ipswich, Massachusetts, was known as Agawam by early European Puritan settlers during the seventeenth century. Agawam was the English name for the Agawam tribe of northeastern Massachusetts. The area of this Native American tribe extended from Cape Ann to the Merrimac River, covering what is now the eastern portion of Essex County. The word Agawam is derived from the native American self-reference to themselves as "Wonnesquamsauke" ("*wonne*" means pleasant, "*asquam*" means water, and "*auke*" means place). See generally, Howard S. Russell, *Indian New England Before the Mayflower* (Lebanon, N.H.: University Press of New England, 1980).

54. J. Stephen Casscles, "Pink and Light Red Heirloom Grape Varieties for the Northeast," appearing concurrently in *Fruit Notes*, Vol. 83, No. 4 (Amherst: A publication of the Stockbridge School of Agriculture at University of Massachusetts, Fall 2018), and *Horticultural News*, Vol. 98, No. 4 (Hackettstown, N.J.: New Jersey State Horticultural Society, Fall 2018), 23–26; J. Stephen Casscles, "Great Cool Climate Heirloom Grape Varieties," *AWS Wine Journal*, Vol. 50, No. 1 (Scranton: American Wine Society, Spring 2019), 14; Observations by the author at his farm Cedar Cliff, Athens, New York, from 2010 to 2021 for Agawam, and tasting notes of wines made from Agawam; Nelson Shaulis, John Einset, and A. Boyd Pack, *Growing Cold-Tender Grape Varieties in New York*, Bulletin No. 821 (New York: Agricultural Experiment Station, Cornell University, August 1968), 5; J. S. Shoemaker, *Sugar, Acidity, and Juice Color Determinations in Grapes*, Ohio Agricultural Experiment Station Bulletin 550 (Wooster, Ohio: Ohio State University, 1935), 5–12.

In the late nineteenth century, grape evaluators and critics from across the country were aggressively evaluating Agawam. Among the main critiques, see Hedrick, *Grapes of New York*, 158–60; *Bushberg Catalogue*, 83–84; Hedrick, *Manual of American Grape-Growing*, 288, 298, 331; State of New York Department of Agriculture, *The Fruit Industry in New York State*, Bulletin No. 79, Part II (Albany, N.Y.: Department of Agriculture, 1915), 1244; S. A. Beach, *Fertilizing Self-Sterile Grapes*, New York Agricultural Experiment Station Bulletin No. 169 (Geneva, N.Y.: New York Agricultural Experiment Station, December 1899), 368; F. H. Hall and S. A. Beach, *Why Some Grapes Fail to Fruit*, New York Agricultural Experiment Station Bulletin No. 157 (Geneva, N.Y.: New York Agricultural Experiment Station, December 1898), 6–7; J. C. Whitten, *The Grape – I. A Study of Types and Varieties*. University of the State of Missouri, College of Agricultural and Mechanic Arts, Agricultural Experiment Station, Bulletin No. 46. (Columbia, M.O.: E. W. Stephens, April 1899), 47; Samuel T. Maynard, Report on Fruits, Hatch Experiment Station of the Massachusetts Agricultural College Bulletin No. 21 (Amherst, Mass.: Press of Carpenter & Morehouse, April 1893), 14; G. W. McCluer, "*Grapes, Test of Varieties*," Agricultural Experiment Station Bulletin No. 28, December 1893, in *Bulletins of the Agricultural Experiment Station, University of Illinois, Vol. 2, Bulletins 17–36, August 1891–August 1894* (Urbana, Ill.: Agricultural Experiment Station, University of Illinois, 1894), 257–58.

55. Patrick Barry (1816–1890) was a nursery owner, pomologist, editor, author, and national leader of the eastern United States fruit industry, especially in western New York. He was born in Belfast, Ireland, in 1816, and came to the United States in 1836. When Barry first came to America, he worked for several years at the Prince Nursery, in Flushing Landing, New York, where he learned much about horticulture, pomology, and plant sciences. Barry, along with George Ellwanger, founded the Mount Hope Nurseries in 1840 in Rochester, New York, and in many respects introduced fruit growing to western New York, which still specializes in the cultivation of pears, apples, and grapes. At his nursery, in addition to a large collection of fruits, he offered ornamental trees and shrubs. Early in the nursery business, he imported many plants from Europe. He was especially fond of pears and exhibited them often under the name Ellwanger & Barry at national and state horticultural exhibitions. From 1844 to 1852, Barry edited the *Genesee Farmer* and succeeded A. J. Downing as the editor of *The Horticulturalist* in 1852, after Downing died from a drowning accident in the Hudson River. *The Horticulturalist* was ultimately relocated to Rochester, which was Barry's home, and was published there until it closed in 1887. Barry authored *Treatise on the Fruit Garden* (1851), and published the book again as *Barry's Fruit Garden* in 1872. Further, he was an author of *The Catalogue of Fruits of the American Pomological Society* that, over the years, listed and described all of the new and noteworthy fruits of America. Barry was the first and long-time vice-president of the Massachusetts Horticultural Society, and one of the founders of the Western New York Horticultural Society, one of the leading horticultural societies in the United States. See Hedrick, *The Pears of New York*, 203–05; U. P. Hedrick, *A History of Agriculture in the State of New York* (Albany, N.Y.: New York State Agricultural Society, 1933), 386–88; Julie S. Higginbotham, "Four Centuries of Planting Progress: A History of the U.S. Nursery Industry," *American Nurseryman*, Vol. 171, No. 12 (Chicago: American Nurseryman Publishing Co., June 15, 1990), 36–40.

56. J. Stephen Casscles, "Blue and Black Heirloom Grape Varieties for the Northeast," appearing concurrently in *Fruit Notes*, Vol. 83, No. 4 (Amherst, Mass.: Stockbridge School of Agriculture, University of Massachusetts Amherst, Fall 2018), and *Horticultural News*, Vol. 98, No. 4 (Hackettstown, N.J.: New Jersey State Horticultural Society, Fall

2018), 19; Observations by the author at his farm Cedar Cliff, Athens, New York, from 2010–2021 for Barry and tasting notes of wines made from Barry, including pollination observations; see also Hedrick, *Grapes of New York*, 177–79; *Bushberg Catalogue*, 88; Hedrick, *Manual of American Grape-Growing*, 288, 298, 335–36; Beach, *Fertilizing Self-Sterile Grapes*, 339–67; Maynard, *Report on Fruits*, 14; McCluer, *Grapes, Test of Varieties*, 252. Hall and Beach, *Why Some Grapes Fail to Fruit*, 6–7; Whitten, *The Grape*, 48.

57. Johann Wolfgang von Goethe (1749–1832) was a prominent German literary figure. Goethe was born in Frankfurt into a wealthy family. As an adult, he moved to Weimer, where he lived for much of his remaining life, under the patronage of his friend Duke Charles Augustus, Duke of Saxe-Weimer. Most of his life was devoted to German literature and as a leader of the Romantic Movement. By 1790, Goethe became interested in botany, and wrote a book on the principles of botany, *Essay in Elucidation of the Metamorphosis of Plants* (1791). Goethe was known for writing the play on the legend of *Faust* and having his poems set to music by the great luminaries of German music composition, including Mozart, Beethoven, Schubert, Mahler, Berlioz, and Mendelson. See generally, John R. Williams, *The Life of Goethe: A Critical Biography* (Oxford: Blackwell Publishers, 1998); and George Henry Lewes, *The Life of Goethe*, 2nd ed. (London: Smith, Elder and Co., 1864).

58. George Beach and L. R. Bryant, *Grape Growing in Colorado*, Colorado Agricultural Experiment Station Bulletin 484 (Fort Collins, Colo.: Colorado State College, 1944), 4–5.

59. Observations by the author at his farm Cedar Cliff, Athens, New York, from 2010–2021 for Goethe and tasting notes of wines made from Goethe, including pollination observations.

60. Hedrick, *Grapes of New York*, 276.

61. *Bushberg Catalogue*, 128.

62. Morton, *Wine Growing in Eastern America*, 178–79.

63. Casscles, "Pink and Light Red Heirloom Grape Varieties," 24–25; Hedrick, *Grapes of New York*, 276–77; *Bushberg Catalogue*, 128; Hedrick, *Manual of American Grape-Growing*, 288, 298, 37; Hall and Beach, *Why Some Grapes Fail to Fruit*, 6–7; Whitten, *The Grape*, 50; Maynard, *Report on Fruits*, 14; McCluer, *Grapes, Test of Varieties*, 260.

64. See generally, Valdinho Pellin, Adriana Vieira, Pinto Vieira, "Contributions of Geographical Indications for Territorial Strengthening in Rural Space: A Case Study in Southern Brazil," *Espacious*, Vol. 36, No. 8, 2015; Mariane Schuck, Flavia Moreira, *et. al.*, "Molecular Identification of the Grapevine 'Goethe' from Urussanga (SC) with Microsatellite Markers," *Revista Brasileira de Fruticultura*, Vol. 32, No. 3, September 2010, 825–31; Adriana Carvalho Pinto Vieira, Melissa Watanabe, Kelly Lissandra Bruch, "Perspectives Vitiviniculture Development in View of the Recognition of Geographical Indication Vales Da Uva Goethe," *Revista GEINTEC*, Vol. 2, No. 4, October 2012, 327–43.

65. Shoemaker, *Sugar, Acidity, and Juice Color Determinations in Grapes*, 5–12.

66. See generally, charts of grape variety genealogy by Carl Camper, Chateau Stripmine Experimental Vineyard, http://chateaustripmine.info/Breeders.htm .

67. Beach and Bryant, *Grape Growing in Colorado*, 4–5; Observations by the author at his farm Cedar Cliff, Athens, New York, from 2010–2021 for Herbert, and tasting notes of wines made from Herbert, including pollination observations.

68. Hedrick, *Grapes of New York*, 291–92.

69. *Bushberg Catalogue*, 135.

70. Ibid., 135.

71. *Shoemaker, Sugar, Acidity, and Juice Color Determinations in Grapes*, 5–12.

72. Hedrick, *Grapes of New York*, 291–293; *Bushberg Catalogue*, 135; Hedrick, *Manual of American Grape-Growing*, 288, 298, 383–84; Maynard, *Report on Fruits*, 14; McCluer, *Grapes, Test of Varieties*, 260; S. M. Tracy, ed., Missouri Horticultural Society, *Reports of the Missouri State Horticultural Society for the years 1880 and 1881* (Jefferson City, Mo.: Tribune Printing Co., 1881), 129–33; Beach, *Fertilizing Self-Sterile Grapes*, 338–67; Hall and Beach, *Why Some Grapes Fail to Fruit*, 6–7.

73. Observations by the author at his farm Cedar Cliff, Athens, New York from 2010–2021 for Lindley, and tasting notes of wines made from Lindley, including pollination observations; Hedrick, *Grapes of New York*, 329–31; *Bushberg Catalogue*, 149–50; N.Y. Dept. of Agriculture, Fruit Industry in New York State, Part II, 1233, 1244; Hedrick, *Manual of American Grape-Growing*, 289, 299, 402–03; Beach, *Fertilizing Self-Sterile Grapes*, 338–65; Maynard, *Report on Fruits*, 15–16; McCluer, *Grapes, Test of Varieties*, 260; Hall and Beach, *Why Some Grapes Fail to Fruit*, 6–7; Whitten, *The Grape*, 51.

74. Shoemaker, *Sugar, Acidity, and Juice Color Determinations in Grapes*, 5–12.

75. Hedrick, *Grapes of New York*, 329–31.

76. *Bushberg Catalogue*, 149.

77. Massasoit Ousamequin (1581–1661) was a sachem of the Wampanog Confederacy, which was located south of Boston, where the Plymouth Puritans first landed and settled. A sachem is a paramount chief of an Algonquin tribe and such tribes lived in Massachusetts in colonial times. Massasoit lived in Pokanoket village which is present day Warren, Rhode Island. As sachem, Massasoit built an alliance with Puritan colonial leaders, such as Governor John Carver, William Bradford, and Miles Standish, that culminated in the peace treaty of 1621. This alliance between the Wampanog Confederacy and the Puritan colonialists prevented the failure of the Plymouth colony by establishing a relatively long and stable balance of power against the ascendant and hostile Narragansett Indians who resided to the west of the Plymouth colony. See generally, Alvin G. Weeks, *Massasoit of the Wampanoags* (Norwood, Mass.: Privately Printed, The Plimpton Press, 1919).

78. Casscles, "Great Cool Climate Heirloom Grape Varieties," 15; Observations by the author at his farm Cedar Cliff, Athens, New York, from 2010–2021 for Massasoit, and tasting notes of wines made from Massasoit, including pollination observations; see also Hedrick, *Grapes of New York*, 343–44; *Bushberg Catalogue*, 154; Hedrick, *Manual of American Grape-Growing*, 289, 299, 408; Beach, *Fertilizing Self-Sterile Grapes*, 369; Maynard, *Report on Fruits*, 15; McCluer, *Grapes, Test of Varieties*, 260–61; Hall and Beach, *Why Some Grapes Fail to Fruit*, 6–7; Whitten, *The Grape*, 51; T. T. Lyon, *Fruit Tests at South Haven*, Bulletin No. 143 (Lansing, Mich., Michigan Agricultural Experiment Station, State Agricultural College, April 1897), 191.

79. Michigan State Horticultural Society, "*Salem Grape*," *Eleventh Annual Report of the Secretary of the State Horticultural Society of Michigan* (Lansing, Mich.: W. S. George & Co., 1882), 223.

80. Observations by the author at his farm Cedar Cliff, Athens, New York, from 2010–2021 for Salem, and tasting notes of wines made from Salem, including pollination observations.

81. Casscles, "Pink and Light Red Heirloom Grape Varieties," 26; Casscles, "Great Cool Climate Heirloom Grape Varieties," 15; Hedrick, *The Grapes of New York*, 397–99; *Bushberg Catalogue*, 175–76; Hedrick, *Manual of American Grape-Growing*, 289, 299, 370, 435; Beach, *Fertilizing Self-Sterile Grapes*, 338–68; Hall and Beach, *Why Some Grapes Fail to Fruit*, 6–7; Maynard, *Report on Fruits*, 15; McCluer, *Grapes, Test of Varieties*, 261–62; F. A. Waugh, *Home-Grown Grapes in Vermont*, Bulletin No. 62 (Montpelier, Vt.: Vermont Agricultural Experiment Station, University of Vermont and State Agricultural College, Vermont Watchman Co., January 1898), 42.

82. R. L. Watts, *Varieties of Grapes, University of Tennessee Agricultural Experiment Station Bulletin*, Vol. IX, No. 4 (Knoxville, Tenn.: The Agricultural Experiment Station, December 1896), 187.

83. This information indicates that Rogers's backyard was not an unchanging and static garden that continued to maintain all of his early first-generation hybrids. Instead, it was a dynamic space where new grape hybrids were being developed and unsuitable ones were removed to make room for new hybrids and breeding activities. See *Bushberg Catalogue*, 86.

84. Observations by the author at his farm Cedar Cliff, Athens, New York, from 2010–2021 for Aminia, including pollination observations; See also Hedrick, *Grapes of New York*, 170–71; *Bushberg Catalogue*, 86; Hedrick, *Manual of American Grape-Growing*, 333; Beach, *Fertilizing Self-Sterile Grapes*, 338–65; McCluer, *Grapes, Test of Varieties*, 252; Hall and Beach, *Why Some Grapes Fail to Fruit*, 6–7; Whitten, *The Grape*, 48.

85. Hedrick, *Grapes of New York*, 263–64; *Bushberg Catalogue*, 125; Beach, *Fertilizing Self-Sterile Grapes*, 369; Maynard, *Report on Fruits*, 14; McCluer, *Grapes, Test of Varieties*, 254; Hall and Beach, *Why Some Grapes Fail to Fruit*, 6–7.

86. Karl Friedrich von Gaertner (1772–1850) was a pioneer in the study of plant hybridization. Gaertner challenged the doctrine of Carl Linnaeus who maintained that a new species of vegetation could arise freely from hybridization. Von Gaertner defended the stability of plant species, and that while transmutation of species was possible, the new species would revert back to its original form. See Herbert F. Roberts, "The Contribution of Carl Friedrich von Gartner to the History of Plant Hybridization," *The American Naturalist*, Vol. 53, No. 628 (New York: The Science Press, September–October 1919), 431–45.

87. Hedrick, *Grapes of New York*, 272–74; *Bushberg Catalogue*, 127; Hedrick, *Manual of American Grape-Growing*, 288, 298, 375–76; Beach, *Fertilizing Self-Sterile Grapes*, 369; Hall and Beach, *Why Some Grapes Fail to Fruit*, 6–7.88.

88. The Merrimac River, now generally referred to as the Merrimack River, is a major river of northeast Massachusetts that flows 110 miles starting north of Concord, New Hampshire, and flowing south to Manchester and Nashua, New Hampshire, and then to Lowell and Lawrence, Massachusetts, and which ultimately empties into the Atlantic Ocean at Newburyport, Massachusetts.

89. Observations by the author at his farm Cedar Cliff, Athens, New York, from 2010–2021 for Merrimac, and tasting notes of wines made from Merrimac, including pollination observations; see also Hedrick, *Grapes of New York*, 345–47; *Bushberg Catalogue*, 155; Hedrick, *Manual of American Grape-Growing*, 289, 299, 410; Beach, *Fertilizing Self-Sterile Grapes*, 338–68; Maynard, *Report on Fruits*, 15; Waugh, *Home-Grown Grapes in Vermont*, 41; Hall and Beach, *Why Some Grapes Fail to Fruit*, 6–7.

90. Observations by the author at his farm Cedar Cliff, Athens, New York, from 2016–2021 for Wilder, including pollination observations; see also Hedrick, *Grapes of New York*, 423–25; *Bushberg Catalogue*, 188; Hedrick, *Manual of American Grape-Growing*, 299, 444; Maynard, *Report on Fruits*, 15; McCluer, *Grapes, Test of Varieties*, 257; Hall and Beach, *Why Some Grapes Fail to Fruit*, 6–7.

CHAPTER SIXTEEN

Other Nineteenth Century Grape Breeders of New England

NEW ENGLAND, ESPECIALLY the Boston area, was a special place for nineteenth century American horticulture and grape breeding because of the high concentration of grape hybridizers, nursery owners, farmers, horticultural writers, botanists, and agricultural/horticultural societies that had a common set of breeding goals. These individuals and groups regularly communicated with each other and shared information about their work. In addition to the aforementioned Rogers, noted grape breeders and writers such as John Fisk Allen, Ephraim Bull, James Clough, Diana Crehore, Charles Hovey, Robert Manning, John B. Moore, and Marshall Wilder gave this region its depth of experience.

This portion of the book covers these, as well as a second tier of lesser-known individuals who contributed their expertise and brought to the area much horticultural knowledge and skill. These growers and breeders, such as Asa Clement, William E. Greene, George Haskell, Francis B. Hayes, Jacob W. Manning, Cyrus Pringle, Epaphroditis Steele, and Nelson White pursued their own breeding and nursery work and helped to evaluate the work of their other local colleagues.

Ephraim Wales Bull. *Very early in his breeding career, Bull was noted for the creation of the Concord grape, but he remained a dedicated grape breeder for the rest of his life. The success he found with the Concord grape, however, was never replicated. Circa 1890.*

The common goal of these New England horticulturalists was to create grapes that were winter hardy, resistant to fungus diseases, productive, and which could produce quality fruit that ripened early enough to be consistently harvested in New England. Some wished to create table grapes and others, multi-purpose grapes that could be used either as table, juice, or wine grapes. Many of these horticulturalists were prosperous New Englanders of English descent who earned their money as merchants or in maritime mercantile shipping. Some were prosperous farmers who were branching out into the new lucrative fruit market to supply New England's growing cities and towns with quality grapes and other fruits.

Some of these individuals had large families and were active in their community's civic and religious affairs and agricultural/horticultural societies. Others, such as E. S. Rogers and George Haskell, never married and lived with their mothers for their entire lives. They successfully engaged in a professional occupation and hired domestic help to take care of their personal needs. Their wealth and lifestyles gave them ample time for horticultural pursuits.

Ephraim Wales Bull

(1806–1895)

Ephraim Bull is considered to be one of the most significant grape breeders in North America. His reputation and legacy centers on the creation of the *labrusca* blue grape variety, Concord. Bull was born in Massachusetts, on March 4, 1806, in a farmhouse located near Washington Street in what is now downtown Boston, near the Boston Common. Ephraim was the eldest son of Epaphras Bull, a silversmith who moved back to Boston from Bullsville, Orange County, New York. His family heritage included Captain Thomas Bull, who came to the Massachusetts Bay Colony from England on the transport ship *Hopewell* in 1635.

In Bull's time, Washington Street was a major avenue into the city, when livestock still roamed the Boston Common. Behind his home was a garden where Bull engaged in his horticultural passions. Here, he started to experiment with grapes, along with the cultivation of other fruits and vegetables. As a lad, Bull did well in school and by 1821 was apprenticed in the trade of beating gold leaf. Ultimately, Bull earned his living in the manufacture, purchase, and sale of gold leaf. Gold leaf was in great demand for the gilding of objects and the production of books. At about this time, his family moved to nearby Dorchester. While he was a goldbeater by trade, Bull continued to devote his spare time to horticulture, especially to grape growing. The varieties that he grew were of the then very limited number of common types, such as Catawba, Isabella, and White Chasselas.

In 1826, Bull set up his own goldbeating workshop and married Mary Ellen Walker, who was related to Harvard College President James Walker. Once married, the Bulls moved back to Boston to a small house on Fayette Street, near the South End. Bull was an active tradesman, who worked long hours in a hot, dusty shop on Cornhill Street (where Government Center is located today). His shop was very near to the first headquarters of the Massachusetts Horticultural Society that was established only three years later, in 1829. During this time, Bull continued to pursue horticulture in the small garden that was behind his house.

By 1836, Bull had developed lung ailments due to the dirty air generated at and near his shop. His physician told Bull to move out of Boston and to live in an area that had fresh air so that his lung condition could improve. In August 1836 Bull purchased seventeen acres of land in Concord, Massachusetts.[1]

Bull's farm and grapery was located near Concord Center on the Lexington Road, the same road that was used by the British Army and Patriots sixty years previously in the battles of Lexington and Concord. The Bulls lived in a little white house where he established a small goldbeaters workshop behind his home. The Bull house is still standing. It is located just to the east of the Alcott's Orchard House, where the novel *Little Women* was written by Louisa May Alcott; and by Waysides, a home that had also been owned by the Alcotts and by Nathanial Hawthorne, the romantic novelist and short story writer who wrote *The Scarlet Letter*, *The House of the Seven Gables*, and many other literary works. Nathaniel Hawthorne, incidentally, was the nephew of Salem pomologist Robert Manning. In addition to his immediate neighbors, the Alcott family, the Hawthorne family, and Ralph Waldo Emerson, Bull's farm was surrounded on two sides by the farm and nursery of Captain John B. Moore, a noted breeder and cultivator of grapes and roses in his own right.

While Bull continued to work as a goldbeater, he liked farming and horticulture more. It is at this point, sometime after 1840, that Bull began to concentrate primarily on cultivating grapes. He continued to grow the more commonly cultivated grapes of the day that he had grown previously in Boston, such as Isabella, Catawba, and White Chasselas. As with most other New England growers, he was not able to consistently ripen his Catawba and Isabella grapes; as both varieties originated in the warmer Southern and Middle

Atlantic states of North and South Carolina, Virginia, and Maryland, which had much longer growing seasons. These late ripening varieties were not conducive to the shorter growing season of New England, even along the warmer ocean coast.

In 1841, Bull purchased the Eben Dow Farm, which was next to his own farm. Here he set out many fruit trees, shrubs, and vines. The farm's soil was sandy, with a south facing slope which possessed a warm micro-climate for Concord Center. Once this land was secured, Bull turned his attention even more so to developing an early-ripening grape that would be hardy to grow in Massachusetts. To this end, he secured as many local early-ripening grapes as he could for his breeding program. As with Rogers's efforts in the next decade, Bull knew of Jean Baptiste Van Mons's success in breeding new hybrid pears from seed and believed that the same process could be used to hybridize new grape varieties that could incorporate the desired characteristics of earlier ripening times, cold hardiness, and fungus disease resistance. To this end, in addition to securing *vinifera* and other cultivated grape varieties, he noticed and collected early-ripening native *labrusca* varieties that were growing in his garden and in the surrounding area.

It is fitting that Bull named Concord, his most famous and consequential grape variety, for the town of its origin. Chapter Seven details the historical and

Ephraim Wales Bull's grave marker in Concord, MA.

cultural significance that the development of the Concord grape had on American viticulture. In addition to propagating the seedling Concord, Bull extensively bred new grape varieties long after the introduction of Concord, communicating with other horticulturalists about them. These include Ester, Rockwood, and Cottage, a productive quality blue grape that the author grows at his farm in Athens, New York. None of these other seedlings came close to attaining Concord's magnitude of success.

Bull made very little money by developing Concord, however, other nurseries and growers quickly propagated and sold the vine without restraint. Bull's last days were not prosperous, and the inscription on his tombstone in the Sleepy Hollow Cemetery in Concord, Massachusetts, reads, "He sowed, others reaped."[2]

CONCORD

Concord is by far the most widely-grown grape variety in the eastern and northern regions of the United States. Details of its fruit and growing characteristics can be found in Chapter Seven, on *Labrusca* Hybrids, pages 72–74.

PARENTAGE
labrusca, vinifera

HARVEST DATE
Mid-season to early late season

A+ A B A+ B-

COTTAGE

Cottage is Bull's second most successful hybrid grape variety after Concord, and in some ways is superior to Concord. Cottage is a seedling of Concord, perhaps a third or fourth generation seedling that Bull raised at his home. Cottage was introduced by Bull in 1869 and placed on the *American Pomological Society Fruit Catalogue* grape list in 1879. Cottage resembles its seedling parent Concord in many ways, but distinguishes itself in others.

PARENTAGE
labrusca

HARVEST DATE
Mid–season

A+ B A A+ B+

Cottage is a very vigorous growing variety, even more so than Concord, with a strong branching root system and thick olive-green leaves similar to Concord. The canes are thinner than Concord and it develops a denser canopy. Its growth habit is slightly upward and then downward to lateral, as gravity brings its canes toward the ground. Cottage is more particular to the soils in which it thrives, unlike Concord which thrives in most soils; hence, this characteristic can limit the productivity of Cottage when compared to Concord. Otherwise, it is very vigorous, healthy, and winter hardy that has moderately good fungus disease resistance, but is not as resistant to fungus diseases as Concord. This heightened susceptibility to fungus diseases may be due to the thicker canopy of Cottage that precludes light and air from entering into the canopy to dry off the fruit and leaves and minimize fungus disease.

Cottage fruit is of better quality than Concord, with a less foxy *labrusca* flavor. It is richer and more delicate in flavor and lends itself to the production of better quality wines. Cottage wines have a more muted foxy *labrusca* flavor profile than Concord, with tannins that are more balanced, and which possess a softer, more approachable, acid structure. The color of the wine is a light red to purple. Cottage wines are flowery and aromatic, with nuances of rose petals, violets, strawberry jam, artificial cherry, and red/black raspberry. The clean body and tannin structure consists of integrated aromas and flavors of flint, wet bricks, and earth that is somewhat musty.

Cottage ripens at about the same time or earlier (mid-season to early late season) as Concord, but tends to ripen unevenly and drops its fruit when ripe. The clusters are somewhat smaller than Concord, with similar sized berries that are a dull

black color, unlike the blue-black berries of Concord. The clusters are moderately broad, cylindrical to slightly tapering and somewhat compact. The flowers are self-fertile, open early to mid-early, with upright stamens; hence, Cottage may be a good pollinator for the mid-season blooming Rogers hybrids. Cottage is not a table grape, but it can make acceptable wines, especially when used as a base for dessert wines such as ports and sherries.[3]

MINOR GRAPE VARIETIES OF BULL

ESTER

labrusca, vinifera?

A white seedling of Concord, it is uncertain if Ester is a purebred seedling of Concord or if it has some *vinifera* heritage. Ester resembles its parent Concord in vine and fruit characteristics. Hence, Ester is a healthy, fungus disease resistant, and winter hardy variety. It is variable in vigor and productiveness, usually hardy, with small light green leaves, with the lower leaf surface tinged with bronze. The fruit ripens with Concord or later, with medium to above-sized clusters. The pale yellowish-white berries are covered with small brown dots, with a thin skin that is inclined to crack. The quality of the fruit is vinous, sweet and of fair to very good flavor. The fruit tends to drop as it becomes ripe. Bull named this variety after his mother, who spelled her name "Ester," in the traditional New England way, and not "Esther" as has been found in other places in grape literature. to the flesh.[4]

LAWSON

labrusca

Exhibited by Bull at the Massachusetts Horticultural Society in 1874. A very minor white grape variety with a large berry and cluster size.[5]

ROCKWOOD

labrusca

A minor black seedling of Concord. It has higher fruit quality than its parent Concord and ripens before it. There is little to distinguish it from Concord in a major way, so this variety did not garner interest from growers or horticulturalists. It was introduced in 1889 by nurseryman George S. Josselyn of Fredonia, New York.[6]

SHEPPERD

labrusca

A minor black grape that was exhibited by Bull at the Massachusetts Horticultural Society in 1874. This black grape has a medium cluster and berry size.[7]

UNA

labrusca

A white Concord seedling bred around 1867. The vine is healthy and vigorous, with a cluster and berry size that is slightly smaller than Concord. Una is as sweet as Concord and ripens one week before Concord.[8]

James Milton Clough

(1820?–1890?)

During the middle-to-late nineteenth century, James Milton Clough lived in Stamford, one of the southernmost towns in Vermont. Little is known of Clough's horticultural work or life, but he did raise one of the most important white wine grape varieties to be found or bred in nineteenth century America. That grape variety has been called, alternatively, Winchell and Green Mountain. The name that was finally settled upon was Winchell, even though there were horticultural societies that pushed to have the variety re-named after its originator Clough.

WINCHELL

Winchell is a superior white grape whose seed was planted by, and the vine raised by, Clough around 1850. The seed came from an unknown purple grape. Winchell was propagated from cuttings by Clough and his neighbors in Vermont; for years it remained a locally-grown grape at its place of origin in southern Vermont. In December 1885, the Ellwanger & Barry Nursery of Rochester, New York, received cuttings of this variety from C. E. Winchell, also of Stamford, Vermont. In 1888, Ellwanger & Barry[9] introduced this newly named grape to the public as "Winchell."

PARENTAGE
labrusca, vinifera? aestivalis?

HARVEST DATE
Early mid–season

A+	A	B+	A+	A

During the same year, Stephen Hoyt's Sons of New Canaan, Connecticut, introduced the same grape variety under the name Green Mountain, after having purchased cuttings of this vine from James M. Paul of North Adams, Massachusetts, earlier in 1885; North Adams being located no more than six miles south of Stamford. While its genetic ancestry is unknown, observers maintain that while it is chiefly a *labrusca* variety, it may have some *vinifera* and *aestivalis* in its genetic background because of its very muted *labrusca* flavors.

Winchell is an important variety for three reasons. First, it is a viticulturally superior grape both in the field and in the wine cellar. Second, Winchell and its progeny were utilized and continue to be extensively used by state agricultural experiment station breeding programs across the United States and by private grape breeders, to impart its superior growing characteristics and fruit quality to future progeny. Third, Winchell's flower is self-fertile with upright stamens. Winchell blooms by mid-season or earlier, at about the same time as when most Rogers hybrids, which are often self-sterile, are blooming. Winchell is therefore a compatible pollinator for most Rogers hybrids and many other self-sterile grape varieties. Planting Winchell as a pollinator close to nearly all other Rogers hybrids will help to increase the fertilization rates and production levels of these infertile Rogers hybrids.

Winchell is a vigorous to very vigorous growing white variety that is adapted to many different soil types and cool climates throughout the United States. It is very winter hardy, healthy, fungus disease resistant, and very productive. Winchell has an upright growth habit with medium thick-sized canes. It has a very dense canopy, so it should be pruned to open its canopy to minimize fungus diseases. While Winchell is not a good table grape because it shells when ripe and its berry skin is too soft and fragile to be easily shipped, it is a fine grape for wine production. The clusters vary between being moderately loose to moderately compact and are moderately large and long. The cluster is slightly cylindrical that tapers, with one prominent shoulder.

The medium-sized berries are greenish to yellow with small brown dots. The variety ripens by late-early to early mid-season and attains sugars of between 18–19° Brix, but its taste is perceived to be sweeter because of the grape's low acidity. Winchell clusters can hang on the vine until they completely ripen by early mid-season to mid-season, but then the berries start to fall off, hence it should be harvested in a timely manner. The variety has a soft, nuanced, and very approachable *labrusca* taste that is muted and subtle. The wines are interesting. They are yellow-green in color, floral, grapey, with very fine soft white *labrusca* flavors complimented by layered notes of honey, apricots, bananas, melons, white peaches, and pineapples, with a body and viscosity that is medium to heavy for a white wine.[10]

Winchell's solid growing attributes in the field and superior winemaking capabilities in the cellar, made this variety very popular with those grape breeders who wanted to impart their new hybrids with these desirable attributes. Winchell and its progeny were used as one of the primary genetic bases by the New York Agricultural Experiment Station (Geneva Experiment Station) breeding program and by the grape breeding programs at the University of Florida and University of Minnesota.

The Geneva Experiment Station introduced the grape variety Ontario (Winchell x Diamond) in 1908, and was one of their primary grape varieties used for table, seedless, and wine grapes. Ontario, whose seed parent is Winchell, is genetically found in a preponderance of the Geneva Experiment Station developed and named grape varieties such as: Wayne (Mills x Ontario), released in 1927; Watkins (Mills x Ontario), released in 1930; Seneca (Lignan Blanc x Ontario), 1930; Yates (Mills x Ontario), 1937; Eden (Ontario x (Triumph x Mills), 1938; Schuyler (Zinfandel x Ontario) 1947; the seedless grape varieties Interlaken Seedless (Ontario x Sultanina), 1947; Romulus (Ontario x Sultanina), 1952; Himrod (Ontario x Sultanina), 1952; and the seeded table grapes Alden (Ontario x Grosse Guillaume), 1952; New York Muscat (Muscat Hamburg x Ontario), 1961; Glenora (Ontario x Russian Seedless), 1977; and the white wine grape Melody (Seyval x GW5 (Pinot Blanc x Ontario), 1985.

Winchell was also used to breed the Geneva varieties Ripley (Winchell x Diamond), 1912, and Brocton (Brighton x (Winchell x Diamond)), 1919. In addition, Winchell, the pollen grandparent of Schuyler (Zinfandel x Ontario), developed in 1947, is the grandfather of the noted white wine grapes Cayuga White (Seyval x Schuyler), 1972; and Horizon (Seyval x Schuyler), 1982. Cayuga White is probably the most commonly grown Geneva white hybrid in the United States.

The University of Florida grape breeding program used many of Winchell's progeny such as: Alden (Ontario x Grosse Guillaume), Glenora (Ontario x Russian Seedless); and Schuyler (Zinfandel x Ontario) in many of its new hybrids. Further, the University of Minnesota extensively used the Winchell progeny hybrids of Alden and Himrod.[11]

Because of its strengths in the field and cellar, Winchell should be more extensively evaluated by growers and state agricultural experiment stations, and ultimately planted in more vineyards in New England, New York State, and other cool and cold climate regions of North America and Northeast Asia.

Diana Ames Crehore

(1796–1885?)

Diana Ames Crehore was one of the few women in the nineteenth century to be publicly interested in grape breeding, and who was active, in her own right, in the Massachusetts Horticultural Society. Crehore's husband, John Shepard Crehore (1763–1819), was a gentleman farmer who made his living manufacturing wooden chairs. The couple were married at the Milton Meeting House in 1817. Diana raised at least two children, John Crehore (1818–1877) and Ann Elizabeth (1834–1917). The Crehore family, originally from Ireland, had an extensive dairy farm in the western part of Milton, Massachusetts, near what was known as Hubbard's Bridge or Paul's Bridge. The farm had been in the Crehore family since the late seventeenth century, and had a considerable dairy operation; they also manufactured cheese, commercially cultivated apples and strawberries, grew other fruits such as grapes, operated greenhouses, and maintained extensive flowerbeds. It is here that Diana Crehore first planted the grape seeds that in 1844 would become the vine that the Massachusetts Horticultural Society named in her honor.

DIANA

The grape Diana was planted in about 1834 from a seed of Catawba that was open pollenated. Crehore had received some Catawba grapes from a friend and planted the seeds the following spring. Of the many seeds that were planted, only a few germinated. The sole vine retained was initially a slow grower, and was neglected at the farm until about 1843 when it was pruned, fruited, and exhibited at the Massachusetts Horticulture Society. Previously, it had been located near the house, until it was moved to a new location sometime before that year.

PARENTAGE
labrusca, vinifera

HARVEST DATE
Late

A-	B	B+	A+	B+

Crehore exhibited the grape in 1843 at the Society to determine if it was a new variety or not. The grape variety was named "Diana" by the Society in honor of its originator after the 1844 showing. In 1850, at the Second Congress of the American Pomological Society, the Diana grape was placed on a trial list of grapes for future cultivation as a new variety that promises well. Diana did not test well in the cool New England climate and soils, and so it did not draw much attention. However, Diana did thrive in the warmer climate and soils of the Mid-Hudson River Valley, where it came to the attention of local nurseryman, horticultural writer, and landscape designer Andrew Jackson Downing of Newburgh, New York, who noticed and admired the growing characteristics and high fruit quality of the grape. Diana was included in the American Pomological Society list of recommended varieties in 1854, and interest in this variety increased.

Sometime around 1850, Diana was planted at Wodenethe, the country estate of Henry Winthrop Sargent (1810–1882). Sargent came from a well-established Yankee family, and was born, raised, and educated in Boston. After graduating from Harvard College, Sargent moved to New York City to engage in banking. In 1841, he retired from banking and moved to his country estate, Wodenethe, which was located on about twenty acres of land just north of Fishkill Landing, now Beacon, New York, on a plateau that overlooked the Hudson River. At Wodenethe Sargent studied to become an expert horticulturalist and landscape gardener. Henry's younger cousin was Charles Sprague

Sargent (1841–1927), the first director of the Arnold Arboretum located in the Jamaica Plain neighborhood of Boston

Diana was one of the first American hybrids to be widely distributed to ascertain its suitability for cultivation throughout the eastern United States. The positive attributes of Diana were its full and pretty cluster, tasty flavor profile, and high sugar content for wine production. It ripens at least ten days before its very late ripening seed parent, Catawba; and about one week to ten days after the mid-season ripening variety Concord. This meant that it was possible to grow Diana, a Catawba-like grape variety, in New England and New York, where Catawba could not generally ripen on a consistent basis.

The vine can be capricious where it grows, and does best on relatively poor, dry, gravelly soil that does not have much compost or nitrogen. Diana takes at least four to five years to establish itself before it can produce an adequate crop. The cluster is more compact than Catawba, with a lighter color of delicate pale red/amber to rose. The vine can be somewhat tender to winter damage, is somewhat resistant to fungus diseases, with medium-sized berries that can ripen unevenly. The skin adheres to the flesh and is not a slip-skin variety. Diana does well in the soils that it likes, being vigorous to very vigorous and productive to very productive, otherwise is variable in its yield. It has a lateral growth habit with medium thin canes. The cluster is medium-large and stubby, being tapering to cylindrical. The flowers are fertile, open by mid-season, with upright stamens, and may be a suitable pollinator for most of the Rogers hybrids. The sugars of Diana can reach 21° Brix, hence it has applications for wine production.

The flavor of Diana is similar to a more delicate version of Catawba, but it has more guava flavors that are rich, spicy, and tangy. The wine is very grapey, with fruits of strawberry, bubble gum, grape-berry, and is very *labrusca*-like, similar to a Pink Catawba, with a guava-like twang at the end. In addition, it can have flavors and aromatics of violets and red raspberries. Diana wines are currently made by the Arbor Hill Grapery and Raymor Estate Cellars of the Finger Lakes, New York.[12]

After 1850, Diana was a grape that generated much interest in the American horticultural community. It was viewed as a replacement wine grape for Catawba that could be successfully grown in the more northern parts of the eastern United States, above Ohio and Pennsylvania. At that time, Catawba was used extensively to produce quality white and sparkling wines in America from the middle to late nineteenth century.[13]

Diana is still a relevant grape variety today as it is most probably the seed parent of Iona. According to its breeder, Dr. Charles W. Grant, Iona originated from a seed of Diana planted around 1855 on Iona Island on the Hudson River in northern Rockland County, New York. Iona is a superior grape for white wine production, and is making a comeback in popularity. Further, Iona is the parent of many other quality grapes such as Jefferson and Diamond, which in turn begat other varieties developed by Geneva breeding program, such as Alden, Buffalo, Cayuga White, Horizon, Melody, Schuyler, and Steuben.[14]

John Fisk Allen

(1807–1876)

John Fisk Allen, a nationally-known horticulturalist from Salem, Massachusetts, was one of the first in the United States to specialize in the cultivation and purposeful breeding of new grape varieties. His goal was to combine the superior flavors of European *vinifera* grape varieties with the more winter hardy and fungus disease resistant characteristics of American *labrusca* grape varieties that ripened early.

Allen was a horticultural generalist who also grew and exhibited amaryllis, nelumbium, and tropical lilies in his greenhouses; he was also one of the charter members of the Massachusetts Horticultural Society. He often exhibited grapes and introduced new grape varieties at local and national horticultural society exhibitions. On a smaller scale, he grew and exhibited forced cherries, peaches, nectarines, figs, pineapples, and pears. In 1843, Allen built a grapery on Dean Street in Salem, Massachusetts, to enclose and cultivate over four hundred varieties of grape vines, peaches, cherries, and other fruits. Not only did he conduct an extensive grape breeding operation, but he also encouraged others to breed new grape varieties that married the finer flavors of European *vinifera* grape varieties with local American *labrusca* grape varieties known for their winter hardiness, fungus disease resistance, big berry and cluster size, early ripening time, and high productivity.

To disseminate information on his grape cultivation and breeding work, Allen wrote *The Culture of the Grape, First ed.*, (Boston, 1847); and expanded third edition of *A Practical Treatise on the Culture and Treatment of the Grape Vine* (New York, 1859). This book was a comprehensive work on growing and breeding grapes. He had at his farm, under glass, approximately 300 European grape varieties, then believed to be one of the largest such collections in the United States.

Through Allen's breeding work, he was one of the first in the United States to usher in the purposeful breeding of grapes to enhance their taste with superior European *vinifera* flavors, and to diminish the off-putting "wild" American *labrusca* grape flavors that were described as being "foxy." Foxy flavors are often described as a combination of grapey and musty flavors, strawberry, or cotton candy, and those heavily perfumed and over the top floral flavors that are often associated with Concord or Niagara grape varieties.[15] Those coarse American foxy flavors were deemed to be inferior to the more subtle flavors of European grape varieties.

The significance of Allen's work is seen after 1830, as American horticulturalists began taking matters into their own hands, purposefully breeding grapes to satisfy the needs of American consumers and growers. Those needs being to create grape varieties that incorporated the beneficial native-American traits of winter hardiness, fungus disease resistance and ripening early so that they could be profitably grown in America's relatively harsh climate when compared to Europe. In addition, these hybrids should have the desired characteristics of a big berry and cluster size of native-American grapes coupled with the finer subtle flavors of European grape varieties.

Allen's greatest horticultural achievement was most likely growing the Victoria Regia, or Great Water Lily.[16] He had just one grape hybrid that was ultimately distributed and widely grown throughout New England, New York, and Canada. It was called, appropriately, Allen's Hybrid.

Early Life

Allen came from a prosperous New England shipping and merchant family that centered its life in Salem. It was precisely because Allen was so prosperous that he had the time, money, and inclination to pursue grape breeding. Early in his business life, Allen made several trips to India as a supercargo and shipmaster, after which he engaged in business with his brother Edward for a few years. He had been a bookkeeper in the office of the distinguished merchant Pickering Dodge; in 1833, he married Dodge's daughter, Lucy Pickering Dodge (1810–1840). They lived at 31–35 Chestnut

Street, Salem, one of the first townhouses built by father-in-law Dodge, Allen's wife Lucy died in Richmond, Virginia, in 1840, leaving one surviving child, Pickering Dodge Allen (1838–1863). In 1842, Allen married Mary Hodges Cleveland (1817–1873); they had one surviving child, Elizabeth Cleveland Allen (1850–1906). For 44 years, Allen was a member of Salem's East India Society, and was secretary of that organization from 1834–1836. He was also an original member of the Essex Institute, and was curator of horticulture from 1864–1867. Both of these organizations ultimately merged and became part of the Peabody Essex Museum of Salem.

MINOR GRAPE VARIETY OF ALLEN

ALLEN'S HYBRID

labrusca, vinifera

Allen's Hybrid is a cross of Isabella and Chasselas de Fontainebleau fertilized by Allen in his grapery in the winter of 1843–1844. Of the many other seedlings that came from Allen's cross fertilization of Isabella and Chasselas de Fontainebleau, the only seedling to be officially named was Allen's Hybrid. It was first exhibited publicly at the Massachusetts Horticultural Society meeting in 1854.

This seedling was one of the first purposeful interspecific hybrids between *vinifera* and *labrusca* grape species made in the United States. After its introduction and for decades afterward, this grape was widely distributed in the United States. This distribution was to encourage its cultivation and to obtain comments from growers from across the United States on its prowess in the field. If nothing else, Allen's work proved that *labrusca-vinifera* hybrid crosses could be made in a practical manner.

Allen's Hybrid is a moderately compact cluster that is medium to large and long. The berry is of medium size with a thin semi-transparent skin the color of which is nearly white, with a slight shade of amber. The flesh is tender with a superior mild Muscat flavor that had few of the perceived disagreeable wild *labrusca* foxy flavors. It was this non-foxy taste that drew horticulturalists and grape breeders to cultivate and evaluate Allen's Hybrid. It ripens by mid-season around the same time as Concord, with a growth habit that is vigorous. It is very susceptible to fungus diseases and is cold sensitive, so it was never widely planted on a commercial basis. The leaf pattern is like its *vinifera* pollen parent Chasselas de Fontainebleau, hence its *vinifera* heritage also transferred to Allen's Hybrid a susceptibility to American fungus diseases.[17]

According to U. P. Hedrick,[18] Allen's Hybrid was grown for at least a generation in the viticultural areas of the United States and Canada. As the variety was distributed, it became more apparent that some first generation *labrusca-vinifera* hybrids such as Allen's Hybrid, may be generally unsuitable for cultivation in the northeastern United States; the reason being that they were still too sensitive for the then very harsh winter temperatures of the northeast United States.[19] It was also very susceptible to the many American fungus diseases, in a manner similar to the European *vinifera* parent. Allen's Hybrid was used by J. H. Ricketts to create Lady Washington (Concord x Allen's Hybrid), which was introduced in 1878, and used in other subsequent hybrids.[20] Ultimately, the cultivation of Allen's Hybrid, and other new hybrid grapes of that time such as Ada (which was bred by Dr. William W. Valk in Flushing, Long Island, and introduced in 1852 (Isabella x Black Hamburg)) was abandoned by commercial growers and horticulturalists alike.[21]

Captain John Brooks Moore

(1817–1887)

Captain John B. Moore of Concord, Massachusetts, was a grape breeder and nursery owner who bred the then widely-planted early blue grape Moore Early. In addition to breeding and cultivating grapes and roses, he operated a nursery in Concord that sold small fruit, grape vines, asparagus, and bedding plants. His property on Lexington Road near Concord Center was immediately to the west of the homes of the Alcott family and Nathanial Hawthorne, the grapery of Ephraim Bull, and what is now the Minute Man National Historical Park. Captain Moore had a sizable farm that was over 115 acres.

Captain Moore married Sarah Augusta Hunt (1817–1849) in 1840 and had one child Caroline Augusta Moore (Lovett) (1840–1892).

In 1851, the widower Moore married his wife's relative Almira Caroline Hunt (1826-1877). In 1885, just two years before his death, he was elected President of the Massachusetts Horticultural Society.[22]

Below are three grape varieties that were either attributed to Moore or which he secured from other breeders to sell at his nursery.

MOORE EARLY

This variety comes from a seed of Concord planted about 1868. In 1871, it was first exhibited at the Massachusetts Horticultural Society by Captain Moore with fifty other seedlings of the same parentage. It was awarded a First Class Certificate of Merit by the Society in 1877, in the same year that it was introduced by Moore himself to the public. In 1881, the *American Pomological Society Fruit Catalogue* placed Moore Early on its list of suitable grapes.

PARENTAGE
labrusca, seedling of Concord

HARVEST DATE
Early-late early

A+ B A A B

Essentially, Moore Early[23] is an early Concord, having many of the same growing characteristics as Concord on the vine. Moore Early is a good, strong grower; it is winter hardy, fungus disease resistant, and adaptable to many different soil types. Its most important differentiating feature is that it ripens ten days to two weeks before Concord, so it is a very early ripening blue grape. Like Concord, its flowers are self-fertile that open in mid-season with upright stamens.

While the vine looks like Concord, it is not as vigorous as Concord, nor as heavy a producer. Moore Early is also more particular on the soils that it thrives on, optimally growing best in rich and well-drained soils that are loose and cultivated often. The cluster and berry look like Concord, except that the clusters tend to be somewhat smaller, looser and have a cluster shape that is cylindrical. The fruit's foxy taste is about the same as Concord and it is better than Hartford, the other popular early-

ripening blue grape of the nineteenth century. At the turn of the twentieth century Moore Early was widely grown and liked by growers as an early blue-black grape, especially by those in Massachusetts and Vermont. It is rarely grown today.[24]

Moore Early is an important grape variety to viticulture because it is the seed parent of Campbell Early (Moore Early x (Belvidere x Muscat Hamburg)), which was bred and introduced by George W. Campbell of Delaware, Ohio, in 1892. While Campbell Early can still be found growing in the eastern United States (more so than Moore Early), it is extremely popular in Northeast Asia.[25]

MINOR GRAPE VARIETIES OF MOORE

EATON

labrusca, seedling of Concord

A seedling of Concord that was originated by Calvin Eaton of Concord, New Hampshire, around 1868. The variety was purchased by John B. Moore and Son of Concord, Massachusetts, in 1882 and introduced by their nursery in 1885. After Captain Moore died in 1887, control of the Eaton grape passed to the nursery of T. S. Hubbard Company of Fredonia, New York.

This blue-black Concord seedling surpasses Concord in appearance, but not in the quality of its fruit, so it was not very popular with either growers or consumers. Similar to Concord in its strong healthy growing characteristics, it ripens unevenly and late to very late, so it can be a shy bearer in cool and short season locations. The flowers are semi-sterile, bloom early, with upright stamens.

As for the Eaton fruit, the flesh is tough and stringy, with the same foxy taste of Concord, but with more juice and less richness in taste. This variety is not grown any longer, chiefly due to its poor fruit quality.[26]

HAYES (MOORE'S NO. 31)

labrusca, seedling of Concord

A white grape that is a seedling of Concord, originating from the same batch of Concord seedlings that created Moore Early. It is a greenish-yellow to amber-yellow hued grape, unlike its blue seed parent Concord. It is a high-quality grape that first fruited in 1872, and was first exhibited in Boston in 1874. In 1880, the Massachusetts Horticultural Society awarded Hayes a Certificate of Merit for its high quality fruit, and it was officially introduced in 1884. The variety was named in honor of Francis B. Hayes (1819–1884),[27] the President of the Massachusetts Horticultural Society from 1879 until his death in 1884. Hayes was somewhat popular before 1900 as it was a good quality, early-ripening white grape, but it fell out of favor due to its defects in the field.

Hayes is hardy, somewhat susceptible to susceptible to fungus diseases, and somewhat vigorous; but it can be a shy bearer. The cluster and berry are variable in size, but generally small to medium-small, and it ripens about one week to a few days before Concord. Its light golden yellow berries are thin-skinned and tender and have small reddish-brown dots, like other high-quality white Concord seedlings, such as Dutchess. It seems that white Concord seedlings that have small reddish-brown dots on its berries are superior quality grapes. The flowers are almost self-sterile, open medium late, with upright stamens. The flavor and texture of the grape is good to excellent, with a vinous delicate taste and texture.

Moore's goal as a breeder was to develop a quality white grape for cool climates. As he wrote, "[i]t is our endeavor to place before the fruit growers of the country, a grape of quality, possessing certain desirable and distinct characteristics which are rare exceptions in grapes of this class. This grape [Hayes] is a pure native possessing hardiness and vigor to a degree seldom attained by other white varieties, and especially adapting it for cultivation in northern and eastern sections." It is sensitive to summer heat, so should be planted where summer heat is comparatively limited.[28]

CYRUS GUERNSEY PRINGLE

(1838–1911)

Dr. Cyrus Guernsey Pringle was a leading American botanist, plant collector, and cataloger of American, Canadian, and Mexican plants. In his early years, he hybridized new varieties of grapes, apples, currants, gooseberries, plums, tomatoes, potatoes, wheat, and oats. Throughout his long and productive career, Pringle was associated with the Smithsonian Institution, American Museum of Natural History, Massachusetts Horticultural Society, Arnold Arboretum, Harvard University, Mexican Medical Institute, Boston Natural History Society, Champlain Valley Horticultural Society, Vermont Botanical Club, and the New England Botanical Club.

Pringle was born in 1838 in the lower Champlain Valley community of East Charlotte, Vermont. He came from a family that had long been interested in fruit cultivation in Vermont's Champlain Valley. In 1859, Pringle attended the University of Vermont, but remained there for only one semester, as his older brother had recently died. Pringle returned home to operate the family farm for his widowed mother. Later on in life, he was awarded an honorary degree of Master of Arts from Middlebury College and a Doctor of Science degree from the University of Vermont.

Cyrus Pringle acquired his longstanding interest in horticulture, fruit growing, and botany from his family. Beginning in 1857, at the age of nineteen, he began to bud apple trees and established his nursery the following year where he propagated pears, currants, cherries, grapes, peaches, and potatoes. Pringle hybridized a new potato, called Snowflake, and new grape varieties, such as Golden Drop, and sold seedlings of lilies, gladioli, and wheat.

Five months after he was married in early 1863, Pringle was drafted into the Union Army, along with two other Vermont Quakers. Being a practicing Quaker and pacifist, he was a declared conscientious objector who opposed any war, including the Civil War. Because of his moral stance against the War, he was subjected to many forms of severe harassment and imprisonment during his time in the U.S. Army until he was finally paroled, which came only after the direct intervention of President Lincoln. Fifty years later, and seven years after his death, the book *The Record of Quaker Conscience: Cyrus Pringle's Diary* (New York, 1918) memorialized his experiences as a pacifist during the Civil War.[29]

In 1868, after returning home to recuperate from injuries inflicted upon him by U.S. Army personnel during the war; he once again started to conduct plant breeding activities. It was during this brief time-period, when he was back in Vermont, that Pringle hybridized several grape varieties, including Golden Drop, Lady Charlotte, and Mansfield.

In 1874, Pringle was appointed to the Vermont Board of Agriculture, which in effect meant that he was awarded an itinerant agricultural professorship. Pringle did have an extensive portfolio of research publications in the areas of botany and horticulture that warranted the Doctor of Science degree. These included scores of articles for *The Country Gentleman*, *Garden & Forest*, and reports for the United State Census Bureau, the Smithsonian Institution, and Harvard University.

By 1875, Pringle started to spread his botanically-inspired wings. He began to collect plant specimens in the field professionally, and lead botanical field surveys and categorize those plants for various herbariums in Vermont, New England, New York City, and Washington, D.C. After working for the State of Vermont, he then worked for Asa Gray[30] of Harvard University, and then for many nationally-known American scientific organizations where he conducted botanical field trips. After 1878, Pringle abandoned plant breeding work in Vermont and concentrated on collecting botanical samples throughout the United States, Canada, and Mexico. Pringle was a quiet and shy person who often botanized alone or with only one or two other close colleagues. He worked closely with Professor Asa Gray, Charles James Sprague of the Boston Natural

History Society, and Professor Charles Sprague Sargent of the Arnold Arboretum.

After 1880, Pringle was appointed as a botanical collector for the American Museum of Natural History in New York City. During his thirty-five years of botanical fieldwork, he collected an estimated 20,000 different plant species, of which more than a few thousand were newly-found plants. By the end of his life, he had collected and amassed one of the largest herbariums in New England that contained over 300,000 sheets of mounted and accessioned plants. Today, the Pringle Herbarium is located at the Plant Biology Department at the University of Vermont. For cool climate viticulturalists, however, he will be best remembered for breeding the grape variety Golden Drop.[31]

MINOR GRAPE VARIETIES OF PRINGLE

GOLDEN DROP

Adirondac x Delaware

A late-early season ripening white grape of superior quality that is resistant to fungus diseases. The seed of this variety was planted in 1869 and introduced by B. K. Bliss & Son of New York about 1880.

The vine is of medium vigor to moderately vigorous and very winter hardy. but not so productive since its cluster and berry size are below average. Its growth habit is upward and then bends slightly downward. The cluster is cylindrical, moderately loose, and of moderate length, less than its pollen parent Delaware. The berry is only medium to medium-small and becomes a deep golden color when ripe.

The quality of Golden Drop's juice for wine is excellent. It has a soft, delicate, and clean *labrusca* flavor profile, like its pollen parent Delaware without its Muscat flavors. It is a clean-tasting fruit that is sweet, achieving a Brix of 17 18° degrees. The leaves can get sun-scaled in August, which can limit its productivity. The flowers are reportedly sterile or nearly so, open by mid-season, with upright stamens. Golden Drop, due to its winter hardiness, quality of juice, and early ripening time, should be considered for shorter season growing areas. My experience is that while the flowers are said to be self-sterile, it will easily pollinate with other mid-season pollinators in mixed variety vineyards.[32]

LADY CHARLOTTE

Delaware x Iona

A vigorous variety that was planted in 1869. Lady Charlotte is a minor grape variety that has a large cluster size that is shouldered and compact. The berry is medium size, with a light green to golden color.[33]

MANSFIELD

Concord x Iona

A minor variety that was raised in 1869 and has not been widely grown. A very vigorous vine with a large compact cluster that is shouldered. The black berry is large with a taste that has some quality. It ripens by mid-season before its seed parent Concord.[34]

Epaphroditis "Paphro" Steele

(1782–1869)

Epaphroditis "Paphro" Steele was a farmer from West Hartford, Connecticut. Paphro's parents were Ebenezer Steele (1753–1805) and Rachel Seymour (1755–1839), also from West Hartford, Connecticut. Paphro and his wife Mary (née Whiting) Steele (1786–1823) had eight children, many of whom did not live past childhood.

Steele is known for identifying the black grape variety first called Hartford Prolific. Its name was later shortened to Hartford. During the nineteenth century, Hartford was so commonly grown that both it and Concord were used as reference grapes in U. P. Hedrick's *The Grapes of New York* to compare its flower blooming and harvest times to other more obscure varieties that were being described.

MINOR GRAPE VARIETY OF STEELE

HARTFORD

chance seedling possibly of
Isabella x wild fox (labrusca) grape

A chance seedling, possibly of Isabella and another wild fox *labrusca* grape, found and identified by Steele in his garden. It first fruited in 1849, was named Hartford Prolific by the Hartford County Horticultural Society, and placed in the *American Pomological Society Fruit Catalogue* in 1862. Hartford was commonly used by American grape breeders in the nineteenth century to impart its beneficial growing characteristics, but not for its faulty fruit qualities.

The vine quality is good, Hartford being vigorous, healthy, very productive, and fungus disease resistant. Unfortunately, its fruit quality is not good, having a very foxy taste, with berries that shell badly when ripe. The black variety ripens by early to late-early season, about one to two weeks before Concord.

Unlike Concord, the variety neither packed, shipped, nor kept well, so it was not considered a suitable table grape. Its advantage was that the vine grows in most places and is productive. The flower is mostly self-fertile, opens in mid-season with upright stamens, but the clusters can be straggly if bad weather occurs during bloom time. The clusters are of above average size, long, tapering, and relatively slender, except when shouldered. The variety was sweet enough at between 17° and 18° Brix that it was used in wine production; the wine quality was below Concord, however.

After 1900, it declined in popularity and was replaced with better quality varieties by growers and nursery owners. Hartford is rarely grown today, but is included here due to its importance during the nineteenth century and its use in developing new hybrids.[35]

Nelson Bonney White

(1824–1914)

Nelson Bonney White was born in Putney, Vermont, in 1824. When he was young, he lived for a time in the grape-growing states of Ohio and New York before settling in West Norwood, Massachusetts. He was a cabinet maker, who came under the influence of E. S. Rogers after the Civil War, at a time when the Rogers hybrids were causing much excitement in New England. It was at this time that White took up grape breeding as his hobby.

By 1908, Hedrick commented that White was probably the oldest grape breeder of note then living, and that he had bred grapes for over fifty years. Like Rogers, White used Black Hamburg primarily as the pollen parent for his hybrids. White also published technical articles for horticultural societies such as the Horticultural Society of New York and the Massachusetts Horticultural Society. These articles detailed his theories and practices in grape breeding.

White's primary breeding goal was to produce superior seeded and seedless table grapes, and to develop new hybrids that were more self-fertile. Creating self-fertile varieties would increase the progenies' cluster size and attractiveness for the market, and thereby increase yields and profitability of such varieties.

The 1895 *Bushberg Catalogue* stated that White's varieties "have not received the attention they deserve; their originator being an unobtrusive, though enthusiastic and careful viticulturalist, working during the last quarter [of the nineteenth] century, on the improvement of the American grape."[36] White's named grape varieties are described below to give an indication of the genetic material that he used in his grape breeding program.[37]

MINOR GRAPE VARIETIES OF WHITE

AUGUST GIANT

early, large red-berried native wild labrusca x Black Hamburg

August Giant has been described as a strong to very strong grower that was bred in 1861. It is a productive variety that has small to large clusters that are long and tapering. For a *vinifera* hybrid, it is quite winter hardy and somewhat fungus disease resistant; the variety, however, needs a long growing season. Its beautiful foliage, luxuriant growth, and strong vigor made it a suitable ornamental vine for home landscaping. The berries are black, large to very large, and somewhat oblong. August Giant ripens late mid-season to late season. The foxy tasting grape is of modest quality.[38]

AMBER QUEEN

Marion x Black Hamburg

Amber Queen was described by White himself as having a large berry and medium-to-large cluster size that is shouldered like Black Hamburg, which is unproductive in the field. It was first exhibited at the Massachusetts Horticultural Society in 1870, but has never been grown much. The berry holds on to the stem, which is amber colored at first, then grows darker over time until it is a deep dark purple to almost black. The leaves are thick with a somewhat pale green underside like Concord or Black Eagle. Some of its clusters do not fill out well, which indicates that it is not completely self-fertile. The flower blooms by mid-season and is sterile or nearly sterile.[39]

BAY-STATE

Marion x Black Hamburg

This obscure variety was reported to be vigorous and hardy, with large leaves. The cluster is of medium size to large that is shouldered. The berry is red, slightly oblong and ripens early which stores well as the berry adheres to the stem. The grape is considered to be a table grape. There seems to be no mention of Bay-State's self-fertility.

NORFOLK

native wild labrusca x Black Hamburg

A vigorous to medium vigorous vine that is winter hardy, and is generally not very productive. The flowers are nearly fertile and open early with upright stamens. Norfolk is an early-ripening variety which looks like Catawba in berry and cluster, with a slight Muscat flavor. Hedrick reports that the clusters are only medium to small, often broad, tapering, and loose. Its flavor is similar to, but not better than, the Rogers hybrid Agawam.[40]

NORWOOD

native wild labrusca x Black Hamburg (according to The Bushberg Catalogue) or (Concord x Black Hamburg, according to Hedrick)

Norwood was first exhibited at the Massachusetts Horticultural Society fall meeting in 1880, where it received a first-class certificate of merit. It has a large to very large berry and cluster size that can produce large, double-sided clusters which ripen a little before Concord. It is, however, not self-fertile to imperfectly self-fertile, and its cluster size and production yields suffer.

The color of the variety is in dispute with *The Bushberg Catalogue* stating that it is a bright amber, with a cluster that is similar to Lindley, while Hedrick and others maintained that it was a purple-black grape. Reports were mixed on whether it is a strong grower in the field, but when grown well it was a productive variety with good quality fruit for the table market.[41]

George Haskell

(1809–1898)

George Haskell conducted a forty-year grape breeding program to create varieties that were winter hardy, resistant to fungus diseases, and which could produce quality fruit that ripened early enough to be consistently harvested in Massachusetts. This book details his work, as a case study, to illustrate the prosperous New Englanders who pursued horticulture in the nineteenth century and how he conducted his work.[42]

Haskell was born in Newburyport, Massachusetts, on August 24, 1809. His father was the son of Deacon Nehemiah Haskell of Newburyport. His mother was Eunice Dodge, daughter of Barnabas Dodge, a land surveyor who laid out Maine's township lines. Haskell's family moved to Ipswich when he was ten years old. His father died soon thereafter.

In 1825, Haskell was apprenticed to a Boston merchant who manufactured and sold dressing cases and pocket books. By the spring of 1831, he started his own manufacturing business in Ipswich. After the economic Panic of 1837, he lost much of his money and quit the manufacturing business.

Even with his financial setback, Haskell had sufficient funds to pursue grape breeding, and to maintain his extensive greenhouses and nursery. In 1837, Haskell began to study the law and was admitted to the Massachusetts Bar in 1843. He opened a general practice law firm in Ipswich and held public office to support himself. In 1838, he was elected to the Massachusetts House of Representatives for two terms.

Politically, Haskell was an anti-slavery Whig who ultimately became a Republican. In 1864, he ran for and lost an election for the U.S. House of Representatives. He was endorsed by the Democratic Party (not by the Republican Party, his party of choice) for his position that the Southern states should be admitted back into the Union, even if slavery was not abolished in the South.

Haskell was a life-long bachelor who lived with his mother until she died in 1851. Before his mother's death, Haskell tended grape vines and fruit trees in her garden for relaxation. After 1830, Haskell read William Robert Prince's *A Treatise on the Vine* and planted 200 vines and 2,000 cuttings of Isabella. After cultivating Isabella for a few years, Haskell realized these vines produced no fruit as the growing season in Massachusetts was too short. At this point, Haskell re-directed his viticultural efforts to create new grape varieties that could consistently ripen earlier than Isabella and Catawba to accommodate the shorter Massachusetts growing season. Around 1856, he built a large home near Heartbreak Hill just outside of Ipswich Center. It was at Heartbreak Hill that Haskell operated his fruit farm, greenhouses, and breeding program.

Haskell's Breeding Program

Haskell's hybrids, like those of Rogers, cross pollinated *vinifera* and local native varieties. Unlike Rogers, who used a local early-ripening native *labrusca* variety, called Carter; Haskell used *riparia* varieties that he found growing wild near his home. Literature from Haskell's era implies that he primarily used wild *labrusca* varieties that he had found. However, Haskell may not have been able to initially distinguish between wild *labrusca* and *riparia* varieties. While he called these wild grapes "fox" (i.e., *labrusca*) grapes, they were indeed *riparias*.

For the first ten to 15 years of his breeding program, Haskell planted the seeds of at least three generations of his earliest and best *vinifera* and *riparia* vines separately, in order to create new generations of seedlings. Simultaneously, Haskell tried to combine the characteristics of the *vinifera* and *riparia* varieties by grafting or inarching: a) the roots and scion, and/or, b) the canes of these different varieties. Then he planted the seeds of the grape clusters derived from the portions of the *vinifera* and *riparia* canes. The resultant seedlings and fruit did not produce satisfactory results. We know now that such efforts were doomed because Haskell's grafting and propagation did not intermix the genetic material of the *vinifera* and *riparia* varieties.

Around 1858, Haskell started breeding hybrids between his *vinifera* and *riparia* varieties by cross pollinization.[43] Over time, Haskell, made approximately eighty crosses of first- and then second-generation *riparia* and *vinifera* seedlings. Unlike Rogers, who consistently used the native local *labrusca* grape variety Carter as the seed parent; Haskell alternatively used *riparia* and *vinifera* varieties as the seed parent; using the alternate *vinifera* and *riparia* variety, as the pollen parent. The *vinifera* varieties that he used were almost exclusively Black Hamburg, White Frontignan, Grizzly Frontignan, and White Chasselas. For the thousands of seedlings that Haskell propagated, he kept records on their progress and parentage. This was done by assigning them hybridization numbers so that he could track their growth habit, disease, and cold resistance, ripening time, and fruit characteristics.

To cull the large number of seedlings that he propagated, Haskell let mother nature take its course. Those seedlings that survived New England's harsh climate were the most winter-hardy and fungus disease resistant of his seedlings.

Haskell studied the fruit quality and early ripening abilities of the remainder of the surviving seedlings. This evaluation—to identify desirable grape flavor profiles and cluster and berry size—took at least five growing seasons. Describing Haskell's hybridization process shows what other breeders covered in this book probably did when conducting their own breeding programs.

By 1869, Haskell selected forty seedlings and sent out scores of sample grape clusters for evaluation. The first to evaluate them were Marshall P. Wilder and Jacob W. Manning. In 1870, eighteen samples were sent to J. B. Garber of Columbia, Pennsylvania. In 1873 and 1874, six to nine samples were sent to the editors of *Gardener's Monthly, The Country Gentleman*, and *Coleman's Rural World, Massachusetts Ploughman* and *New England Journal of Agriculture*, as well as to William Saunders, Superintendent of the U.S. Government Experimental Grounds. Finally, in 1875, a select sampling of Haskell's new grape varieties was sent to Dr. Robert Hogg of the *London Journal of Horticulture*, and again to Marshall P. Wilder. To Wilder, he asked for the American Pomological Society, through its Fruit Committee, to publicly evaluate them.[44]

In 1877, Haskell published an 18-page grape catalog that detailed his four decades of hybridization work and gave descriptions of the forty hybrids that he offered for sale; the sale price being $100 for a selection of thirty of his vines. In his catalog's introduction, Haskell offered this dower assessment of his work:

> "I am sorry the success has not been more obvious and decided; but, thinking it will be a long time before any other individual will be so unwise as to spend his money, time and thought in such efforts, or will do so with a better chance of success, I have concluded to offer to the public the fruits of these labors, such as they are. I had hoped to obtain a reimbursement of my money-outlay in this matter, but see no chance for even that. ... Whether I obtain any pecuniary recompense for the products of these labors, may be a question of justice; but it is not by any means a matter of necessity with me. I hope the fruits obtained will be a benefit to the public, and that the future will show them to be of great value to the country, in both an economic and commercial aspect."[45]
>
> "If any grape-grower or nurseryman desires to purchase such an assortment of thirty vines, I shall be glad to supply him with them at a very low price, considering their cost to me; if they are not desired it may excite my regret, but it will do me no harm, nor, indeed, would it were I to annihilate the whole brood of vines, as I, at times, have been almost tempted to do."[46]

For decades after the Civil War, Haskell wrote extensively about his horticultural experiences in *The Country Gentleman*, *The Horticulturalist*, *The Magazine of Horticulture*, and publications issued by the Massachusetts Horticultural Society. In 1896, he self-published an autobiography that contained his grape breeding practices and experiments. It is in these writings that Haskell's contribution to American horticulture and grape breeding can be found.

MINOR GRAPE VARIETY OF HASKELL

RUBY

labrusca

Haskell was known for only one grape hybrid, named Ruby. In *The Grapes of New York* Hedrick described Ruby as hardy, variable in vigor; somewhat subject to rot; bunch medium, loose; berry above medium, round, dark ruby red; very good quality; ripens about with Hartford; stamens reflexed."[47] In *The Bushberg Catalogue*, Ruby had a minor entry indicating that Samuel Miller of Bluffton, Missouri, had conducted a two-year trial of Ruby and intended to propagate it. Ruby was a "vine hardy and very vigorous, bears a most excellent grape; above medium in bunch and berry; a regular ruby in color (hence the name) and in quality unsurpassed." While this description was glowing, Ruby was not embraced by nursery owners or growers.[48]

Asa Clement Jr.

(1813–1894)

Asa Clement Jr. was born to Asa Clement (1784–1851), who was a captain in the militia. He married Elizabeth Wilson in 1812 and operated a fruit farm in Dracut, Massachusetts. His son, Clement Jr. married Hannah J. Peabody of Methuen (1818–1898) in 1837 and they had five children.

Upon inheriting the family farm, Clement Jr. became interested in horticulture. He was a successful fruit grower who raised apples and grapes. He was one of the original members of the Middlesex North Agricultural Society, serving as president (1867–1869), and was a delegate to the Massachusetts State Board of Agriculture (1862–1869). He participated in the Farmers' Institute and read essays on farming before the Institute. Clement is mentioned in a few exhibitions and covered in the Massachusetts Horticultural Society annual reports for receiving prizes for his apples. He was an active member of the Pawtucket Church, serving as deacon and Sunday school superintendent.[49]

MINOR GRAPE VARIETY OF CLEMENT

DRACUT AMBER

labrusca

This variety came from a seed of a native reddish *labrusca* vine that grew near a Catawba vine. Clement planted the seed around 1855, and the resulting vine was called Amber. Later on, the variety was commercially introduced by nursery owner Jacob W. Manning,[50] of the Reading Nursery, as Dracut Amber. The vine is vigorous, hardy, and productive, but somewhat subject to attacks by leafhoppers. The cluster is medium in size, cylindrical, and rather broad. The berry is medium to large in size, of a dull pale red to dark amber color. The fruit is very foxy and musky of inferior quality. The flowers are sometimes on a plan of six, variably slightly sterile to fertile; they open late-early to mid-season, with stamens of variable length. To ensure good fruit set, suitable self-fertile pollinators such as Cottage, Croton, Delaware, Empire State, Iona, and Winchell should be used. Dracut Amber ripens late early season just before Concord. Due to its inferior flavors and texture, the variety was abandoned by 1910.[51]

William Ellery Greene

(1810–1886)

The Greene family were longtime residents of New England in general, and the Vergennes area of Vermont, in particular. Many paternal and maternal ancestors of William E. Greene lived in New England well before 1670. Greene married Anna Jane Sherrill (1817–1893) in Vergennes, where they lived, raised their family, and died. The couple had eight children, some of whom survived to adulthood. Little is known of Greene's horticultural interests, but sometime after the Civil War he did find and identify in his garden the grape variety that would later be named Vergennes, in honor of his hometown.

MINOR GRAPE VARIETIES OF GREENE

VERGENNES

chance seedling, labrusca

This red table grape is a chance seedling that was found in the garden of William E. Greene in 1874. Vergennes is located in the lower Lake Champlain Valley. The grape is not commonly grown in New England today, but is cultivated extensively by Arbor Hill Grapery and Winery of Naples, New York, to produce a white wine. Vergennes wines, both dry and semi-dry, are quality white *labrusca*-type wines that have become a specialty of this winery.

The vine is generally winter hardy and productive to very productive, but can overbear if not properly cluster thinned in select years. Vergennes produces a large-berried, attractive, bright-colored red grape. The clusters are of medium size and length, broad and cylindrical to tapering, and of variable compactness, but generally are loose. Overall, the vine is healthy and somewhat resistant to fungus diseases. Vergennes, while variable in its yield from year to year, is consistently able to produce an ample crop; it can also be variable in its vigorousness, which will affect its productivity and ripening times. Vergennes flowers by mid-season to late season, whose flowers are semi-sterile, with upright stamens. For this reason this variety may not be a good pollinator for the Rogers hybrids, but is suitable for self-fertile pollinators such as Concord, Diamond, Jefferson, and Niagara. Hedrick stated that Vergennes readily transmits its beneficial characteristics to its offspring, indicating that this variety is of value for grape breeding. The variety ripens late about one to two weeks after Concord and about the same time as Empire State, Chelois, or Chambourcin.[52] In the end, it is a safe and productive variety for marginal vineyard sites.

Vergennes produces a very light green colored wine that is grapey, very *labrusca*-like, clean, and aromatic like the native grape variety Diamond. This fruit-forward wine has flavors of pink grapefruit, bananas, lemons, and butternut, melding with the softer fruits of pears, melons, white peaches, and some spice. It is an easy-to-drink wine that can be herbal, has elements of Moscato, that is balanced with sufficient acidity to produce a semi-dry wine.[53]

This grape should be re-evaluated to determine if it is suitable for future cultivation in New England, New York, and other cool climates of North America and Northeast Asia.

EARLY DELMONICO

labrusca, vinifera

A minor grape bred by Greene that was not widely planted. It has a medium-large and long conical cluster of medium sized red berries. Greene stated that it ripened very early and was superior to Vergennes, but its lack of popularity diminishes that claim by Greene.[54]

NOTES

1. Edmund A. Schofield, "'He Sowed; Others Reaped': Ephraim Wales Bull and the Origins of the 'Concord' Grape," *Arnoldia*, Vol. 48, No 4, (Boston: The Arnold Arboretum of Harvard University, Fall 1988); 7–10.

2. Schofield, "He Sowed; Others Reaped," 7–10; For more on Ephraim Bull, see Chapter 7, Note 8, page 80; Also Thomas Pinney, *A History of Wine in America: From the Beginnings to Prohibition, Vol. 1* (Berkeley: University of California Press, 2007), 211–14; U. P. Hedrick, *The Grapes of New York. Report of the New York Agricultural Experiment Station for the Year 1907*. Department of Agriculture, State of New York. Fifteenth Annual Report, Vol. 3 – Part II. (Albany, N.Y.: Lyon Company, 1908), 221.

3. J. Stephen Casscles, "Blue and Black Heirloom Grape Varieties for the Northeast," appearing concurrently in *Fruit Notes*, Vol. 83, No. 4 (Amherst, Mass.: Stockbridge School of Agriculture, University of Massachusetts Amherst, Fall 2018), and *Horticultural News*, Vol. 98, No. 4 (Hackettstown, N.J.: New Jersey State Horticultural Society, Fall 2018), 19; J. Stephen Casscles, "Great Cool Climate Heirloom Grape Varieties," *AWS Wine Journal*, Vol. 50, No. 1 (Scranton: American Wine Society, Spring 2019), 17; Observations by the author at his farm Cedar Cliff, Athens, NY from 2010 to 2021 for Cottage and tasting notes of wines made from Cottage; Hedrick, *Grapes of New York*, 222–24; *The Bushberg Catalogue, A Grape Manual* (Bushberg, Jefferson City, Mo.: Bush and Son & Meissner, 1895), 108–09; U. P. Hedrick, *A Manual of American Grape-Growing* (New York: The Macmillan Company, 1919), 288, 298, 354; Lucie T. Morton, *Wine Growing in Eastern America: An Illustrated Guide to Viniculture East of the Rockies* (Ithaca, N.Y.: Cornell University Press, 1985), 139, 178; U. P. Hedrick, *Cyclopedia of Hardy Fruits* (New York: The Macmillan Company, 1922), 239–40; Samuel T. Maynard, Report on Fruits, Hatch Experiment Station of the Massachusetts Agricultural College Bulletin No. 21 (Amherst, Mass.: Press of Carpenter & Morehouse, April 1893), 14; G. W. McCluer, "*Grapes, Test of Varieties*," Agricultural Experiment Station Bulletin No. 28, in *Bulletins of the Agricultural Experiment Station, University of Illinois, Vol. 2, Bulletins 17–36, August 1891–August 1894* (Urbana, Ill.: Agricultural Experiment Station, University of Illinois, December 1893), 253; J. C. Whitten, *The Grape – I. A Study of Types and Varieties*. Bulletin No. 46. University of the State of Missouri, College of Agricultural and Mechanic Arts, Agricultural Experiment Station (Columbia, Mo.: E. W. Stephens, April 1899), 49.

4. Hedrick, *Grapes of New York*, 264; *Bushberg Catalogue*, 125; Maynard, Report on Fruits, 14.

5. Hedrick, Grapes of New York, 479.

6. Ibid., 389; *Bushberg Catalogue*, 107; T. T. Lyon, *Fruit Tests at South Haven, Bulletin No. 143* (Lansing, Mich.: Michigan Agricultural Experiment Station, State Agricultural College, April 1897), 162.

7. Hedrick, *Grapes of New York*, 512.

8. Ibid., 519.

9. The Ellwanger & Barry Nursery, also known as the Mount Hope Nursery, was founded in Rochester, New York, in 1840. Patrick Barry, along with George Ellwanger, founded the Mount Hope Nurseries and in many respects introduced fruit growing to western New York, which still specializes in the cultivation of pears, apples, grapes, stone fruits, and small berries to this day. At this nursery, in addition to a large collection of fruits, they offered ornamental trees and shrubs. Early in the nursery business, he imported many plants from Europe. See U. P. Hedrick, *The Pears of New York* (Albany, N.Y.: Department of Agriculture, State of New York, 1921), 203–05; U. P. Hedrick, *A History of Agriculture in the State of New York*, (Albany, N.Y.: New York State Agricultural Society, 1933), 386–88; Julie S. Higginbotham, "Four Centuries of Planting and Progress: A History of the U.S. Nursery Industry," *American Nurseryman*, Vol. 171, No. 12 (Chicago: American Nurseryman Publishing Co., June 15, 1990), 36–40.

10. J. Stephen Casscles, "Selected White Heirloom Grape Varieties for the Northeast," appearing concurrently in Fruit Notes. Vol. 83, No. 4 (Amherst, Mass.: Stockbridge School of Agriculture, University of Massachusetts Amherst, Fall 2018), and Horticultural News. Vol. 98, No. 4 (New Brunswick, N.J.: New Jersey State Horticultural Society, Fall 2018), 15–18; Casscles, "Great Cool Climate Heirloom Grape Varieties," 13; Observations by the author at his farm Cedar Cliff, Athens, New York, from 2010 to 2021 for Winchell and tasting notes of wine made from Winchell; Todd Trzaskos, *Wines of Vermont: A History of Pioneer Fermentation*, (Charleston, S.C.: American Palate, The History Press, 2015), 81–82; Hedrick, *Grapes of New York*, 425–27; Hedrick, *Manual of American Grape-Growing*, 445–46; *Bushberg's Catalogue*, 130–31.

11. Chateau Stripmine, http://www.chateaustripmine.info/Breeders/Cornell.html; See generally Grape Varieties Named at the New York State Agricultural Experiment Station, Geneva, NY, http://www.hort.cornell.edu/reisch/grapegenetics/nyreleases.html; Hudson Cattell, *Wines of Eastern North America: From Prohibition to the Present: A History and Desk Reference* (Ithaca, N.Y.: Cornell University Press, 2014), 238–44.

12. J. Stephen Casscles, "Pink and Light Red Heirloom Grape Varieties for the Northeast," appearing concurrently in *Fruit Notes*, Vol. 83, No. 4 (Amherst, Mass.: Stockbridge School of Agriculture, University of Massachusetts Amherst, Fall 2018), and *Horticultural News*, Vol. 98, No. 4 (Hackettstown, N.J.: New Jersey State Horticultural Society, Fall 2018), 23–26; Casscles, "Great Cool Climate Heirloom Grape Varieties," 14; Observations by the author at his farm Cedar Cliff, Athens, New York, from 2010 to 2021 for Diana and tasting notes of wines made from Diana.

13. Casscles, "Pink and Light Red Heirloom Grape Varieties," 23–24; Casscles, "Great Cool Climate Heirloom Grape Varieties, 14; Hedrick, *Grapes of New York*, 238–41; Bushberg Catalogue, 116–17; C. M. Hovey, ed., *The Magazine of Horticulture, Botany, and All Useful Discoveries and Improvements in Rural Affairs*, Vol. X (Boston: Hovey and Co., 1844), 242–43; A. K. Teele, ed., *The History of Milton, Mass.,1640–1887* (Milton, MA: Town of Milton, 1887), 149–50, 177–78, 561–62; Anthony M. Sammarco, "The History of the Diana Grape Has Roots in Milton," *The Milton Historical Society Newsletter* (Milton, Mass.: The Milton Historical Society, Fall 2008), http://www.miltonhistoricalsociety.org/Sampler/DianaGrape.html

14. There are reports that Iona may be a seedling of Catawba. However, since Diana is also claimed to be a seedling of Catawba, it would be difficult to differentiate the characteristics of the parents to determine Iona's real parents. See the entry on Iona in Chapter Eight, pages 98–99; also Hedrick, *Grapes of New York*, 304.

15. See the chapter on *Labrusca Hybrids*, Chapter Seven, specifically pages 69–70.

16. See *Transactions of the Massachusetts Horticultural Society for the Year 1876, Part II, Resolution adopted on November 4, 1876 on the Death of John Fisk Allen* (Boston: Printed for the Massachusetts Horticultural Society, 1877), 8–10; John Fisk Allen, *Victoria Regina or the Great Water Lily of America* (Boston: Dutton and Wentworth, 1854); Frank Cousins and Phil M. Riley, *The Colonial Architecture of Salem* (Boston: Little, Brown and Company, 1919), reprinted by Dover Publications, Mineola, N.Y., 2000), Plate XLV; Philip M. Wagner, *A Wine-Grower's Guide* (San Francisco: The Wine Appreciation Guild, 1985), 33; Maine Agricultural Experiment Station, "Orchard Notes," *Bulletin No. 128*, March 1906, 167–68.

17. Hedrick, *Grapes of New York*, 165–66; *Bushberg Catalogue*, 84.

18. Dr. Ulysses Prentice Hedrick (1870–1951) was a horticultural writer, educator, and longtime Director of the New York State Agricultural Station. For more on Hedrick, see Chapter One, page 17, notes 1 and 2.

19. It is important to mention that the temperatures in the northeastern United States in the 1850s were much colder than it is today. This was due to the "Little Ice Age" which was just coming to an end, having started in northern America and Europe in the 1350s. See generally Brian Fagan, *The Little Ice Age: How Climate Made History 1300–1850* (New York: Basic Books, Perseus Books Group, 2000).

20. Hedrick, *Grapes of New York*, 327–28.

21. Pinney, *History of Wine I*, 206; Philip M. Wagner, *American Wines & Wine Making* (New York: Alfred A. Knopf, 1965), 36; Wagner, *Wine Grower's Guide*, 32–33; Hedrick, *Grapes of New York*, 165–66.

22. Dr. F. M. Hexamer, ed., "Exhibitions and Societies, Massachusetts Horticultural Society," *The American Garden, A Monthly Journal of Practical Gardening, Vol. VI, No. 2* (New York: B. K. Bliss & Sons, February 1885), 32; E. H. Libby, ed., "Two Notables Gone – John B. Moore," *The American Garden and Floral Cabinet, A Monthly Journal of Practical Gardening, Vol. III, No. 11* (New York: B. K. Bliss & Sons, November 1887), 345; H. E. Hoard, Sec., *Annual Report of the Minnesota State Agricultural Society for the Year 1887, Executive Documents of the State of Minnesota, for Fiscal Year 1887-8*, Vol. II (St. Paul, Minn.: J. W. Cunningham & Co., 1888), 576.

23. A note here about the grape variety known as Moore Early, and at times referred to as Moore's Early. Initially, when introduced by the American Pomological Society it was listed as Moore's Early, and it continued to be so by the various horticultural societies of the day. Bushberg also referred to the grape as Moore's Early as late as 1895. But by 1908, Hedrick was calling this variety Moore Early. This change in grape naming reflected the trend that grape variety names should not include a possessive in them. In 2012, Jancis Robinson referred to this grape as "Moore Early" clearly signifying that the Moore's Early designation is no longer used. See Jancis Robinson, Julia Harding, and Jose Vouillamoz, *Wine Grapes: A Complete Guide to 1,368 Vine Varieties, Including Their Origins and Flavor* (New York: HarperCollins, 2012), 136, 181.

24. Hedrick, *Grapes of New York*, 353–54; *Bushberg Catalogue*, 158–59; Hedrick, *Manual of American Grape-Growing*, 289, 299, 414–15; Hedrick, *Cyclopedia of Hardy Fruits*, 254; Maynard, *Report on Fruits*, 15–16; McCluer, *Grapes, Test of Varieties*, 256; R. L. Watts, *Varieties of Grapes*, Vol. IX, No. 4, University of Tennessee Agricultural Experiment Station Bulletin No. 38 (Knoxville, Tenn.: Tennessee Agricultural Experiment Station, December 1896), 189; F. A. Waugh, *Home-Grown Grapes in Vermont, III. Variety Notes*, Bulletin No. 62 (Montpelier, Vt.: Vermont Agricultural Experiment Station, University of Vermont and State Agricultural College, Vermont Watchman Co., January 1898), 41–42.

25. Campbell Early is the most widely-planted grape variety in South Korea, covering over 31,000 acres or approximately 70% of all grapes grown in Korea. Campbell Early is used to produce table grapes, wines, and juices in Korea. In Japan, Campbell Early covers over 3,700 acres of ground. Unlike in Korea, in Japan Campbell Early is used almost exclusively as a table grape as is the other American grape Delaware. The attraction to Campbell Early in Korea and Japan is its early ripening time (to avoid very heavy late summer and early fall rains), big and full cluster size, large berry size, attractive blue color, high productivity, and fungus disease resistance. See generally, Australia Department of Agriculture, Fisheries and Forestry, "Biosecurity Australia (2011) Final Non-Regulated Risk Analysis Report for Table Grapes from the Republic of Korea (Canberra, Australia: Australia Department of Agriculture, Fisheries and Forestry, 2011), 15–23; Minas K. Papademetriou and Frank J. Dent, "Grape Production in the Asia-Pacific Region," *RAP Publication*, 2000/13, (Bangkok, Thailand: Food and Agriculture Organization of the United Nations Regional Office for Asia and the Pacific, 2001), Chapters 6 and 7.

26. Hedrick, *Grapes of New York*, 252–54; *Bushberg Catalogue*, 123; Hedrick, *Manual of American Grape-Growing*, 288, 298, 365–66; Hedrick, *Cyclopedia of Hardy Fruits*, 243; Maynard, *Report on Fruits*, 14; McCluer, *Grapes, Test of Varieties*, 254; Watts, *Varieties of Grapes*, 178–80.

27. Francis B. Hayes (1819–1884) was born in South Berwick, Maine. He graduated from Harvard College and Law School and became a prosperous corporate attorney who managed railroads. He was on the board of directors of the Boston & Maine and the Atlantic & Pacific Railroad. He was elected to the Massachusetts House of Representatives in 1873, and to the State Senate in 1874. In 1861, he established in Lexington, Massachusetts, a four-hundred-acre estate where he grew roses, azaleas, rhododendrons, and chrysanthemums. He was President of the Massachusetts Horticultural Society from 1879 until his death in 1884. See Hamilton D. Hurd, *History of Middlesex County, Massachusetts, with Biographical Sketches of Many of its Pioneers & Prominent Men, Vol. 1* (Philadelphia: J. W. Lewis Co., 1899), 636; Old Berwick Historical Society, South Berwick, Maine: http://www.oldberwick.org/

28. The quote about Hayes is from *Bushberg Catalogue*, 133; On Hayes, see also Hedrick, *Grapes of New York*, 286–87; Hedrick, *Manual of American Grape-Growing*, 381; Hedrick, *Cyclopedia of Hardy Fruits*, 246; Lyon, *Fruit Tests*, 161; Maynard, *Report on Fruits*, 14; McCluer, *Grapes, Test of Varieties*, 264.

29. See generally Cyrus Guernsey Pringle, *The Record of a Quaker Conscience: Cyrus Pringle's Diary* (New York: Macmillan & Co., 1918).

30. Asa Gray (1810–1888), was one of the great American academic leaders of botanical studies who based most of his career at Harvard University. Gray was born in Sauquoit, near Utica, New York. In 1831, he graduated from Fairfield Academy, near Little Falls, New York, to become an apprentice physician, but he always preferred to study botany and to collect and classify plants. By 1832, Gray quit medicine, and became associated with Professor John Torrey of New York City, and jointly wrote with Torrey the landmark book, *A Flora of North America*, Vols. I and II (New York: Wiley & Putnam, 1838–40), which was continually updated until his death in 1888. As an educator, he wrote many textbooks on the subject of botany for those in high school and college including: *A Manual of Botany of the Northern United States* (1848); *First Lessons in Botany and Vegetable Physiology* (1857); *How Plants Grow: A Simple Introduction to Structural Botany* (1858); *The Elements of Botany for Beginners and for Schools* (1887); and many more texts.

In 1864, Gray donated over 200,000 plant specimens and over 2,000 books to Harvard University that ultimately became the base of the Gray Herbarium and the Arnold Arboretum. He had a long-standing working relationship with Dr. George Engelmann of St. Louis, Missouri, who specialized in the study of native grape species. Gray also managed the Harvard Botanic Garden, that ultimately became the Arnold Arboretum. See generally, A. Hunter Dupree, *Asa Gray: American Botanist, Friend of Darwin* (Baltimore: John Hopkins University Press, 1988).

31. See generally Helen Burns Davis, *Life and Work of Cyrus Guernsey Pringle* (Burlington, VT: University of Vermont, Free Press Printing Co., 1936); http://www.uvm.edu/vtnaturalhistory/pringle-herbarium.html; Rob Nicholson, "The Splendid Haul of Cyrus Guernsey Pringle," *Arnoldia*, Vol 61, Issue 1, (Boston: The Arnold Arboretum of Harvard University, 2001), 2–9; Trzaskos, *Wines of Vermont*, 81–82.

32. Casscles, "White Heirloom Grape Varieties," 16; Observations by the author at his farm Cedar Cliff, Athens, New York, from 2010 to 2021 for Golden Drop and tasting notes of wines made from Golden Drop; Hedrick, *Grapes of New York*, 281, 464; *Bushberg Catalogue*, 130; Trzaskos, *Wines of Vermont*, 81–82; Maynard, *Report on Fruits*, 15.

33. Hedrick, *Grapes of New York*, 477; *Bushberg Catalogue*, 147.

34. Hedrick, *Grapes of New York*, 484; *Bushberg Catalogue*, 151.

35. Hedrick, *Grapes of New York*, 284–85; Hedrick, *Manual of American Grape-Growing*, 288, 298, 380; *Bushberg Cata-*

logue, 132; Hedrick, *Cyclopedia of Hardy Fruits*, 246–47; Maynard, *Report on Fruits*, 14; McCluer, *Grapes, Test of Varieties*, 254; Waugh, *Home-Grown Grapes* III, 41.

36. *Bushberg Catalogue*, 186.

37. Hedrick, *Grapes of New York*, 166–67, 172–73, 364–65, 369, 437; *Bushberg Catalogue*, 85, 86, 88, 89, 163–65, 186; N. B. White, "On Grape Hybrids," *Memoirs, Vol. I, Proceedings: International Conference on Plant Breeding and Hybridization, September 30, October 1 and 2, 1902* (New York: Horticultural Society of New York, 1902), 221–23; Whitten, *The Grape – I*, 51; McCluer, *Grapes, Test of Varieties*, 252, 256, 258.

38. Maynard, *Report on Fruits*, 14; McCluer, *Grapes, Test of Varieties*, 252.

39. Maynard, Ibid.; McCluer, *Grapes, Test of Varieties*, 258.

40. McCluer, Ibid., 261.

41. Ibid., 256.

42. See generally George Haskell, *An Account of Various Experiments for the Production of New and Desirable Grapes, and a Description of Forty Varieties Obtained by Hybridization* (Ipswich, Mass.: printed by the author, 1877); and George Haskell, *A Narrative of the Life, Experience and Work of an American Citizen* (Ipswich, Mass.: Chronicle Publishing Company, 1896).

43. See Chapter Fifteen on Rogers's hybridization process, pages 240–244.

44. For more on George Haskell, see his previously mentioned autobiography, *A Narrative of the Life, Experience and Work of an American Citizen*; Hedrick, *Grapes of New York*, 467, 508; *Bushberg Catalogue*, 132, 175; Thomas Meehan, ed., "New or Rare Fruits: Haskell's Grapes," *The Gardener's Monthly and Horticulturist, Vol. XXI, No. 245*, (Philadelphia: Charles H. Marot, May 1879), 149; Pinney, *History of Wine I*, 207.

45. Haskell, *Account of Various Experiments*, 6–7.

46. Ibid., 13.

47. Hedrick, *Grapes of New York*, 508.

48. *Bushberg Catalogue*, 175.

49. Silas R. Coburn, *History of Dracut, Massachusetts* (Lowell, Mass.: Press of the Courier-Citizen Co., 1922), 378–79.

50. Jacob Warren Manning (1826–1904) was born and raised in Bedford, New Hampshire. The Mannings come from old New England stock, descending from William Manning, born in England in 1620, an early settler of Cambridge, Massachusetts. The father of Jacob, born in Chelmsford, Massachusetts, was a farmer, fruit grower, lumbered wood, and active in church and town affairs. In 1847, Manning worked for Amos Carleton of Chelmsford at his fruit farm and nursery business. In 1849, he became the superintendent of the Winnesemitt Nursery in Chelsea. This nursery was owned by S. W. Cole, the editor of the *New England Farmer* and the *Boston Cultivator*, and author of *The American Fruit Book* (Boston: John P. Jewett & Co., 1850). Cole also engaged in the breeding of new fruit varieties, especially small fruits.

From 1850 to 1854, Manning was employed as a gardener for various country homes in Massachusetts and Vermont. In the early summer of 1854, Manning moved permanently to Reading, Massachusetts, to establish his own nursery and fruit farm, where he raised and propagated large and small fruits, ornamental trees, and shrubs. In 1859, Manning bred the Cutter Seedling strawberry. Manning contacted many other growers and amateur fruit breeders in New England and purchased the rights to their new fruit varieties, including the Dracut Amber, and introduced them to the public for sale at his nursery. Manning was a long-standing member of the Massachusetts Horticultural Society, its prestigious Fruit Committee, and the Middlesex Agricultural Society, American Pomological Society, American Nurserymen's Association, Massachusetts Fruit Growers Association, Massachusetts Agricultural Society, and the American Forest Association. Jacob Manning married Lydia Brooks Chandler in 1858 and had five sons. All of the Manning sons were noted for their own work in landscape architecture, gardening, or nursery work. The Mannings, both father and sons, worked closely with Charles Downing and his nursery of Newburgh, New York. See "Jacob Warren Manning," *Biographical Sketches of Representative Citizens of the Commonwealth of Massachusetts, American Series of Popular Biographies, Massachusetts Edition* (Boston: Graves and Steinbarger, 1901), 816–20.

51. Hedrick, *Grapes of New York*, 244–45; Hedrick, *Manual of American Grape-Growing*, 288, 298, 362; *Bushberg Catalogue*, 117; Hedrick, *Cyclopedia of Hardy Fruits*, 241–42; Jacob W. Manning, *The National Nurseryman. For Growers and Dealers in Nursery Stock, Vol. V, No. 2* (Rochester, N.Y.: The National Nurseryman Publishing Company, March 1897), 21; Hall and Beach, *Grapes Fail to Fruit*, 7; McCluer, Grapes, 259–60.

52. Hedrick, *Grapes of New York*, 416–18; Hedrick, *Manual of American Grape-Growing*, 289, 299, 442–43; *Bushberg Catalogue*, 184; State of New York Department of Agriculture, *The Fruit Industry in New York State, Bulletin 79, Part II* (Albany, N.Y.: Department of Agriculture, 1915), 1235; Hedrick, *Cyclopedia of Hardy Fruits*, 260; the author's discussion with Mr. Brahm of Arbor Hill Grapery and Winery, in 2015, on the cultivation of Vergennes. Mr. Brahm cultivated several acres of Vergennes, loved to talk about how he cultivated grapes, and made lovely wines from them.

Mr. Brahm and his family are generous growers who share their extensive knowledge on grape growing and winemaking. Mr. Brahm gave this author his cuttings of Dutchess, which is now a fine vineyard of Dutchess vines; Trzaskos, *Wines of Vermont*, 81–82; Maynard, *Report on Fruits*, 15; McCluer, *Grapes, Test of Varieties*, 262; Waugh, *Home-Grown Grapes III*, 42. Watts, *Varieties of Grapes*, 189; Whitten, The Grape – I, 53; F. H. Hall and S. A. Beach, *Why Some Grapes Fail to Fruit*, New York Agricultural Experiment Station Bulletin No. 157 (Geneva, N.Y.: New York Agricultural Experiment Station, December 1898), 7.

53. Casscles, "Pink and Light Red Heirloom Grape Varieties," 26; Casscles, "Great Cool Climate Heirloom Grape Varieties," 15; Observations by the author and tasting notes of wines made from Vegennes.

54. Hedrick, *Grapes of New York*, 456.

BIBLIOGRAPHY

Adams, Leon D. *The Wines of America*. Boston: Houghton Mifflin Company, 1973.

American Wine Society. *The Complete Handbook of Wine-Making*. Ann Arbor, Mich.: G. W. Kent, Inc., 1993.

Amerine, M. A. *The Technology of Winemaking*. Westport, Conn.: AVI Publishing, 1980.

Anthony, R. D., and U. P. Hedrick. *Vinifera Grapes in New York*. New York Agricultural Experiment Station, Cornell University, Bulletin no. 432 (April). Geneva, N.Y.: New York Department of Agriculture, 1917.

Armstead, Myra B. Young. *Freedom's Gardener: James F. Brown, Horticulture, and the Hudson Valley in Antebellum America*. New York: New York University Press, 2012.

Australia Department of Agriculture, Fisheries and Forestry. *Final Non-Regulated Risk Analysis Report for Table Grapes from the Republic of Korea*. Canberra: Australia Department of Agriculture, Fisheries and Forestry, August 2011.

Bailey, Liberty Hyde. *Annals of Horticulture in North America for the Years 1889: A Witness of Passing Events and a Record of Progress*. New York: Rural Publishing Company, 1889.

——. The Principles of Fruit Growing. New York: MacMillan Company, 1910.

——. *The Standard Cyclopedia of Horticulture*, Vol. 3. New York: Macmillan, 1915.

Beach, George, and L. R. Bryant. *Grape Growing in Colorado*, Colorado Agricultural Experiment Station Bulletin 484. Fort Collins, Colo.: Colorado State College, 1944.

Beach, S. A. *Fertilizing Self-Sterile Grapes*, New York Agricultural Experiment Station Bulletin No. 169. Geneva, N.Y.: New York Agricultural Experiment Station, December 1899.

Bedford, Robert. "Dr. Fritz Zweigelt," *The Zweigelt Project*, http://zweigeltproject.com/all-things-zweigelt/ dr-fritz-zweigelt.

——. *The Story of Brotherhood, America's Oldest Winery*. Coxsackie, N.Y.: Flint Mine Press, 2014.

Bowers, John E. and Carole P. Meredith. "The Parentage of a Classic Wine Grape, Cabernet Sauvignon," *Nature Genetics* vol. 16, April 30, 1997.

Brooks, Reid Merrifield, and Harold Paul Olmo. *The Brooks and Olmo Register of Fruit and Nut Varieties* (3rd ed.). American Society for Horticultural Sciences. Alexandria, Va.: ASHS Press, 1997.

Burns Davis, Helen. *Life and Work of Cyrus Guernsey Pringle*. Burlington, Vt.: University of Vermont, Free Press Printing Co., 1936.

Burroughs, John. *My Boyhood, with a Conclusion by His Son Julian Burroughs*. Garden City, N.Y.: Doubleday, Page & Company, 1922.

Bush & Son & Meissner. *Bushberg Catalogue: A Grape Manual. An Illustrated Descriptive Catalogue of American Grape Vines* (4th ed.). A Grape Growers Manual by Bush & Son & Meissner, Viticulturalists and Proprietors of Bushberg Vineyards and Grape Nursery. Jefferson County, Mo.: Bush & Son & Meissner, 1895.

Casscles, J. Stephen. "Blue and Black Heirloom Grape Varieties for the Northeast," *Fruit Notes*, Vol. 83, No. 4. Amherst, Mass.: Stockbridge School of Agriculture, University of Massachusetts Amherst, Fall 2018, and *Horticultural News*, Vol. 98, No. 4. Hackettstown, N.J.: New Jersey State Horticultural Society, Fall 2018, 19–21.

———. "Elmer Swenson: Pioneer Cool Climate Grape Breeder (1913–2004)," *AWS Wine Journal*, Vol. 52, No. 4. Scranton, Pa.: American Wine Society, Winter 2020, 6–9.

———. "E. S. Rogers and the Origins of American Grape Breeding," *Arnoldia*, Vol. 77, Issue 2. Boston: The Arnold Arboretum of Harvard University, 2019, 30–39.

———. "Great Cool Climate Heirloom Grape Varieties," *AWS Wine Journal*, Vol. 50, No. 1. Scranton: American Wine Society, Spring 2019, 12–17.

———. "Pink and Light Red Heirloom Grape Varieties for the Northeast," *Fruit Notes*, Vol. 83, No. 4 Amherst, Mass.: A publication of the Stockbridge School of Agriculture at University of Massachusetts, Fall 2018, and *Horticultural News*, Vol. 98, No. 4. Hackettstown, N.J.: New Jersey State Horticultural Society, Fall 2018, 23–26.

———. "Selected White Heirloom Grape Varieties for the Northeast," *Fruit Notes*, Vol. 83, No. 4. Amherst, Mass.: Stockbridge School of Agriculture at University of Massachusetts, Fall 2018, and *Horticultural News*, Vol. 98, No. 4. Hackettstown, N.J.: New Jersey State Horticultural Society, Fall 2018, 15–18.

Cattell, Hudson. *Wines of Eastern North America: From Prohibition to the Present: A History and Desk Reference*. Ithaca, N.Y.: Cornell University Press, 2014.

Cattell, Hudson, and H. Lee Stauffer. *The Wines of the East: The Hybrids*. Lancaster, Pa.: Eastern Wine Publications, L & H Photojournalism, 1978.

Cedar Cliff Nursery. *Cedar Cliff Nursery: Grape Vines for Grape Growers*. J. Stephen Casscles, Proprietor. Catskill, N.Y.: Cedar Cliff Nursery, 2013.

Chilberg, Joe, and Bob Baber. *New York Wine Country, A Tour Guide*. Utica, N.Y.: North Country Books, 1986.

Child, Hamilton. "The History of Esopus, NY," *Gazetteer and Business Directory of Ulster County, N.Y., for 1871–1872*. Syracuse, N.Y., Printed at the Journal Office, 1871: 80–86.

Childers, Norman F. *Modern Fruit Science* (6th ed.). New Brunswick, N.J.: Rutgers University Press, 1975.

Christman, Henry. "Iona Island and the Fruit Growers' Convention of 1864," *New York History, The Quarterly Journal of the New York State Historical Association*, volume 48, 1967: 332–51.

The City of Newburgh City Directory. Newburgh, N.Y.: L. P. Wait & Company, 1870 through 1890.

Clarke, Oz. *Oz Clarke's Encyclopedia of Grapes: A Comprehensive Guide to Varieties and Flavors*. New York: Harcourt, 2010.

Clyne, Patricia Edwards. *Hudson Valley Faces and Places*. New York: Overlook Press, 2005.

Cole, Rev. David. *History of Rockland County, New York with Biographical Sketches of Its Prominent Men*. New York: J. B. Beers & Company, 1884.

Cornett, Peggy. "Encounters with America's Premier Nursery and Botanic Garden: The Prince Nursery of New York City," *Twinleaf Journal*. Charlottesville, Va.: Th. Jefferson Monticello, January 2004, 1–12.

Cox, J. *From Vines to Wines: The Complete Guide to Growing Grapes and Making Your Own Wine*. Pownal, Vt.: Storey Books, 1994.

De Vito, Carlo. East Coast Wineries: *A Complete Guide from Maine to Virginia*. New Brunswick, N.J.: Rutgers University Press, 2004.

Dean, John Ward. *New England Bibliopolist: Containing the 'Book Notice' in the Historical Genealogical Register, Notices of Books on American History*, Vol. 1, No. 1. Boston: David Clapp & Son, January 1880.

Dial, Tom. *The Wines of New York*. Utica, N.Y.: North Country Books, 1986.

Downing, A. J. *Cottage Residences: or a Series of Designs for Rural Cottages and Cottage Villas, and Their Gardens and Grounds, Adapted to North America*. New York: Wiley and Putnum, 1842.

———. *A Treatise on the Theory and Practice of Landscape Gardening, Adapted to North America: With a View to the Improvement of Country Residences*. New York: Wiley and Putnum, 1841.

Downing, A. J. *The Fruits and Fruit Trees of America: Or the Culture, Propagation and Management in the Garden and Orchard of Fruit Trees Generally; with Descriptions of All the Finest Varieties of Fruit, Native and Foreign in this Country*. (New York: Wiley and Putnam, 1845).

Downing, A. J., and Charles Downing. *The Fruits and Fruit Trees of America* (2nd ed.). New York: John Wiley & Sons, 1869.

Dupree, A. Hunter. *Asa Gray: American Botanist, Friend of Darwin*. Baltimore: John Hopkins University Press, 1988.

Einset, J., and W. B. Robinson. *Cayuga White, the First of a Finger Lakes Series of Wine Grapes for New York*. Cornell University, Bulletin no. 22. Geneva, N.Y.: New York State Agricultural Experiment Station, 1972.

Gager, C. Stuart, ed., "The First Botanic Garden on Long Island," *Brooklyn Botanic Garden Record*, Vol. 1, No. 4. New York: Press of L. Middleditch Co., October, 1912, 97–99.

Galet, Pierre. *Cépages et vignobles de France*. Montpellier: P. Déhan, 1956-1964.

———. *Dictionnaire encyclopédique des cépages*. Paris: Hachette Pratique, 2000.

———. *Les maladies et les parasites de la vigne*. Montpellier: Impr. Du Paysan du Midi, 1977 & 1982.

———. *Précis d'Ampélographie Pratique*. Montpellier: Impr. De P. Déhan, 1952.

Galet, Pierre, and Lucie T. Morton. *A Practical Ampelography: Grapevine Identification*. Ithaca, N.Y., and London: Cornell University Press, 1979.

Gennery-Taylor, Mrs. *Easy to Make Wine, with Additional Recipes for Cocktails, Cider, Beer, Fruit Syrups, and Herb Teas*. New York: Gramercy Publishing Company, 1963.

Gladwin, F. E. "A History of Grape-Growing in the Eastern United States: Part VI," *Rural New Yorker*, Vol. XC, No. 5161, May 23, 1931. New York: Rural Publishing Co., 1931, 613–15.

"The Grapes of New York," *The Country Gentleman*, Vol. 74, No. 2933. Albany, N.Y.: Luther Tucker & Son, April 15, 1909, 376.

Graves and Steinbarger, eds. *American Series of Popular Biographies, Massachusetts Edition: Biographical Sketches of Representative Citizens of the Commonwealth of Massachusetts*. Boston: Graves & Steinbarger, 1901.

Hall, F. H., and S. A. Beach, *Why Some Grapes Fail to Fruit*, New York Agricultural Experiment Station Bulletin No. 157. Geneva, N.Y.: New York Agricultural Experiment Station, December 1898.

Hansen, Melissa. "Grapes for Puget Sound," *Good Fruit Grower*, June 2011. www.goodfruit.com.

Hardwick, Homer. *Winemaking at Home*. New York: Wilfred Funk, 1954.

Haskell, George. *An Account of Various Experiments for the Production of New and Desirable Grapes, and a Description of Forty Varieties Obtained by Hybridization*. Ipswich, Mass.: printed by the author, 1877.

———. *A Narrative of the Life, Experience and Work of an American Citizen*. Ipswich, Mass.: Chronicle Publishing Company, 1896.

Headley, Russel. *The History of Orange County, New York*. New York: Van Deusen & Son, 1908.

Hedrick, U. P. *Fruits for the Home Garden*. London, New York, and Toronto: Oxford University Press, 1944.

———. *Grape Stocks for American Grapes*. New York Agricultural Experiment Station, Bulletin no. 355. Geneva, N.Y.: Department of Agriculture, 1912.

———. *Grapes and Wines from Home Vineyards*. London: Oxford University Press, 1945.

———. *The Grapes of New York*. Report of the New York Agricultural Experiment Station for the Year 1907. Department of Agriculture, State of New York. Fifteenth Annual Report, Vol. 3 – Part II. Albany, N.Y.: J. B. Lyon Company, 1908.

———. *A History of Agriculture in the State of New York*. The New York State Agricultural Society. Albany, N.Y.: J. B. Lyon Company, 1933.

———. *A History of Horticulture in America to 1860*. New York: Oxford University Press, 1950.

———. *Manual of American Grape-Growing*. New York: The Macmillan Company, 1919.

———. *The Pears of New York*. Albany, N.Y.: Department of Agriculture, State of New York, 1921.

———. *The Plums of New York: Report of the New York State Agricultural Experiment Station for the Year 1910, II*. Albany, N.Y.: J. B. Lyon Co., 1911.

———. *The Small Fruits of New York*. Albany, N.Y.: Department of Agriculture, State of New York, 1925.

———. *The Vegetables of New York*, vols. 1, parts I–IV. Albany, N.Y.: Education Department, State of New York, 1928, 1931, 1934, 1937.

Hexamer, Dr. F. M., ed. "Exhibitions and Societies, Massuchusetts Horticultural Society," *The American Garden, A Monthly Journal of Practical Gardening*, Vol. VI, No. 2. New York: B. K. Bliss & Sons, February 1885, 32.

Higginbotham, Julie S. "Four Centuries of Planting Progress: A History of the U.S. Nursery Industry," *American Nurseryman*, Vol. 171, No. 12. Chicago: American Nurseryman Publishing Co., June 15, 1990, 36–42.

"History of the University of Minnesota grape-breeding program," *Enology Blog, University of Minnesota*, 2012. http://enology.umn.edu/history/.

Hoard, H. E., ed. *Annual Report of the Minnesota State Agricultural Society for the Year 1887, Executive Documents of the State of Minnesota, for Fiscal Year 1887–8*, Vol. II. St. Paul, Minn.: J. W. Cunningham & Co., 1888, 576.

Hovey, C. M., ed., *The Magazine of Horticulture, Botany, and All Useful Discoveries and Improvements in Rural Affairs*, Vol. XXII (Vol. II, Third Series); Vol. XXIII (Vol. III, Third Series), January 1857, 6, and February 1857, Vol. X, 1844. Boston: Hovey and Co.

Huber, F., F. Röckel, F. Schwander, E. Maul, R. Eibach, P. Cousins, and R. Töpfer. "A view into American grapevine history: *Vitis vinifera* cv. 'Sémillon' is an ancestor of 'Catawba' and 'Concord,'" *Vitis – Journal of Grapevine Research*, vol. 55, no. 2 (2016),

Geilweilerhof: Julius Kühn-Institut, Institute for Grapevine Breeding, , 53–56.

Hurd, Hamilton D. *History of Middlesex County, Massachusettes, with Biographical Sketches of Many of its Pioneers & Prominent Men*, Vol. I. Philadelphia: J. W. Lewis Co., 1899.

Hutchinson, B. June "A Taste for Horticulture," *Arnoldia*, Vol. 40, Issue 1. Boston: The Arnold Arboretum of Harvard University, 1980, 31–48.

Hutchinson, Ralph, Richard Figiel, and Ted Meredith. *A Dictionary of American Wines*. New York: Beech Tree Books, William Morrow, 1985.

Jackisch, Philip. *Modern Winemaking*. Ithaca and London: Cornell University Press, 1985.

Jacobellis v. Ohio, 378 U.S. 184, 197; 84 SCT 1676, 1683 (1964).

Jacobsen, Abigail and J. Dustin Williams. "Tales from the Archives, Prince Family Nurseries (circa 1737–post 1851)," *Bulletin of the Hunt Institute for Botanical Documentation*, Vol. 21, No. 1. Pittsburgh: Carnegie Mellon University, Spring 2009, 4–7.

Jameson, J. Franklin. *Narratives of New Netherland 1609–1664. A Short Account of the Mohawk Indians, by Reverend Johannes Megapolensis*. New York: Charles Scribners Sons, 1909.

Johnson, Hugh. *Wine*. New York: Simon & Schuster, 1969.

———. *The World Atlas of Wine*. New York: Fireside Book, Simon & Schuster, 1971.

Johnson, Hugh, and Hubecht Duijker. *The Wine Atlas of France, and Traveller's Guide to the Vineyards*. New York: Simon & Schuster, 1987.

Johnson, Victoria. *American Eden: David Hosack, Botany, and Medicine in the Garden of the Early Republic*. New York: W. W. Norton & Co., 2018.

Jordan, T. D, R. M. Pool, T. J. Zabadal, and J. P. Tomkins. *Cultural Practices for Commercial Vineyards*. New York State Agricultural Experiment Station, Cornell University, Misc. Bulletin no. 111. Geneva, N.Y.: 1981.

Kains, M. G. *Five Acres and Independence: A Practical Guide to the Selection and Management of the Small Farm*. New York: Greenberg Publisher, Inc., 1942.

Kracen, Emilie. "Grant Family Paper, 1817–1869," Litchfield Historical Society. www.litchfieldhistoricalsociety.org.

Kowsky, Francis R. *Country Park & City: The Architecture and Life of Calvert Vaux*. New York: Oxford University Press, 1998.

Larkin, William J., John J. Bonacic, Vincent L. Leibell, Thomas P. Morahan, Stephen M. Saland, James L. Seward, and J. Stephen Casscles. *The New York State Senate Task Force for Hudson Valley Fruit Growers. Tending the Vineyards: Renewed Growth for the Hudson Valley's Grape and Wine Industry*. Final Report: May 10, 2006. Albany, N.Y.: New York State Senate Task Force for Hudson Valley Fruit Growers, 2006.

Larkin Jr., State Senator William J., and J. Stephen Casscles. "Fruit Cultivation in the Hudson Valley," *New York Fruit Quarterly* vol. 13, no. 2 (Summer 2005).

———. "A Proposal for Renewed Growth of the Hudson Valley's Grape and Wine Industry," *New York Fruit Quarterly* vol. 14, no. 4 (Winter 2006): 12.

———. "Recommendations by the NYS Senate Task Force for Hudson Valley Fruit Growers to Promote Fruit Farming," *New York Fruit Quarterly* vol. 13, no. 2 (Summer 2005): 9.

Larkin Jr., William J., J. Stephen Casscles, and the Senate Majority Steering Committee. *The New York State Senate Task Force for Hudson Valley Fruit Growers. Action Plan Final Report*. Albany, N.Y.: New York State Senate Task Force for Hudson Valley Fruit Growers, December 2004.

Libby, E. H., ed. "Two Notables Gone – John B. Moore," *The American Garden and Floral Cabinet, A Monthly Journal of Practical Gardening*, Vol. III, No. 11. New York: B. K. Bliss & Sons, November 1887, 345.

Lord, E. L. "Grape Varieties," *Proceedings of the Thirty-Fifth Annual Meeting of the Florida State Horticultural Society, Held at Lakeland, Fla., May 2, 3, 4, 5, 1922*, Vol. 35. Deland, Fla.: The Society Office of Publication, 1922.

Lyon, T. T. *Fruit Tests at South Haven*, Bulletin No. 143. Lansing, Mich.: Michigan Agricultural Experiment Station, State Agricultural College, April 1897.

Maine Agricultural Experiment Station. "Orchard Notes," *Bulletin No. 128*, March 1906, 167–70.

Maniscalco, Matt. "Brianna Grape Is Midwest's New Tropical Fruit," *Midwest Wine Press*, May 3, 2012. http://midwestwinepress.com/

Marquis Who's Who. *Who Was Who in American History – Science and Technology: A Component of Who's Who in American History*. Chicago: Marquis Who's Who, 1976.

Marshall P. Wilder Collection 1848–1886. Special Collections & University Archives, University of Massachusetts Amherst Libraries. Call No. R6 002/3 W55.

Martell, Alan R., and Alton Long, with Laury A. Egan, photographer. *The Wines and Wineries of the Hudson River Valley*. Woodstock, Vt.: The Countryman Press, 1993.

Massachusetts Horticultural Society. *For the Year 1887, Part 1, Memorial of Marshall P. Wilder by Pres. Walcott*. Boston: Printed for the Society, 1887.

———. *History of the Massachusetts Horticultural Society, 1829–1878*. Boston: Rand, Avery & Co. for the Massachusetts Horticultural Society, 1880.

———. *Transactions of the Massachusetts Horticultural Society for the Year 1876, Part II, Resolution adopted on November 4, 1876 on the Death of John Fisk Allen*. Boston: Printed for the Society, 1877, 8–10.

Maynard, Samuel T. *Report on Fruits*, Hatch Experiment Station of the Massachusetts Agricultural College Bulletin No. 21. Amherst, Mass.: Press of Carpenter & Morehouse, April 1893.

McCluer, G. W. "*Grapes, Test of Varieties*," Agricultural Experiment Station Bulletin No. 28, December 1893, in *Bulletins of the Agricultural Experiment Station, University of Illinois, Vol. 2, Bulletins 17–36, August 1891–August 1894*. Urbana, Ill.: Agricultural Experiment Station, University of Illinois, 1894.

McGrew, J. R., J. Loenholdt, T. Zabadal, A. Hunt, and H. Amberg. *The American Wine Society Presents: Growing Wine Grapes*. Ann Arbor, Mich.: G. W. Kent, Inc., 1993.

McLaughlin III, Edward J. *Around the Watering Trough: A History of Washingtonville*. Washington Centennial Celebration. Washingtonville, N.Y.: Washingtonville Centennial Celebration, Inc., 1994.

McLeRoy, Sherrie, and Roy E. Renfro Jr. *Grape Man of Texas: Thomas Volney Munson & the Origins of American Viticulture*. San Francisco: The Wine Appreciation Guild, 2008.

Meehan, Thomas, ed. "New or Rare Fruits: Haskell's Grapes," *The Gardener's Monthly and Horticulturist*, Vol. XXI, No. 245. Philadelphia: Charles H. Marot, May 1879, 149.

Meisel, Anthony, and Sheila Rosenzweig. *American Wine*. New York: Excalibur Books, 1983.

Michigan State Horticultural Society, "Salem Grape," *Eleventh Annual Report of the Secretary of the State Horticultural Society of Michigan*. Lansing, Mich.: W. S. George & Co., 1882, 223.

Miller, Mark. *Benmarl Vineyards: Manual for Home Wine-makers*. Marlboro, N.Y.: Benmarl Vineyards, circa 1977.

———. *Wine—A Gentleman's Game, the Adventures of an Amateur Winemaker Turned Professional*. New York: Harper & Row, 1984.

Morton, Lucie T. *Winegrowing in Eastern America: An Illustrated Guide to Viniculture East of the Rockies*. Ithaca, N.Y.: Cornell University Press, 1985.

Munsell, W. W. *History of Queens County, New York, with Illustrations, Portraits, & Sketches of Prominent Families and Individuals*. New York: Press of George MacNamara, 1882.

Nanovic, John. *The Complete Book of Wines, Vineyards & Labels*. Baltimore, Md.: Ottenheimer Publishers, 1979.

Neiles, Edward. *The New York Red Book*. Albany: New York Legal Publishing Corp., 2011.

Neirynck, Baudouin. *The Grapes of Wine: The Fine Art of Growing Grapes and Making Wine*. Garden City, N.Y.: SquareOne Publishers, 2009.

New York Fruit Tree and Vineyard Survey. Albany, N.Y.: United States Department of Agriculture, National Agricultural Statistical Services and the New York State Department of Agriculture and Markets, 2006.

New York State Statistical Yearbook (31st ed., rev. and exp.). Albany, N.Y.: The Nelson A. Rockefeller Institute of Government, State University of New York, 2006.

Nicholson, Rob. "The Splendid Haul of Cyrus Guernsey Pringle," *Arnoldia*, Vol. 61, Issue 1. Boston: The Arnold Arboretum of Harvard University, 2001, 2–9.

Nutt, John J. *Newburgh: Her Institutions, Industries, and Leading Citizens*. Newburgh, N.Y.: Ritchie & Hull, 1891.

"Obituary of Edward Rogers, Originator of the Rogers Grape," 3. Boston: The Boston Transcript Company, March 30, 1899.

Pambianchi, Daniel. *Techniques in Home Winemaking*. Montreal: Véhicule Press, 2002.

Papademetriou, Minas K. and Frank J. Dent. "Grape Production in the Asia-Pacific Region," *RAP Publication*, 2000/13. Bangkok: Food and Agriculture Organization of the United Nations Regional Office for Asia and the Pacific, July 2001.

Pardee, R. G. "Iona, Pelham, and Wodenethe," *The Horticulturalist and Journal of Rural Art and Rural Taste*, vol. 14. New York: C.M. Saxton, Barker & Co. 1859.

Parker, Mark. "Legend of Hermann Jaeger," April 2009. www.MissouriRuralist.com.

Pawley, Emily. "Cataloging Nature: Standardizing Fruit Varieties in the United States, 1800–1860," *Business History Review*, Vol. 90, No 3. Cambridge: Cambridge University Press, Fall 2016, 405–29.

Peynaud, Emile, and Alan F. Spencer. *Knowing and Making Wine*. New York: John Wiley & Sons, 1984.

Pinney, Thomas. *A History of Wine in America: From the Beginnings to Prohibition*, vol. 1. Berkeley: University of California Press, 2007.

Plocher, Tom. "Swenson Vineyard Preservation Project," *Northern Wine Work*, November 2003. http://northernwinework.com/cms/index.php?page=swenson-vineyard-preservation.

———. "Tasting Grapes and Wines at Geilweilerhof, Notes From the North," *Minnesota Grape Growers Association, Quarterly Newsletter*, Winter 1999.

Plocher, Tom, and Bob Parke. *Northern Winework: Growing Grapes and Making Wine in Cold Climates*. Hugo, Minn.: Northern Winework, Inc., 2001.

Pool, Robert M., J. Einset, K. H. Kimball, and J. P. Watson. *Vineyard and Cellar Notes, 1958–1973*. Special Report no. 22-A., Ithaca, N.Y.: New York State Agricultural Experiment Station, 1976.

Pool, Robert M., et al. *Growing Vitis Vinifera Grapes in New York State, I: Performance of New and Interesting Varieties*. Geneva, N.Y.: New York Agricultural Experiment Station, Cornell, University, 1990.

Potanovic, Joanne. *The Chapel of St. John the Divine, Our History, Tompkins Cove, NY*. Tompkins Cove, N.Y.: The Chapel of St. John the Divine, 2012.

Powell, E. P. *The Orchard and Fruit Garden*. New York and Garden City: Doubleday, Page & Company, 1914.

Present, Jess J., and Joseph S. Casscles. *The New York State Grape and Wine Industry: A Discussion of Mandate Relief and Regulatory Reform to Stabilize and Promote the State's Grape and Wine Industry*. Albany, N.Y.: New York State Senate Majority Program Development Committee, June 1992.

Pringle, Cyrus Guernsey. *The Record of a Quaker Conscience: Cyrus Pringle's Diary*. New York: Macmillan & Co., 1918.

Reisch, Bruce I., and R. S. Luce. *Corot Noir*. Cornell University, Bulletin no. 159. Geneva, N.Y.: New York State Agricultural Experiment Station, 2006.

———. *Noiret Grape*. Cornell University, Bulletin no. 160. Geneva, N.Y.: New York State Agricultural Experiment Station, 2006.

Reisch, Bruce I., and Robert M. Pool. *Chardonel Grape*. Cornell University, Bulletin no. 132. Geneva, N.Y.: New York State Agricultural Experiment Station, 1990.

———. *Melody Grape*. Cornell University, Bulletin no. 112. Geneva, N.Y.: New York State Agricultural Experiment Station, 1985.

———. *Traminette Grape*. Cornell University, Bulletin no. 149. Geneva, N.Y.: New York State Agricultural Experiment Station, 1996.

Reisch, Bruce I., Robert M. Pool, David V. Peterson, Mary-Howell Martens, and Thomas Henick-Kling. *Wine and Juice Grape Varieties for Cool Climates*. New York State Agricultural Experiment Station, Cornell University, Information Bulletin no. 233. Geneva, N.Y.: Cornell Cooperative Extension, 1993.

Reisch, Bruce I., W. B. Robinson, K. Kimball, Robert M. Pool and J. Watson. *Horizon Grape*. Cornell University, Bulletin no. 96. Geneva, N.Y.: New York State Agricultural Experiment Station, 1982.

Renehan, Edward J. Jr. *John Burroughs: An American Naturalist*. Post Mills, Vt.: Chelsea Green Publishing Company, 1992.

Robards, Terry, *The New York Times Book of Wine*. New York: Avon Books, 1976.

———. Terry Robard's *New Book of Wine*. New York: G. P. Putnam's Sons, 1976.

Roberts, Herbert F. "The Contribution of Carl Friedrich von Gartner to the History of Plant Hybridization," *The American Naturalist*, Vol. 53, No. 628. New York: The Science Press, September–October 1919, 431–45.

Robinson, Jancis. *The Oxford Companion to Wine* (3rd ed.). New York: Oxford University Press, 2006.

———. *Vines, Grapes, and Wines*. New York: Alfred A. Knopf, 1986.

Robinson, Jancis, Julia Harding, and Jose Vouillamoz. *Wine Grapes: A Complete Guide to 1,368 Vine Varieties, Including Their Origins and Flavor*. New York: Ecco/HarperCollins, 2012.

Röckel, F., O. Trapp, E. Zyprian, L. Hausmann, D. Migliaro, S. Vezzulli, R. Töpper and E. Maul. "A 'Regent' pedigree update: ancestors, offspring and their confirmed resistance loci," *Vitis, Journal of Grapevine Research*, Vol. 60, No 4, October 2021: 189–93.

Roe, Edward Payson. *Success with Small Fruits*. New York: Dodd, Mead & Company, 1880.

Rogers Family Papers, 1792–1901, MSS 87, Box 8, Phillips Library at the Peabody Essex Museum, Salem, Massachusetts.

Ronbough, Lon. *The Grape Grower: A Guide to Organic Viticulture*. White River Junction, Vt.: Chelsea Green Publishing, 2002.

Russell, Howard S. *Indian New England Before the Mayflower*. Lebanon, N.H.: University Press of New England, 1980.

Ruttenber, E. M., and L. H. Clark. *A History of Orange County, New York, with Illustrations and Biographical Sketches of Its Many Pioneer and Prominent Men*. Philadelphia: Everts & Peck, 1881.

Sammarco, Anthony M. "The History of the Diana Grape Has Roots in Milton." *The Milton Historical Society Newsletter*. Milton, Mass.: The Milton Historical Society, Fall 2008. http://www.miltonhistoricalsociety.org/Sampler/DianaGrape.html.

Schofield, Edumund A. "'He Sowed; Others Reaped': Ephraim Wales Bull and the Origins of the 'Concord' Grape" *Arnoldia*, Vol. 48, No 4. Boston: The Arnold Arboretum of Harvard University, Fall 1999, 7–10.

Schoonmaker, Frank. *Frank Schoonmaker's Encyclopedia of Wine*. New York: Hastings House, 1969.

Schuyler, David. *Apostle of Taste: Andrew Jackson Downing*, 1815–1852. Baltimore: Johns Hopkins University Press, 1996.

Sears, Fred Coleman. *Productive Small Fruit Culture*. Philadelphia and London: J. B. Lippincott Company, 1920.

Shaulis, Nelson, John Einset, and A. Boyd Pack. *Growing Cold-Tender Grape Varieties in New York. New York*. Agricultural Experiment Station, Cornell University, Bulletin no. 821 (August). Geneva, N.Y.: Cornell University Agricultural Experiment Station, 1968.

Shoemaker, James Sheldon. *Small-Fruit Culture—A Text for Instruction and Reference Work and a Guide for Field Practice*. (2nd ed.) Philadelphia and Toronto: Blakiston Company, 1950.

———. *Sugar, Acidity, and Juice Color Determinations in Grapes*. Ohio Agricultural Experiment Station Bulletin 550. Wooster, Ohio: Ohio State University, 1935, 5–13.

Slate, George L., John Watson, and John Eisnset. *Grape Varieties Introduced by the New York State Agricultural Experiment Station, 1928–1961*. Cornell University, Bulletin 794 (February). Geneva, N.Y.: New York State Agricultural Experiment Station, 1962.

Smiley, Lisa. *Brianna*. Iowa State University, Dept. of Horticulture bulletin, 2008. http://viticulture.hort.iastate.edu/cultivars/cultivars.html.

———. *Kay Gray*. Iowa State University, Dept. of Horticulture bulletin, 2008. http://viticulture.hort.iastate.edu/cultivars/cultivars.html.

———. *Lemberger*. Iowa State University, Dept. of Horticulture bulletin, 2008. http://viticulture.hort.iastate.edu/cultivars/cultivars.html.

———. *Sabrevois*. Iowa State University, Dept. of Horticulture bulletin, 2008. http://viticulture.hort.iastate.edu/cultivars/cultivars.html.

———. *St. Croix*. Iowa State University, Dept. of Horticulture bulletin, 2008. http://viticulture.hort.iastate.edu/cultivars/cultivars.html.

———. *St. Pepin*. Iowa State University, Dept. of Horticulture bulletin, 2008. http://viticulture.hort.iastate.edu/cultivars/cultivars.html.

Smith, Bruce. "History of the Early Swenson Hybrids," A presentation at the Nebraska Grape and Wine Conference, November 11, 2006. http://agronomy.unl.edu/c/document.

Subden, Ronald E. "The Odyssey of DeChaunac (Seibel 9549)," *American Wine Society Journal* vol. 12, no. 4 (Winter 1980): 81-84.

Stalter, Elizabeth "Perk." *Doodletown: Hiking Through History in a Vanished Hamlet on the Hudson*. Bear

Mountain, N.Y.: Palisades Interstate Park Commission Press, 1996.

State of New York Department of Agriculture. *The Fruit Industry in New York State*, Bulletin No. 79, Part II. Albany, N.Y.: Department of Agriculture, 1915.

Taylor, Allan. *What Everybody Wants to Know About Wine*. New York: Alfred A. Knopf, 1934.

Taylor, Walter S. and Richard P. Vine. *Home Winemaker's Handbook*. New York: Harper & Row, 1968.

Thorton, Tamara Plakins. *Cultivating Gentlemen: The Meaning of Country Life Among the Boston Elite, 1785–1860*. New Haven, Conn.: Yale University Press, 1989.

Trzaskos, Todd. *Wines of Vermont: A History of Pioneer Fermentation*. Charleston, S.C.: American Palate, The History Press, 2015.

Underhill, Sarah Gibbs. "Tales from Croton Point," A Brief History of Croton-on-Hudson, from the perspective of the Great Hudson River Brick Industry at Croton Landing and Croton Point, *Brick Collecting*, http://brickcollecting.com/croton.htm.

"Understanding Earthworms," *Organic Gardening*, www.organicgardening.com/learn-and-grow/understanding-earthworms.

United States Department of Agriculture, *Report of the Commissioner of Agriculture for the Year 1864*. Washington, D.C.: Government Printing Office, 1865.

———. *Report of the Commissioner of Agriculture for the Year 1875*. Washington, D.C.: Government Printing Office, 1876.

Wagner, Philip M. *American Wines and Wine-Making*. New York: Alfred A. Knopf, 1969.

———. *Grapes Into Wine*. New York: Alfred A. Knopf, 1976.

———. *A Wine-Grower's Guide*. New York: Alfred A. Knopf, 1985.

Waldron, Rodney K. "Prince's Plants," *The Call Number*, Vol. 20, No. 1. Portland, Ore.: University of Oregon Library, Fall 1958, 7–12;

Watts, R. L. *Varieties of Grapes, University of Tennessee Agricultural Experiment Station Bulletin*, Vol. IX, No. 4. Knoxville, Tenn.: The Agricultural Experiment Station, December 1896.

Waugh, F. A. *Home-Grown Grapes in Vermont, III. Variety Notes*, Bulletin No. 62. Montpelier, Vt.: Vermont Agricultural Experiment Station, University of Vermont and State Agricultural College, Vermont Watchman Co., January 1898.

Weeks, Alvin G. *Massasoit of the Wampanoags*. Norwood, Mass.: Privately Printed, The Plimpton Press, 1919.

Wellhouse, Walter, ed. "Transactions of the Kansas State Horticultural Society," *The Proceedings of the Forty-Second and Forty-Third Annual Meetings*, vol. XXX, December 1908 and 1909. Topeka, Kan.: State Printing Office, 1910.

White, Charles A. *Memoir of George Engelmann, 1809–1884*. A paper presented before the National Academy of Sciences, April 1896. Washington, D.C.: Judd & Detweiler, Printers, 1896.

White, N. B. "On Grape Hybrids," *Memoirs, Vol. I, Proceedings: International Conference on Plant Breeding and Hybridization, September 30, October 1 and 2, 1902*. New York: Horticultural Society of New York, 1902, 221–23.

Whitten, J. C. *The Grape – I. A Study of Types and Varieties*. University of the State of Missouri, College of Agricultural and Mechanic Arts, Agricultural Experiment Station, Bulletin No. 46. Columbia, Mo.: E. W. Stephens, April 1899.

Wilder, Marshall B. *Address at the Twentieth Session of the American Pomological Society, held in Grand Rapids, MI, September 9–11, 1885*. Boston: American Pomological Society, 1885.

Woolsey, C. M. *History of the Town of Marlborough, Ulster County, NY*. Albany, N.Y.: J. B. Lyons Company, 1908.

Wolf, Tony K., et al. *Wine Grape Production Guide for Eastern North America*. Ithaca, N.Y.: Natural Resource Agriculture and Engineering Service (NRAES) Cooperative Extension, 2008.

Yang, Dan Yi, Yukio Kakuda, and Ronald E. Subden, "Higher Alcohols, Diacetyl, Acetoin, and 2,3-Butanediol Biosynthesis in Grapes Undergoing Carbonic Maceration," *Food Research International* vol. 39, issue 1, January 2006: 112-116.

Zraly, Kevin. *Windows on the World Complete Wine Course*. New York: Dell Publishing, 1985.

———. *Windows on the World Complete Wine Course* (25th Anniversary ed.). New York: Sterling Publishing Co., 2009.

PERIODICALS AND CATALOGS

American Wine Society Journal. American Wine Society, Englewood, Ohio.

Benmarl Vineyards Catalog, *Supplies for Home Winemakers*. Catalogue BV-1. Marlboro, N.Y.: Mark and Dene Miller, Benmarl Vineyards, circa 1975.

Boordy Nursery. *Boordy Nursery Grape Book. The Boordy Nursery Grape Vine Catalogue, Grape Wines for Wine Growers*. Philip and Jocelyn Wagner, Proprietors. Riderwood, Md.: Boordy Nursery, 1993.

Double A Vineyards Nursery Catalog 1989/1990 through 2009/2010, 20th Anniversary. Double A Vineyards, Inc., 10277 Christy Road, Fredonia, N.Y. 14063. www.doubleavineyards.com.

Fosters Concord Nursery Catalog (1998 through 2000). Fosters Concord Nursery, 10175 Mile Block Road, North Collins, N.Y.

Fruit Testing Association Nursery, Inc. *A Catalog of New and Noteworthy Fruits*, 1978 through 1995. New York State Fruit Testing Cooperative Association, Geneva, N.Y.

Fulkerson's Winery and Juice Plant: Price List 2000 through 2009. Sayre and Nancy Fulkerson, Proprietors, 5576 Route 14, Dundee, N.Y. Dundee, N.Y.: Fulkerson's Winery and Juice Plan, 2010.

A Guide to American and French Hybrid Grape Varieties. Foster Nursery Co., Fredonia, N.Y.

The Horticulturalist and Journal of Rural Art and Rural Taste. Edited by A. J. Downing. Albany, N.Y.: Luther Tucker, 1846.

J. E. Miller's Nursery Catalog, 1989/1990 through 2009/2010. J. E. Miller, 5060 County Road 16, Canandaigua, N.Y.

Practical Winery and Vineyard Journal. Don Neel, publisher, San Rafael, Calif.

Presque Isle Wine Cellars Catalog. Presque Isle Wine Cellars, 9440 W. Main Rd., North East, Pa.

Vineyard & Winery Management News. J. William Moffett and Hope Merletti, founders and publishers, Emeritas, Watkins Glen, N.Y. and Santa Rosa, Calif.

Wine East. L & H Photojournalism, 620 N. Pine Street, Lancaster, Pa.

Wines and Vines. 1800 Lincoln Avenue, San Rafael, Calif.

PHOTO CREDITS

The author and publisher would like to thank the following for their kind permission to reproduce their photographs.

Al Nowak/On Location Studios, Poughkeepsie, NY: Cover (bottom). American Museum of Natural History, New York City, NY: 9 (all). Archives, Flint Mine Press, Coxsackie, NY: Cover (top), iv, xii, xxi, xxiv, 2, 3, 5, 6, 7, 8, 11, 12, 14, 15 (right), 19, 34, 42, 56 (inset, background), 58, 59, 60, 62, 63, 68 (inset, background), 71, 73 (all), 75, 82 (all), 84, 91, 92, 95 (bottom), 94 (top), 96, 97, 101, 107, 108, 109, 111, 112 (top, right), 113, 115 (all), 119, 128 (all), 129, 130, 136, 145 (left, center), 148, 163, 175 (top), 176, 239, 243, 249, 252, 256, 259, 271, 277, 280 (bottom), 283, 289. Dave Rosenberger, Cornell's Hudson Valley Lab, Highland, NY: 191. Eric Miller, West Chester, PA: 15 (left). Irving Geary, Minnesota Grape Growers Association, Red Wing, MN: 194 (inset). Maïté Laberiotte, Chairman of the Cultural Center of Pays d'Orthe, the Pommies Family, Daniel Dufau, Mayor of Bélus, and Bernard Matabos, Saint-Lon-Les-Mine, France: 151. Marlboro Free Library, Marlboro, NY: 22, 88. Mary McTamaney, City of Newburgh Historian, Newburgh, NY: 4, 85, 117, 118 (all). Massachusetts Horticultural Society, Wellesley, MA: 234, 241, 268, 270, 280 (top), 282, 288. Millbrook Vineyards & Winery, Millbrook, NY: 218. Minnesota Agricultural Experiment Station (MAES), University of Minnesota, St. Paul, MN: 194 (background), 198. New York State Agricultural Experiment Station, Cornell University, Geneva, NY: 174 (bottom). New York State Archives, Albany, NY: 16. Niagara Falls Public Library, Niagara Falls, Ontario: 136 (bottom). Phillips Library, Peabody Essex Museum, Salem, MA: 278. Randall Tagg Photography, www.Taggphotography.com: 168, 171. Robert "Woody" Underhill, The Underhill Society of America, Oyster Bay, NY: 112 (left). Special Collections, National Agricultural Library, Beltsville, MD: 135. Steve Casscles, Athens, NY: 13 (all), 145 (right), 251, 272, 275, 317.

Color Insert: Minnesota Agricultural Experiment Station (MAES), University of Minnesota, St. Paul, MN: Frontenac Gris, Marquette. New York State Agricultural Experiment Station, Geneva, NY: Seyval Blanc. Getty Images: Riesling. Palaia Vineyards, Highland Mills, NY: Lemberger. Millbrook Vineyards & Winery, Millbrook, NY: Veraison. All other photos, except last page, were taken at Cedar Cliff farm in Athens, NY: Archives, Flint Mine Press, Coxsackie, NY.

Every effort has been made to identify and accurately credit the photographers and copyright holders of the images in this book. The author and publisher apologizes for any unintentional inaccuracies, errors, or omissions, which will be corrected in future editions of this work.

INDEX OF MAJOR AND MINOR GRAPE VARIETIES

INDEX

Page references followed by *fig* indicate an illustration or photograph.

G

H

S

T

U

ABOUT THE AUTHOR

J. STEPHEN CASSCLES, ESQ., comes from a fruit growing family that has been in the Hudson River Valley since the 1820s. Beginning in 1972, Stephen gained fruit-growing experience by working at local orchards and vineyards in Marlboro and Middle Hope, New York. Stephen worked at Benmarl Vineyards in Marlboro from 1973 to 1986 where he became fascinated with growing grapes and making wine from heritage and French-American hybrid grape varieties.

After working as a municipal attorney for two years in Walden, New York, Stephen had a long career as an attorney for the New York State Senate and the New York State Department of Health. For 33 years he worked for six state senators, two of whom, Jess J. Present and William J. Larkin Jr., represented the grape growing areas of western New York and the Hudson Valley. Stephen has authored at least 27 laws related to the production, distribution, and sale of wine, spirits, beer, and cider, in addition to laws related to his legal specialty of insurance, health care financing, and racing law.

In 1990, Stephen established Cedar Cliff Vineyards and Nursery, in Athens, New York, and since 2008 has concentrated on evaluating and propagating nineteenth century heritage grape varieties that were developed in the Hudson Valley, New York, and Cape Ann, Massachusetts. Stephen was a winemaker at Hudson-Chatham Winery in Ghent, New York, from 2007 to 2020, where he specialized in making wines from heritage grape varieties. In 2021, he was appointed director/winemaker of Milea Estate Vineyard's Hudson Valley Heritage Grape Project, which seeks to conserve and promote historically relevant local grape varieties for future generations.

He operates a small nursery that propagates rare nineteenth-century heritage grape varieties and own rooted chance hybrids identified at his farm. He is an award-winning winemaker whose wines and work has been covered by *The New York Times*, *The Wall Street Journal*, *The Washington Post*, *New York Post*, *Wine Enthusiast*, *Forbes*, *Hugh Johnson's Pocket Wine Book* (2021), the *Times-Union*, and the *Hartford Courant*.

Stephen is a frequent lecturer on wine, grape cultivation, nineteenth-century American horticulture, and landscape architecture at botanical gardens and historical societies throughout New York and New England.

In addition to his full-length works on grape varieties, grape cultivation, and nineteenth-century horticulture, he is a contributor to academic and trade journals such as *Arnoldia* of the Arnold Arboretum of Boston, MA, *Fruit Notes* of the University of Massachusetts Amherst, *Horticultural News*, *New Jersey State Horticultural Society*, *Wine Journal* of the American Wine Society, *New York Fruit Quarterly* of the NYS Horticultural Society, and *Hudson Valley Wine Magazine*.

As a culmination of his horticultural pursuits, Stephen advises and lectures at the Fermentation Sciences Program at SUNY at Cobleskill and has a working relationship with many in the Korean grape and wine industry.

Stephen lives with his wife Lilly Casscles at Cedar Cliff Vineyards, Athens, New York. They have three adult children, Benjamin, Noah, and Grace, their dog Nixon, and Peebles the turtle.

This book was typeset in Goudy, a font designed by type designer Frederic W. Goudy (1865–1947) who lived and worked in Marlborough-on-Hudson, New York, from 1924 until his death in 1947.

Goudy (and Goudy Old Style) is one of 124 typefaces designed by Frederic Goudy, and considered to be among the most legible and readible serif typefaces for use in print. Goudy's home and studio workshop, a converted pre-Revolutionary mill he christened "Deepdene" was on the Old Post Road property in Marlboro, just across from the Caywood family's boarding house. The house no longer exists, and the lot is now wooded.